THE MOTION OF BUBBLES AND DROPS IN REDUCED GRAVITY

Orbiting spacecraft provide a valuable laboratory for experiments on physical and biological systems in a reduced gravity environment. Materials processing experiments have commonly involved the growth of crystals from the melt or solution and the processing of alloys and composites. Biological experiments have been performed on a variety of subjects, including protein crystal growth, bioreactors, and the adaptation of humans to extended periods of weightlessness. In these studies, fluid masses containing bubbles and drops are encountered routinely. This book is the first to provide a clear, thorough review of the motion of bubbles and drops in reduced gravity, particularly motion caused by variations in interfacial tension arising from temperature gradients on their surfaces. The emphasis is on theoretical analysis from first principles; experimental results are discussed and compared with predictions where appropriate. Students and researchers interested in fluid mechanics in reduced gravity will welcome this state-of-the-art reference.

R. Shankar Subramanian is Professor of Chemical Engineering at Clarkson University.

R. Balasubramaniam is Staff Scientist at the National Center for Microgravity Research on Fluids and Combustion, NASA John H. Glenn Research Center.

THE MOTION OF BUBBLES AND DROPS IN REDUCED GRAVITY

R. SHANKAR SUBRAMANIAN

Clarkson University

R. BALASUBRAMANIAM

National Center for Microgravity Research
on Fluids and Combustion
at NASA John H. Glenn Research Center

CAMBRIDGE
UNIVERSITY PRESS

CAMBRIDGE UNIVERSITY PRESS
Cambridge, New York, Melbourne, Madrid, Cape Town, Singapore, São Paulo

Cambridge University Press
The Edinburgh Building, Cambridge CB2 2RU, UK

Published in the United States of America by Cambridge University Press, New York

www.cambridge.org
Information on this title: www.cambridge.org/9780521496056

First published 2001
This digitally printed first paperback version 2005

A catalogue record for this publication is available from the British Library

Library of Congress Cataloguing in Publication data
Subramanian, R. Shankar (Ram Shankar), 1947–
     The motion of bubbles and drops in reduced gravity / R. Shankar Subramanian, R. Balasubramaniam.
        p.   cm.
     Includes bibliographical references.
     ISBN 0-521-49605-5
     1. Drops—Effect of reduced gravity on.   2. Bubbles—Effect of reduced gravity on.
     3. Fluid mechanics.   I. Balasubramaniam, R. (Ramaswamy), 1957–   II. Title.
     QC183 .S83   2001
     530.4′27 – dc21   00-041411

ISBN-13 978-0-521-49605-6 hardback
ISBN-10 0-521-49605-5 hardback

ISBN-13 978-0-521-01948-4 paperback
ISBN-10 0-521-01948-6 paperback

Dedicated to our parents

Sita Lakshmi and R. K. Ramasubramaniam

and

Saraswathi and D. Ramaswamy

Contents

Errata

The correct versions in each case are given below.

Page 8, line 20: proportional to the square of the distance

Page 26, line 2 below Equation (2.2.8): dimensions of force per unit volume

Page 41, Figure 2.6.3: upper case Φ should be replaced by lower case ϕ

Page 116, line 2 below Equation (4.7.5): Ferrers' functions

Page 172, line 3 from the bottom: for T_1', but ... equation for T_1' of

Page 231, lines 4 and 5 below Equation (5.1.5): spindle-shaped surfaces when $\eta > \dfrac{\pi}{2}$, apple-shaped surfaces when $\eta < \dfrac{\pi}{2}$,

Page 430, Equation (9.6.24): The correct form is given below.

$$G_1 = Be^{-\chi Pr\eta}\left(1+\sum_{n=1}^{\infty}C_n e^{-n\chi\eta}\right) \qquad (9.6.24)$$

Prepared by R. S. Subramanian and R. Balasubramaniam

October 28, 2004

Preface

This monograph is principally about the motion of bubbles and drops caused by variations in interfacial tension arising from temperature gradients on their surfaces. We have attempted to provide a reasonably comprehensive picture of the progress and the current status of research on this subject. It is our opinion that, in the long run, this driving force for the motion of bubbles and drops will prove to be as ubiquitous in a reduced gravity environment as gravity is on the surface of Earth.

The book is divided into four parts. In Part One, we introduce the reader to the role of gravity and interfacial tension in the motion of bubbles and drops in Chapter One and cover the governing equations in Chapter Two. Part Two is devoted to the motion of isolated bubbles or drops and contains two chapters. Some important aspects of the motion of bubbles and drops due to gravity, which is a familiar body force, are treated in Chapter Three. This is done for the purpose of completeness in coverage and to provide contrast where needed with features of the motion driven by the interface, which is discussed in Chapter Four. In Part Three, which is composed of three chapters, we discuss the interactions of bubbles and drops with each other and with neighboring boundaries. General solutions are given in Chapter Five and are then used in Chapters Six and Seven, which are devoted to body-force-driven motion and motion driven by the interface, respectively. In Part Four, two chapters cover topics that are closely related to the main theme. Chapter Eight deals with mass transport to bubbles and drops in reduced gravity conditions, and Chapter Nine is devoted to motion that occurs in a body of fluid due to interfacial tension gradients on its free surface. Although the emphasis in this work is on theoretical analysis, we have presented and discussed experimental results wherever appropriate and possible.

We hope that scholars who choose to work on bubbles and drops, on fluid mechanics in reduced gravity, and on interfacial phenomena will find this book useful. We have employed a level that is suitable for advanced students in engineering and science with the expectation that some of this material may be used in courses dealing with transport phenomena associated with motion driven by the interface. Also, the topics covered should be of interest to scientists studying the processing of materials in reduced gravity.

We have used the following system for numbering equations, figures, and tables. In each section, equations are numbered sequentially, beginning with 1. The identification number assigned to an equation also includes the chapter number and the section number, separated by periods. Thus, an equation number has the form $C.S.N$, where C designates the chapter number, S is the section number, and N stands for the sequential number within the section. Regarding symbols, we have used a uniform convention throughout the book, to the extent possible. In some instances, however, it has been necessary to use the same symbol with different connotations in different parts of the book. Definitions of the symbols used commonly are given when they are first introduced and repeated when needed.

We are grateful to our mentors, to our present and former students, and to our colleagues, too numerous to mention individually, who have offered valuable suggestions along the way. We alone are responsible for any errors and omissions. We would appreciate readers informing us about any errors that they may find. We wish to thank Florence Padgett of Cambridge University Press for her consistent encouragement and support, Nancy Mieczkowski of GRAFIXWORKS.COM, Lorain, Ohio, for preparing the drawings, and Erin Subramanian, Potsdam, New York, for preparing the bulk of the equations. We are indebted to our spouses and children for the patience and understanding they have displayed during the years when we devoted time toward the preparation of this work. Finally, we wish to express our appreciation to the National Aeronautics and Space Administration for steady support of our research program in numerous ways and to the European Space Agency for their kind hospitality during the years when we collaborated with them on the design and conduct of the flight experiments.

PART ONE

INTRODUCTION

CHAPTER ONE

The Role of Gravity and Interfacial Tension in the Motion of Bubbles and Drops

1.1 Bubbles and Drops and the Influence of Gravity

The principal subject of this book is the motion of bubbles and drops in a reduced gravity environment. A bubble is a fluid object that contains a gas or vapor, and a drop is a similar object which instead contains liquid. A gas or vapor bubble can exist only in a liquid, whereas drops can be present within liquids or gases. Bubbles and drops are encountered in everyday life. For instance, we see a collection of gas bubbles formed when we pour a carbonated beverage into a glass, and we see vapor bubbles when water boils in a pot. Oil drops may be observed in salad dressing and rain drops in the air surrounding us. Bubbles and drops also are industrially important. In the manufacture of a variety of chemicals which we use routinely, contacting two immiscible fluids is often necessary to purify one of them or to remove some useful substance dissolved in one fluid by transferring it to the other fluid. Both fluids can be liquids, or one of these fluids can be a gas or vapor. If we contact one fluid in a container with a layer of the other, the surface area between the two is relatively small, and the exchange will be slow. Therefore, we break one fluid into small bubbles or drops, allowing relatively rapid exchange of material or energy between the two fluids. Operations such as distillation, absorption, and liquid-liquid extraction that are used in the oil, chemical, and pharmaceutical industries employ this technique effectively. In another example, certain alloys contain a fine dispersion of one substance in another in the solid state. This dispersion is formed by heating the materials involved to a high temperature to form a single liquid phase and then cooling the liquid. At a certain temperature, this liquid spontaneously separates into two immiscible liquids, and one of these is precipitated in the form of fine drops. These alloys are used in numerous applications. Of course, there are situations in which bubbles or drops can be a nuisance and need to be eliminated to make a useful product. An example can be given from the glass industry in which air entrapped between the grains of the raw material, gases dissolved in the grains, and gases produced by chemical reactions during the glass-making process all lead to gas bubbles in the molten glass when it is formed. These bubbles must be removed because they affect the strength of glass, as well as its optical properties where such properties are important. Another example is in the casting of metal alloys, wherein unwanted gas bubbles, formed in a mixture of the solidified alloy and the liquid called "the mushy zone," need to be removed.

Relative motion between a bubble or drop and the fluid in which it is suspended enhances the transfer of dissolved substances between the two phases by continuously bringing the bubble or drop into fresh fluid. We take such motion for granted, but it arises mainly from the action of Earth's gravitational field. A bubble or drop experiences a downward force due to gravity that is equal to its mass times the acceleration due to gravity. It also is subject to an upward force exerted by the surrounding fluid. This force, called the buoyancy force, is a consequence of the increase of pressure with depth in this fluid, also caused by Earth's gravitational pull on it. The net force on the bubble or drop is the algebraic sum of these two forces, and causes its acceleration from a state of rest when it is first introduced into a second fluid. If this were the only force on a bubble, it would continue to accelerate, but bubbles and drops move at steady velocity. This is because a moving object experiences a force from the surrounding fluid resisting its movement, which is termed the *hydrodynamic force* on the object. Motion at constant velocity ensues when the forces on the bubble or drop are in balance, leading to zero net force. Some body forces other than gravity, such as those resulting from electric and magnetic fields, also can cause the motion of a bubble or drop under suitable conditions.

Another common consequence of Earth's gravitational field on fluids is *buoyant convection,* also called *free* or *natural* convection. This is motion of a fluid caused by a difference in density when under the influence of gravity. Buoyant convection occurs in everyday situations, such as when heated air, which is less dense than cooler air, rises in a room. It is encountered in larger scale systems, causing atmospheric and oceanic circulation patterns, and in industrial processes in which heat and mass transfer are promoted by it. The physical origin of the buoyant force on a mass of fluid, of a density different from that of the neighboring fluid, is the same as that on a gas bubble or liquid drop, which was discussed in the previous paragraph.

The magnitude of the acceleration due to gravity on Mars is approximately 0.38 of its value on Earth; the value on the surface of our moon is approximately one-sixth of that on Earth. These may be regarded as reduced gravity environments when compared with that on Earth. A much larger reduction occurs within spacecraft that are nearly in free fall, however. Satellites, several hundred of which orbit Earth at present, are a good example of such vehicles. Many of them carry fuel to power thrusters. The need to manage fuel and propulsion systems aboard these satellites required the development of a knowledge base regarding the behavior of fluids – and bubbles and drops suspended in them – under reduced gravity conditions. This is an early example of an activity that prompted a study of phenomena occurring in fluids when gravitational levels are reduced.

Under the right conditions, the steady residual gravity level aboard a vehicle such as the space shuttle can be as much as six orders of magnitude smaller than that on the surface of Earth. This provides an attractive laboratory setting for experiments on materials processing aimed at either taking advantage of the low gravity levels or at unraveling the role of gravity in similar experiments on Earth. The physical science areas most studied in such orbiting laboratories have been the growth of crystals for electronic and biological applications, processing of alloys and composites, and container-less processing. Also, there have been experiments in the biological sciences, both with a view to understanding the changes in biological systems when gravitational effects are reduced and from the perspective of "processing." The materials processing experiments have been designed to take advantage of the near absence of sedimentation of objects of a different density from that of the fluid in which they are present, the near absence of buoyant convection, and the ability to suspend large liquid bodies without the use of a container. Even the tiny levels of background gravitational acceleration aboard the space shuttle may be too large for certain sensitive experiments, however. Examples are experiments that cannot tolerate buoyant convection caused by density variations resulting from temperature or concentration gradients if the container for the fluid is too large.

Virtually all the experimental systems studied in reduced gravity conditions involve fluid masses. In many of them, bubbles, drops, or particles suspended in a fluid are encountered. In space excursions involving humans, it is necessary to use separation processes for recycling oxygen and water. These processes also may involve drops and bubbles for facilitating extraction of the desired species from one phase into another. For these reasons, a study of the behavior of bubbles and drops in a reduced gravity environment is a useful endeavor.

1.2 The Reduced Gravity Environment Aboard Spacecraft

The use of spacecraft provides relatively long durations of free-fall conditions. Shorter durations of reduced gravity conditions can be obtained on the surface of Earth by using other techniques. One such method uses drop towers or drop tubes, which either are evacuated or use some way of reducing the drag exerted by the surrounding gas on the experiment package. The entire experiment package is dropped from the top. In the few seconds available before the package falls to the bottom, it is in free fall in the prevailing gravitational field. In the reference frame attached to the falling experiment, gravitational levels can become small, typically between 10^{-5} and $10^{-6}g$ where g is the acceleration due to gravity at the surface of Earth. A fall distance of 150 m, which is a practically realizable distance, provides a free-fall time of approximately 5.5 s. In some instances, this time is increased by propelling the experiment upward first. Once released, the package is in free fall even as it moves upward, stops, and falls down. The gravity level is not precisely zero because of imperfections such as restraints that guide the fall and because of drag from the surrounding air. Even though the fall occurs commonly in an evacuated tube, there is still some residual air contributing to such drag. The low gravity time can be increased to a value between 15 and 60 s by using aircraft flying ballistic trajectories, but the quality of the low gravity levels is not as good, varying from $0.001g$ to $0.1g$. Sounding rockets that go well above the atmosphere are in near free fall during the coasting phase. The times of free fall are between four and seven minutes, and the gravitational levels to which the experiments are subjected are of the order of $10^{-5}g$.

With regard to the gravitational environment aboard orbiting spacecraft, whether it is an orbiting space vehicle such as the space shuttle or an outer space probe traveling to a planet or another solar system, spacecraft are in a state of free fall in the local gravitational field. In the following, we focus on orbiting spacecraft, which generally are the type of vehicle used for reduced gravity experiments. Some examples are satellites, the U.S. Skylab, and, more recently, the space shuttle, the Russian Mir station, and the international space station that is scheduled to become functional near the beginning of the 21st century. In a reference frame attached to the spacecraft, the acceleration experienced by objects within the spacecraft is not exactly zero for reasons discussed in depth by Hamacher, Fitton, and Kingdon (1987) and summarized in Alexander (1990). First, any object not located at the center of mass of the spacecraft will be subject to a residual acceleration with respect to the spacecraft because of the difference in the distances from the center of mass of Earth to the object and to the center of mass of the spacecraft. This is known as the gravity gradient effect. It leads to acceleration levels of the order of 0.1 to 0.5 μg per meter of distance from the center of mass of the spacecraft, where the symbol μg refers to "micro-g" and means $10^{-6}g$. The actual magnitude of the acceleration level from the gravity gradient effect will depend on the direction of the position vector of the point in question referred to the center of mass of the spacecraft and on the altitude of the orbit. This residual gravitational level can be time dependent or steady, depending on whether the spacecraft maintains a fixed absolute orientation or an orientation fixed with respect to its position vector from the center of mass of Earth, respectively. In addition, atmospheric drag exerts a force on the spacecraft that makes a contribution to the acceleration, which is estimated to vary from a small fraction of a μg at an altitude of 500 km above the surface of Earth to a value as large as 10 μg at an altitude of 250 km, depending on the type of spacecraft. There are other small contributions arising from rotation of the spacecraft and solar radiation pressure. If the spacecraft rotates, moving objects also experience a Coriolis force that will need to be included. Furthermore, the movement of astronauts, the operation of pumps, the dumping of waste, and regular thruster firings made for correcting the spacecraft's attitude, along with the resulting vibrations of the spacecraft, all make time-dependent contributions to the gravitational acceleration experienced by objects within the spacecraft. The time-dependent acceleration covers a broad range of frequencies, and its instantaneous amplitude can be larger than that of the residual steady gravity level by 3 to 4 orders of magnitude. It is difficult to estimate the magnitude of these fluctuations, and usually information about them is obtained from accelerometer measurements made at several locations aboard the spacecraft. A sample acceleration record from the Life and Microgravity Spacelab (LMS) mission of the space shuttle, which took place in summer 1996, from Hakimzadeh et al. (1997), is given in Figure 1.2.1.

The data displayed in the figure were obtained by an instrument called the Orbital Acceleration Research Experiment (OARE). It was designed to measure quasi-steady accelerations between $10^{-8}g$ and $2.5 \times 10^{-3}g$, with a cut-off frequency of 1 Hz. The measured signal was digitized at ten samples per second and processed by using an adaptive filter to exclude transient acceleration of higher magnitude from the quasi-steady data. The x-coordinate is directed from the tail to the nose of the shuttle. In the plane normal to it, the y-direction points from the port wing to the starboard wing, and the z-direction points from the top of the fuselage to the orbiter belly. The relatively quiet periods in the data correspond to times during which the crew members were asleep. The higher acceleration levels correspond to periods of crew and orbiter activity, and the latter is likely the cause of the differences in acceleration levels between the x-direction and those in the other two directions. Additional details can be found in Hakimzadeh et al. (1997). Transient acceleration data at higher frequencies were measured by another instrument, called the Space Acceleration Measurement System (SAMS). Analysis of the role of such transient gravitational fields in affecting fluid motion and the motion of bubbles and drops is an interesting exercise, but too little is known at present to include it here.

From this brief discussion, we see that orbiting laboratories provide for the longest periods of experimentation under reduced gravity conditions. Even in the relatively short period that humans have been able to use such facilities, a good number of studies have been performed, and a vigorous effort is ongoing at this time. The gravitational environment aboard such spacecraft is complex as seen from Figure 1.2.1. Nevertheless, the steady gravity level relative to the spacecraft is typically 5 to 6 orders of magnitude smaller than that on the surface of Earth. This leads one to consider what mechanisms might be active in causing the motion of bubbles and drops in such an environment.

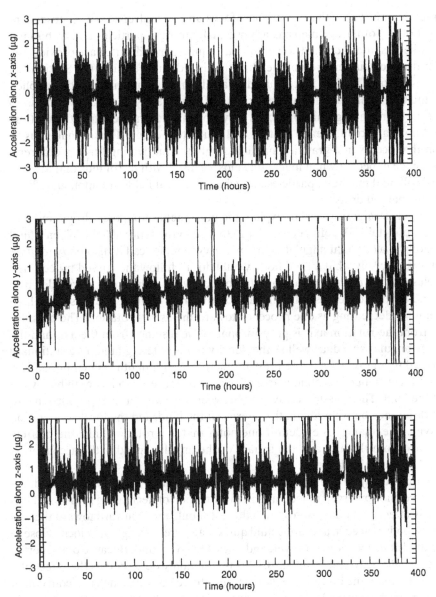

Figure 1.2.1   Typical acceleration environment aboard the space shuttle *Columbia* during the Life and Microgravity Spacelab (LMS) mission. (Figure reproduced with permission of NASA from Technical Memorandum 107401, January, 1997. Roshanak Hakimzadeh, Kenneth Hrovat, Kevin M. McPherson, Milton E. Moskowitz, and Melissa J. B. Rogers. Permission by NASA does not constitute an official endorsement, either expressed or implied, by the National Aeronautics and Space Administration.)

1.3   Mechanisms That Can Cause the Motion of Bubbles and Drops in the Absence of Gravity

Mechanisms independent of gravity that will cause a bubble or drop to move include other body forces induced by electrical or magnetic fields, phoretic phenomena such as electrophoresis, thermophoresis, or diffusiophoresis, rotation or vibration of the container (which, in some cases, can be another mass of fluid floating freely), and interfacial tension gradients. The last of these is our main focus, and we shall discuss it in subsequent sections. All of the other mechanisms apply equally to rigid particles and to bubbles and drops, whereas interfacial tension gradients require a mobile interface provided by a bubble or drop to cause motion of the object. This is discussed further in Section 1.5.

When a body force, such as that induced by an electrical or magnetic field, is applied to a drop, the drop will respond by accelerating. If it is initially at rest, it will begin to move, and the surrounding fluid will exert a hydrodynamic drag resisting the motion. If the conditions are right, steady motion will ultimately ensue. This is analogous to the motion that occurs in a gravitational field. Only the motivating force is different. Subtleties can be expected when magnetic fields are used with ferrofluids that are suspensions of colloidal magnetic particles and when electrical fields are applied to fluids containing uncharged drops.

It is relatively straightforward to explain the movement of bubbles, drops, and particles in a rotating fluid. The effect is similar to that of a gravitational field. When a fluid rotates steadily at a constant angular velocity everywhere, we call it *rigid body rotation* because this is precisely what a rigid body would do. To follow the curved trajectories, fluid particles must be pushed inward by a force everywhere. This force is provided by a pressure gradient that develops naturally in a rotating mass of fluid. The pressure is high at the periphery, decreasing as one proceeds toward the axis of rotation. The pressure difference from the rotation axis is proportional to the distance from the axis and the square of the angular speed, as well as the density of the rotating fluid. In a stationary fluid subjected to a gravity field, the pressure increases with depth, as mentioned earlier, because the weight of the fluid above a location increases with its depth below the surface of the fluid. The pressure difference between any two points is proportional to the height difference between them, the density of the fluid, and the local acceleration due to gravity. The analogy is evident if one notes that in a rotating fluid, the product of the distance from the axis and the square of the angular speed is the acceleration experienced by the fluid elements.

In a rotating fluid, just as in the case of a stationary fluid in a gravitational field, the buoyancy force on submerged objects acts in the direction of decreasing pressure. If this were the only effect on a body in a rotating fluid, it should move inward toward the axis of rotation. A body placed in a rotating fluid quickly acquires an angular velocity close to that of the fluid at its location, however, and begins to go around. Because of its inertia, its tendency would be to spiral outward. This is commonly described as the effect of a "centrifugal" force on the body, which is a fictitious force but nevertheless a convenient way to recognize this particular manifestation of inertia. The force is proportional to the mass of the body and the product of its distance from the axis and the square of its angular speed. The ultimate consequence is that the body will move away from the rotation axis if it is more dense than the fluid and toward the rotation axis if it is less dense, reminiscent of the behavior of objects in a gravitational field. This motion will be resisted by a hydrodynamic force exerted by the fluid. Because the effective gravity experienced by the body depends on its distance from the rotation axis even in a steadily rotating fluid, the velocity will continually change, and there is no equivalent of a steady settling situation as that obtained in Earth's gravitational field. Rotating fluid flows are interesting in themselves, and a good discussion of phenomena that can occur in a rotating body of fluid can be found in the book by Greenspan (1969).

The behavior of drops and bubbles in a rotating liquid has been the subject of many studies. Some of the early impetus appears to have come from a desire to understand the motion of bubbles in tanks containing liquid propellant (see, for example, Clifton, Hopkins, and Goodwin, 1969) in a low gravity environment. Applications also exist in industrial situations on Earth, where bubbles are eliminated by rotating a liquid body. This action forces them to the axis of rotation, where they can coalesce and be removed more easily. The motion of bubbles and drops in rotating fluids has been investigated by Schrage and Perkins (1972), Siekmann and Johann (1976), Siekmann and Dittrich (1977), Annamalai, Subramanian, and Cole (1982b), Annamalai and Cole (1983, 1986), Ruggles et al. (1988), and Ruggles, Cook, and Cole (1990). For a discussion of the motion of rigid particles in a rotating fluid, the reader is referred to Herron, Davis, and Bretherton (1975).

Annamalai et al. (1982a) have suggested that, in reduced gravity, the rotation of a freely suspended drop containing one or more bubbles may be a useful way to move the bubbles to desired locations. One possibility is that small bubbles can be made to coalesce into larger ones after migrating to the rotation axis. Such larger bubbles can be removed from the liquid mass by applying local heating to the surface, causing interfacial tension gradient-driven flows and bubble motion. The rotation of a drop can be achieved by using an acoustic field as discussed by Wang, Saffren, and Elleman (1974). A device that employs this technique has been built and flown on the space shuttle a few times with the objective of studying fluid mechanical problems of relevance to containerless processing. The term *containerless processing* implies that one is working with materials in the fluid state, suspended without a container. This can be achieved on Earth for objects that weigh very little, but large, heavy drops cannot be levitated easily against gravity. In reduced gravity, even massive drops can be floated freely, but one must prevent the drop from drifting off and colliding with a container wall because of residual gravitational effects. This is why an acoustic field is used to create a gentle potential well in which the drop can be held. The same acoustic field can be manipulated to produce drop rotation or oscillation. Containerless processing is a useful endeavor because it permits one to study the possible effects of crystallization induced by container walls in glass-forming systems. At one time, it was suggested as a means of extending the range of glass-forming compositions in rare earth oxide mixtures because it was believed that the container wall promoted crystallization. Also, containerless processing allows the measurement of the physical properties of materials in pure form, when these properties can be affected by contamination from the container.

The term *phoretic phenomena* refers to the motion of a particle in a field that interacts with the surface of the particle. Anderson (1989) has provided a good discussion of this topic. Some examples of phoretic motion include electrophoresis, which is caused by the action of an applied electric field on a particle that carries a surface charge, thermophoresis, which is driven by a temperature gradient in the surrounding fluid, and diffusiophoresis, which is caused by solute composition gradients in the fluid. *Photophoresis* is a term used to designate motion that occurs because of nonuniform absorption of electromagnetic radiation by a particle but that is really a thermophoretic consequence of the resulting temperature gradient in the fluid surrounding the particle, usually a gas.

In phoretic motion, if one includes fluid in a thin region surrounding the particle, the external field imposes no force or torque on the particle taken together with this fluid. The thickness of this region depends on the nature of the field involved and the physical characteristics of the system. It is the mobility of the fluid in this region near the particle that leads to relative motion between the particle and the fluid as a whole.

Phoretic motion does not require that the interface itself be mobile and applies equally well to rigid particles as it does to bubbles and drops. In fact, the familiar example of the soot deposit formed inside the glass cover of an oil lamp is attributed to thermophoresis, which drives the soot particles from the hot region of the flame to the relatively cool glass surface. Thermophoresis is used in air cleaning operations and is known to play a role in the process used for forming the precursors for optical waveguides. Thermophoresis has been discussed by Rosner et al. (1992), who also provided several useful references. Electrophoresis results directly from the interaction of an applied electric field and the double layer of electrical charge on a particle, and much has been written about it. Various aspects of electrophoresis are discussed in the books by Shaw (1969), Russel, Saville, and Schowalter (1989), and Masliyah (1994). Diffusiophoresis is a term used to describe the movement of particles due to gradients in the concentration of solute. To cause it, there must be an attractive or repulsive interaction between the particle and the solute molecules. A simplistic picture of the physical mechanism can be obtained (for example, in the case of an attractive potential) by noting that the solute in the higher concentration side of the particle will exert a greater pull on it than that on the lower concentration side, resulting in particle movement toward regions of higher solute concentration. The review by Anderson (1989) is a good starting point for learning more about this and related phenomena.

1.4 Interfacial Tension

1.4.1 Elementary Concepts

Central to our subject is the concept of interfacial tension and its dependence on temperature. We also need to appreciate that interfacial tension additionally depends on the concentration of adsorbed surface active species known as *surfactants*, which are discussed later. It is not our object to devote a large discussion to the subject of interfacial tension because detailed treatments already are available in articles by Ono and Kondo (1960) and Buff (1960). Various aspects of the interface also are discussed in books by Davies and Rideal (1963), Defay and Prigogine (1966), Gaines (1966), Adamson (1976), Jaycock and Parfitt (1981), Miller and Neogi (1985), and Lyklema (2000).

Briefly, when two immiscible fluids are brought into contact, an apparent surface of separation, which we call the interface, is observed. Such an interface can be formed between a gas and a liquid or between two liquids. Molecules in the bulk are surrounded in all directions by molecules of the same species and do not experience a net force on a time-average basis. This is not the situation in the case of molecules in the neighborhood of an interface with a second fluid. This is the origin of interfacial tension. The interface is an active region with molecules constantly moving in and out of it, even though it may appear quiescent to the human eye. The term *interface* usually refers to a region a few molecular diameters thick within which the density and composition change rapidly from one side to the other. In a system at thermodynamic equilibrium, these quantities approach the values prevalent in the bulk phases outside of this region. There are a variety of levels at which the interfacial region is handled in modeling. In dealing with transport phenomena in fluids, it is typical to treat the interface as a geometrical surface of zero thickness, at which certain boundary conditions are applied on the field variables. This of course ignores the structure of the interfacial region but is consistent with the approach of ignoring the molecular structure that is taken regarding the bulk phases, as discussed in Chapter Two. Such an approach is sufficient to deal with the class of problems with which we are concerned. In Chapter Two we introduce the boundary conditions commonly employed at the interface between two fluids. Here, a few more words are in order regarding the interface and the behavior of the interfacial tension.

When an interface exists between two immiscible fluids, molecules of a substance in the interfacial region experience a pull toward the bulk phase of that region. As a result, the interface displays a tendency to minimize the area it occupies. This explains why small gas bubbles, liquid drops, and soap bubbles are spherical in shape. As a result of its tendency to contract, the interface appears to be in a state of tension. The interfacial tension is a force exerted per unit length of an imaginary line on the interface by interfacial fluid on one side of this line on the interfacial fluid on the other side. The sense of the force is to pull on the line from both sides. In view of the units, it also is common to think of this as an energy per unit area of the interface and refer to it as the excess free energy associated with forming the interface. This numerical equality with an interfacial free energy does not hold in multicomponent systems but is true, for example, for a pure liquid in equilibrium with its own vapor.

Ono and Kondo (1960) have described a fundamental approach to the calculation of interfacial tension, based on its origin from the interactions among molecules. The approach is based on writing the potential energy function as a superposition of pairwise potentials of interaction between molecules and using a distribution function to describe the variation of the density of molecules with distance from a given molecule. The distribution functions depend on whether the system is in equilibrium or not. The potential energy function is related to the Helmholtz free energy via the partition function. In the equilibrium statistical thermodynamics approach, as well as in the statistical mechanical approach that applies when the system is not in equilibrium, the same integral expression is obtained for the interfacial tension. The result for the interfacial tension consists of a thermal contribution, as well as a contribution from the intermolecular potential and the distribution function for pairs of molecules. Jaycock and Parfitt (1981) have pointed out that the practical utility of such methods is limited even for calculating the interfacial tension of a pure liquid because of a lack of knowledge of the precise distribution functions in the interfacial region. X-ray diffraction data, which are typically used to obtain the bulk distribution functions used in the models, are not applicable in the interfacial region.

We now discuss a macroscopic description of the interfacial region and give the definition of the interfacial tension from that description. The thermodynamics of interfaces is discussed in the books by Guggenheim (1967) and Hill (1963/1964). Summaries of the relevant thermodynamic results also are given in several references cited earlier.

The brief discussion here follows Guggenheim's (1967) development. If the interface is curved, the picture is more complex when the thickness of the interfacial region is not small compared with the radius of curvature. Therefore, this discussion is limited to a planar interface between two phases in equilibrium. It is convenient to divide the region straddling the interface into an interfacial region ω and bulk regions α and β on either side as shown in Figure 1.4.1. The distinction is that within the region ω, properties such as composition, internal energy, and entropy change in a direction normal to the planar bounding surfaces that intersect the plane of the diagram in the straight lines AA' and BB', whereas these properties are uniform in the regions α and β. Properties in all three regions are assumed not to vary in the direction normal to the plane of the diagram nor along the direction parallel to AA' and BB'. The normal force acting

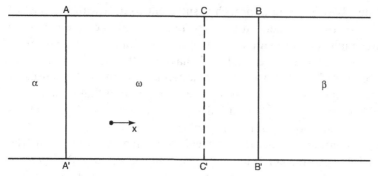

Figure 1.4.1   Schematic of the interfacial region.

on an element of area will be isotropic in the bulk phases, and this force per unit area is called the pressure P. This means that regardless of the orientation of the area element, the force is the same. This is not true in the interfacial region. Here, area elements parallel to the bounding surfaces will experience the same normal force P regardless of their location. Other area elements inclined with respect to such elements will experience different normal forces, however, depending on their orientation and location. If we construct a line of length h normal to AA' and BB' that extends from one bulk phase to another (and therefore passes through the entire interfacial region) and consider the area formed by this line and another of width ℓ normal to the plane of the figure, the normal force on this area will be $(Ph - \sigma)\ell$ where σ is defined as the interfacial tension. In fact, the following definition of σ can be developed from this picture. Consider a coordinate x measured from a straight line parallel to AA' and located somewhere within the interfacial region. Construct area elements using a width ℓ normal to the plane of the figure and a straight line on the plane of the figure, taken either parallel or perpendicular to AA'. Designate the normal force per unit area on such area elements with a side parallel to AA' as P, and the normal force per unit area on area elements oriented perpendicular to these elements as p. Whereas P is independent of x, p will be a function of x. The interfacial tension is then given by the following integral:

$$\sigma = \int_{-\infty}^{\infty} (P - p)\, dx. \tag{1.4.1}$$

The limits on the integral are formally an infinite distance away from the interface in opposite directions, providing for an asymptotic approach of $P - p$ to zero as $|x| \to \infty$. Of course, in practice, the difference between P and p will become negligible once $|x|$ exceeds a few molecular diameters.

For developing the thermodynamics of the interface, we consider the interfacial region occupying a width ℓ normal to the plane of Figure 1.4.1. By examining the consequences of a differential change in the volume of this region while keeping its material content unchanged, one can obtain an expression for the resulting work done on ω by the forces acting on it as $-P dV^{\omega} + \sigma dA$, where the superscript identifies the region, V is the volume, and A is the area of the interfacial region. This is different from the work associated with a similar volume change in the bulk phase α, for example, which is given by $-P dV^{\alpha}$. From here, one can define thermodynamic functions such as the internal energy, enthalpy, and the Helmholtz and Gibbs free energy in the interfacial region. The enthalpy function in the interfacial region is defined slightly differently from that in the bulk, so that changes in the Gibbs free energy function represent net work. Note that in this equilibrium system, the chemical potentials of each species remain the same throughout the three regions.

It is possible to treat the problem from a different perspective, originally put forth by Gibbs according to Guggenheim (1967). The results are entirely equivalent. Gibbs proposed the concept of a geometrical dividing surface of zero thickness, shown as the dashed line CC' in Figure 1.4.1, placed somewhere in the interfacial region. Assume that the bulk properties on either side extend all the way up to this imaginary dividing surface. Then, for extensive properties, one can define a surface excess as follows. Consider a region enclosing the interfacial region bounded by AA' and BB' and extending into both bulk phases. An accounting is made of the entity in question in the real system. Now, the total amount of the same entity is calculated for the hypothetical system with bulk phases on either side extending right up to the dividing surface, recalling that the properties are uniform within each bulk phase at equilibrium. The excess of the former over the latter (which can be negative or positive) is termed the *surface excess* of the entity. The amounts of solute and solvent in a liquid in contact with its vapor can serve as useful examples of this concept. Using unit area of the interface as a basis, one can define a surface excess concentration of each at the interface, the implication being a positive or negative adsorption of the species at the interface, depending on the sign of the surface excess. Of course, one problem is the vagueness regarding the location of this dividing surface. The surface excess quantities just defined will vary depending on the actual location. Gibbs chose to position this hypothetical surface such that the surface excess of the solvent is zero, leading only to a surface excess of solute. For a given region straddling the interface, this can then be defined as the excess of solute per unit area of surface over that which would be present in a bulk region containing the same number of moles of solvent as does the selected region. One can make other choices for the dividing surface, as detailed in several references mentioned earlier. The thermodynamic quantities are defined such that they are independent of the location of the dividing surface. For a planar interface, we can write the Gibbs-Duhem equation as follows:

$$d\sigma = -S^{(\sigma)}dT - \sum_i \Gamma_i d\mu_i. \qquad (1.4.2)$$

Here, $d\sigma$ is a differential change in interfacial tension that corresponds to differential changes in the surface excess concentrations Γ_i of the species present and a differential change in temperature dT. The symbol T represents the thermodynamic temperature measured on an absolute scale. The symbols μ_i represent the chemical potentials of the species present, and $S^{(\sigma)}$ is the surface excess entropy. The summation in Equation (1.4.2) is carried out over the number of species present. For a binary system at constant temperature, choosing the dividing surface so that the surface excess of the solvent A is zero, we obtain

$$d\sigma = -\Gamma_B d\mu_B \qquad (1.4.3)$$

for solute B. From here, relating the chemical potential of the solute to its activity a_B, it is possible to obtain the following result, which is known as the Gibbs adsorption isotherm:

$$\Gamma_B = -\frac{a_B}{RT}\frac{d\sigma}{da_B}. \qquad (1.4.4)$$

The symbol R stands for the gas constant. The idea is that when the temperature is held fixed, this result permits one to make a connection between the rate of change in interfacial tension with solute activity and the surface excess of that solute. From Equation (1.4.4), it is clear that when a solute lowers the interfacial tension, it will be positively adsorbed onto the interface. In the case of an ideal solution, the activity coefficient is unity so that the activity a_B can be replaced by the mole fraction of the solute. In the dilute solution limit, the mole fraction is proportional to the concentration of solute c_B so that we can simplify Equation (1.4.4) to the following useful form:

$$\Gamma_B = -\frac{c_B}{RT}\frac{d\sigma}{dc_B}. \tag{1.4.5}$$

Equation (1.4.5) can be used to evaluate the surface excess of a solute, given the dependence of the surface tension on the concentration of the solute, provided the simplifying assumptions used in obtaining it are satisfied.

1.4.2 The Dependence of Interfacial Tension on Adsorbed Surfactant Concentration

Certain molecules are composed of some segments that prefer to be within one fluid and other segments that do not. An example is a long chain organic fatty acid or alcohol placed in water. The nonpolar hydrocarbon chain in the molecule is hydrophobic (water hating), whereas the polar end group is hydrophilic (water loving). The hydrophobic part resists dissolution in water, whereas the polar group promotes it. If the chain is sufficiently short, as in the case of ethyl alcohol, the organic material is perfectly miscible with water. As the chain length is increased, however, the solubility decreases, and the organic molecule displays more of a tendency to adsorb onto the air-water interface, where the polar end group is surrounded by water while the hydrocarbon tail sticks out of the water. When a species adsorbs onto the interface and significantly reduces the interfacial tension, it is considered surface active, or a *surfactant*, for the interface in question. Many surface-active agents also can ionize in solution and are classified as anionic or cationic depending on whether the surface active hydrophobic chain has a negative or positive charge, respectively. A surface-active agent that does not ionize is termed *nonionic*.

Because surface-active species lower the interfacial tension, an interface possessing a high value of this tension is more susceptible to their adsorption than one with a low interfacial tension. These are, of course, relative terms. Water, with a surface tension $\sigma \approx 72.8$ mN/m against air, is very susceptible to contamination by small amounts of organic material that display surface activity. On the other hand, certain liquids such as silicone oils, which are siloxane polymers with alkyl groups attached to the carbon atoms in the backbone, have a relatively low value of the surface tension against air of $\sigma \approx 21$ mN/m. Even though several chemicals, such as the polymerization initiators, terminators, and the monomer, must be dissolved in these silicone oils, fluid mechanical experiments performed using them reveal no sign that these substances are surface active in the silicone oils.

It is possible for the adsorbed species to form a film one molecule thick, which is called a *Gibbs monolayer*. Monolayers are discussed in great detail, for example, in the book by Gaines (1966). Because we shall be doing some modeling later that involves the influence of monolayers on the motion of drops, we briefly present the necessary thermodynamic concepts here. The main idea is to describe the influence of the concentration of surfactant in the adsorbed monolayer on the interfacial tension. There are many models for doing this, and the simplest one is the ideal gas model. First we introduce a useful concept called *surface pressure* for which the symbol Π is used. This is a term used to identify the reduction in surface tension caused by a surfactant film adsorbed on the interface. If σ_0 designates the interfacial tension at a surfactant-free interface, and $\sigma(\Gamma_B)$ is the interfacial tension corresponding to an adsorbed surfactant concentration Γ_B, then

$$\Pi = \sigma_0 - \sigma(\Gamma_B). \tag{1.4.6}$$

It can be measured directly as shown by Langmuir (1917) by using a suitable device that permits compression of this film gradually while measuring the reduction in interfacial tension. In this way, the surface pressure can be related to the adsorbed surfactant concentration. When only a small number of surfactant molecules are present on the interface, each molecule has a relatively large area in which to move along the surface before encountering another molecule. In this respect, we can construct an analogy to the ideal gas, ignoring interactions of these molecules with each other and with the liquid. Because there are two degrees of freedom for a molecule moving about on the surface, this leads to the following equation of state:

$$\Pi A_M = kT. \tag{1.4.7}$$

Here, A_M is the area assigned on the surface to each molecule, and k is the Boltzmann constant. Converting to a molar basis, we can write

$$\sigma_0 - \sigma = \Pi = \Gamma_B RT, \tag{1.4.8}$$

where we identify the inverse of the area per mole of adsorbed species as the concentration of the adsorbed species per unit area on the interface, Γ_B. This ideal gas model of the adsorbed film is only useful when the concentration of adsorbed species is small. One must use more involved equations of state to account for interaction effects among the molecules and with the solvent. These are discussed in Gaines (1966), as well as in some of the other references mentioned earlier. We do not devote space to them here because we do not consider situations that necessitate such equations of state.

1.4.3 The Dependence of Interfacial Tension on Temperature

We shall be concerned mostly with motion in a fluid that is driven by the variation of interfacial tension with temperature. Hence, it is useful to provide a few comments on this subject. The interfacial tension of a liquid consisting of a single component against its own vapor will decrease with temperature. It can be shown that the rate of change of the interfacial tension with temperature $\sigma_T = -S^{(\sigma)}$. The reader will recall that $S^{(\sigma)}$ stands for the surface excess entropy, which is positive. Such decrease of the interfacial tension with temperature is not always to be expected in multicomponent systems. As a practical matter, in most liquid-gas systems, the interfacial tension does decrease with temperature. At the liquid-gas interface, the temperature coefficient σ_T is practically constant over a temperature range extending over several tens of Kelvin. The interfacial tension at liquid-liquid interfaces behaves in a more complex way with temperature, and it is more common to find the coefficient σ_T varying with temperature, even over moderate ranges in temperature. Additional discussion may be found in several references cited earlier, which deal with the thermodynamics of interfaces in detail.

The surface tension of most liquid elements, as a function of temperature, can be found in the CRC Handbook of Chemistry and Physics (1999). Reid, Prausnitz, and Poling (1987) have devoted a chapter in their book to the estimation of surface tension. They recommended using a compilation of data by Jasper (1972) for organic liquids. Jasper has collected data on the variation of surface tension with temperature for approximately two thousand pure liquids from various sources, and most of the data have been fitted to a straight line. Reid et al. provided two techniques for estimating σ for pure liquids when experimental data are not available. If the liquid is non-hydrogen-bonded, a result based on corresponding states is suggested as optimal, whereas for hydrogen-bonded liquids, a correlation attributed to Macleod and Sugden is recommended. When the reduced temperature, which is the ratio of the absolute temperature to the critical temperature, lies between 0.4 and 0.7, the results from the correlations are well approximated by a linear relationship between surface tension and temperature. In the case of liquid mixtures, Reid et al. discussed nonaqueous and aqueous solutions separately. For nonaqueous systems, a correlation obtained from thermodynamics, which requires a knowledge of activity coefficients, is recommended for accurate evaluations. Less accurate results are achieved by using an extension of the Macleod–Sugden or the corresponding states correlations. For aqueous solutions of organics, Reid et al. recommended a correlation based on modifying the Macleod–Sugden result. The interested reader should consult the book by Reid et al. for more information.

Many measurements of σ_T, assumed independent of temperature, have been reported in the recent literature, especially for silicone oils and fluorocarbons. These are displayed in Table 1.4.1 for several pairs of fluids. Unless otherwise stated, the values of σ are at room temperature, which can be taken to be $T = 25°C$. The values of σ_T are based on measurements of σ at temperatures in the vicinity of room temperature. Because the temperature coefficient is small, to obtain reliable estimates of σ_T, such interfacial tension measurements are usually carried out over a few tens of Kelvin. In the case of the silicone oils, we have included trade names used by different manufacturers in Table 1.4.1. Uncertainty estimates, as reported by the authors of the cited references, are included in the table where available. The reader should consult the original article for the method by which the uncertainty was evaluated. In Table 1.4.2, we provide the surface tension as a function of temperature, from Jasper (1972), for a few selected fluids. A typical value of the uncertainty reported by Jasper is 0.1 mN/m. The term *vapor* in Table 1.4.2 implies the pure vapor of the same liquid.

1.5   Description of Interfacial Tension Driven Motion

Now, we turn to a brief explanation of how changes in interfacial tension can cause the movement of bubbles or drops suspended in a fluid. Because the most common way to change the interfacial tension is through a temperature variation in the fluid, we use this for illustration. Note that a gradient of concentration of a substance that affects the interfacial tension, along the interface, will lead to similar consequences. A film by Trefethen (1963) demonstrates several fluid mechanical situations in which motion is driven by interfacial tension gradients.

When a bubble or drop is placed in a fluid in which the temperature changes from one place to another, temperature variations can be expected to arise at the interface. The consequence is a variation of the local interfacial tension along the interface. This variation causes elements of fluid in the interfacial region to experience a pull from

Table 1.4.1 Interfacial Tension and Its Temperature Coefficient for Some Fluid Pairs

Fluid Pair	σ mN/m	σ_T mN/(m·K)	Reference
Ethylene glycol/Air	46.9	-0.077 ± 0.019	Thompson (1979)
Fluorinert FC-40/Air	17.5 ± 0.09	-0.075 ± 0.0017	Burkersroda et al. (1994)
Fluorinert FC-70/Air	18.6 ± 0.11	-0.067 ± 0.002	"
Fluorinert FC-75/Air	15.1 ± 0.07	-0.083 ± 0.0013	"
Potassium chloride/Air	97.5 @ 790°C	-0.073	Velten et al. (1991)
Silicon/Argon	874 @ 1450°C	-0.279 ± 0.004	Hardy (1984)
Silicone oil KF-96L 1 cs/Air	16.2	-0.0459	Kurosaki et al. (1989)
Silicone oil Hüls 2 cs/Air	18.2 ± 0.12	-0.062 ± 0.0022	Burkersroda et al. (1994)
Silicone oil Hüls 10 cs/Air	19.7 ± 0.07	-0.060 ± 0.0013	"
Silicone oil DC-200 10 cs/Air	19.0 ± 0.1	-0.061 ± 0.003	Hadland et al. (1999)
Silicone oil KF-96 10 cs/Air	20.1	-0.0375	Kurosaki et al. (1989)
Silicone oil DC-200 50 cs/Air	20.1 ± 0.1	-0.0623 ± 0.002	Balasubramaniam et al. (1996)
Silicone oil AK-50 50 cs/Air	19.9	-0.059	Schwabe et al. (1990)
Silicone oil Hüls 100 cs/Air	20.7 ± 0.17	-0.054 ± 0.0033	Burkersroda et al. (1994)
Silicone oil KF-96 100 cs/Air	20.9	-0.0516	Kurosaki et al. (1989)
Silicone oil DC-200 150 cs/Air	21	-0.05	Schwabe et al. (1990)
Sodium nitrate/Air	117.5 @ 360°C	-0.055	Velten et al. (1991)
Tetracosane/Air	26 @ 90°C	-0.067	"
Butylbenzoate/Bidistilled water	24.4 ± 0.2	-0.0322 ± 0.003	Ma (1998)
Castor oil/Silicone oil DC-200 50 cs	3.6 ± 0.1	-0.0148 ± 0.0019	"
Fluorinert FC-40/Silicone oil Hüls 2 cs	4.5 ± 0.052	-0.035 ± 0.0009	Burkersroda et al. (1994)
Fluorinert FC-40/Silicone oil Hüls 10 cs	6.1 ± 0.063	-0.037 ± 0.0011	"
Fluorinert FC-40/Silicone oil Hüls 100 cs	7.0 ± 0.084	-0.026 ± 0.0017	"
Fluorinert FC-70/Silicone oil Hüls 2 cs	5.4 ± 0.093	-0.032 ± 0.0015	"
Fluorinert FC-70/Silicone oil Hüls 10 cs	6.9 ± 0.091	-0.029 ± 0.0014	"
Fluorinert FC-70/Silicone oil Hüls 100 cs	7.8 ± 0.047	-0.026 ± 0.0008	"
Fluorinert FC-75/Silicone oil Hüls 2 cs	3.2 ± 0.125	-0.038 ± 0.0024	"
Fluorinert FC-75/Silicone oil Hüls 10 cs	4.9 ± 0.133	-0.024 ± 0.0023	"
Fluorinert FC-75/Silicone oil DC-200 10 cs	4.6 ± 0.1	-0.036 ± 0.002	Hadland et al. (1999)
Fluorinert FC-75/Silicone oil DC-200 50 cs	5.3 ± 0.1	-0.0320 ± 0.0012	Balasubramaniam et al. (1996)
Fluorinert FC-75/Silicone oil Hüls 100 cs	5.9 ± 0.055	-0.030 ± 0.0010	Burkersroda et al. (1994)
Vegetable oil/Silicone oil DC-200 5 cs	1.2 ± 0.15	-0.0088 ± 0.001	Rashidnia and Balasubramaniam (1991)

Table 1.4.2 Surface Tension as a Function of Temperature for Some Fluid Pairs. (Data from Jasper, J.J. 1972. The Surface Tension of Pure Liquid Compounds. *J. Phys. Chem. Ref. Data* 1, No. 4, 841–1009.)

Fluid Pair	σ mN/m	
Acetone/Air	$26.26 - 0.1120T \pm 0.15$	$25 \le T(C) \le 50$
Benzaldehyde/Air	$40.72 - 0.1090T \pm 0.5$	$10 \le T(C) \le 100$
Carbon tetrachloride/Air	$29.49 - 0.1224T$	$15 \le T(C) \le 105$
Cyclohexane/Air	$27.62 - 0.1188T \pm 0.1$	$5 \le T(C) \le 70$
Decane/Nitrogen	$25.67 - 0.0920T \pm 0.1$	$10 \le T(C) \le 120$
Deuterium oxide/Vapor	$80.62 - 0.2201T \pm 0.1$	$100 \le T(C) \le 215$
Dibutyl ether/Air	$24.78 - 0.0934T \pm 0.1$	$15 \le T(C) \le 90$
Diethylene glycol/Vapor	$46.97 - 0.0880T \pm 0.14$	$20 \le T(C) \le 140$
Dodecane/Nitrogen	$27.12 - 0.0884T \pm 0.1$	$10 \le T(C) \le 120$
Ethanol/Air	$24.05 - 0.0832T \pm 0.1$	$10 \le T(C) \le 70$
Ethyl acetate/Air	$26.29 - 0.1161T \pm 0.1$	$10 \le T(C) \le 100$
Ethylene glycol/Vapor	$50.21 - 0.0890T \pm 0.14$	$20 \le T(C) \le 140$
Formamide/Air	$59.13 - 0.0842T \pm 0.2$	$25 \le T(C) \le 120$
Heavy water/Air	$74.64 - 0.1082T^{1.1} \pm 0.18$	$10 \le T(C) \le 75$
Helium II/Vapor	$0.352 - 0.0069T^{\frac{7}{3}}$	$0.5 \le T(K) \le 2.1$
Hexane/Nitrogen	$20.44 - 0.1020T \pm 0.1$	$10 \le T(C) \le 60$
Isopropyl alchohol/Air	$22.90 - 0.0789T \pm 0.1$	$10 \le T(C) \le 80$
Mercury/Vapor	$490.6 - 0.2049T \pm 2.0$	$5 \le T(C) \le 200$
Methanol/Air	$24.00 - 0.0773T \pm 0.1$	$10 \le T(C) \le 60$
Octane/Nitrogen	$23.52 - 0.0951T \pm 0.1$	$10 \le T(C) \le 120$
Toluene/Air	$30.90 - 0.1189T \pm 0.1$	$10 \le T(C) \le 100$
Water/Air	$75.83 - 0.1477T \pm 0.1$	$10 \le T(C) \le 100$
Xenon/Vapor	$0.03703(289.74 - T)^{1.287}$	$165 \le T(K) \le 285$

C = degrees Celsius; K = Kelvin.

the lower interfacial tension side toward the side with higher interfacial tension. This pull, expressed as tangential force per unit area, is often termed the *thermocapillary* stress because its origin is from temperature gradients and capillary effects, the latter being a term used to describe effects associated with the interface. The consequence of this stress is to cause the fluid adjacent to the interface on either side to move. The stress transmitted to fluid within the drop cannot cause any force to be exerted on the drop. The stress transmitted to the exterior fluid will cause a reaction on the drop of the same magnitude, but opposite in direction. If the drop is held fixed by attaching it to a wire, for example, it would experience this reaction force. If it is free to move, it will begin to move in the direction of this reaction force when it is placed in the continuous phase. Because the interfacial tension in many systems decreases as the temperature increases, bubbles and drops are found to swim toward warmer regions in the fluid. In thermocapillary motion of a drop, there is no external force acting on the drop, as is the case with gravitationally driven motion. This has some interesting consequences, as we shall see in Chapters Four and Seven. A somewhat analogous situation exists with respect to phoretic phenomena. In electrophoresis of a charged colloidal particle, the electric field exerts a force on the particle. The field also exerts an opposite force on the double layer of counterions adjacent to the particle. If one examines the situation from outside an envelope surrounding the particle and the double layer, which is very thin, there is no net force on the contents. This affects the way the flow decays away from the particle, which is similar to that observed in comparable thermocapillary migration problems involving drops.

It also is possible for a drop, initially stationary in isothermal surroundings, to move because of the action of a uniform source of heat within the drop, or uniform generation of heat at the surface of the drop due to chemical reaction, because of the thermocapillary effect. In this situation, the source of energy leads to a uniform temperature on the surface of the drop, which can be different from that of the undisturbed continuous phase. In the absence of temperature variations along the interface between the two fluids, there can be no thermocapillary movement of the drop. The symmetry can be broken, however, by a disturbance that causes slight movement of the drop. Such movement will lead to a temperature variation on the surface of the drop because of nonuniform heat transport between the drop and the surrounding fluid. Under the right conditions, the drop will continue to move in the same direction because of the action of the thermocapillary stress, sustaining the motion indefinitely. Although this phenomenon has not been experimentally observed yet, it was predicted by Ryazantsev (1985). Subsequent analyses can be found in Golovin, Gupalo, and Ryazantsev (1986), and Rednikov, Ryazantsev, and Velarde (1994a). We do not consider this topic further, but the interested reader can consult a review provided by Rednikov, Ryazantsev, and Velarde (1994b). On the general subject of thermocapillary migration of drops and bubbles, three reviews are available. They are authored by Wozniak, Siekmann, and Srulijes (1988), Subramanian (1992), and most recently, Subramanian, Balasubramaniam, and Wozniak (2001).

When a gravitational force also acts on the drop at the same time, the relative importance of the thermocapillary stress in driving drop motion will depend on the size of the object. The net force exerted by gravity on a drop immersed in a fluid is proportional to the volume of the drop, whereas the thermocapillary stress acts on the surface, and the force arising from its action on the entire drop is proportional to the interfacial area of the drop. Therefore, one can conclude that thermocapillary effects will be prominent in influencing the motion of drops of sufficiently small size on Earth. As an example, a gas bubble of radius 20 μm, placed in a temperature gradient of 1 K/mm in a silicone oil, will move with approximately the same velocity that it would have if rising under the influence of buoyancy. If one uses a liquid drop and adjusts the density difference to be almost zero, then thermocapillary effects can be observed on Earth at much larger drop sizes. In an environment where the gravitational level is about $10^{-6}g$, even under a gentle gradient of 0.01 K/mm, a bubble must be as large as 0.2 m in radius for the gravitational contribution to its motion to be comparable to that of thermocapillarity.

The movement of objects in a temperature gradient is consistent with thermodynamic principles. The systems involved are not equilibrium systems because the temperature is not uniform. Nonetheless, in most cases we can assume local equilibrium in applying these principles. The thermocapillary velocity of a drop can be calculated using the idea that the work done by the thermocapillary stress in moving fluid at the interface is irreversibly converted to heat by viscous dissipation in the bulk fluids. We use this approach when needed. In a similar manner, when an object moves because of a body force such as that arising from the action of gravity, the work done by the object on the fluid, which exerts a drag on it, must equal the heat generated by viscous dissipation. This concept has been used on occasion to calculate the hydrodynamic drag exerted by a fluid on a moving object.

The same thermocapillary stress, which causes the motion of bubbles and drops, also will lead to motion in a fluid held in a container when an interface between two fluids is present. Some examples are a pool of liquid exposed to a gas or vapor or multiple layers of liquid such as those encountered in making photographic film or in encapsulated crystal growth. In these situations, thermocapillary motion becomes important compared with that caused by buoyancy in a gravitational field when the ratio of surface area to volume becomes sufficiently large. This is the case in shallow liquid layers.

The presence of fluids on both sides of an interface is crucial to motion driven by the interface. One can define an interfacial tension for the interface between a rigid object and a fluid, but a stress arising from a gradient of interfacial tension along the surface will be communicated entirely to the rigid object. No stress will be transmitted to the adjacent fluid and, as a consequence, the object itself will not move. In the reverse situation, where a drop is located within a rigid solid, thermocapillary stress arising from a gradient of interfacial tension will be communicated entirely to the solid in which the drop is submerged, and the drop will not move because of the fluid mechanical consequences of this stress. Interestingly, drops and bubbles do move in a solid under the action of a temperature gradient, as discussed by Anthony and Kline (1972), but the mechanism is not of fluid mechanical origin.

We shall not consider the motion of rigid particles. In Section 1.3, we mentioned various mechanisms independent of gravity that can cause the movement of rigid particles, as well as that of drops and bubbles. These mechanisms have been studied for a relatively long time, and much has already been written about them.

In materials processing, it is common to work with fluids in which the temperature varies from one location to the next. The resulting temperature gradients provide a natural way for interfacial tension gradients to arise and cause bubbles and drops to move in a reduced gravity environment. As mentioned earlier, materials processing is an area in which substantial research is being carried out in reduced gravity. Therefore, it is our hope that the contents of this book will be of some use to the community of scientists working in various aspects of this endeavor.

Finally, we mention the availability of two books that deal with fluid behavior in low gravity conditions. Myshkis et al. (1987) have considered the stability of the equilibrium states of bodies of liquid with free surfaces and their oscillations and also have provided a discussion of thermocapillary motion in liquid bodies. Antar and Nuotio-Antar (1993) have discussed a variety of transport problems in the context of reduced gravity conditions and have included a chapter devoted to drops and bubbles.

1.6 Scope of the Book

We now briefly describe the contents of the remainder of the book. In Chapter Two, the governing equations and boundary conditions used in modeling transport problems encountered in the remaining chapters are presented and discussed. Also, we discuss the physical assumptions that will be made, and several dimensionless parameters that will be encountered, in the subsequent chapters. At the end of Chapter Two, we provide tables of the governing equations in common coordinate systems. Part Two is devoted to a discussion of various aspects of the motion of isolated drops. The term *isolated* implies that there are no neighboring objects to interfere with the drop in question, nor are there any neighboring surfaces with which it can interact. Although no physical system is truly isolated, it is a useful exercise to treat this problem. The reason is that, in thermocapillary migration, the velocity and temperature disturbances introduced by such a drop only penetrate a relatively small distance of a few drop radii into the neighboring fluid before becoming too weak to be significant. Although the disturbance decays more slowly in the case of motion driven by a body force, in a sufficiently large container, such motion can be regarded as being isolated. Part Two is divided into two chapters, with Chapter Three focusing on motion driven by a body force and Chapter Four dealing with thermocapillary migration. In Chapter Four, we also include a discussion of some situations wherein a drop moves under the combined action of gravity and thermocapillarity. In Part Three, we consider the problem of interactions of moving drops with neighboring drops and surfaces that are sufficiently close and present results that are currently available. This subject is still one of active investigation at the present time, and much remains to be done. First, we provide a preliminary discussion. Some general solutions are reported in Chapter Five. Then, Part Three is divided into two subsequent chapters in which body force driven motion and thermocapillary migration, respectively, are considered. Finally, in Part Four, we discuss two related topics. Chapter Eight deals with mass transfer to and from drops and bubbles. There, we focus primarily on stationary objects and objects that are propelled by interfacial tension gradients. We do not consider mass transfer from objects moving because of a body force; this topic is discussed in depth in other available books. In Chapter Nine, we consider the related problems of motion driven in a body of liquid by gradients in interfacial tension. Much of Chapter Nine is devoted to problems involving rectangular pools of liquid, but we also cover elementary cases in cylindrical and spherical geometries. Although this is not strictly the motion of these objects but rather motion in them, we have included some simple solutions of problems on this subject because some readers might find them useful.

CHAPTER TWO

The Governing Equations

2.1 Introduction

The purpose of this chapter is to introduce the reader briefly to the differential equations that govern the behavior of fluids and to discuss the assumptions commonly made in simplifying them. When modeling any specific problem, we shall see that it is necessary to begin with an appropriate subset of these equations, write all the applicable boundary conditions, and then proceed to solve them. The usual result from this process is a detailed description of velocity, pressure, temperature, and concentration fields as appropriate to the problem at hand, as well as answers for the velocity of a bubble or a drop (or the force on it) and its shape.

The governing equations are those that state the principles of conservation of each chemically distinct species, momentum, and energy. Experience has shown these principles to hold true for each of the entities mentioned. Much has been written regarding the origin and derivation of these equations. Therefore, very little space will be devoted to a detailed discussion of those topics. The interested reader is referred to the books by Bird, Stewart, and Lightfoot (1960), Aris (1962), and Batchelor (1967) for further information. In the following material, it is assumed that the reader is familiar with elementary vector and tensor operations and with the meaning of symbols such as the gradient operator and the Laplacian. A good introduction to such operations and symbols can be found in the book by Aris.

A central idea in modeling the behavior of fluids is the continuum assumption. Although in reality all matter is made up of subatomic particles and mostly empty space, on the common length and time scales of observation and measurement, fluids behave as though the velocity, temperature, and species concentration are continuous functions of position and time. Where needed, suitable derivatives of these quantities of the desired order also are found to behave in this way. This is a consequence of the averaging that is done over an enormous number of the individual atoms or molecules making up the substance in any volume that is considered a point for measurement purposes. A good discussion of this idea is given in the book by Leal (1992), which is an excellent resource on methods to solve problems encountered in fluid mechanics. The continuum that comprises the model of the fluid consists of point particles called *material particles* from hereon. Note that these particles occupy no volume and are simply a useful construct in discussing fluid behavior.

The derivation of the governing equations from the conservation principles involves writing the principle in mathematical form for a control volume that can be assumed to be either fixed in space or moving and deforming with the fluid. The former approach is used in Bird et al. (1960), whereas examples of the power of the latter point of view may be found in the book by Aris (1962).

A concept that is useful when deriving and using conservation equations is that of the material derivative. Imagine that at some instant, which we arbitrarily designate as the origin of time, we identify the position of a given material particle by the position vector ξ. In a fluid executing motion, at later times this particle would, in general, occupy some position given by the position vector $\mathbf{x}$. The paths followed by the material particles comprising the fluid then can be described formally by the equation

$$\mathbf{x} = \mathbf{x}(\xi, t). \tag{2.1.1}$$

That is, we say that the position vector $\mathbf{x}$ of any material particle is a function of its initial position ξ and time t. Starting with different material particles, different curves can be drawn through space describing the paths followed by those material particles. Such curves are called *pathlines*. The initial coordinates of material particles are called *material coordinates*, and usually the particle itself is identified by its initial position, that is, one often refers to it as the particle ξ.

We can follow changes with time of a given property of a fluid, such as the density or temperature at a fixed point in space, as different material particles traverse through that point. This would yield the usual partial derivative with respect to time, holding spatial coordinates fixed. On the other hand, we can travel with a material particle and investigate how the property changes with time in an infinitesimal neighborhood of that particle. In this case, it is the initial position of the particle in question or its material coordinates that are being held fixed in the time differentiation. This time derivative is known as the *material derivative* or the *substantial derivative* and is designated by a standard total derivative symbol, $\frac{d}{dt}$. It is now straightforward to define the velocity vector $\mathbf{v}$ in a fluid as simply the material derivative of the position vector of material particles:

$$\mathbf{v} = \frac{d\mathbf{x}}{dt}. \tag{2.1.2}$$

The material derivative of any field variable, such as the density, temperature, or species concentration, can be related in a simple way to the usual partial derivative with respect to time, $\frac{\partial}{\partial t}$, which implies holding the spatial coordinates fixed. If the variable being considered is a scalar field, such as mass density $\rho(\mathbf{x}, t)$, then

$$\frac{d\rho}{dt} = \frac{\partial \rho}{\partial t} + (\mathbf{v} \bullet \nabla)\rho, \tag{2.1.3}$$

which can be rewritten in rectangular Cartesian coordinates, (x, y, z) (abbreviated as rectangular coordinates throughout the rest of this book), as follows:

$$\frac{d\rho}{dt} = \frac{\partial \rho}{\partial t} + v_x \frac{\partial \rho}{\partial x} + v_y \frac{\partial \rho}{\partial y} + v_z \frac{\partial \rho}{\partial z}. \tag{2.1.4}$$

In the above results, we have introduced the gradient operator ∇ whose definition in rectangular coordinates is

$$\nabla = \mathbf{i}\frac{\partial}{\partial x} + \mathbf{j}\frac{\partial}{\partial y} + \mathbf{k}\frac{\partial}{\partial z}. \tag{2.1.5}$$

Here, the orthogonal unit vectors $(\mathbf{i}, \mathbf{j}, \mathbf{k})$ correspond respectively to the spatial directions (x, y, z). Suitable expressions in other common coordinate systems can be found in Bird et al. (1960). The gradient of a scalar field points in the direction of maximum change of that field variable at every point, and its component in any given direction gives the rate of change of the field variable in that direction. Hence, the gradient is normal to surfaces on which the variable is constant, a concept that will prove useful.

2.2 Conservation Equations

2.2.1 Conservation of Total Mass

Now, we can begin to write down the governing conservation equations. It is customary to start with the equation of conservation of mass. This is usually called the continuity equation and is written as follows:

$$\frac{d\rho}{dt} + \rho(\nabla \bullet \mathbf{v}) = 0. \tag{2.2.1}$$

When the density of an element of the fluid does not change with changing pressure, the fluid is considered incompressible. The pressure changes encountered in problems to be considered in this book will always be small compared with the actual value of the pressure itself, and in these cases, the error made by assuming the density to be unaffected by pressure variations is negligible. Often we shall consider problems where the temperature of an element of fluid varies as it moves because of the conduction of heat from neighboring elements, leading to a change in the density of the element. In multicomponent systems, the composition can vary because of diffusion, leading to similar consequences. As pointed out by Batchelor (1967) in his book, however, the variation of density with temperature and composition can almost always be ignored in considering the continuity equation. The assumption that the density of each fluid element remains constant leads to the vanishing of the material derivative of the density in Equation (2.2.1), leading to the following *incompressible* version of the continuity equation:

$$\nabla \bullet \mathbf{v} = 0. \tag{2.2.2}$$

A vector field, the divergence of which vanishes, is called a *solenoidal field;* therefore, in incompressible flow, the velocity field is solenoidal.

2.2.2 Conservation of Momentum

For the equation of conservation of momentum, we first state the following form, which is known as Cauchy's equation of motion:

$$\rho\frac{d\mathbf{v}}{dt} = \nabla \bullet \mathbf{\Pi} + \rho\,\mathbf{f}. \tag{2.2.3}$$

There are two new symbols used in the right side of the above equation. The first is the tensor stress field $\mathbf{\Pi}$ in the fluid, and the second is the body force per unit mass $\mathbf{f}$ at any given point in the fluid. A good discussion of the reason that stress can be described by a second order tensor is given by Aris (1962). The corresponding force on any given area element $d\mathbf{S}$ is $d\mathbf{S} \bullet \mathbf{\Pi}$. In the absence of body or surface couples, the stress tensor is symmetric, a result that can be proved using the principle of conservation of angular momentum. As pointed out by Leal (1992), there is no evidence that real fluids exhibit surface torque, but body couples can exist in ferrofluids, for example. We shall always assume a symmetric stress tensor in problems considered in this book. The most common example of a body force is that due to gravity. In this case, $\mathbf{f}$ would be replaced by the more common symbol $\mathbf{g}$, which stands for the acceleration due to gravity. Henceforth, we shall use this symbol in the body force term.

Cauchy's equation, as written above, has too many unknowns in it. The standard approach for proceeding further is to consider a constitutive model of the fluid that permits one to relate the state of stress to the state of motion in the fluid. The only type of constitutive model that we shall use in this book is the Newtonian model, which assumes a linear connection between stress and the rate of deformation. This is written as follows:

$$\mathbf{\Pi} = [-p + \lambda(\nabla \bullet \mathbf{v})]\mathbf{I} + 2\mu\,\mathbf{D}. \tag{2.2.4}$$

Here, $\mathbf{I}$ is the unit tensor, and the rate of deformation of the fluid $\mathbf{D}$ is a symmetric tensor related to the gradient of the velocity field by

$$\mathbf{D} = \frac{1}{2}[\nabla\mathbf{v} + (\nabla\mathbf{v})^t]. \tag{2.2.5}$$

In the above, the superscript t refers to the transpose.

The symbol λ appearing in the constitutive equation is related to the coefficient of bulk viscosity, defined as $\lambda + \frac{2}{3}\mu$, and μ is known as the coefficient of shear viscosity. When the term viscosity or dynamic viscosity is used without qualification, it always refers to the coefficient of shear viscosity. The coefficient of bulk viscosity, which characterizes dissipative processes during compression and expansion, is difficult to measure, and there appears to be a long history of misconceptions regarding it. Because the only flows we consider are those in which the velocity field is solenoidal, the value of λ is irrelevant for our purposes.

The third new symbol encountered in Equation (2.2.4) is the scalar field $p(\mathbf{x}, t)$, which is the thermodynamic pressure in the fluid. In a stationary fluid, it can be seen that the stress is isotropic and equal to $-p\mathbf{I}$. In a moving fluid, however, because of the prospect of contributions to the normal force on an area element from the flow, the normal surface force is not the same on area elements pointing in different directions at the same point. In this case, it is customary to define a mean pressure as minus one-third the trace of the stress tensor. For incompressible Newtonian flow, it can be seen that this mean pressure is equal to the pressure p that appears in the constitutive equation. We shall use the symbol τ to refer to the viscous part of the stress tensor. For incompressible flow, $\tau = \mathbf{\Pi} + p\mathbf{I}$.

When the constitutive model is introduced into Cauchy's equation, and we assume that the coefficient of shear viscosity is constant and that the flow is incompressible, we arrive at the commonly used Navier–Stokes equation:

$$\rho\left[\frac{\partial\mathbf{v}}{\partial t} + (\mathbf{v} \bullet \nabla)\mathbf{v}\right] = \rho\mathbf{g} - \nabla p + \mu\nabla^2\mathbf{v}. \tag{2.2.6}$$

In the left side of the above equation, we have replaced the material derivative in time in terms of the usual partial derivative holding spatial coordinates fixed and the convective contribution. In the right-hand side, the symbol for the Laplacian ∇^2, which can be formally written as $\nabla \bullet \nabla$, is introduced.

Because the hydrostatic variation in pressure does not contribute to fluid motion, it is useful to define a hydrodynamic (or simply dynamic) pressure P as follows:

$$-\nabla P = -\nabla p + \rho \mathbf{g}. \tag{2.2.7}$$

The dynamic pressure, then, is uniform in a stationary fluid. Almost exclusively, we shall use the dynamic pressure in this book and, for consistency in notation with common usage, employ p to designate it. With this interpretation of the symbol p, the Navier–Stokes equation may be rewritten as

$$\rho\left[\frac{\partial \mathbf{v}}{\partial t} + (\mathbf{v} \bullet \nabla)\mathbf{v}\right] = -\nabla p + \mu \nabla^2 \mathbf{v}. \tag{2.2.8}$$

It is worthwhile recapitulating the significance of the various terms that appear in the above equation. All the terms have dimensions of momentum per unit volume. The derivation makes it evident that Equation (2.2.8) is the statement of Newton's law applied to a continuum; the left-hand side symbolizes inertia, and the right-hand side includes forces that act on the fluid to accelerate it. Another useful perspective can be gained by first multiplying every term by an elemental volume dV located at some arbitrary position $\mathbf{x}$ and time t. In this case, the term involving the partial derivative with respect to time represents the time rate of change of momentum of fluid occupying that elemental volume at that given instant. Several items contribute to this time rate of change. First, from a result from vector calculus known as the divergence theorem, the term $\rho \mathbf{v} \bullet \nabla \mathbf{v} dV$ can be shown to be the net efflux of momentum from this volume arising from the flow of fluid into and out of it. This is usually termed *convective transport of momentum* because momentum moves in and out of the volume as a result of bulk movement of the fluid. The pressure gradient term gives the pressure force on the fluid occupying the elemental volume dV. The viscous term $\mu \nabla^2 \mathbf{v} dV$ represents the viscous force exerted by the adjoining fluid on the fluid occupying dV. This force actually arises from interactions at the atomic and molecular level, and therefore is considered molecular transport of momentum. Because momentum is transported within the fluid by the two mechanisms discussed above, the relative importance of convective transport when compared with molecular transport is assessed from the value of the Reynolds number, which will be defined in Section 2.5. It is important to know about this, because the types of simplifications one can make in analyzing a problem depend on whether one or the other mechanism is dominant.

We conclude this section by noting that the ratio of the dynamic viscosity to the density of the fluid is sufficiently important in its own right that it is called the kinematic viscosity $v = \frac{\mu}{\rho}$. The dimensions of this quantity can be seen to be length squared per time. The kinematic viscosity can be envisioned as a diffusivity of momentum in the fluid. The usefulness of this interpretation will be seen later. Viewed from the perspective of kinematic viscosity, air at room temperature is far more viscous than water. Therefore, if a surface is suddenly moved in either fluid, the effect will be felt much farther in air than in water after the lapse of the same amount of time. Such physical pictures can be helpful in making the right assumptions about a situation.

2.2.3 Conservation of Energy

Next we turn to the equation of conservation of energy. The most common use of this equation is in determining temperature distributions in a moving fluid and heat transfer rates. Various versions of the energy equation can be found in textbooks. From a mathematical statement of the principle that total energy is conserved, by subtracting the balance of mechanical energy that may be obtained by taking the inner (dot) product of the Navier–Stokes equation with the velocity vector, one proceeds to an equation for thermal energy. As with the principle of conservation of momentum, the principle of conservation of energy leads first to an equation with too many unknowns. Therefore, a constitutive equation is needed which relates the heat flux $\mathbf{q}$ to the temperature field $T(\mathbf{x}, t)$ itself. The most commonly used one is Fourier's law, which assumes a linear connection between the heat flux and the temperature gradient, namely

$$\mathbf{q} = -k\nabla T. \tag{2.2.9}$$

Here, the symbol k stands for the thermal conductivity of the fluid at the point in question. The above version assumes conduction to occur isotropically in the medium, which is adequate for describing the behavior of common fluids. Fourier's law is used to describe material behavior, even when substantial temperature variations occur, by permitting the thermal conductivity to depend on the temperature. When it is inserted into the conservation equation for thermal energy and the assumption is made that the density and the thermal conductivity are independent of temperature, the following version of the energy equation is obtained:

$$\frac{\partial T}{\partial t} + \mathbf{v} \bullet \nabla T = \kappa \nabla^2 T. \tag{2.2.10}$$

In this result, κ is the thermal diffusivity of the fluid. It may be written as $\kappa = \frac{k}{\rho C_p}$ where C_p is the specific heat of the fluid at constant pressure. The dimensions of κ are the same as those of the kinematic viscosity, namely length squared per time. Several assumptions besides the ones mentioned already are inherent in this equation. The most important are that the contributions from viscous dissipation are ignored, radiant transport of energy within the fluid is completely neglected, and heat sources or sinks are not present. These assumptions usually are satisfactory in the context of problems considered in this book. Heat generation or consumption from chemical reactions appears naturally in a more general version of the energy equation, when it is formulated for multicomponent systems. Other types of heat sources, such as electrical heating of the fluid or radiation that is absorbed by the fluid, can be included by adding a term to the right-hand side that represents the time rate of heat generation per unit volume.

Just as in the Navier–Stokes equation, two mechanisms for the transport of energy in a moving fluid are represented in Equation (2.2.10). One is convective transport of energy, exemplified by $\rho C_p \mathbf{v} \bullet \nabla T$, and the other is conduction represented by $k\nabla^2 T$. Once again, the relative importance of one mechanism, when compared with the other, plays a role in the choice of a solution technique. A ratio that represents the relative importance of convective transport to conduction is the Péclet number, which will be defined in Section 2.4. Also important is the dimensionless group Prandtl number $Pr = \frac{\nu}{\kappa}$. The value of the Prandtl number provides a gauge of the relative ability of a fluid to transport momentum by molecular means when compared with its ability to transport energy by similar means. Its magnitude is approximately unity in gases and larger than that in liquids. An exception is the case of liquid metals, which are excellent conductors of heat. The Prandtl number of a liquid metal is typically of the order of magnitude 0.01.

When the temperature varies in a fluid, material properties, such as the density, viscosity, specific heat, and thermal conductivity, will vary as well. The simplicity of the governing equations is lost if one were to attempt to permit such variations in them, however, and there are very few situations where analytical solutions can be written. This has led to the use of the *constant property* equations, although they do not provide a precise description. A brief discussion of this topic is given in Section 2.4.

2.2.4 Conservation of Species

Now we turn to the principle of conservation of species. It is an inherent aspect of molecular diffusion of species through each other that the very act of molecular diffusion can result in net average motion of the fluid. In constructing the constitutive equation, one has to be careful to subtract out the convective flux arising from this average motion. In cases where, however, one is dealing with a dilute solution of one or more species in a solvent, this motion can be neglected as a first approximation, and one can solve for the velocity field independently of the concentration distribution of species. This known velocity field can be used in the conservation equations for species to determine concentration distributions and mass transfer rates. Another approximation in the dilute limit involves using a binary approach to describe the problems of multiple species, assuming that interactions among the species diffusing in the solvent are negligible. An important aspect to note is that the driving force for molecular transport is the gradient in chemical potential and not simply the concentration gradient. An introduction to these and other matters is given by Bird et al. (1960), and an in-depth discussion may be found in the book by Taylor and Krishna (1993).

The general form of the equation of conservation of species contains too many unknowns in the same manner as the general forms of the equations of motion and energy. Therefore, a constitutive equation becomes necessary. One equation that is widely used in describing the diffusion of species is Fick's law. In a binary system containing only two species, A and B, one form of Fick's law can be written as follows:

$$\mathbf{n}_A = \omega_A(\mathbf{n}_A + \mathbf{n}_B) - \rho \mathcal{D}_{AB} \nabla \omega_A. \tag{2.2.11}$$

Here, $\mathbf{n}_A$ is the mass flux of species A as measured in a laboratory reference frame, and ω_A and ω_B represent the mass fractions of A and B. The symbol $\mathcal{D}_{AB}$ is the binary diffusivity associated with the pair A-B. Its dimensions are length squared per time.

When Fick's law is inserted into the equation of conservation of species and the density and mass diffusivity are assumed constant, the following result is obtained:

$$\frac{\partial \rho_A}{\partial t} + \mathbf{v} \bullet \nabla \rho_A = \mathcal{D}_{AB} \nabla^2 \rho_A + r_A. \tag{2.2.12}$$

Here, ρ_A is the mass concentration of species A in solution. The symbol r_A stands for the rate of production of A per unit volume of the fluid by chemical reaction. Equation (2.2.12) is the starting point for the modeling of mass transport problems in dilute liquid solutions. In a manner analogous to the situation with regard to the transport of energy, one can identify the convective and diffusive transport terms in the above equation.

A suitable ratio of the rates of transport by these two mechanisms also is termed the Péclet number. The ratio of the material properties that reflects the ability of the fluid to transport momentum by molecular means when compared with its ability to transport a given species by the same mechanism is defined as the Schmidt Number, $Sc = \frac{\nu}{D_{AB}}$.

2.3 Boundary Conditions

2.3.1 General Comments

When constructing a mathematical model of a given problem, we shall begin by writing down the applicable conservation equations and simplifying them, using physically plausible assumptions. These are partial differential equations, however, and to determine solutions applicable to a given situation it is necessary to specify additional conditions at the boundaries of the system. These boundaries may be solid-fluid interfaces or fluid-fluid interfaces. Also, it is possible to pose problems in unbounded domains, for example, when describing the motion of a bubble or drop in a fluid, even though in reality there always are boundaries in a problem. This is acceptable because the relevant field variables, such as velocity or temperature, in an unbounded domain sensibly decay to those prevailing in the undisturbed state of the fluid as one moves away from the bubble or drop, long before the boundaries of the system are encountered. In problems involving time, it is necessary to specify the initial distribution of the relevant field variables at some value of time; in these cases, it is necessary to specify boundary conditions at all values of time.

In Chapter One, we devoted some discussion to the fact that an interface is not a mathematical surface of zero thickness, but one that is conveniently assumed to be so on the length scales of observation. This means that one must regard boundary conditions merely as plausible statements based on experience that permit one to obtain results for velocities, temperatures, and concentrations that are consistent with experimental observations. A good discussion of the physical basis of boundary conditions to be presented here can be found in the books by Batchelor (1967) and Leal (1992).

We can make a broad distinction between fluid-solid and fluid-fluid interfaces. On the solid side of the fluid-solid interface, we usually do not seek solutions for field variables. Therefore, conditions typically involve specifying the field variable, or its spatial derivatives, on the fluid side of the boundary. In contrast, at fluid-fluid interfaces, it is often necessary to solve for the fields in both fluids. In this case, usually neither the field variable nor its derivatives are known at the interface. Rather, we can only write connections between such quantities on either side of the interface.

In stating the boundary conditions relevant to the momentum and energy equations, we shall ignore the consequences of mass transfer across the interface. Significant complications are introduced by the movement of species across an interface, and for the most part, we shall not be concerned with such situations. When species are transferred across an interface, the interface cannot be considered a material surface, and the normal components of the fluid velocity at the interface on either side are not in general equal to the velocity of the interface normal to itself. In Chapter Eight, we specifically deal with mass transport problems. Therefore, we shall introduce the reader to the relevant jump condition on the species mass flux when discussing boundary conditions associated with the mass conservation equation in Section 8.2. The generalization of other conditions to accommodate mass transfer will be made as needed in Chapter Eight. Jump balances including mass transfer effects are discussed in the book by Slattery (1972).

2.3.2 Fluid-Solid Interface

First, let us consider the interface between a fluid and a solid. At each point on the interface, the fluid is assumed to have the velocity of the solid at that point. Therefore, if the solid is stationary in the reference frame that is used, the fluid also must be stationary at the interface. Usually, this is broken into two parts. One is a purely kinematical consequence of the fact that the fluid does not penetrate the solid unless one is considering a porous wall. Therefore, in a reference frame attached to the solid, the normal velocity of the fluid must vanish at the interface unless there is significant mass transport to or from the interface. If the unit normal to the interface is designated $\mathbf{n}$, one can write

$$\mathbf{v} \bullet \mathbf{n} = 0. \tag{2.3.1}$$

The other component of the velocity is, of course, the tangential component that lies on the surface. This is assumed to be the tangential component of the velocity of the solid, which would be zero in a reference frame attached to the solid. This is called the no-slip condition and is written as

$$\mathbf{v} \bullet \mathbf{t} = 0, \tag{2.3.2}$$

where $\mathbf{t}$ is any vector tangent to the surface. The correctness of this condition is borne out by experience with numerous problems, wherein the solutions obtained using it are found to be consistent with observation. As a matter of interest, there was some controversy in the eighteenth and nineteenth centuries regarding the correct condition to impose at a rigid boundary, as is evident from the summary given by Goldstein (1938). Today, the no-slip condition is taken for granted, with some exceptions that are not pertinent to us in the subject matter of this book.

In the case of energy transport, the temperature field may be prescribed at a solid boundary. Another option is to prescribe the heat flux into the fluid. Because there is no normal velocity relative to the boundary, the heat flux must be solely due to conduction. Invoking Fourier's law then leads to a prescription of the normal component of the temperature gradient at the boundary. As a specific example, if the solid is a good insulator compared with the fluid, one might set this heat flux to zero. Other options are possible wherein the temperature at the boundary is related to the heat flux. In problems wherein neither the temperature field nor the heat flux is prescribed at a solid-fluid interface, one would normally have to solve for the temperature field in the fluid and in the solid. In this case, one would require the temperature and the heat flux to be continuous across the interface. Discontinuities in heat flux can arise when there is a phase change process at the interface because the latent heat must be accommodated.

The boundary conditions to be used in conjunction with the equation of species conservation are similar to those discussed above. If the solid is soluble in the fluid, it is assumed that thermodynamic equilibrium exists locally at every point on the interface between the solid and the fluid. This translates to equality of the chemical potential of each species on either side of the interface. In practice, one usually prescribes the concentration of the dissolving species as the equilibrium value corresponding to the local temperature.

2.3.3 Fluid-Fluid Interface

The situation is more involved when a fluid-fluid interface is considered. In this case, one often has to solve the conservation equations for the fluids on either side of the interface. To further complicate matters, the shape of the interface is not known a priori and must be determined as part of the solution of the problem. The interface can move and deform with the passage of time. All of this makes the solution of problems involving fluid-fluid interfaces more challenging. The usual practice has been to simplify the problems as much as possible using physical insight. For example, bubbles and drops are observed to be nearly spherical, and the surfaces of films of liquid in rectangular troughs are found to be nearly flat. This permits the shape of the interface to be assumed as given at least as a first approximation. By the use of perturbation theory, one can then construct better approximations to the shape when solving the problem, if desired.

As before, we begin with conditions associated with the velocity field. If the interface can be described by

$$F(\mathbf{x}, t) = 0, \tag{2.3.3}$$

then we can write $\frac{dF}{dt} = 0$, where $\frac{d}{dt}$ continues to stand for the material derivative. Let us designate the fluid on one side of the interface as Fluid I and that on the other side as Fluid II and choose the unit normal vector to the interface $\mathbf{n}$ to point into Fluid I. The unit normal to the surface at any point can be obtained from

$$\mathbf{n} = \frac{\nabla F}{|\nabla F|}. \tag{2.3.4}$$

The velocity and stress fields in fluid I are identified as $(\mathbf{v}, \mathbf{\Pi})$ respectively, and those in Fluid II as $(\mathbf{v'}, \mathbf{\Pi'})$. We shall use the convention of employing a prime to designate fields in Fluid II in the context of temperature and concentration fields as well. In the absence of mass transfer, the kinematic condition at every point on the interface can be stated as

$$\mathbf{v} \bullet \mathbf{n} = \mathbf{v'} \bullet \mathbf{n} = -\frac{1}{|\nabla F|} \frac{\partial F}{\partial t}. \tag{2.3.5}$$

Note the appearance on the right-hand side of the usual partial derivative with time holding spatial coordinates fixed.

In most situations encountered in this book, the interface location does not change with time. In this case, the partial derivative with respect to time vanishes, yielding the simpler kinematic condition

$$\mathbf{v} \bullet \mathbf{n} = \mathbf{v'} \bullet \mathbf{n} = 0. \tag{2.3.6}$$

The tangential velocity components on the two sides of the interface also can be set equal to each other. Taken together with the statement that the normal velocity components must be continuous in the absence of significant mass transfer effects, the vector velocity is continuous across a fluid-fluid interface

$$\mathbf{v} = \mathbf{v'}. \tag{2.3.7}$$

Because the above result does not specify the velocity itself, but rather relates the fields on either side of the interface, it is necessary to look for additional conditions which one can write at the interface. Based on physical considerations, it can be established that the stress vectors must be discontinuous across an interface because of the effect of interfacial tension. The usual jump condition written on the stress fields is given below.

$$\mathbf{n} \bullet (\mathbf{\Pi} - \mathbf{\Pi}') = \sigma \mathbf{n}(\nabla \bullet \mathbf{n}) - \nabla_s \sigma. \tag{2.3.8}$$

Here, σ is the interfacial tension and ∇_s is the surface gradient. This is related to the spatial gradient via $\nabla_s = \nabla - \mathbf{n}(\mathbf{n} \bullet \nabla)$. Examination of the stress condition reveals that both the normal and the tangential components of the stress suffer discontinuities across the interface. In the case of the normal stress, the discontinuity is proportional to the local curvature whereas the tangential stress discontinuity is proportional to the local gradient of interfacial tension. For a flat surface, the curvature is zero and the normal stress would be continuous. In a similar manner, the tangential stress would be continuous across an interface on which the interfacial tension is uniform.

More complicated descriptions of the interface than the one used here exist. These involve postulating dissipative processes to occur in the interface, and there is some experimental evidence supporting such a picture when surface active agents are adsorbed at the interface. Treating situations that involve complex interfacial rheology is beyond the scope of this work, however. Later, in Section 4.9, we analyze an example problem involving Newtonian rheology at the interface. The reader interested in more involved interfacial rheological models may wish to consult the book by Edwards, Brenner, and Wasan (1991).

It is useful to write the divergence of the unit normal to a surface at a given point in terms of the principal radii of curvature at that point, R_1 and R_2:

$$\nabla \bullet \mathbf{n} = 2H = \frac{1}{R_1} + \frac{1}{R_2}. \tag{2.3.9}$$

Here, H is the mean curvature of the surface at that point.

Now examining the discontinuity in the tangential component of the stress on either side of the interface, we see that it is equal to the local value of the surface gradient of the interfacial tension. Because the interfacial tension is assumed to be the equilibrium tension corresponding to the local thermodynamic state at the point in question, it can depend on the temperature and the concentrations of any surface-active solutes at that point. Therefore, gradients in interfacial tension can be related to gradients in these quantities, as shown in the example below.

$$\nabla_s \sigma = \sigma_T \nabla_s T + \sigma_\Gamma \nabla_s \Gamma. \tag{2.3.10}$$

The symbols σ_T and σ_Γ represent the partial derivatives of the interfacial tension with respect to temperature and surfactant concentration, respectively. In writing Equation (2.3.10), we have assumed that only a single surface active solute is present. Its surface concentration is represented by the symbol Γ. When a temperature or surfactant concentration gradient is present along an interface, the fluid adjoining it must move because the viscous stress is precisely zero in fluids at rest. It would not be possible to satisfy the jump balance in tangential stress at the interface if the fluids on both sides of the interface are quiescent. Because a knowledge of the temperature and the concentration field at the interface is necessary for describing the interfacial tension gradient, it is clear that the velocity distribution depends on these fields. In turn, because convective transport of energy influences temperature fields, a knowledge of the velocity distribution is necessary to obtain the temperature field. This means that in problems in which convective transport is not negligible, the velocity and temperature fields are bidirectionally coupled. Analogous considerations apply to the velocity and composition distributions in systems involving surfactant transport.

The boundary conditions with respect to the temperature fields at a fluid-fluid interface also are more involved than those at a fluid-solid interface. It is common to ignore the contributions arising from stretching and shrinking of elements of fluid on the interface with little error and write the following conditions on the temperature field. Also, it is assumed that no phase change processes occur that will introduce a discontinuity in the heat flux across the interface due to a contribution from latent heat.

$$T = T' \tag{2.3.11}$$

$$\mathbf{n} \bullet \mathbf{q} = \mathbf{n} \bullet \mathbf{q}'. \tag{2.3.12}$$

Here, the heat flux $\mathbf{q} = -k\nabla T$ in each phase. In subsequent chapters, we shall almost always use the above boundary conditions. When the thermal conductivity of Fluid II is negligible compared with that of Fluid I, Equation (2.3.12) can be simplified to

$$\mathbf{n} \bullet \mathbf{q} = 0. \tag{2.3.13}$$

We note that in certain situations, we must include the effects of the stretching and shrinkage of interface elements. For example, when a bubble or drop moves in another fluid, area elements in the front half are stretched as they move toward the equator. Energy must be supplied by the neighboring fluid to provide the additional interfacial internal energy for this increasing area. This fluid therefore must cool as a result. The opposite is true in the rear half. This leads to a jump in the heat flux across the interface. The boundary condition given below is adequate for accommodating such effects:

$$(\sigma - e_s)\nabla_s \bullet \mathbf{v} = \mathbf{n} \bullet (\mathbf{q} - \mathbf{q}'). \tag{2.3.14}$$

The surface divergence of the velocity field on the interface appears here. A definition can be constructed using the definition of ∇_s given earlier. The quantity e_s appearing in this equation is the internal energy of the interface per unit area and is different from the interfacial tension, which measures the free energy of the interface per unit area for a pure substance. For a single component liquid, the two are related by the thermodynamic identity

$$e_s = \sigma - T\sigma_T, \tag{2.3.15}$$

where T is the absolute temperature. Note that this is the only place where T is defined in this strict way. For instance, in the equation of conservation of energy, the scale used for measuring temperature is irrelevant as long as it is consistently used everywhere. A detailed derivation of Equation (2.3.14), along with the assumptions employed, can be found in Torres and Herbolzheimer (1993).

Now, we discuss boundary conditions on the species concentration fields at fluid-fluid interfaces. First, we shall consider species that are not surface active. In this case, one writes conditions pertaining to local thermodynamic equilibrium between the two phases at the given point. These would be statements that the chemical potential of each species takes on the same value in the fluids on each side of the interface. In practice, instead of working with chemical potentials, we use concentrations in mass or molar units. Therefore, the condition becomes one of assuming equilibrium concentrations to prevail in the two fluids on either side of the interface at that point. Of course, the concentrations can vary along the interface if the thermodynamic variables that determine the equilibrium compositions change. An example would be a variation of temperature along the interface. The second condition on each species is that the mass flux is continuous across the interface. For species A, this jump mass balance can be written as follows:

$$\mathbf{n}_A \bullet \mathbf{n} - \rho_A v_i = \mathbf{n}'_A \bullet \mathbf{n} - \rho'_A v_i. \tag{2.3.16}$$

Here, v_i is the velocity of the interface normal to itself. The flux $\mathbf{n}_A$ is given by the sum of the convective and diffusive fluxes in Equation (2.2.11) for a binary system. When necessary, dilute multicomponent systems can be approximated as though each solute and the solvent form a binary system.

The situation is somewhat more complicated when one considers species that can adsorb on the interface. As mentioned in Chapter One, such chemicals are called surface active agents or surfactants. Typically a chemical that is dissolved in one of the fluids in bulk, if it acts as a surfactant for the pair, will diffuse through the fluid and adsorb on the interface. It may or may not cross the interface to the other side, depending on whether it is soluble in the other liquid. Connections have to be made between the volumetric concentration in the fluid adjoining the interface and that in the adsorbed surface layer through a suitable adsorption isotherm or, if nonequilibrium conditions exist, a more involved rate expression. The mass flux of surfactant is not, in general, continuous across an interface because the surface can serve as a source or sink for the adsorbing species.

The only type of problems involving surfactants that we shall address in this book are those in which surfactants are insoluble in the bulk. They reside only on the interface. This may seem artificial at first sight because a surfactant normally needs to be present in one of the two fluids to reach the interface, especially in the case of a liquid-liquid interface. The "insoluble surfactant" case can be envisioned as a physical limit of very slow adsorption/desorption or very slow diffusion in the bulk. In this case, over the time scales of the experiment, one can assume that all the surfactant present at the surface resides there, and there is no sensible exchange with the bulk. As a consequence, surfactant is conserved at the interface, and its surface concentration Γ satisfies the following equation:

$$\frac{\partial \Gamma}{\partial t} + \nabla \bullet (\mathbf{v}_s \Gamma) = \mathcal{D}_s \nabla_s^2 \Gamma. \tag{2.3.17}$$

The symbol $\mathbf{v}_s$ represents the velocity vector on the interface, and $\mathcal{D}_s$ represents the diffusivity of the surfactant on the interface. The derivation of Equation (2.3.17) is straightforward and follows along lines similar to those used in deriving the bulk phase conservation equations. A surface version of Fick's law is used in relating the diffusive flux to the gradient in surfactant concentration, and the surface diffusivity of surfactant is assumed constant. We shall have occasion to use this equation in Chapter Four. Equation (2.3.17) can be modified easily to accommodate situations where the surfactant is also soluble in one or both fluids by adding a term representing the net flux of surfactant to the interface to the right-hand side. Such a result more closely resembles the jump balances for stress and energy written earlier in this section.

In closing this section, a brief comment is appropriate regarding homogeneous and nonhomogeneous boundary conditions. A linear equation is considered homogeneous when a constant times the dependent variable also satisfies the same equation.

In practical terms, this usually means a condition in which a field variable and its spatial derivatives appear in some linear combination set equal to zero. Because the governing equations we shall use also are homogeneous, if all the boundary conditions are homogeneous, one can usually expect only a trivial answer. It is inhomogeneity in a boundary condition that will lead to a nontrivial situation. As an example, if the temperature and composition fields in a fluid are uniform, in the absence of gravity, a drop present in such a fluid would not move, and no fluid motion will ensue. A variation of temperature in the fluid, on the other hand, will serve to force motion of the drop by causing variation of the interfacial tension on the drop surface. As an alternative, a nonzero heat flux imposed on the drop surface, perhaps from a laser beam, can be used to generate temperature variations on its surface and therefore cause its motion. There are certain classes of homogeneous problems containing a parameter that can exhibit nontrivial solutions when the parameter takes on characteristic values, also known as eigenvalues. Such problems are rare in the context of the motion of drops and bubbles.

2.4 Physical Assumptions

In subsequent chapters, we shall be using the conservation equations and boundary conditions stated in the previous sections to set up mathematical models of various problems. There will be certain common themes in the modeling that we discuss at this stage.

Many problems considered here involve the movement of a bubble or drop in a continuous phase due to the action of a temperature gradient. The temperature gradient at the interface can arise because of the presence of temperature variations in the fluids on either side or on both sides, as well as from external sources such as radiation absorbed at the interface. In all of these cases, one can expect temperature variations in the fluid. An important consequence of these temperature variations is a concomitant variation of physical properties that depend on temperature. The list includes transport properties, such as the viscosity, thermal conductivity, and diffusivity, as well as thermodynamic properties, such as the density, specific heat, and of course, the interfacial tension. In our modeling, however, we shall always treat these properties as constant, with the exception of the interfacial tension. In that case, we shall assume that the rate of change of the interfacial tension with temperature or surfactant concentration is constant. These assumptions might, at first glance, seem to be so restrictive that they would compromise the utility of the various models to be developed here. We shall offer convincing evidence, based on comparison with experimental observations, that it is reasonable to make these assumptions if they are used properly. Here, we discuss the basis for such assumptions.

To take a specific example, if we were to model the settling of a fluid drop in another fluid that is isothermal and of infinite extent, we would note that a drop initially at rest would accelerate, but after some time, it would reach an asymptotic state wherein it would settle at constant velocity known as the terminal settling velocity. This is possible because the weight of the drop minus the buoyant force on it is independent of time, and the drag on the drop is steady, given constant physical properties and a steady velocity. Therefore, it is possible to reach a condition wherein the net force on the drop is zero, and it experiences no acceleration.

Let us consider the thermocapillary analog of the same problem. Here, a drop is placed in a liquid in which there is a gradient of temperature. For simplicity, assume that there is no body force such as gravity acting on the system. If the drop is initially at rest, the variation of temperature on its surface will lead to a variation of interfacial tension, and the resulting imbalance of tangential stresses between the two fluids at the interface will lead to motion in the fluids as noted earlier. As a consequence, the drop will experience a hydrodynamic force that will lead to acceleration of the drop from the initial state. As the drop begins to move, however, it will come into contact with fluid at different temperatures, and the temperatures within the drop itself will begin a process of adjustment to the surroundings. The physical properties cataloged earlier will change continually as the drop moves, both within the drop and in the fluid surrounding it. As a consequence, the hydrodynamic force on the drop will always keep changing and lead to a nonzero acceleration at all times. Thus, the drop can never really be in steady motion, as in the analogous case of the settling drop in a gravitational field.

At first glance, it appears that we have a difficult problem. If we assume that the drop moves sufficiently slowly, however, we can recover a quasi-steady situation here as well. Imagine that the movement is so slow that, for any given drop location, the velocity and temperature fields in the affected region correspond to steady versions of these fields if the drop were moving at that velocity but literally staying in place. We realize that this is physically impossible, but it can be approximated if the fields can be established relatively quickly. The affected region can be of the order of several drop radii in situations where molecular transport dominates, or it can be only a very thin region called a *boundary layer* when convective transport is dominant. Situations also exist wherein the transport of momentum is dominated by one and that of energy is dominated by the other. Therefore, in specific problems, one should attempt to make an order of magnitude estimate of the relevant time scales for transport processes to establish quasi-steady fields and compare these with the time scales for the drop to move appreciably. This usually provides an indication of when the approximation of quasi-steady fields is reasonable.

Among the relevant properties, it is the viscosity that is usually most sensitive to temperature. For instance, the dynamic viscosity of water approximately halves in magnitude when the temperature is increased from 20° to 55°C. Some other fluids show even steeper change with temperature. With all the other properties involved, the change over such a temperature range might be of the order of several percent; usually a simple average, such as the value corresponding to the temperature in the undisturbed fluid on a plane normal to the prevalent temperature gradient that is a symmetry plane for the drop, is adequate for use. In the case of viscosity, this is usually the first approximation used in the absence of a precise model accounting for its variation with temperature.

The behavior of the interfacial tension gradient with temperature was briefly mentioned in Chapter One. At a gas-liquid surface, σ_T is sensibly constant over temperature ranges as large as 50 K unless one is near the critical point. In the case of liquid-liquid interfaces, the interfacial tension exhibits more complex dependence, particularly when one considers partially miscible fluids. Here, the assumption of constant σ_T would not work as well over as wide a temperature range.

Temperature variations also cause an added complication if a gravitational field is present. It is well known from everyday experience that when the density changes such that the component of the density gradient normal to the gravitational vector is nonzero, a fluid will move. As noted in Chapter One, such motion is known as natural, free, or buoyant convection. Even when the density gradient is vertical, if the density increases with height, one has a potentially unstable stratification of the fluid. In this situation, buoyant convection can set in at a critical density gradient, the value of which will depend on the system. A proper analysis of problems in which both a temperature gradient and a gravitational field are present would incorporate such buoyant convection effects. Analysis of such problems is not simple, however, and we do not attempt to treat them in this book. Thus, we always make the assumption that convection due to buoyancy will be ignored.

In general, the comments made above in the context of temperature variations and their impact on properties apply to species concentration variations as well. In dilute solutions, however, one can expect the effects to be somewhat less important than those caused by temperature differences.

2.5   Important Dimensionless Parameters

We now provide a brief discussion of the physical significance of dimensionless parameters encountered in this book. These naturally arise when one scales the governing conservation equations and boundary conditions using suitable reference quantities. One purpose of scaling, also known as nondimensionalization, is to identify the minimum number of parameters relevant to a given problem. In fluid mechanics, one begins this process by selecting a reference length, L, in the problem of interest and a reference velocity, V. For a drop settling due to gravity, the length might be the radius or the diameter of the drop. It is not critical which is chosen as long as one is consistent. We shall always choose the radius of the drop. A logical choice for the velocity scale would be the settling velocity of the drop. The scale for pressure is the same as that used for stresses. In problems wherein viscous forces dominate inertia, a viscous scale $\mu \frac{V}{L}$ is chosen for stresses and pressure. If inertia were to be dominant, one would instead choose ρV^2. The idea is to choose the scale that would be more likely to normalize the variable in question, that is, make the dimensionless variable range from 0 to 1. The choice of a time scale is a bit more involved, depending on the phenomenon one wishes to highlight. We shall first comment on the Navier–Stokes equation with one choice of time scale, namely that for molecular transport of momentum over the reference length. This is given by $\frac{L^2}{\nu}$.

When the above reference values are used to nondimensionalize velocity, distances, pressure, and time, the Navier–Stokes Equation (2.2.8) can be rewritten for the scaled variables in the following form:

$$Re\left[\frac{\partial \mathbf{v}}{\partial t} + (\mathbf{v} \bullet \nabla)\mathbf{v}\right] = -\nabla p + \nabla^2 \mathbf{v}. \tag{2.5.1}$$

Here, we have retained the same symbols for the scaled variables as for the physical counterparts for convenience. The operator ∇ now refers to differentiation in scaled distance variables. The dimensionless quantity multiplying the inertia terms on the left side is the Reynolds number, Re, which is defined as follows:

$$Re = \frac{LV}{\nu}. \tag{2.5.2}$$

The physical significance of the Reynolds number is that it measures the relative importance of inertial effects when compared with viscous effects in the flow. For small Reynolds number, it is common to ignore the entire left-hand side of the above equation and write

$$\nabla^2 \mathbf{v} = \nabla p. \tag{2.5.3}$$

This is commonly known as Stokes's equation. Being linear, unlike the original Navier–Stokes equation, it is solved relatively more easily, and a substantial literature can be found on *Stokes problems*. The book by Happel and Brenner (1965) is a good starting point for the reader interested in exploring such problems. Two facts are worthy of note. First, the dynamic pressure field in Stokes flow satisfies Laplace's equation. Second, Stokes flows are reversible. This means that in a given problem, reversing the flow direction simply changes the signs of the velocities and dynamic pressures, preserving the same absolute magnitudes everywhere. This will not hold if inertial effects are included. In this book, although we shall provide solutions for certain problems at high Reynolds number, most of our attention will be devoted to problems with negligible inertia. As a rule of thumb, for the settling of particles, drops, or bubbles, one can safely ignore inertial corrections to the velocity when the Reynolds number is less than 0.01, and such corrections are relatively small even when the Reynolds number is as large as 0.1. Nonetheless, examination of order of magnitude estimates of inertial and viscous forces at a given distance from a settling particle shows that inertial effects become comparable to viscous effects at scaled distances of the order $\frac{1}{Re}$ from the particle, no matter how small the Reynolds number might be. Therefore, a standard perturbational approach to solving problems with small inertial effects will fail unless proper account is taken of this fact. We briefly consider this problem in Section 3.2.

A comment regarding flows at high Reynolds number past particles, drops, and bubbles is in order. We might assume that ignoring viscous effects in the flow would be a reasonable approximation, and this is the case over most of the flow domain. It is possible to show that an inviscid flow in a body of fluid will remain irrotational if it starts that way, which means that the vorticity $\nabla \times \mathbf{v} = 0$. This implies that the velocity $\mathbf{v}$ can be written as the gradient of a scalar field ϕ, known as the velocity potential. If the flow is incompressible, it follows from Equation (2.2.2) that ϕ satisfies Laplace's equation:

$$\nabla^2 \phi = 0. \tag{2.5.4}$$

The Navier–Stokes equation in this case is used to infer the pressure distribution using the velocity field obtained from $\mathbf{v} = \nabla \phi$. Such flows are known as potential flows. The difficulty with this approach is that steady potential flows do not exert any drag on objects. Because the continuity equation, which is used to infer the velocity field, is lower in order than the Navier–Stokes equation, we lose the ability to satisfy relevant boundary conditions on the surfaces of objects. Thus, a potential flow would slip right over the surface of a rigid object, and in the case of fluid drops and bubbles, the tangential stress would not be continuous across the interface. Prandtl (1905) developed boundary layer theory to deal with this difficulty in the context of rigid surfaces. He envisioned a thin boundary layer near a rigid surface in which the velocity from potential flow should rapidly decay to zero at the surface, leading to large viscous forces that become comparable to inertia in the boundary layer. For our purposes, it suffices to note that potential flow still provides a useful approximation of the velocity field past fluid spheres, with discontinuities only in the stress at the surface. We shall return to this subject later in the book.

In a manner similar to that used in nondimensionalizing the Navier–Stokes equation, we can nondimensionalize the equation of conservation of energy. In this case, we encounter a dimensionless group known as the Péclet number, Pe. This group multiplies the convective transport terms, which play a role analogous to that of inertia in the Navier–Stokes equation. The Péclet number is defined as

$$Pe = \frac{LV}{\kappa}. \tag{2.5.5}$$

It follows that the Péclet number in heat transport provides information on the relative importance of convective transport of energy when compared with molecular transport. In this book, we shall consider problems in which the Péclet number is negligible, as well as those in which it is relatively large. The Péclet number can be seen to be the product of the Reynolds number and the Prandtl number. When a drop moves because of the action of interfacial tension gradients caused by temperature variations, we shall encounter a dimensionless group known as the Marangoni number, Ma, when we scale the energy equation. As noted in Section 4.1, this group plays the role of the Péclet number in such problems.

In the case of mass transport, the Péclet number can be defined in an analogous manner to that in heat transport:

$$Pe = \frac{LV}{D_{AB}}. \tag{2.5.6}$$

We see that the the Péclet number in mass transport is the product of the Reynolds number and the Schmidt number.

One must consider the boundary conditions and any other applicable equations, such as a force balance, to determine if additional dimensionless parameters emerge. We shall examine this subject when we analyze specific problems. Most of the boundary conditions that we use do not yield any additional parameters; however, the normal stress balance, which can be inferred from Equation (2.3.8), does lead to a new parameter. If the scale for stresses is the viscous stress mentioned earlier, this parameter is the Capillary number, Ca, defined as

$$Ca = \frac{\mu V}{\sigma}. \tag{2.5.7}$$

The magnitude of the Capillary number determines the relative importance of the normal stress imbalance in deforming a fluid-fluid interface. If the difference in normal stress across the interface is uniform over the surface of a drop, the curvature would be uniform, leading to a spherical shape. It is variations in this difference that lead to deformation from the spherical shape. In a similar manner, with a uniform normal stress difference, a liquid layer in a long and wide trough would assume a flat surface except in the region where contact is made with the solid walls. The Capillary number provides a measure of the relative importance of the deforming forces when compared with the interfacial tension, which acts to maintain the shape the interface would assume if there were no fluid motion. In problems wherein inertial effects dominate over viscous effects, stress would be nondimensionalized with an inertial scale. The dimensionless group determining the relative importance of the deforming forces in that situation is the Weber number, which is product of the Capillary and Reynolds numbers. The definition of the Weber number, We, is

$$We = \frac{\rho L V^2}{\sigma}. \tag{2.5.8}$$

We shall be working with scaled variables and dimensionless parameters in this book, with only a small number of exceptions. We shall use an asterisk to distinguish physical entities with dimensions from their nondimensional counterparts, where appropriate.

2.6   Conservation Equations in Common Coordinate Systems

The conservation equations are given in invariant notation in the earlier sections. When solving specific problems, however, it is useful to begin with a form that involves components in a certain coordinate system that is suited to the compact description of the important boundaries in a physical situation. For example, the motion of fluid in a rectangular trough usually would be solved by using a rectangular coordinate system, whereas a problem involving a drop or bubble is best posed in spherical polar coordinates. In subsequent chapters, we shall be modeling such problems, as well as those involving a pair of bubbles or drops, collections of drops, compound drops that consist of a droplet suspended within a drop that itself is suspended in another fluid, a drop moving in the vicinity of a rigid or fluid surface, and flow in a cylindrical container. Some of these configurations require special coordinate systems to be used. For example, the problem of a pair of spheres, or a sphere and a plane surface, can be handled by using bipolar coordinates, discussed in Section 5.1. In some other cases, no simple coordinate system can be used in which all the relevant boundaries can be described in a simple manner. The most common coordinate systems for which we shall find substantial use are the rectangular, cylindrical polar, and spherical polar coordinates. Each is an orthogonal system of coordinates, as can be seen from the definition sketches given in Figures 2.6.1, 2.6.2, and 2.6.3.

In each coordinate system, a triad of unit basis vectors can be constructed at any given point. These vectors point in the three coordinate directions at that point and are mutually orthogonal. The basis set is labeled $(\mathbf{i}, \mathbf{j}, \mathbf{k})$ in rectangular coordinates, $(\mathbf{i}_r, \mathbf{i}_\theta, \mathbf{i}_z)$ in cylindrical polar coordinates, and $(\mathbf{i}_r, \mathbf{i}_\theta, \mathbf{i}_\phi)$ in spherical polar coordinates. Rectangular coordinates have the simplicity that the basis set $(\mathbf{i}, \mathbf{j}, \mathbf{k})$ does not change with position, whereas in the other two systems, the triad of basis vectors rotates as one moves about, so that some derivatives of these basis vectors with respect to position do not vanish. One of the consequences is the appearance of certain additional terms in the governing equations when written in component form.

Table 2.6.1 provides the continuity equation in rectangular, cylindrical polar, and spherical polar coordinates. Similarly, Tables 2.6.2 and 2.6.3 list the Navier–Stokes equations and the components of the viscous stress tensor, respectively, in these coordinate systems. In a like manner, Tables 2.6.4, 2.6.5, and 2.6.6 provide the equations of conservation of energy, species, and insoluble surfactant on an interface, respectively, in these coordinate systems. These tables should suffice for the reader of this book. Any additional information that is needed can be found in Bird et al. (1960).

Table 2.6.1   The Equation of Continuity for Incompressible Flow

Rectangular coordinates (x, y, z)

$$\frac{\partial v_x}{\partial x} + \frac{\partial v_y}{\partial y} + \frac{\partial v_z}{\partial z} = 0$$

Cylindrical polar coordinates (r, θ, z)

$$\frac{1}{r}\frac{\partial}{\partial r}(r v_r) + \frac{1}{r}\frac{\partial v_\theta}{\partial \theta} + \frac{\partial v_z}{\partial z} = 0$$

Spherical polar coordinates (r, θ, ϕ)

$$\frac{1}{r^2}\frac{\partial}{\partial r}(r^2 v_r) + \frac{1}{r\sin\theta}\frac{\partial}{\partial \theta}(v_\theta \sin\theta) + \frac{1}{r\sin\theta}\frac{\partial v_\phi}{\partial \phi} = 0$$

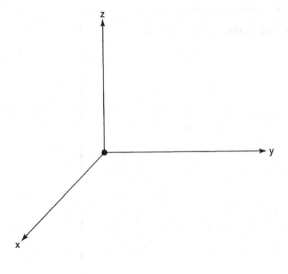

Figure 2.6.1 Sketch of the rectangular coordinate system.

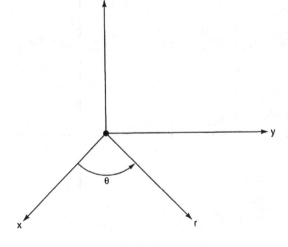

Figure 2.6.2 Sketch of the cylindrical polar coordinate system.

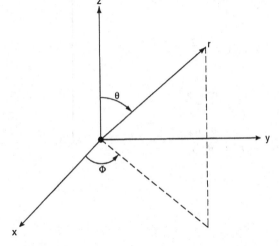

Figure 2.6.3 Sketch of the spherical polar coordinate system.

Table 2.6.2 The Navier–Stokes Equations (for Incompressible Newtonian Flow with Constant Viscosity μ)

Rectangular coordinates (x, y, z)

x-component

$$\rho\left[\frac{\partial v_x}{\partial t} + v_x\frac{\partial v_x}{\partial x} + v_y\frac{\partial v_x}{\partial y} + v_z\frac{\partial v_x}{\partial z}\right] = -\frac{\partial p}{\partial x} + \mu\left[\frac{\partial^2 v_x}{\partial x^2} + \frac{\partial^2 v_x}{\partial y^2} + \frac{\partial^2 v_x}{\partial z^2}\right] + \rho g_x$$

y-component

$$\rho\left[\frac{\partial v_y}{\partial t} + v_x\frac{\partial v_y}{\partial x} + v_y\frac{\partial v_y}{\partial y} + v_z\frac{\partial v_y}{\partial z}\right] = -\frac{\partial p}{\partial y} + \mu\left[\frac{\partial^2 v_y}{\partial x^2} + \frac{\partial^2 v_y}{\partial y^2} + \frac{\partial^2 v_y}{\partial z^2}\right] + \rho g_y$$

z-component

$$\rho\left[\frac{\partial v_z}{\partial t} + v_x\frac{\partial v_z}{\partial x} + v_y\frac{\partial v_z}{\partial y} + v_z\frac{\partial v_z}{\partial z}\right] = -\frac{\partial p}{\partial z} + \mu\left[\frac{\partial^2 v_z}{\partial x^2} + \frac{\partial^2 v_z}{\partial y^2} + \frac{\partial^2 v_z}{\partial z^2}\right] + \rho g_z$$

Cylindrical polar coordinates (r, θ, z)

r-component

$$\rho\left[\frac{\partial v_r}{\partial t} + v_r\frac{\partial v_r}{\partial r} + \frac{v_\theta}{r}\frac{\partial v_r}{\partial \theta} - \frac{v_\theta^2}{r} + v_z\frac{\partial v_r}{\partial z}\right]$$

$$= -\frac{\partial p}{\partial r} + \mu\left[\frac{\partial}{\partial r}\left(\frac{1}{r}\frac{\partial}{\partial r}(rv_r)\right) + \frac{1}{r^2}\frac{\partial^2 v_r}{\partial \theta^2} - \frac{2}{r^2}\frac{\partial v_\theta}{\partial \theta} + \frac{\partial^2 v_r}{\partial z^2}\right] + \rho g_r$$

θ-component

$$\rho\left[\frac{\partial v_\theta}{\partial t} + v_r\frac{\partial v_\theta}{\partial r} + \frac{v_\theta}{r}\frac{\partial v_\theta}{\partial \theta} + \frac{v_r v_\theta}{r} + v_z\frac{\partial v_\theta}{\partial z}\right]$$

$$= -\frac{1}{r}\frac{\partial p}{\partial \theta} + \mu\left[\frac{\partial}{\partial r}\left(\frac{1}{r}\frac{\partial}{\partial r}(rv_\theta)\right) + \frac{1}{r^2}\frac{\partial^2 v_\theta}{\partial \theta^2} + \frac{2}{r^2}\frac{\partial v_r}{\partial \theta} + \frac{\partial^2 v_\theta}{\partial z^2}\right] + \rho g_\theta$$

z-component

$$\rho\left[\frac{\partial v_z}{\partial t} + v_r\frac{\partial v_z}{\partial r} + \frac{v_\theta}{r}\frac{\partial v_z}{\partial \theta} + v_z\frac{\partial v_z}{\partial z}\right]$$

$$= -\frac{\partial p}{\partial z} + \mu\left[\frac{1}{r}\frac{\partial}{\partial r}\left(r\frac{\partial v_z}{\partial r}\right) + \frac{1}{r^2}\frac{\partial^2 v_z}{\partial \theta^2} + \frac{\partial^2 v_z}{\partial z^2}\right] + \rho g_z$$

Spherical polar coordinates (r, θ, ϕ)

r-component

$$\rho\left[\frac{\partial v_r}{\partial t} + v_r\frac{\partial v_r}{\partial r} + \frac{v_\theta}{r}\frac{\partial v_r}{\partial \theta} + \frac{v_\phi}{r\sin\theta}\frac{\partial v_r}{\partial \phi} - \frac{v_\theta^2 + v_\phi^2}{r}\right]$$

$$= -\frac{\partial p}{\partial r} + \mu\left[\mathcal{L}v_r - \frac{2}{r^2}v_r - \frac{2}{r^2}\frac{\partial v_\theta}{\partial \theta} - \frac{2}{r^2}v_\theta\cot\theta - \frac{2}{r^2\sin\theta}\frac{\partial v_\phi}{\partial \phi}\right] + \rho g_r$$

θ-component

$$\rho\left[\frac{\partial v_\theta}{\partial t} + v_r\frac{\partial v_\theta}{\partial r} + \frac{v_\theta}{r}\frac{\partial v_\theta}{\partial \theta} + \frac{v_\phi}{r\sin\theta}\frac{\partial v_\theta}{\partial \phi} + \frac{v_r v_\theta}{r} - \frac{v_\phi^2\cot\theta}{r}\right]$$

$$= -\frac{1}{r}\frac{\partial p}{\partial \theta} + \mu\left[\mathcal{L}v_\theta + \frac{2}{r^2}\frac{\partial v_r}{\partial \theta} - \frac{v_\theta}{r^2\sin^2\theta} - \frac{2\cos\theta}{r^2\sin^2\theta}\frac{\partial v_\phi}{\partial \phi}\right] + \rho g_\theta$$

Table 2.6.2 *Continued*

ϕ-component

$$\rho\left[\frac{\partial v_\phi}{\partial t} + v_r\frac{\partial v_\phi}{\partial r} + \frac{v_\theta}{r}\frac{\partial v_\phi}{\partial \theta} + \frac{v_\phi}{r\sin\theta}\frac{\partial v_\phi}{\partial \phi} + \frac{v_\phi v_r}{r} + \frac{v_\theta v_\phi}{r}\cot\theta\right]$$

$$= -\frac{1}{r\sin\theta}\frac{\partial p}{\partial \phi} + \mu\left[\mathcal{L}v_\phi - \frac{v_\phi}{r^2\sin^2\theta} + \frac{2}{r^2\sin\theta}\frac{\partial v_r}{\partial \phi} + \frac{2\cos\theta}{r^2\sin^2\theta}\frac{\partial v_\theta}{\partial \phi}\right] + \rho g_\phi$$

In the above equations, $\mathcal{L}$ is the Laplacian operator defined as follows:

$$\mathcal{L} = \frac{1}{r^2}\frac{\partial}{\partial r}\left(r^2\frac{\partial}{\partial r}\right) + \frac{1}{r^2\sin\theta}\frac{\partial}{\partial \theta}\left(\sin\theta\frac{\partial}{\partial \theta}\right) + \frac{1}{r^2\sin^2\theta}\left(\frac{\partial^2}{\partial \phi^2}\right)$$

Table 2.6.3 **Components of the Viscous Stress Tensor (for Newtonian Constitutive Model and Incompressible Flow)**

Rectangular coordinates (x, y, z)

$$\tau_{xx} = 2\mu\frac{\partial v_x}{\partial x} \qquad\qquad \tau_{xy} = \tau_{yx} = \mu\left[\frac{\partial v_x}{\partial y} + \frac{\partial v_y}{\partial x}\right]$$

$$\tau_{yy} = 2\mu\frac{\partial v_y}{\partial y} \qquad\qquad \tau_{yz} = \tau_{zy} = \mu\left[\frac{\partial v_y}{\partial z} + \frac{\partial v_z}{\partial y}\right]$$

$$\tau_{zz} = 2\mu\frac{\partial v_z}{\partial z} \qquad\qquad \tau_{zx} = \tau_{xz} = \mu\left[\frac{\partial v_z}{\partial x} + \frac{\partial v_x}{\partial z}\right]$$

Cylindrical polar coordinates (r, θ, z)

$$\tau_{rr} = 2\mu\frac{\partial v_r}{\partial r} \qquad\qquad \tau_{r\theta} = \tau_{\theta r} = \mu\left[r\frac{\partial}{\partial r}\left(\frac{v_\theta}{r}\right) + \frac{1}{r}\frac{\partial v_r}{\partial \theta}\right]$$

$$\tau_{\theta\theta} = 2\mu\left(\frac{1}{r}\frac{\partial v_\theta}{\partial \theta} + \frac{v_r}{r}\right) \qquad\qquad \tau_{\theta z} = \tau_{z\theta} = \mu\left[\frac{\partial v_\theta}{\partial z} + \frac{1}{r}\frac{\partial v_z}{\partial \theta}\right]$$

$$\tau_{zz} = 2\mu\frac{\partial v_z}{\partial z} \qquad\qquad \tau_{zr} = \tau_{rz} = \mu\left[\frac{\partial v_z}{\partial r} + \frac{\partial v_r}{\partial z}\right]$$

Spherical polar coordinates (r, θ, ϕ)

$$\tau_{rr} = 2\mu\frac{\partial v_r}{\partial r} \qquad\qquad \tau_{r\theta} = \tau_{\theta r} = \mu\left[r\frac{\partial}{\partial r}\left(\frac{v_\theta}{r}\right) + \frac{1}{r}\frac{\partial v_r}{\partial \theta}\right]$$

$$\tau_{\theta\theta} = 2\mu\left(\frac{1}{r}\frac{\partial v_\theta}{\partial \theta} + \frac{v_r}{r}\right) \qquad\qquad \tau_{\theta\phi} = \tau_{\phi\theta} = \mu\left[\frac{\sin\theta}{r}\frac{\partial}{\partial \theta}\left(\frac{v_\phi}{\sin\theta}\right) + \frac{1}{r\sin\theta}\frac{\partial v_\theta}{\partial \phi}\right]$$

$$\tau_{\phi\phi} = 2\mu\left(\frac{1}{r\sin\theta}\frac{\partial v_\phi}{\partial \phi} + \frac{v_r}{r} + \frac{v_\theta\cot\theta}{r}\right) \qquad \tau_{\phi r} = \tau_{r\phi} = \mu\left[\frac{1}{r\sin\theta}\frac{\partial v_r}{\partial \phi} + r\frac{\partial}{\partial r}\left(\frac{v_\phi}{r}\right)\right]$$

Table 2.6.4 The Equation of Conservation of Energy (for Constant Density and Thermal Conductivity, Neglecting Viscous Dissipation and Sources)

Rectangular coordinates (x, y, z)

$$\frac{\partial T}{\partial t} + v_x \frac{\partial T}{\partial x} + v_y \frac{\partial T}{\partial y} + v_z \frac{\partial T}{\partial z} = \kappa \left[\frac{\partial^2 T}{\partial x^2} + \frac{\partial^2 T}{\partial y^2} + \frac{\partial^2 T}{\partial z^2} \right]$$

Cylindrical polar coordinates (r, θ, z)

$$\frac{\partial T}{\partial t} + v_r \frac{\partial T}{\partial r} + \frac{v_\theta}{r} \frac{\partial T}{\partial \theta} + v_z \frac{\partial T}{\partial z} = \kappa \left[\frac{1}{r} \frac{\partial}{\partial r} \left(r \frac{\partial T}{\partial r} \right) + \frac{1}{r^2} \frac{\partial^2 T}{\partial \theta^2} + \frac{\partial^2 T}{\partial z^2} \right]$$

Spherical polar coordinates (r, θ, ϕ)

$$\frac{\partial T}{\partial t} + v_r \frac{\partial T}{\partial r} + \frac{v_\theta}{r} \frac{\partial T}{\partial \theta} + \frac{v_\phi}{r \sin \theta} \frac{\partial T}{\partial \phi}$$

$$= \kappa \left[\frac{1}{r^2} \frac{\partial}{\partial r} \left(r^2 \frac{\partial T}{\partial r} \right) + \frac{1}{r^2 \sin \theta} \frac{\partial}{\partial \theta} \left(\sin \theta \frac{\partial T}{\partial \theta} \right) + \frac{1}{r^2 \sin^2 \theta} \frac{\partial^2 T}{\partial \phi^2} \right]$$

Table 2.6.5 The Equation of Conservation of Species (for Constant Density and Diffusivity)

Rectangular coordinates (x, y, z)

$$\frac{\partial \rho_A}{\partial t} + v_x \frac{\partial \rho_A}{\partial x} + v_y \frac{\partial \rho_A}{\partial y} + v_z \frac{\partial \rho_A}{\partial z} = D_{AB} \left[\frac{\partial^2 \rho_A}{\partial x^2} + \frac{\partial^2 \rho_A}{\partial y^2} + \frac{\partial^2 \rho_A}{\partial z^2} \right] + r_A$$

Cylindrical polar coordinates (r, θ, z)

$$\frac{\partial \rho_A}{\partial t} + v_r \frac{\partial \rho_A}{\partial r} + \frac{v_\theta}{r} \frac{\partial \rho_A}{\partial \theta} + v_z \frac{\partial \rho_A}{\partial z} = D_{AB} \left[\frac{1}{r} \frac{\partial}{\partial r} \left(r \frac{\partial \rho_A}{\partial r} \right) + \frac{1}{r^2} \frac{\partial^2 \rho_A}{\partial \theta^2} + \frac{\partial^2 \rho_A}{\partial z^2} \right] + r_A$$

Spherical polar coordinates (r, θ, ϕ)

$$\frac{\partial \rho_A}{\partial t} + v_r \frac{\partial \rho_A}{\partial r} + \frac{v_\theta}{r} \frac{\partial \rho_A}{\partial \theta} + \frac{v_\phi}{r \sin \theta} \frac{\partial \rho_A}{\partial \phi}$$

$$= D_{AB} \left[\frac{1}{r^2} \frac{\partial}{\partial r} \left(r^2 \frac{\partial \rho_A}{\partial r} \right) + \frac{1}{r^2 \sin \theta} \frac{\partial}{\partial \theta} \left(\sin \theta \frac{\partial \rho_A}{\partial \theta} \right) + \frac{1}{r^2 \sin^2 \theta} \frac{\partial^2 \rho_A}{\partial \phi^2} \right] + r_A$$

Table 2.6.6 The Equation of Conservation of Insoluble Surfactant on an Interface (for Constant Surface Diffusivity)

Rectangular surface coordinates (x, y) (Surface at constant $z = c$)

$$\frac{\partial \Gamma}{\partial t} + \frac{\partial}{\partial x} (v_x \Gamma) + \frac{\partial}{\partial y} (v_y \Gamma) = D_s \left[\frac{\partial^2 \Gamma}{\partial x^2} + \frac{\partial^2 \Gamma}{\partial y^2} \right]$$

Cylindrical polar surface coordinates (θ, z) (Surface at constant $r = R$)

$$\frac{\partial \Gamma}{\partial t} + \frac{1}{R} \frac{\partial}{\partial \theta} (v_\theta \Gamma) + \frac{\partial}{\partial z} (v_z \Gamma) = D_s \left[\frac{1}{R^2} \frac{\partial^2 \Gamma}{\partial \theta^2} + \frac{\partial^2 \Gamma}{\partial z^2} \right]$$

Spherical polar surface coordinates (θ, ϕ) (Surface at constant $r = R$)

$$\frac{\partial \Gamma}{\partial t} + \frac{1}{R \sin \theta} \frac{\partial}{\partial \theta} (v_\theta \Gamma \sin \theta) + \frac{1}{R \sin \theta} \frac{\partial}{\partial \phi} (v_\phi \Gamma) = D_s \left[\frac{1}{R^2 \sin \theta} \frac{\partial}{\partial \theta} \left(\sin \theta \frac{\partial \Gamma}{\partial \theta} \right) + \frac{1}{R^2 \sin^2 \theta} \frac{\partial^2 \Gamma}{\partial \phi^2} \right]$$

THE MOTION OF ISOLATED BUBBLES AND DROPS

In Part Two, we consider problems that arise in the motion of isolated bubbles or drops. To avoid repetition, we use the word *drop* generically for both bubbles and drops except where it is important to distinguish the special limiting case of a gas bubble. The word *isolated* has the following implication: The behavior of the drop is unaffected by the presence of any other drops or boundaries of the container within which the drop moves. Mathematically, the problem would be posed as though the fluid body surrounding the drop extends to infinity in all directions as one moves away from the drop. The only other boundary exterior to the drop on which conditions would be imposed is that at infinity. Usually, these conditions are statements that certain behavior be approached asymptotically by the fields as the distance from the center of the drop approaches infinity.

CHAPTER THREE

Motion Driven by a Body Force

3.1 Motion When Inertial Effects Are Negligible

3.1.1 Introduction

This chapter deals with traditional problems that arise when a drop moves because of a body force. Gravity is used as the model body force. It is uniform over space and constant in time over the distances and time periods of interest here. Consideration of other body forces, such as those due to electric or magnetic fields, with the attendant prospect of variation in space and time, is possible but is beyond the scope of this book. In Chapter Four, we treat problems involving the motion of a drop due to the action of interfacial tension gradients.

Consider the motion of a drop of radius, R, and constant dynamic viscosity, μ', which retains its spherical shape while settling at constant velocity, U, through a second fluid of constant dynamic viscosity, μ, which is infinite in extent. Inertial effects are assumed negligible. The viscosities of fluids are usually sensitive to temperature so that their constancy can be ensured only under isothermal conditions. This is the analog of the problem analyzed by Stokes (1851) for the motion of a rigid sphere under similar conditions. The problem for the case of drops was solved independently by Hadamard (1911) and Rybczyński (1911) and thus is known as the Hadamard–Rybczyński problem. Of course, drops also can deform in shape, and the actual shape must be obtained as part of the final solution of the problem. In this problem, the spherical shape happens to be correct as long as inertia is neglected. In general, deformation from the spherical shape can be ignored if the interfacial tension is sufficiently large to maintain nearly uniform curvature in spite of deforming forces that arise from the spatial variation of the normal stress difference across the interface. This statement can be cast in the context of a suitable dimensionless group, such as the Capillary number, but we postpone doing that until later. The neglect of the inertial terms in the governing equations implies a negligible value of the Reynolds number. It is not possible to say at the outset what numerical value of the Reynolds number or any other dimensionless group can be considered negligible. In practice, this is determined either from experimental observation or analysis that includes the effects of the neglected groups.

Besides being the simplest problem one can analyze, there is some practical merit in considering the Hadamard–Rybczyński problem. Situations arise in industrial operations and in day-to-day applications in which small drops are encountered. One example can be found in separation processes where two phases are contacted to transfer one or more species between them. To achieve efficient operation, one of the phases is broken up into a fine dispersion of drops or bubbles. Another reason for using the Stokes settling of a drop as a model case is to introduce the notation as well as several ideas that will be used in the rest of the book.

The physical problem involves the motion a drop executes after it is released into a second fluid. A standard question asked is: What is the terminal settling velocity of the drop, given the relevant physical properties and geometrical information? Other questions include: What is the detailed velocity field outside the drop and within it? How long does it take for the drop to achieve its terminal settling velocity? How far does a boundary or another drop have to be for its influence to be judged negligible? Although the analysis in this section can help answer the first two questions, answers for the last two can only be estimated. We make such estimates at the end of this section.

3.1.2 Analysis

The starting point of the analysis is the choice of a coordinate system and the making of suitable assumptions to simplify the governing equations. A sketch showing the drop and its surroundings, along with some useful information, is given in Figure 3.1.1. The spherical polar coordinates to be used here (r, θ, ϕ) are marked on the sketch, as well as cylindrical polar coordinates (ω, z). This figure also will serve as a reference for several problems posed in subsequent sections.

The assumptions include incompressible Newtonian flow with a constant viscosity, as well as steady velocity and pressure fields in a reference frame attached to the drop. By assuming steady fields, the initial transient period after which the drop will achieve terminal settling conditions is neglected. Assuming negligible Reynolds number permits one to neglect the convective transport of momentum. The making of these assumptions

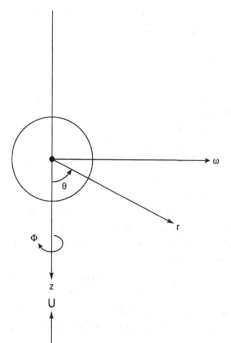

Figure 3.1.1 Sketch of the system in a reference frame attached to the drop, showing spherical polar and cylindrical polar coordinates.

precludes the generation of answers to some of the questions raised. For instance, without doing an analysis of the unsteady problem, one cannot determine how long it will take for the drop to reach terminal velocity to within a specified tolerance. We address this subject toward the end of this section.

For convenience at a later stage, we shall use a reference frame attached to the moving drop. In such a reference frame, the drop appears stationary, and fluid appears to be approaching the drop at the terminal settling velocity U in the negative z-direction as shown in Figure 3.1.1. In a reference frame that moves at constant velocity with respect to a laboratory fixed on Earth, the governing Navier–Stokes equation remains unchanged. If the reference frame were to be accelerating with respect to such a laboratory, one would need to modify the Navier–Stokes equation by making the necessary transformation to the new reference frame. It is best to work with scaled variables as far as possible. The reference length is the radius of the drop, R, and the reference velocity is the settling velocity, U. Stresses, including pressure, are scaled by using the viscous stress, $\mu\frac{U}{R}$, as a reference.

In the absence of inertial effects, and for incompressible Newtonian flow, the governing equations for the scaled velocity $\mathbf{v}$ and the scaled hydrodynamic pressure, p, are linear and known as the Stokes (or alternatively Stokes's) equations. In the continuous phase, they are written as follows:

$$\nabla \bullet \mathbf{v} = 0 \tag{3.1.1}$$

and

$$\nabla^2 \mathbf{v} = \nabla p. \tag{3.1.2}$$

The same equations, using symbols with a prime for the scaled velocity and pressure, apply inside the drop, where the symbols with a prime are used to distinguish these variables within the drop from those outside of it. The only change is the appearance of a factor α in front of the viscous term in the left side of Equation (3.1.2) if we use the same reference pressure within the drop phase. The symbol $\alpha(= \frac{\mu'}{\mu})$ stands for the ratio of the viscosity of the drop phase to that of the continuous phase. We define all similar ratios of properties in subsequent sections in the same manner, namely, the value of the property in the fluid inside the drop divided by its value in the exterior fluid.

The problem is axially symmetric. That is, the velocity and pressure fields are independent of the azimuthal coordinate ϕ, and there is no velocity component in the ϕ direction. This permits some immediate and important simplifications. For axially symmetric velocity fields, the Stokes streamfunction, ψ, can be introduced. The value of ψ at a point is physically related to the volumetric flow rate through a surface obtained by joining the point in question to the axis by any curve and rotating this curve by an angle of 2π about the axis. Because the velocity field is solenoidal, this result holds regardless of the actual point on the axis at which the curve meets it. This can be seen by applying the divergence theorem to a volume enclosed by two surfaces obtained by the same process using two different points on the axis of symmetry. The sign convention is such that the physical stream function represents the volumetric flow rate in the negative z-direction in Figure 3.1.1, divided by 2π. Of course, the scaled streamfunction will have no units. The reference quantity for the streamfunction is UR^2. Our notation below parallels that of Happel and Brenner (1965), who discussed in detail numerous problems of motion at low Reynolds number. We also have learned much from their book in formulating the subsequent development in this and several other sections. The scaled streamfunction is related to the velocity field as follows:

$$\mathbf{v} = \frac{1}{\omega}\mathbf{i}_\phi \times \nabla\psi.$$

(3.1.3)

From this definition, it is possible to write the connection between the velocity components in spherical polar coordinates and the streamfunction:

$$v_r = -\frac{1}{r^2\sin\theta}\frac{\partial\psi}{\partial\theta},$$

(3.1.4)

$$v_\theta = \frac{1}{r\sin\theta}\frac{\partial\psi}{\partial r}.$$

(3.1.5)

By taking the curl of both sides of Equation (3.1.2), it is possible to eliminate pressure as a dependent variable at the cost of increasing the order of the equation. Then, inserting the definition of ψ leads to a fourth-order partial differential equation, which is known as Stokes's equation for the streamfunction:

$$E^4\psi = 0.$$

(3.1.6)

The operator E^4 may be formally written as $E^2(E^2)$. E^2 is closely related to the Laplacian ∇^2 encountered in Chapter Two. In spherical polar coordinates, E^2 is written as follows:

$$E^2 = \frac{\partial^2}{\partial r^2} + \frac{\sin\theta}{r^2}\frac{\partial}{\partial\theta}\left(\frac{1}{\sin\theta}\frac{\partial}{\partial\theta}\right).$$

(3.1.7)

General solutions of $E^4\psi = 0$ are available in spherical polar coordinates, as well as in some other coordinate systems used in axisymmetric problems. They commonly are written in the form of expansions using a suitable orthogonal set of basis functions, obtained by the method of separation of variables. They are then specialized by applying boundary conditions relevant to the problem in question. Happel and Brenner (1965) give the following solution in spherical polar coordinates, attributing its origin to Sampson (1891):

$$\psi(r, s) = \sum_{n=0}^{\infty}(A_n r^n + B_n r^{-n+1} + C_n r^{n+2} + D_n r^{-n+3})C_n(s)$$

$$+ \sum_{n=2}^{\infty}(a_n r^n + b_n r^{-n+1} + c_n r^{n+2} + d_n r^{-n+3})\mathcal{H}_n(s).$$

(3.1.8)

In Equation (3.1.8), $s = \cos\theta$. Here, and in the rest of the book, we use the set of independent variables (r, s) interchangeably with the set (r, θ). In some results involving θ, we use s and θ in the same expression if it is convenient to do so. For economy, we do not designate the equivalent functions of (r, s) and (r, θ) by different symbols. In other words, both $\psi(r, s)$ and $\psi(r, \theta)$ are used to represent the same dependent variable. Of course, when differentiations or integrations with respect to θ or s are involved, one must be careful in converting the dependence to the correct variable before carrying out the operation. In Equation (3.1.8), the symbols A_n, B_n, C_n, and D_n, as well as the corresponding lower case symbols, refer to sets of arbitrary constants.

Note the appearance of the functions $C_n(s)$ and $\mathcal{H}_n(s)$ in the solution for ψ. To avoid clutter, we have deleted the superscript $-\frac{1}{2}$ which must be used with these functions, but it is implied everywhere. The functions $C_n(s)$ and $\mathcal{H}_n(s)$ are known as Gegenbauer functions of the first and second kind, respectively, and are closely related to the Legendre functions of the first and second kind. This is not surprising given the close connection between the Stokes and Laplace operators. The subscript, n, is the order of the Gegenbauer function, and the omitted superscript $-\frac{1}{2}$ is the degree. When the order n is an integer, the functions $C_n(s)$ are polynomials.[†] Some properties of Gegenbauer functions are provided in the book by Happel and Brenner (1965). Additional information can be found in the Handbook of Mathematical Functions by Abramowitz and Stegun (1965).

The solution can be specialized further by using the fact that neither the streamfunction nor the velocity component in the θ-direction, v_θ, can be allowed to become unbounded on the axis of symmetry. This rules out the entire set of Gegenbauer functions of the second kind, as well as the contributions from the terms corresponding to $n = 0$ and $n = 1$ in the rest of the solution. The result, given below, is our starting point here and in other related problems. For completeness, we also have written the corresponding result for the streamfunction field within the drop phase fluid.

$$\psi(r, s) = \sum_{n=2}^{\infty} (A_n r^n + B_n r^{-n+1} + C_n r^{n+2} + D_n r^{-n+3}) C_n(s), \tag{3.1.9}$$

$$\psi'(r, s) = \sum_{n=2}^{\infty} (A_n' r^n + B_n' r^{-n+1} + C_n' r^{n+2} + D_n' r^{-n+3}) C_n(s). \tag{3.1.10}$$

The two nonzero velocity components, v_r and v_θ, and their counterparts within the drop, can be obtained by suitable differentiation of Equations (3.1.9) and (3.1.10).

$$v_r = -\sum_{n=2}^{\infty} (A_n r^{n-2} + B_n r^{-n-1} + C_n r^n + D_n r^{-n+1}) P_{n-1}(s), \tag{3.1.11}$$

$$v_\theta = \sum_{n=2}^{\infty} (n A_n r^{n-2} - (n-1) B_n r^{-n-1} + (n+2) C_n r^n - (n-3) D_n r^{-n+1}) \frac{C_n(s)}{\sin\theta}, \tag{3.1.12}$$

$$v_r' = -\sum_{n=2}^{\infty} (A_n' r^{n-2} + B_n' r^{-n-1} + C_n' r^n + D_n' r^{-n+1}) P_{n-1}(s), \tag{3.1.13}$$

$$v_\theta' = \sum_{n=2}^{\infty} (n A_n' r^{n-2} - (n-1) B_n' r^{-n-1} + (n+2) C_n' r^n - (n-3) D_n' r^{-n+1}) \frac{C_n(s)}{\sin\theta}. \tag{3.1.14}$$

Note the appearance of the Legendre Polynomials $P_n(s)$[†] in the results for v_r and v_r'. Many important properties of these polynomials, as well as those of Legendre functions of the first and second kind, can be found in the book by MacRobert (1967).

Specialization of the general solution requires the use of boundary conditions. As mentioned in Chapter Two, these must be obtained from physical grounds. Although we do not normally go through the detailed application of each boundary condition in every problem, it is instructive to do a little bit of that in this first example case. We shall find it convenient to apply most of the boundary conditions directly on the velocity components, even though one can equally well apply them to the results for the streamfunctions. The complete set of boundary conditions applicable to this problem follows.

As $r \to \infty$, the velocity field must approach the uniform stream shown in the sketch. That is, because $\mathbf{v} \to -\mathbf{k}$ as $r \to \infty$, v_r must approach $-\cos\theta$, and v_θ must approach

---

[†] The Gegenbauer Polynomials $C_n(s)$ and the Legendre Polynomials $P_n(s)$ satisfy the orthogonality relations given below:

$$\int_{-1}^{+1} \frac{C_m(s) C_n(s)}{(1-s^2)} ds = \frac{2}{n(n-1)(2n-1)} \delta_{mn} \qquad \int_{-1}^{+1} P_m(s) P_n(s) ds = \frac{2}{(2n+1)} \delta_{mn}$$

$\sin\theta$. Using Equations (3.1.4) and (3.1.5), it can be established by integration that the streamfunction ψ must satisfy

$$\psi \to \frac{1}{2}r^2(1-s^2) \quad \text{as} \quad r \to \infty. \tag{3.1.15}$$

At the center of the drop, the velocity components must remain bounded.

$$|v_r'| < \infty, \quad |v_\theta'| < \infty \quad \text{as} \quad r \to 0 \tag{3.1.16}$$

At the drop surface, several conditions must be satisfied. The drop surface is fixed in this reference frame. Therefore, the kinematic condition can be written as

$$v_r(1, s) = v_r'(1, s) = 0. \tag{3.1.17}$$

The velocity field is continuous across the interface. The continuity of radial velocities is already stated in Equation (3.1.17). The continuity of tangential velocities yields

$$v_\theta(1, s) = v_\theta'(1, s). \tag{3.1.18}$$

Next, we use the stress balance. Because the shape of the drop has been assumed, it generally is not possible to satisfy the balance of normal stresses. We comment on this subject later in this section. In the absence of any interfacial tension gradients, the balance of tangential stresses at the interface reduces to

$$\tau_{r\theta}(1, s) = \tau_{r\theta}'(1, s), \tag{3.1.19}$$

or, in terms of velocity components,

$$\left[\frac{\partial}{\partial r}\left(\frac{v_\theta}{r}\right)\right]_{r=1} = \alpha\left[\frac{\partial}{\partial r}\left(\frac{v_\theta'}{r}\right)\right]_{r=1}, \tag{3.1.20}$$

because the terms involving $\frac{\partial v_r}{\partial\theta}$ and $\frac{\partial v_r'}{\partial\theta}$ are zero as a consequence of the kinematic condition.

When the condition as $r \to \infty$ is used with the solution for the streamfunction in the continuous phase, we see that all the constants $C_n = 0, n \geq 2$, and further that $A_n = 0, n \geq 3$ and $A_2 = 1$. This still leaves us with the infinite set of constants B_n and D_n, which cannot yet be determined. Similarly, because the velocity components must remain bounded everywhere within the drop and therefore as $r \to 0$, it can be seen that $B_n' = D_n' = 0, n \geq 2$. After this, the remaining conditions are applied. Note that some textbooks assume the solution for ψ to be of the form $f(r)\sin^2(\theta)$, thereby not including contributions from the remaining members of the basis set of Gegenbauer polynomials at the outset. This is usually justified by the fact that the field as $r \to \infty$ is of this form in its dependence on θ; however, as seen from the above, the correct application of the condition does not eliminate the contributions from other members of the basis set. It is the application of the remaining conditions at the surface of the drop that leads to their elimination, leaving only $C_2(s)$ as the survivor from the infinite set in the expansion for the streamfunction. This is simply $\frac{\sin^2\theta}{2}$. The final results for the various fields are reported below.

$$\psi = \frac{1}{2}\left[r^2 - \frac{2+3\alpha}{2(1+\alpha)}r + \frac{\alpha}{2(1+\alpha)}\frac{1}{r}\right]\sin^2\theta, \tag{3.1.21}$$

$$v_r = -\left[1 - \frac{2+3\alpha}{2(1+\alpha)}\frac{1}{r} + \frac{\alpha}{2(1+\alpha)}\frac{1}{r^3}\right]\cos\theta, \tag{3.1.22}$$

$$v_\theta = \left[1 - \frac{2+3\alpha}{4(1+\alpha)}\frac{1}{r} - \frac{\alpha}{4(1+\alpha)}\frac{1}{r^3}\right]\sin\theta, \tag{3.1.23}$$

$$p = \frac{2+3\alpha}{2(1+\alpha)}\frac{1}{r^2}\cos\theta + c, \tag{3.1.24}$$

$$\psi' = \frac{1}{4(1+\alpha)}(r^4 - r^2)\sin^2\theta, \tag{3.1.25}$$

$$v_r' = \frac{1}{2(1+\alpha)}(1 - r^2)\cos\theta, \tag{3.1.26}$$

$$v_\theta' = -\frac{1}{2(1+\alpha)}(1 - 2r^2)\sin\theta, \tag{3.1.27}$$

$$p' = -\frac{5\alpha}{(1+\alpha)}r\cos\theta + c'. \tag{3.1.28}$$

Along with the streamfunction and velocity fields, we have included the results for the scaled hydrodynamic pressure fields to within an arbitrary constant in each case. These constants, c and c', are related to each other. The relationship can be established by considering the balance of normal stresses that determines the shape of the drop. Recall that we assumed the shape to be a sphere at the beginning. It was shown by Taylor and Acrivos (1964) that for drops settling because of the action of gravity at negligible Reynolds number, the shape is indeed a sphere regardless of the value of the interfacial tension and that to cause deformation from that shape, inertial effects must be included. This topic is discussed in more detail in Taylor and Acrivos (1964), and a good summary of the case of negligible Reynolds number can be found in Leal (1992).

The results for the streamfunctions and velocity fields in the continuous phase can be considered as the consequence of superposing solutions corresponding to a uniform flow and those corresponding to simple singularities located at the center of the drop. Consider the streamfunction in the outer fluid. The first term in the right side corresponds to the uniform flow coming from far away and will disappear if we revert to the laboratory reference frame. The second term is called a *Stokeslet* and is the flow driven by a scaled point force $\mathbf{F} = 2\pi\frac{2+3\alpha}{(1+\alpha)}\mathbf{k}$. This contains all the vorticity in this flow. The third term corresponds to a potential flow created by a dipole or a doublet, which means a source-sink pair located at the center of the drop. We have discussed potential flows briefly in Section 2.5. Usually, the potential flow approximation is made when viscous effects are small. Of course, the flow considered here is very viscous flow, but that does not prevent the solution for the velocity from being the same as that in potential flow. Naturally, the corresponding pressure distribution would be different for the two types of flows. Kim and Karrila (1991) have illustrated the utility of singularity solutions in a variety of problems involving the motion of isolated particles, as well as particles interacting with each other and with boundaries.

The flow inside the drop is worthy of discussion. Clearly, the fluid within the drop must recirculate because it cannot leave the drop. The solution given above for the viscous flow problem also happens to satisfy the complete Navier–Stokes equation inside and is known as Hill's spherical vortex. This flow is not irrotational. It can be shown that the vorticity vector points in the ϕ direction at all locations within the drop. The contours of constant magnitude of the vorticity are coaxial cylinders, with their axis coinciding with the axis of symmetry.

3.1.3 Results

Now that we have the solution for the velocity fields and the streamfunction both outside and inside the drop, what can we do with them? First, if one were interested in problems of heat or mass transport, one can use these known velocity distributions in approaching the relevant transport problem. Second, by plotting level curves for the streamfunction, called streamlines, one can develop a mental picture of the flow. It is not difficult to show that the velocity vector is everywhere tangent to curves on which the streamfunction is constant. In unsteady flow, these curves will change with time because the velocity field changes with time. In the present problem, because the drop is fixed in the chosen reference frame, the flow is steady. In this case, pathlines, which are curves traced by material particles as they move about, also coincide with streamlines. Furthermore, streaklines, which are curves traced by "dye" that might be emitted continuously from a point in the domain, also coincide with streamlines in steady flows. Therefore, streamlines are instructive devices that help us learn about the nature of a given flow. In Figure 3.1.2, we have provided some sample sketches of streamlines in a meridian plane (a plane that

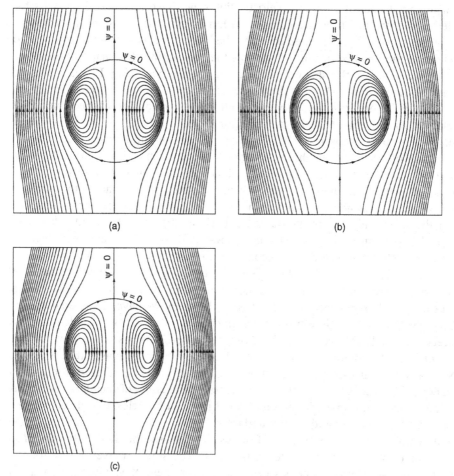

Figure 3.1.2 Streamlines in a reference frame attached to the drop (a) $\alpha = 0$, $\Delta\psi = \frac{1}{19}$, and $\Delta\psi' = -\frac{1}{144}$; (b) $\alpha = 1$, $\Delta\psi = \frac{13}{304}$, and $\Delta\psi' = -\frac{1}{288}$; (c) $\alpha = 10$, $\Delta\psi = \frac{29}{836}$, and $\Delta\psi' = -\frac{1}{1584}$; $\psi' = 0$ on the surface of the drop, and $\Delta\psi$ and $\Delta\psi'$ are increments measured away from this surface.

contains the axis of symmetry) for a few values of the viscosity ratio, α. Because of the symmetry of the problem, it is necessary to provide the drawings only in one quadrant of the plane; the results for the other quadrants can be immediately inferred. Nonetheless, we have shown the entire drop for aesthetic reasons. The streamfunction is zero on the axis of symmetry, as well as on the surface of the drop. Streamlines are shown outside with equal increments of ψ to a maximum that corresponds to the value on the equatorial plane at $r = 2$. Within the drop, equal increments in ψ', which are negative in this case, are used proceeding inward from the surface of the drop to the internal stagnation point. The range of values of ψ' used within the drop is from 0 to $\psi'_{min} = -\frac{1}{16(1+\alpha)}$. When streamlines are plotted with equal increments in the streamfunction, the variation in their spatial density in a given region can be used to infer the variation in the magnitude of the velocity in that region.

It also is of interest to record details of the instantaneous flow field as seen by an observer in a fixed reference frame, known as the laboratory reference frame. Although flow in this reference frame will be unsteady because the region affected by the motion of the drop always changes with time, knowledge of the velocity field still provides a general idea of the direction and intensity of the motion of fluid at various locations in the neighborhood of the moving drop. The streamfunctions in the laboratory reference frame can be obtained from Equations (3.1.21) and (3.1.25) by subtracting $\frac{1}{2}r^2 \sin^2 \theta$, which corresponds to the uniform stream, from the expressions on the right side. Figure 3.1.3 shows sample streamlines in the laboratory reference frame. In this case, streamlines are shown with equal increments from 0 to the maximum value that is reached on the equator at $r = 2$. Note that $\psi = 0$ along the axis of symmetry, but the drop surface is no longer a streamsurface. The streamlines intersect the surface of the drop and continue from the continuous phase smoothly into the drop phase. This apparent smoothness is a consequence of the continuity of the first derivatives of the streamfunction. Discontinuities in the second derivatives cannot be discerned in such a drawing. Finally, when $\alpha = 10$, the figure shows that fluid motion inside the drop is so weak that the drop almost appears to move as a rigid object.

We have not yet worked out answers to some of the questions raised earlier. Perhaps the most important question is that of the terminal velocity of the drop. Even though Equations (3.1.21–3.1.28) appear to contain no unknowns, included in them is the velocity scale U, which is the quantity we seek. Therefore, it appears that we have not used all the information about the physical situation here. We now invoke Newton's law of motion for the drop to see if it yields some useful information. Newton's law states that the acceleration experienced by the drop must equal the ratio of the force acting on it to its mass. The drop is moving at a steady velocity and must therefore experience zero net force. In the following discussion, physical quantities with dimensions are used for clarity.

For convenience, the force on the drop can be envisioned as the sum of the force due to gravity acting on the drop and the force exerted by the continuous phase on it at the interface. The former is simply the weight of the drop and is given by $\frac{4}{3}\pi R^3 \rho' g\mathbf{k}$, where ρ' is the density of the drop phase and g is the magnitude of the acceleration due to gravity. The force exerted by the continuous phase is customarily divided into a static part that would arise even if there were no motion in the fluid and a hydrodynamic contribution solely due to the motion in the fluid. The static force, arising because of the hydrostatic pressure variation in the continuous phase, also is known as the *buoyancy force* on the drop. It can be shown to be the weight of the displaced continuous phase

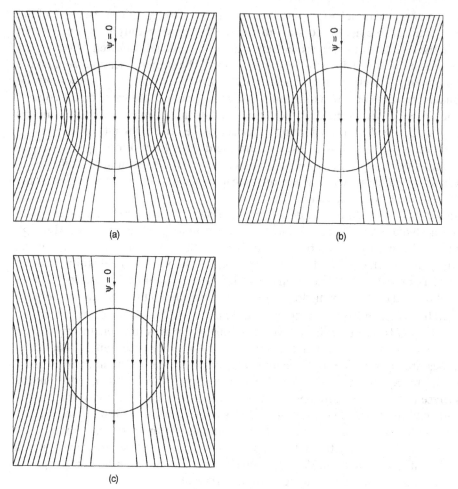

Figure 3.1.3   Streamlines in the laboratory reference frame (a) $\alpha = 0$, $\Delta\psi = \Delta\psi' = -\frac{1}{19}$; (b) $\alpha = 1$, $\Delta\psi = \Delta\psi' = -\frac{1}{16}$; (c) $\alpha = 10$, $\Delta\psi = \Delta\psi' = -\frac{59}{836}$; $\psi = \psi' = 0$ along the symmetry axis, and $\Delta\psi$ and $\Delta\psi'$ are increments measured away from this axis.

fluid and acts upward. When this is combined with the weight of the drop, the net hydrostatic force on the drop, $\mathbf{F}_H$, can be written as $\frac{4}{3}\pi R^3(\rho' - \rho)g\mathbf{k}$, where ρ is the density of the continuous phase fluid. The calculation of the hydrodynamic force, $\mathbf{F}_D$, can be performed in a variety of ways. In principle, the approach is to obtain the stress tensor Π in the continuous phase at an arbitrary point on the interface and take the inner (dot) product with an area element $d\mathbf{S}$, which will yield the force on that area. The result is then integrated over the entire drop surface to obtain the total hydrodynamic force. That is,

$$\mathbf{F}_D = \int_S d\mathbf{S} \bullet \Pi. \tag{3.1.29}$$

The hydrodynamic pressure must be used in the stress, and not the total pressure, because the hydrostatic contribution already has been included in $\mathbf{F}_H$. Performing the integration in Equation (3.1.29) is somewhat tedious, but not difficult. Happel and Brenner (1965) have discussed some other ways of calculating the hydrodynamic force.

They have given the result for the hydrodynamic force when the streamfunction is represented by a solution of the form in Equation (3.1.9). In the present notation, it is

$$\mathbf{F}_D = 4\pi \mu U R D_2 \mathbf{k}. \tag{3.1.30}$$

Because of symmetry, the force $\mathbf{F}_D$ points in the direction of motion of the approaching fluid and is also known as the *drag*, which accounts for our choice of subscript for this force. From the solution, it can be seen that D_2 is given by

$$D_2 = -\frac{2 + 3\alpha}{2(1 + \alpha)} \tag{3.1.31}$$

so that the hydrodynamic force on the drop can be written as

$$\mathbf{F}_D = -2\pi \mu U R \frac{2 + 3\alpha}{1 + \alpha} \mathbf{k}. \tag{3.1.32}$$

The terminal settling velocity is now obtained by setting the sum of the two forces to zero. This yields

$$U = \frac{2(1 + \alpha)}{3(2 + 3\alpha)} \frac{(\rho' - \rho)g R^2}{\mu}. \tag{3.1.33}$$

The contribution from the hydrodynamic pressure force to the total hydrodynamic force is one-third of the total, regardless of the viscosity ratio. The remaining two-thirds is made up of a fraction $\frac{4}{3(2+3\alpha)}$ from the normal viscous stress and the balance of $\frac{2\alpha}{2+3\alpha}$ from the tangential viscous stress.

In the above development, it was necessary to invoke Newton's law for the drop to obtain its velocity. If we had used the normal stress balance as part of the problem statement and calculated the shape of the drop as well, it would have been superfluous to write Newton's law for the drop separately because it is already implied in the set of governing equations being used.

We made no assumptions about the property ratios that appear in the development above. It is possible to recover the limiting results for a gas bubble by simply setting the viscosity ratio α to zero. For instance, the hydrodynamic force on a bubble would be $-4\pi \mu U R \mathbf{k}$, and its terminal rise velocity would be $\frac{\rho' g R^2}{3\mu}$. Naturally, the results for the velocity and streamfunction fields also can be specialized in this way. Note the disappearance of the potential dipole contribution in the flow outside in this limiting case. Therefore, the Stokes flow outside a rising gas bubble is simply the same as that produced by a Stokeslet or point force at its center.

In the limit as the viscosity of the drop phase becomes very large compared with that of the continuous phase, the drop phase would represent a rigid object; therefore, it should be possible to recover Stokes's results for the motion of a rigid sphere in the exterior fluid, including the hydrodynamic force. This corresponds to the limit $\alpha \to \infty$.

We conclude this section by returning to some of the questions posed near the beginning. The two unanswered questions are those concerning the time taken to achieve a steady settling velocity and the distance at which a boundary or another drop must be located for its influence to be judged negligible. The question of time scales can be answered correctly by considering unsteady analysis that was performed by Chisnell (1987). The issue might be separated into two questions. First, if we consider a situation where a drop moves at fixed velocity $\mathbf{U}$ through a fluid at rest, we can envision, in a reference frame attached to the drop, that the velocity field will take some time to achieve a steady distribution in both fluids. This time is estimated approximately as follows. For flow at negligible Reynolds number, both in the exterior and in the interior fluids, viscous effects dominate in transporting momentum. Therefore, the time taken to achieve steady state flow in the external fluid would be proportional to the viscous relaxation time $\frac{R^2}{\nu}$, where ν is the kinematic viscosity of that fluid, whereas that in the fluid within the drop would be proportional to $\frac{R^2}{\nu'}$, where ν' stands for the kinematic viscosity of the drop phase. These time scales are much smaller than the time it takes for the drop to move a distance equal to its own radius, which is $\frac{R}{U}$, because the ratio of the latter time scale to either of the viscous relaxation time scales is the Reynolds number for that phase.

If we impart the final velocity to the drop instantaneously, the velocity fields and the drag on the drop would achieve their steady values before the drop has had a chance to move too far. Because the drag is balanced by the net weight of the drop for steady settling in a gravitational field, one might guess that a drop released from a state of rest will be moving at terminal settling velocity before it moves a distance equal to its own radius; but the situation is not that simple. In the case of a drop with a very large density compared with that of the continuous phase, one must consider the acceleration of this drop from a state of rest to its terminal settling condition, which introduces an additional time scale from Newton's law applied to the drop. This time scale is $\gamma \frac{R^2}{\nu}$, where $\gamma = \frac{\rho'}{\rho}$ is the ratio of the density of the drop phase to that of the continuous phase. This can, of course, be much larger than the viscous relaxation time scale alone when γ is large. For a water drop falling in air, Chisnell (1987) presented numerical results showing that it takes a time of approximately $500 \frac{R^2}{\nu}$ for the velocity to approach 90% of its terminal velocity. It takes much longer for closer approach because, in Stokes flow, the difference between the actual settling velocity and the asymptotic terminal settling velocity decays only as the square root of time. For a rigid sphere, when inertial effects are included, Lovalenti and Brady (1993) show that the decay is more rapid; when the sphere moves from rest, the difference between the two velocities approaches zero as the inverse square of time. The temporal development of the flow within the drop is more complex with two separate time scales. The streamline structure corresponding to Hill's spherical vortex is achieved quickly, whereas the actual magnitudes of the velocities take the same length of time to achieve asymptotic values as do the hydrodynamic force and the settling velocity.

In the reverse situation of a gas bubble accelerated from rest in a liquid, the significant contribution to the time required is not that needed for accelerating the bubble itself, but that required to accelerate the liquid surrounding the bubble. According to an unsteady-state analysis performed by Sy, Taunton, and Lightfoot (1970), the time required to achieve steady bubble rise for negligible Reynolds number works out to be of the order of viscous relaxation time in the liquid.

Regarding the question of distances at which a neighboring drop or surface can be located and assumed not to influence the drop in question, we need to consider the decay of the velocities as one moves away from the drop. Because the disturbance emanating from the drop decays as $\frac{1}{r}$, one might roughly estimate that a neighboring surface can be located 10 radii away if disturbances of the order of 10 to 20% are acceptable. Note that we are only making very gross estimates at this point. Intuitively, one might expect that a large surface will exert a greater influence than another drop located at the same distance will exert, even though this cannot be established based on what we have developed in this section. We shall return to the subject of interactions with neighbors in Part Three.

3.2 Motion Accounting for Small to Moderate Inertial Effects

3.2.1 Introduction

In the previous section, inertial effects were completely ignored. In mathematical terms, the Reynolds number was set equal to zero in the Navier–Stokes equation. In this section, we consider the problem in the context of small inertial effects because the Reynolds number is never equal to zero in physical situations involving an object moving in a fluid.

The problem has an interesting history. Originally, Stokes (1851) obtained the solution for the steady motion of a rigid sphere when the Reynolds number is zero. Whitehead (1889) used this known velocity field to approximate the inertia term and attempted to solve the resulting equation, which is rendered linear by this process, for the velocity. He was unable to satisfy the boundary condition at infinity using the solution, however. The reason for this was explained correctly by Oseen (1910). The problem, accounting for small amounts of inertia, was solved by Proudman and Pearson (1957), who used the method of matched asymptotic expansions.

In the present section, we first introduce the reader to the language of asymptotic expansions. Then we provide a discussion of available results for the motion of drops accounting for small inertial effects, concluding with a brief discussion of the literature on numerical solutions, which are the only recourse when inertial effects become more important.

3.2.2 Asymptotic Expansions

Whitehead's procedure, mentioned in the introduction, is equivalent to writing the scaled velocity field $\mathbf{v}(\mathbf{x})$ in the form of the following asymptotic series:

$$\mathbf{v}(\mathbf{x}) = \mathbf{v}_0(\mathbf{x}) + Re\,\mathbf{v}_1(\mathbf{x}) + \cdots. \tag{3.2.1}$$

Here, the velocity is scaled using the velocity of the sphere, U, as a reference. The Reynolds number $Re = \frac{\rho R U}{\mu}$, where ρ stands for the density of the continuous phase, R is the radius of the sphere, and μ is the dynamic viscosity of the continuous phase.

Properties of asymptotic expansions such as the one above are discussed in books by Van Dyke (1975) and Kevorkian and Cole (1981). The reader is referred to them for mathematical details. Asymptotic series need not be convergent to be useful, and usually one can conveniently calculate only the first few terms. For the present, we shall only introduce the reader to certain basic terminology used with such expansions, postponing detailed illustration of the application of asymptotic expansions to a later section.

When a function $y(x, \epsilon)$ depends on a small parameter, ϵ, and the solution of the governing equations is known when this parameter is zero, a perturbation method is often tried for solving the equations for $y(x, \epsilon)$, especially when the equations are nonlinear and no general techniques are available for exact solution. The usual approach is to try a power series, which is an especially simple version of a perturbation expansion:

$$y(x, \epsilon) = y_0(x) + \epsilon y_1(x) + \epsilon^2 y_2(x) + \cdots. \tag{3.2.2}$$

This series is inserted into the governing equations, and coefficients of like powers of ϵ are grouped to obtain a series of equations for the coefficient functions in

Equation (3.2.2). Typically, the first equation is solved for $y_0(x)$, and this solution for $y_0(x)$ is used in the equations for $y_1(x)$, which are then solved for $y_1(x)$, and so on. The resulting series for $y(x, \epsilon)$ need not converge for any ϵ, however, and often does not. Nevertheless, it is useful because for small values of ϵ a few terms still can give an excellent approximation of the function $y(x, \epsilon)$ in the domain of interest. The error in the approximation of the function by a fixed number of terms becomes smaller as the value of ϵ is decreased, and herein lies the utility of the series, which is called an *asymptotic series* because it represents the function in the asymptotic limit as ϵ approaches 0.

The symbols O and o are commonly used in asymptotic analysis. They are used to describe the behavior of a function of ϵ as $\epsilon \to 0$. If we have two different functions of ϵ, say $f(\epsilon)$ and $g(\epsilon)$, we say that

$$f(\epsilon) = O[g(\epsilon)] \quad \text{if} \quad \lim_{\epsilon \to 0} \frac{f(\epsilon)}{g(\epsilon)} < \infty. \tag{3.2.3}$$

If the limit is zero, then the symbol o is used. That is,

$$f(\epsilon) = o[g(\epsilon)] \quad \text{if} \quad \lim_{\epsilon \to 0} \frac{f(\epsilon)}{g(\epsilon)} = 0. \tag{3.2.4}$$

It is assumed that we intuitively know the behavior of powers of epsilon, so that often the symbol O is used to describe the behavior of a function of ϵ in terms of a power of ϵ. Thus, $\sin \epsilon$ is $O(\epsilon)$. Of course, so also are $\tan \epsilon$ and $\log(1 + \epsilon)$. To this order, all three functions have the same representation, namely, ϵ. The symbol $\sim$ is used to mean "asymptotically equal to" as in $\sin \epsilon \sim \epsilon$.

Power series are just one type of asymptotic series, and the general asymptotic series is of the form

$$y(x, \epsilon) = \sum_{n=0}^{N} f_n(\epsilon) y_n(\epsilon), \tag{3.2.5}$$

where the summation is carried out to a desired upper limit N. When the upper limit is infinity, convergence is *not* implied. The functions $f_n(x)$ must satisfy

$$\lim_{\epsilon \to 0} \frac{f_{n+1}(\epsilon)}{f_n(\epsilon)} = 0, \quad n = 0, 1, 2 \cdots. \tag{3.2.6}$$

This means that each successive member of the set approaches zero more rapidly than the previous member as $\epsilon \to 0$. This provides a method for calculating any given coefficient function uniquely.

$$y_j(x) = \lim_{\epsilon \to 0} \frac{y(x, \epsilon) - \sum_{n=0}^{j-1} f_n(\epsilon) y_n(\epsilon)}{f_j(\epsilon)}, \quad j = 1, 2, 3 \ldots, \tag{3.2.7}$$

and

$$y_0(x) = \lim_{\epsilon \to 0} \frac{y(x, \epsilon)}{f_0(\epsilon)}. \tag{3.2.8}$$

By choosing different asymptotic sequences, one can write different asymptotic series for the same function, but in any given sequence, the representation is unique. On the other hand, it is possible for two entirely different functions to have exactly the same asymptotic series representation up to a point. As simple examples, the functions $\sin \epsilon$, $\tan \epsilon$, and $\log(1 + \epsilon)$ all have the same representation in powers of ϵ to $O(\epsilon)$.

Asymptotic series for $\sin \epsilon$ are written below illustrating the use of the two order symbols. The use of the symbol O is superior, indicating knowledge of the order of the next term in the series.

$$\sin \epsilon \sim \epsilon - \frac{\epsilon^3}{6} + O(\epsilon^5) \tag{3.2.9}$$

$$\sin \epsilon \sim \epsilon - \frac{\epsilon^3}{6} + o(\epsilon^3) \tag{3.2.10}$$

Finally, the symbols $\ll$ and $\gg$ have special meaning in asymptotic analysis. If we write $f(\epsilon) \ll g(\epsilon)$, we imply that

$$\lim_{\epsilon \to 0} \frac{f(\epsilon)}{g(\epsilon)} = 0 \tag{3.2.11}$$

Alternatively, we can write $g(\epsilon) \gg f(\epsilon)$ in this example. For additional information, the reader is encouraged to consult the books by Van Dyke (1975) and Kevorkian and Cole (1981). Now, we return to the main topic of this section.

3.2.3 The Role of Small Inertial Effects

The problem with the approach taken by Whitehead is that the expansion represented in Equation (3.2.1) is not uniformly valid at all spatial locations. One can see this by estimating the order of magnitude of the inertial terms that can be approximated by $\frac{\rho U^2}{r^*}$ and that of the viscous terms which is $\frac{\mu U}{r^{*2}}$. Physical quantities are used in these estimates, wherein we have employed r^* to designate the actual radial coordinate in a spherical polar coordinate system with its origin at the center of the sphere. The ratio of inertial to viscous terms then is $\frac{\rho U r^*}{\mu}$. One can write this ratio in terms of the Reynolds number and the dimensionless radial coordinate $r = \frac{r^*}{R}$ as $r\,Re$. It becomes clear that no matter how small a value of Reynolds number one chooses, it is possible to find a scaled distance r in an unbounded domain such that the product $r\,Re$ becomes as large as one wishes. In particular, the inertial terms are of comparable order of magnitude to the viscous terms at scaled distances of approximate magnitude $\frac{1}{Re}$. The reason for writing the asymptotic expansion given in Equation (3.2.1) is to obtain a result that is good in the limit as $Re \to 0$. Thus, when Stokes neglected inertia terms, the implicit assumption was that these terms are small compared with viscous terms everywhere in the domain. It is evident now that this is not true. Therefore, the expansion in Equation (3.2.1) is not uniformly valid over all space, but only good when $r\,Re \ll 1$.

It is worthwhile examining the Stokes solution to see if it exhibits any peculiar behavior. The velocity field given in Section 3.1 in the continuous phase surrounding the drop appears to behave in a reasonable manner. Its magnitude decays to zero with distance away from the drop in a laboratory frame of reference. In such a reference frame, the fluid far from the moving drop is quiescent. The streamfunction field given in Equation (3.1.21), however, when written in the laboratory reference frame, contains a term that decays inversely with distance r from the center of the drop, corresponding to the potential dipole, and a term that grows linearly with r, arising from the Stokeslet. The growth in the streamfunction is not physically acceptable because the streamfunction represents the volumetric flow rate through a surface formed by rotating a curve from the point in question to the axis by an angle of 2π about the axis, divided by 2π. Surely, one cannot expect the motion of a small drop through an infinite expanse of fluid to cause an infinitely large volumetric flow rate in that fluid. This unacceptable behavior of the streamfunction is another symptom of the breakdown of the assumption of uniformly negligible inertial effects everywhere in the domain when compared with viscous effects.

The reasons for Whitehead's paradox were discussed by Oseen (1910), who also suggested a way to deal with the problem. In a reference frame traveling with the sphere, in the region where inertia becomes important, the velocity is only slightly different from the uniform stream **U**. Therefore, Oseen's idea was to use the following approximate version of the Navier–Stokes equation in which physical variables $\mathbf{v}^*$ and p^* are used, along with the gradient operator ∇^*, which has dimensions of length^{-1}, to emphasize the physical nature of the assumption advocated by Oseen.

$$\rho(\mathbf{U} \bullet \nabla^* \mathbf{v}^*) = \mu \nabla^{*2} \mathbf{v}^* - \nabla^* p^* \tag{3.2.12}$$

Note that the equation is now linear in the velocity field. Oseen gave an approximate solution of this equation for the streamfunction in the case of the steady settling of a rigid sphere that can be written in dimensionless form, in a reference frame attached to the moving sphere, as follows:

$$\psi = \frac{1}{2}\left(r^2 + \frac{1}{2r}\right)\sin^2\theta - \frac{3}{2Re}(1 - \cos\theta)\left\{1 - \exp\left[-\frac{1}{2}Re\,r(1 + \cos\theta)\right]\right\}. \tag{3.2.13}$$

Here, θ is the polar angle measured from the forward stagnation streamline. As Proudman and Pearson (1957) pointed out, this result is consistent with the degree of approximation used in writing Equation (3.2.12). At first sight, it appears that this solution does not satisfy the no-slip boundary condition on the surface of the rigid sphere. If, however, one expands the right side for small Re and retains only the leading terms, the solution will be seen to reduce precisely to the Stokes solution. Therefore, the solution of the Oseen equation can be considered a uniformly valid first approximation of the solution of the Navier–Stokes equation for this problem when the Reynolds number is small. The most important difference of the Oseen solution from that of Stokes is the lack of fore-aft symmetry in the flow, which is illustrated in Figure 3.2.1 for a Reynolds number of 0.4. The figure displays streamlines in a laboratory reference frame. Far from the sphere, the flow is radially outward as though the sphere acts as a source of fluid, with the exception of a narrow wake behind the sphere in which the fluid moves toward it. The interested reader can find more details in Batchelor (1967).

An ordered sequence of approximations was developed by Proudman and Pearson (1957), who used the method of matched asymptotic expansions (see also Kaplun and Lagerstrom (1957)), that permitted them to calculate the correction to the drag on a rigid sphere for small inertial effects. The idea behind this method is to develop separate expansions, each of which is valid in a certain part of the domain. Assuming the domains of validity of these expansions overlap, the expansions are matched by using straightforward principles outlined by Van Dyke (1975). The expansion that is good in a domain near the sphere is used to evaluate the drag on the sphere. Using this, Proudman and Pearson calculated the following result for the drag on a rigid sphere:

$$F_D = 6\pi\mu U R\left(1 + \frac{3}{8}Re + \frac{9}{40}Re^2\log Re\right). \tag{3.2.14}$$

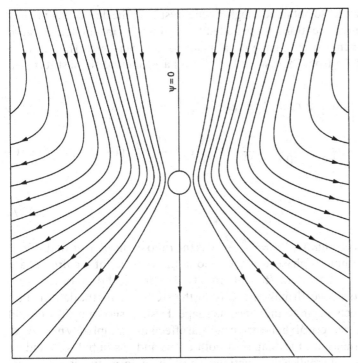

Figure 3.2.1 Streamlines in the laboratory reference frame from the Oseen solution, including a small inertial effect, $Re = 0.4$, $\Delta\psi = 0.3182$; $\psi = 0$ along the symmetry axis, and $\Delta\psi$ stands for the increment measured away from this axis.

We observe that Proudman and Pearson obtained a higher-order term beyond $O(Re)$ that is of $O(Re^2 \log Re)$. The logarithm of the perturbation parameter invariably appears in similar problems. The next higher order term is $O(Re^2)$, as was determined by Chester and Breach (1969), who extended the series up to $O(Re^3 \log Re)$. Fraenkel (1969) pointed out that one has to be careful in matching when such intermediate terms are encountered. The correct way is to calculate all the terms involving a certain power of the perturbation parameter. Nevertheless, the calculation of Chester and Breach revealed the coefficient obtained by Proudman and Pearson for the term involving $O(Re^2 \log Re)$ to be correct. Chester and Breach themselves stopped at $O(Re^3 \log Re)$, without calculating the $O(Re^3)$ term. In any case, it happens that the additional corrections to the drag are not particularly useful because the series is of practical value only in the limited range $0 \leq Re \leq 0.5$, as concluded by Chester and Breach. That is, inclusion of additional terms does not improve the performance of the series beyond $Re \approx 0.5$.

Taylor and Acrivos (1964) analyzed the analogous problem for the settling of a drop due to gravity by using the method of matched asymptotic expansions. As noted earlier, these authors pointed out that the spherical shape is consistent with the assumption of negligible inertial effects. Their principal focus was to calculate the deformation of the drop when that deformation is small; in performing the analysis, Taylor and Acrivos assumed inertial effects to be small both within and outside the drop. They also obtained a result for the drag on the drop, including an inertial correction. Actually, after calculating the leading order deformation, the authors calculated the drag on the slightly deformed drop and also obtained a higher order result for the deformation. They then noted that in practical situations, these added corrections are incomplete because of terms of equal order that should have been included, but were not. Therefore, in giving the principal results from their work below, we have omitted the last terms in each case:

$$F_D = 2\pi \mu U R_0 Z \left[1 + \frac{1}{8} Z\, Re + \frac{1}{40} Z^2\, Re^2 \log Re \right], \tag{3.2.15}$$

$$\frac{R}{R_0} = 1 - \frac{We\, P_2(s)}{4(1+\alpha)^3} \left[\frac{3}{4} + \frac{103}{40}\alpha + \frac{57}{20}\alpha^2 + \frac{81}{80}\alpha^3 + \frac{(1-\gamma)}{12}(1+\alpha) \right], \tag{3.2.16}$$

and

$$Z = \frac{2+3\alpha}{1+\alpha}. \tag{3.2.17}$$

Here, $s = \cos\theta$, the Weber number $We = \frac{\rho R_0 U^2}{\sigma}$, γ is the ratio of the density of the drop phase to that of the continuous phase, α is the ratio of the viscosity of the drop phase to that of the continuous phase, and R_0 is the radius of a spherical drop of the same volume. A few comments are worth making regarding the prediction for the deformation of a drop as inertial effects begin to influence its shape. First, as shown by Taylor and Acrivos, a drop retains its spherical shape when inertial effects are completely neglected. Second, for small deformations, the shapes of both drops and gas bubbles should be oblate spheroids in physically realizable situations. Third, the deformation is insensitive to the ratio of viscosities. In a later paper, Pan and Acrivos (1968) extended the results of Taylor and Acrivos to situations in which the Reynolds number within the drop is not restricted to being small. For gas bubbles, the authors found that the additional effect due to inertia within the bubble, even at a Reynolds number of $O(10^2)$, was negligible and that the results of Taylor and Acrivos were sufficient in this case. They also presented experimental results on air bubbles rising in a viscous lubricating oil, which supported their predictions of bubble shapes. Brignell (1973) extended the calculations of Taylor and Acrivos for the deformation by providing two additional terms, so that the error in the new result is of the order of the square of the magnitude of the relative deformation.

3.2.4 Numerical Solutions for the Case of Intermediate Values of the Reynolds Number

When the Reynolds number increases above approximately 0.5, numerical methods must be used to obtain results for the velocity field, the drag, and the shape of a drop or bubble. Although these results can be interesting in their own right, we do not devote much space to details of such numerical solutions. In the case of rigid spheres, useful results can be found in Dennis and Walker (1971), who covered the range $0.05 \leq Re \leq 20$. These authors solved the axisymmetric problem by using expansions of the stream-function and vorticity in basis sets involving Gegenbauer Polynomials and associated Legendre Polynomials respectively. The series are substituted into the governing equations, and this leads to a set of ordinary differential equations for the coefficient functions that depend on the radial coordinate. This set is truncated and solved numerically. The authors reported good agreement with known analytical solutions for small values of the Reynolds number and with experimental results available at that time. Brabston and Keller (1975) addressed the motion of spherical bubbles in the range $0.05 \leq Re \leq 100$ by using a similar series truncation technique, but they employed a different numerical scheme for solving the resulting equations. These authors reported that their results confirmed the correctness of an asymptotic result for large Reynolds numbers provided by Moore (1963) when $Re \geq 20$. The solution of Moore is discussed in detail in the next section.

Numerical calculations in which the shape deformation of a drop or bubble is taken into account have only become possible in the past two decades when computers with sufficient power became generally available. Examples of such calculations may be found in Ryskin and Leal (1984a, 1984b), Dandy and Leal (1989), and Leal (1989). Leal (1989) reviewed a variety of numerical methods used to solve free-boundary problems in fluid mechanics and provided results for the motion and deformation of bubbles and drops for nonzero Reynolds numbers. For problems in which the Reynolds number is precisely zero or infinity, boundary integral techniques for solution appear to be attractive. The problem is cast in the form of an integral equation for the interfacial velocity distribution, for a specified shape of the interface. The computed interfacial velocity field is then used to update the shape via the kinematic boundary condition. This process is used repeatedly until the solution converges. When the Reynolds number is neither zero nor infinity, the full nonlinear Navier–Stokes equations within and outside the drop need to be solved, along with the equation of continuity applicable to each phase. A crucial step in incorporating the boundary conditions correctly is to ensure that the interface coincides with a coordinate surface of the chosen orthogonal coordinate system. This is indeed possible, and methods are available to generate such coordinate grids numerically. The Navier–Stokes equations are then solved numerically by a finite-difference method in these boundary-fitted coordinates. For three dimensional simulations, the mapping techniques needed to transform the grids are more difficult than their two-dimensional counterparts. Furthermore, Leal (1989) noted that the orthogonality of the coordinate system in three-dimensional problems cannot be preserved while simultaneously requiring the coordinate system to be boundary fitted. A disadvantage of boundary-fitted coordinates is that the grids must be reconstructed when the shape of the drop changes. Thus the simulations are computationally intensive. An alternative method, which avoids the use of boundary-fitted coordinates, was proposed by Unverdi and Tryggvason (1992) and extended to the motion of drops due to thermocapillarity by Nas (1995) and Haj-Hariri, Shi, and Borhan (1997). Here, the interfacial boundary conditions are incorporated directly in the formulation by representing interfacial tension as a body force in the momentum equation. This body force acts only where the interface is located. Thus, it is proportional to a Dirac delta function, with the distance from the interface as its argument. The evolution of the interface is tracked either by using the kinematic condition or by recognizing that the boundary is demarcated by sharp changes in a property such as the density or an artificially introduced *color function* that assumes different values on either side of the interface.

The main features of the numerical solution for the rise of gas bubbles, accounting for shape deformation, are discussed by Leal (1989). When the Reynolds number is relatively low, as the Weber number increases, the shape of the bubble varies smoothly from a sphere to one with a spherical cap. When the Reynolds number is around 200, the bubble is egg shaped for low Weber numbers and becomes flattened when the Weber number is increased to approximately 10. Another interesting feature is that there is almost always a strong recirculating wake downstream of the bubble. This recirculating wake is absent both in the solution for zero Reynolds number provided in Section 3.1 and that for large Reynolds number where deformation from the spherical shape is neglected, provided in Section 3.3. Similar recirculating wakes, which are detached, are predicted for the motion of a viscous drop. These recirculating wakes have been observed experimentally by many investigators (see, for example, Hnat and Buckmaster (1976) and Bhaga and Weber (1981)) and appear to be consistent with numerical predictions. It is often observed experimentally that bubbles and drops moving because of gravity undergo a transition from steady rectilinear motion to time-dependent motion when they are sufficiently large. The precise values of the Reynolds and Weber numbers at transition to time-dependent motion appear to depend on the properties of the liquids and their purity, as noted by Bhaga and Weber (1981) and McLaughlin (1996). Nonetheless, one can state that the transition generally appears at Reynolds numbers of the order 100 and Weber numbers of about 3. Leal (1989) conjectured that the transitions are of two types. One is due to an instability of the recirculating wake and vortex shedding at moderate Weber numbers, whereas the other is a shape instability for larger Weber numbers.

Clift, Grace, and Weber (1978) and Sadhal, Ayyaswamy, and Chung (1997) have provided ample details regarding results from the numerical solutions, as well as experimental data on motion at intermediate Reynolds number. These authors also discuss heat and mass transfer from particles, bubbles, and drops when they execute such motion. Also, two chapters in Churchill (1988) are devoted to a discussion of available results both from theory and experiment. Churchill has provided some general correlations for the drag, represented in dimensionless form as the drag coefficient. Another useful source is the book by Leal (1992), in which problems amenable to analysis are considered in good detail.

In the following sections, we cover certain aspects of motion at large Reynolds number and deformation when the Weber number is large. The reader will find that the treatment of the motion of bubbles and drops at large Reynolds number is useful in subsequent analyses of thermocapillary migration problems at large values of the Marangoni number that exhibit parallel mathematical structure.

3.3 Motion of a Spherical Bubble at Large Reynolds Number – Asymptotic Analysis

3.3.1 Introduction

In this section, we analyze the motion of an isolated spherical bubble in a liquid when inertial effects are predominant. The Reynolds number, Re, defined in Section 3.2, is assumed to be large compared with unity. When inertia is important, the variation of the hydrodynamic pressure in the liquid over the surface of the bubble is more important in deforming it than the variation of the viscous stresses over the surface. The ratio of the variation in the pressure force over the surface to the force of surface tension that resists deformation is given by the Weber number, which is a product of the Reynolds and the Capillary numbers. It is assumed that the value of the Weber number is small compared with unity and that the deformation of the bubble can be neglected. This problem has been analyzed by Levich (1962), Chao (1962), and Moore (1963). Moore as well as Harper (1972) commented about the inadequacies of the earlier treatments. The analysis of the momentum boundary layer that exists near the surface of the bubble given here closely follows the work of Moore (1963). The goal is to determine the velocity field in the boundary layer, the drag experienced by the bubble, and its terminal velocity, when the bubble rises because of the influence of gravity.

The motion of a spherical bubble at large values of the Reynolds number, Re, is an unusual problem in fluid mechanics where the leading order drag on the object can be determined without solving for the velocity and pressure fields within the boundary layer. Knowledge of the velocity field obtained by solution of the potential flow equations outside the boundary layer proves sufficient for this purpose, a fact first noted by Levich (1949, 1962). In the absence of interfacial tension gradients, we infer from the tangential stress balance at the surface of the bubble that this stress is vanishingly small in the liquid because the viscosity of the gas is small compared with that of the liquid. When the Reynolds number is large, a boundary layer exists near the bubble surface where the shear stress changes from its typical value in the region of potential flow outside the boundary layer to a value that is zero on the surface. When distances are scaled by the radius of the bubble and velocities by its rise velocity, there is an $O(1)$ change in the scaled velocity gradient in the boundary layer. The change in the scaled velocity is only $O(\epsilon)$ where ϵ is a measure of the thickness of the momentum boundary layer that can be shown to be proportional to $\frac{1}{\sqrt{Re}}$. Thus, to leading order, the velocity field everywhere, including that in the momentum boundary layer, is given by the potential flow velocity field. The correction to the potential flow field within the boundary layer is important only at higher order in ϵ. This picture is in contrast to that in a boundary layer near a solid body, where the change in the velocity within the boundary layer is $O(1)$, and the change in the velocity gradient is $O(\epsilon^{-1})$. As Leal (1992) has pointed out, for both a solid body and a gas bubble, the vorticity generated by the object is confined within the boundary layer with a thickness that scales as $\frac{1}{\sqrt{Re}}$ in both cases. The difference between the two cases lies in the magnitude of the vorticity. Near a rigid body, the vorticity is greater than that near the bubble surface by a multiplicative factor of $\sqrt{Re}$. In the former case, vorticity is generated on account of a lack of slip between the fluid and the solid, whereas in the case of a gas bubble, vorticity is generated by rotation of fluid elements that follow the curved surface of the bubble.

3.3.2   Analysis at Leading Order

Levich (1962) obtained the drag exerted on the bubble at leading order by considering the energy dissipated in the liquid. After the initial transient period subsides, the rate at which work is done on the liquid (that equals the product of the hydrodynamic force and the steady bubble velocity) is equal to the rate of dissipation of kinetic energy by the viscous stresses in the liquid. Because the volume of the liquid in the momentum boundary layer is small compared with the total volume of the liquid, the rate of dissipation of energy at leading order can be obtained from the potential flow solution. Nondimensionalizing the velocity field by the bubble velocity U and scaling distances by the radius of the bubble R, the potential flow velocity field in a reference frame moving with the bubble can be written in a spherical polar coordinate system as follows:

$$V_r = -\left(1 - \frac{1}{r^3}\right)\cos\theta \tag{3.3.1}$$

and

$$V_\theta = \left(1 + \frac{1}{2r^3}\right)\sin\theta. \tag{3.3.2}$$

Here, the polar angle θ is measured from the forward stagnation streamline, and upper-case symbols are used for this outer velocity field. The rate at which energy is dissipated is scaled by $\mu U^2 R$. This nondimensional dissipation rate, represented by the symbol $\dot{E}$, is given below:

$$\dot{E} = \int_V \Phi \, d\mathcal{V} = \int_V |\nabla \times \mathbf{V}|^2 \, d\mathcal{V} + 2 \int_S \left[\mathbf{n} \bullet (\mathbf{V} \times \nabla \times \mathbf{V}) - \frac{1}{2} \frac{\partial}{\partial n} (V^2) \right] dS.$$

(3.3.3)

Here, Φ is the scaled viscous dissipation function, $\mathcal{V}$ denotes the volume of the fluid scaled by R^3, $\mathbf{V}$ is the scaled velocity vector, S denotes the surface of the bubble, and $\mathbf{n}$ is the unit outward normal vector on this surface. Kang and Leal (1988) and Stone (1993) have shown that the expression for $\dot{E}$ generally may be recast in terms of the vorticity distribution exterior to the translating bubble and on its surface, even if it is deformed. For a spherical bubble, $\dot{E}$ depends only on the surface vorticity distribution. Equating $\dot{E}$ with the product of F_D, the hydrodynamic force on the bubble scaled by $\mu U R$, and the scaled bubble velocity, which is unity, F_D may be determined to be

$$F_D = 12\pi.$$

(3.3.4)

Thus, the drag exerted on the bubble in the inertial regime is three times its value in Stokes flow. Equating the sum of the hydrodynamic and gravitational forces on the bubble to zero, its steady rise velocity can be obtained:

$$U = \frac{1}{9} \frac{g R^2}{\nu}.$$

(3.3.5)

Here, g is the magnitude of the acceleration due to gravity, and ν is the kinematic viscosity of the continuous phase fluid. Using this result for U, it can be seen that the Reynolds number is large when $\frac{g R^3}{\nu^2} \gg 1$, and the Weber number is small when $\frac{\rho g^2 R^5}{\sigma \nu^2} \ll 1$. The results for the drag and the rise velocity are good to $O(1)$ because energy dissipation in the boundary layer, as well as in the wake, has been neglected. It will be shown below that in the boundary layer, the correction to the potential flow occurs at $O(Re^{-\frac{1}{2}})$, leading to corrections to Equation (3.3.4) at this order. Moore (1963) has evaluated the contribution to the energy dissipation in the wake.

We now turn to the analysis of the momentum boundary layer near the surface of the bubble. Such a boundary layer exists because V_θ given in Equation (3.3.2) does not lead to a vanishing shear stress at $r = 1$, i.e., $r \frac{\partial}{\partial r} (\frac{V_\theta}{r})(1, \theta) \neq 0$. Although Moore solved for the correction velocity field in the boundary layer, the complete velocity field is considered below. The method of matched asymptotic expansions is used.[‡] The small parameter is $\epsilon = \frac{1}{\sqrt{Re}}$. The outer variables are (r, θ), and the inner variables are (x, θ) where $x = \frac{r-1}{\epsilon}$. The outer velocity vector field $\mathbf{V}$ in a reference frame moving with the bubble, and the pressure field P, scaled by ρU^2, are expanded as follows:

$$\mathbf{V} = \mathbf{V_0} + \epsilon^2 \mathbf{V_2} + \cdots$$

(3.3.6)

and

$$P = P_0 + \epsilon^2 P_2 + \cdots.$$

(3.3.7)

The outer expansion does not contain terms at $O(\epsilon)$ because the equations and boundary conditions are homogeneous at this order. The outer equations at leading order are the following:

$$\nabla \bullet \mathbf{V_0} = 0$$

(3.3.8)

[‡] This method is illustrated with examples in Van Dyke (1975). More details about the method can be found in Section 4.12 in the context of the thermocapillary motion of a drop.

and

$$\mathbf{V_0} \bullet \nabla \mathbf{V_0} = -\nabla P_0. \tag{3.3.9}$$

The boundary conditions are given below.

$$V_{r_0}(r \to \infty, \theta) \to -\cos\theta, \tag{3.3.10}$$

$$V_{\theta_0}(r \to \infty, \theta) \to \sin\theta, \tag{3.3.11}$$

$$V_{r_0}(1, \theta) = 0, \tag{3.3.12}$$

$$P_0(r \to \infty, \theta) \to P_\infty. \tag{3.3.13}$$

Equation (3.3.12) states that the bubble is not penetrated by the exterior fluid. Strictly speaking, conditions at the surface of the bubble must be imposed on the inner fields. Nonetheless, the outer problem is not fully specified unless one condition is imposed on the bubble surface. Equation (3.3.12) is obtained from requirements of matching with the inner solution. The solution for $\mathbf{V_0}$ is given by Equations (3.3.1) and (3.3.2). The solution for P_0 is obtained by using Bernoulli's theorem as

$$P_0 = P_\infty + \frac{1}{2}(1 - \mathbf{V_0} \bullet \mathbf{V_0}) = P_\infty + \cos^2\theta\left(\frac{1}{r^3} - \frac{1}{2r^6}\right) - \sin^2\theta\left(\frac{1}{2r^3} + \frac{1}{8r^6}\right). \tag{3.3.14}$$

A discussion of Bernoulli's theorem can be found in Batchelor (1967). We now consider the inner velocity and pressure fields. They are expanded as follows:

$$v_r(x, \theta) = \epsilon v_{r_0}(x, \theta) + \epsilon^2 v_{r_1}(x, \theta) + \cdots, \tag{3.3.15}$$

$$v_\theta(x, \theta) = v_{\theta_0}(x, \theta) + \epsilon v_{\theta_1}(x, \theta) + \cdots, \tag{3.3.16}$$

and

$$p(x, \theta) = p_0(x, \theta) + \epsilon p_1(x, \theta) + \epsilon^2 p_2(x, \theta) + \cdots. \tag{3.3.17}$$

The inner equations at leading order are given below.

$$\frac{\partial v_{r_0}}{\partial x} + \frac{1}{\sin\theta}\frac{\partial}{\partial\theta}(v_{\theta_0}\sin\theta) = 0, \tag{3.3.18}$$

$$v_{r_0}\frac{\partial v_{\theta_0}}{\partial x} + v_{\theta_0}\frac{\partial v_{\theta_0}}{\partial\theta} = -\frac{\partial p_0}{\partial\theta} + \frac{\partial^2 v_{\theta_0}}{\partial x^2}, \tag{3.3.19}$$

$$\frac{\partial p_0}{\partial x} = 0. \tag{3.3.20}$$

The boundary conditions are

$$\frac{\partial v_{\theta_0}}{\partial x}(0, \theta) = 0, \tag{3.3.21}$$

$$v_{\theta_0}(x \to \infty, \theta) = V_{\theta_0}(r \to 1, \theta) \to \frac{3}{2}\sin\theta, \tag{3.3.22}$$

$$p_0(x \to \infty, \theta) = P_0(r \to 1, \theta), \tag{3.3.23}$$

$$v_{\theta_0}(x, 0) = 0, \tag{3.3.24}$$

and

$$v_{r_0}(0, \theta) = 0. \tag{3.3.25}$$

Equation (3.3.21) is the result, at leading order, from the condition that the tangential stress vanishes at the surface of the bubble. Equations (3.3.22) and (3.3.23) represent matching requirements. Equation (3.3.24) arises from symmetry requirements, and Equation (3.3.25) represents the condition of impenetrability at the bubble surface. The solution is

$$v_{\theta_0}(x, \theta) = V_{\theta_0}(1, \theta), \tag{3.3.26}$$

$$v_{r_0}(x, \theta) = x \frac{\partial V_{r_0}}{\partial r}(1, \theta), \tag{3.3.27}$$

and

$$p_0(x, \theta) = P_0(1, \theta). \tag{3.3.28}$$

The leading-order inner field is thus the potential flow field (the outer field) evaluated in the vicinity of the surface of the bubble. Thus there is no correction to the potential flow field in the boundary layer at leading order.

3.3.3 Higher-Order Analysis

We now examine the inner velocity field at the next order. Such an analysis provides the correction to the potential flow field within the boundary layer, and also demonstrates how the shear stress behaves in the boundary layer. The shear stress vanishes on the surface of the bubble and must match that given by the potential flow field at the edge of the boundary layer. The equations for the velocity and pressure fields at $O(\epsilon)$ are given below.

$$\frac{\partial v_{r_1}}{\partial x} + \frac{1}{\sin \theta} \frac{\partial}{\partial \theta}(v_{\theta_1} \sin \theta) + 2v_{r_0} - \frac{x}{\sin \theta} \frac{\partial}{\partial \theta}(v_{\theta_0} \sin \theta) = 0, \tag{3.3.29}$$

$$v_{r_0} \frac{\partial v_{\theta_1}}{\partial x} + v_{\theta_0} \frac{\partial v_{\theta_1}}{\partial \theta} + v_{r_1} \frac{\partial v_{\theta_0}}{\partial x} + v_{\theta_1} \frac{\partial v_{\theta_0}}{\partial \theta} - x v_{\theta_0} \frac{\partial v_{\theta_0}}{\partial \theta} + v_{r_0} v_{\theta_0}$$

$$= -\frac{\partial p_1}{\partial \theta} + \frac{\partial^2 v_{\theta_1}}{\partial x^2} + x \frac{\partial p_0}{\partial \theta} + 2 \frac{\partial v_{\theta_0}}{\partial x}, \tag{3.3.30}$$

$$\frac{\partial p_1}{\partial x} = v_{\theta_0}^2. \tag{3.3.31}$$

Integrating Equation (3.3.31) and matching the inner pressure field to $O(\epsilon)$ with the outer pressure field in Equation (3.3.14), p_1 may be determined to be

$$p_1 = x v_{\theta_0}^2 = \frac{9}{4} x \sin^2 \theta. \tag{3.3.32}$$

The inner pressure field to $O(\epsilon)$ is merely the outer pressure field expressed in inner variables. Thus, the boundary layer correction to the pressure field must occur at $o(\epsilon)$. Equation (3.3.30) may be simplified by using Equations (3.3.26), (3.3.28), and (3.3.32) to

$$\frac{3}{2} \frac{\partial}{\partial \theta}(v_{\theta_1} \sin \theta) - 3x \cos \theta \frac{\partial v_{\theta_1}}{\partial x} = \frac{\partial^2 v_{\theta_1}}{\partial x^2}. \tag{3.3.33}$$

The boundary conditions on v_{θ_1} are given below.

$$\frac{\partial v_{\theta_1}}{\partial x}(0, \theta) = v_{\theta_0}(0, \theta) = \frac{3}{2} \sin \theta, \tag{3.3.34}$$

$$v_{\theta_1}(x, 0) = 0, \tag{3.3.35}$$

$$v_{\theta_1}(x \to \infty, \theta) \to -\frac{3}{2} x \sin \theta. \tag{3.3.36}$$

Equation (3.3.34) is a consequence of the vanishing of the tangential stress at the bubble surface. Equation (3.3.35) follows from symmetry considerations, and Equation (3.3.36) is established by matching the inner and outer tangential velocity fields. The solution for v_{θ_1} can be obtained by using a similarity transformation and is given below.

$$v_{\theta_1}(x, \theta) = -\frac{3}{2}x \sin\theta - 6\frac{\xi^{\frac{1}{2}}}{\sin\theta}\left[\frac{1}{\sqrt{\pi}}e^{-\zeta^2} - \zeta \operatorname{erfc}\zeta\right] \tag{3.3.37}$$

Here, the similarity variable is $\zeta = \frac{x \sin^2\theta}{2\sqrt{\xi}}$ and $\xi = (2/9)(2 - 3\cos\theta + \cos^3\theta)$. The radial velocity v_{r_1} that satisfies the condition $v_{r_1}(0, \theta) = 0$ may be determined from Equation (3.3.29) to be

$$v_{r_1}(x, \theta) = 6x^2 \cos\theta + 2(1 - \operatorname{erfc}\zeta) + 6\frac{\xi}{\sin^4\theta}$$

$$\times \left(-1 + \frac{2}{\sqrt{\pi}}\zeta e^{-\zeta^2} - 2\zeta^2 \operatorname{erfc}\zeta + \operatorname{erfc}\zeta\right). \tag{3.3.38}$$

The variation of the shear stress across the boundary layer can be inferred from Equation (3.3.37). It can be shown that the magnitude of the shear stress vanishes at the surface of the bubble and increases monotonically to the value given by the potential flow field at the edge of the boundary layer. Expanding the r-momentum equation for small ϵ and retaining terms to $O(\epsilon^2)$, the following equation for p_2 is obtained:

$$\frac{\partial p_2}{\partial x} = -v_{r_0}\frac{\partial v_{r_0}}{\partial x} - v_{\theta_0}\frac{\partial v_{r_0}}{\partial \theta} + 2v_{\theta_0}v_{\theta_1} - xv_{\theta_0}^2. \tag{3.3.39}$$

The solution for p_2 that matches the outer pressure field in Equation (3.3.14) can be written as

$$p_2 = -\frac{45}{8}x^2 \sin^2\theta - \frac{9}{2}x^2 \cos^2\theta + 36\frac{\xi}{\sin^2\theta}$$

$$\times \left(\frac{1}{4}\operatorname{erfc}\zeta - \frac{\zeta}{2\sqrt{\pi}}e^{-\zeta^2} + \frac{\zeta^2}{2}\operatorname{erfc}\zeta\right) + f(\theta), \tag{3.3.40}$$

where $f(\theta)$ is unknown and cannot be obtained by matching with the outer pressure field given in Equation (3.3.14). It can only be determined by matching the inner and outer pressure fields to $O(\epsilon^2)$. Physically, $f(\theta)$ represents the effect of the displacement of the outer region by the presence of the boundary layer. Moore (1963) did not include such a contribution to the pressure in the boundary layer.

3.3.4 Hydrodynamic Force on the Bubble

It is seen from Equations (3.3.37), (3.3.38), and (3.3.40) that the boundary layer corrections to the velocity and pressure fields, excluding the contribution from $f(\theta)$, are singular at the rear stagnation point, $\theta = \pi$. The scaled leading-order drag exerted on the bubble can be written as

$$F_D = 2\int_0^\pi \left[Re\ p(0, \theta) - 2\frac{\partial v_r}{\partial x}(0, \theta)\right]\cos\theta \sin\theta\ d\theta$$

$$= 2\int_0^\pi \left[p_2(0, \theta) - 2\frac{\partial v_{r_0}}{\partial x}(0, \theta)\right]\cos\theta \sin\theta\ d\theta. \tag{3.3.41}$$

It can be shown that the contribution to the drag from the pressure field given by Equation (3.3.40), excluding the $f(\theta)$ term, is nonintegrable. Thus, the drag cannot be determined by direct integration of the stress on the surface of the bubble unless $f(\theta)$ is determined, and the singularity in the pressure field at $\theta = \pi$ is relieved. The contribution to F_D from the viscous normal stress can be obtained to be 8π. Because F_D has been determined earlier in this section to be 12π by using energy dissipation arguments, it follows that the contribution to the scaled drag from the pressure correction in the boundary layer is 4π, which is one-third of the total. Curiously, as mentioned in Section 3.1, the pressure field also makes a one-third contribution to the total drag exerted on a drop when the Reynolds number is small compared with unity.

Kang and Leal (1988) have mentioned the importance of the displacement thickness effects in the determination of p_2. The presence of the boundary layer near the bubble affects the velocity and pressure distribution in the outer region and makes an important contribution near the rear stagnation point to relieve the singular behavior in p_2. Kang and Leal obtained the pressure field on the bubble surface alternatively in terms of the vorticity distribution and showed that this pressure field is indeed integrable, and that the expression for F_D in Equation (3.3.4) can be obtained from Equation (3.3.41).

By considering the dissipation of energy in the boundary layer, as well as in the wake, Moore (1963) showed that Equation (3.3.4) can be written as

$$F_D = 12\pi \left(1 - \frac{2.2}{\sqrt{Re}} \right) + O(Re^{-\frac{5}{6}}). \tag{3.3.42}$$

The boundary layer correction to the tangential velocity at the surface of the bubble is opposite in sign to the potential flow velocity there. Because this correction is predicted to be large near the rear stagnation point in Equation (3.3.37), it is tempting to speculate that the tangential velocity will reverse its sign and that the flow will separate near $\theta = \pi$. Harper (1972) argued against this conclusion and noted that the boundary layer does not separate when the Reynolds number is large. The reason is that the boundary layer analysis given here breaks down in a small region near the rear stagnation point. Harper provides an analysis that is valid in this region and shows that the boundary layer correction is negligible compared with the tangential velocity given by potential flow.

When the Weber number is not negligible, it can be shown that the bubble deforms to an oblate spheroid (see Leal 1992). Moore (1965) has extended the boundary layer analysis to include small perturbations to the shape of the bubble. He obtained a result similar to Equation (3.3.42). The consequence of the deformation is an increase in the drag experienced by the bubble over that on an equivalent spherical bubble.

3.4   Motion of a Spherical Liquid Drop at Large Reynolds Number – Asymptotic Analysis

3.4.1   Introduction

We now consider the steady motion of a spherical liquid drop in a second fluid with which it is immiscible. We define the Reynolds number in the continuous phase, Re, in the same manner as in Section 3.2 and that in the drop phase, Re', in a similar manner, using the properties of the drop phase. Both Re and Re' are assumed to be large compared with unity. Therefore, inertia plays a dominant role over viscous forces in both fluids. As in Section 3.3, where we analyzed the limiting case of a gas bubble, the Weber number is assumed to be sufficiently small that the deformation from the spherical shape can be neglected. The case of a liquid drop is different from that of a gas bubble because the internal flow within the drop is important. It controls the magnitude of the velocities in the continuous phase in the vicinity of the interface and, therefore, the hydrodynamic force that is exerted on the drop.

As in Section 3.3, the hydrodynamic force exerted on a drop in steady motion can be obtained by equating the rate at which kinetic energy is dissipated by viscous means in the two fluids to the work done, namely, the product of the hydrodynamic force on the drop and its steady velocity. This requires a knowledge of the velocity fields in the two fluids. When the Reynolds numbers are large, the fields may be obtained to a first approximation from the solution of the Navier–Stokes equation completely neglecting viscosity, because the contribution to the viscous dissipation from the fluid in the viscous boundary layers adjacent to the surface is small. As noted in Chapter Two, if the flow is also irrotational, the velocity field can be written as the gradient of a scalar potential. The potential flow solution in the continuous phase, given in Section 3.3, is still applicable. The inviscid solution for the flow within the drop, which is kinematically consistent at the drop surface with the potential flow in the continuous phase, is Hill's spherical vortex. We have encountered this flow in Section 3.1 in the context of Stokes motion of a drop in a second fluid. Hill's vortex is not irrotational and is indeed a solution of the complete Navier–Stokes equation, including the viscous term.

3.4.2 Inviscid Solutions for the Velocity Field

We use spherical polar coordinates (r, θ) with the origin at the center of the drop. The radial coordinate has been scaled by the radius of the drop, R. The polar angle θ is measured from the forward stagnation streamline. Velocities are scaled by that of the drop, U. In a reference frame attached to the moving drop, the solution of the inviscid equations of motion is given below.

$$V_r = -\left(1 - \frac{1}{r^3}\right)\cos\theta, \tag{3.4.1}$$

$$V_\theta = \left(1 + \frac{1}{2r^3}\right)\sin\theta, \tag{3.4.2}$$

$$V_r' = \frac{3}{2}(1 - r^2)\cos\theta, \tag{3.4.3}$$

$$V_\theta' = 3\left(r^2 - \frac{1}{2}\right)\sin\theta. \tag{3.4.4}$$

A result analogous to that given in Equation (3.3.3) can be written for the rate at which energy is dissipated by viscous means in the fluids. The hydrodynamic force exerted on the drop, scaled by $\mu U R$, is denoted by F_D. Here, μ is the viscosity of the continuous phase. Accounting for the viscous dissipation within the drop, F_D can be written as follows:

$$F_D = 12\pi\left(1 + \frac{3}{2}\alpha\right). \tag{3.4.5}$$

Here, α is the ratio of the viscosity of the drop to that of the continuous phase. The limit of a gas bubble analyzed in Section 3.3 is correctly recovered from Equation (3.4.5) in the limit $\alpha \to 0$.

3.4.3　Outer Velocity and Pressure Fields

It is evident that viscous boundary layers must exist adjacent to the surface of the drop on both sides because the tangential stress, obtained from the inviscid solution in Equations (3.4.1) to (3.4.4), is not continuous at $r = 1$. Using arguments similar to those given in Section 3.3, it can be shown that the correction to the velocity field in the boundary layers occurs at $O(\epsilon)$, where $\epsilon = \frac{1}{\sqrt{Re}}$. The analysis of the boundary layers in the two fluids has been performed by Harper and Moore (1968). Although Harper and Moore solved for the correction velocity field in the boundary layers, we consider the complete velocity field. It is assumed that α is $O(1)$. The variables $\mathbf{V}$ and $\mathbf{V}'$ are used to denote the outer velocity fields in the continuous phase and the drop phase, respectively, and $\mathbf{v}$ and $\mathbf{v}'$ are the corresponding inner fields. The outer pressure fields in the two fluids are P and P', and the corresponding inner pressure fields are p and p'. The outer variables are (r, θ), and the inner variables are (x, θ) for the boundary layer in the continuous phase and (x', θ) for the boundary layer within the drop. Here, $x = \frac{r-1}{\epsilon}$ and $x' = (1-r)/(\sqrt{\frac{\alpha}{\gamma}}\epsilon)$, where $\gamma = \frac{\rho'}{\rho}$ is the ratio of the density of the drop phase to that of the continuous phase.

The outer expansion for the velocity and pressure fields in the continuous phase is assumed to be as follows:

$$V_r = V_{r_0} + \cdots, \tag{3.4.6}$$

$$V_\theta = V_{\theta_0} + \cdots, \tag{3.4.7}$$

and

$$P = P_0 + \cdots. \tag{3.4.8}$$

It can be established that the correction to the leading order outer fields occurs at $o(\epsilon)$. The equations and boundary conditions for $\mathbf{V_0}$ and P_0 are identical to Equations (3.3.8) to (3.3.13). The solutions for V_{r_0} and V_{θ_0} are given in Equations (3.4.1) and (3.4.2), respectively, and that for P_0 is given in Equation (3.3.14). We do not attempt to obtain the correction fields here.

The outer expansion within the drop is written as follows:

$$V'_r = V'_{r_0} + \epsilon V'_{r_1} + \cdots, \tag{3.4.9}$$

$$V'_\theta = V'_{\theta_0} + \epsilon V'_{\theta_1} + \cdots, \tag{3.4.10}$$

and

$$P' = P'_0 + \epsilon P'_1 + \cdots. \tag{3.4.11}$$

The governing equations for V'_{r_0}, V'_{θ_0} and P'_0 are the same as those for the continuous phase at this order. The boundary conditions are given below.

$$V'_{r_0}(1, \theta) = 0, \tag{3.4.12}$$

$$V'_{\theta_0}(1, \theta) = V_{\theta_0}(1, \theta), \tag{3.4.13}$$

$$P_0'(1, \theta) = K_0 + P_0(1, \theta), \tag{3.4.14}$$

$$|V_{r_0}'| < \infty, |V_{\theta_0}'| < \infty \text{ as } r \to 0. \tag{3.4.15}$$

Strictly speaking, the kinematic condition at the drop surface, represented by Equation (3.4.12), and the continuity of tangential velocities at the surface, represented by Equation (3.4.13), should be written on the inner field. It can be shown that these conditions on the outer field are correct, however, and that they arise from matching requirements with the boundary layer solution. In Equation (3.4.14), K_0 is an unknown constant whose value cannot be established without considering the normal stress balance at the surface of the drop. The solutions for V_{r_0}' and V_{θ_0}' are as given in Equations (3.4.3) and (3.4.4), respectively, and that for the pressure is as follows:

$$P_0' = K_0 + \frac{9}{8}[(r^4 - r^2)\sin^2\theta + (2r^2 - r^4)\cos^2\theta]. \tag{3.4.16}$$

We now consider the correction to the leading order fields within the drop. Unlike in the exterior fluid, the first correction can occur at $O(\epsilon)$. Harper and Moore showed that this correction field is given by Hill's spherical vortex as well, with a magnitude of the vorticity that is to be determined. The existence of the $O(\epsilon)$ correction in the outer field is subject to verification by matching with the inner field. The governing equations and boundary conditions at $O(\epsilon)$ are as follows:

$$\nabla \bullet \mathbf{V_1'} = 0, \tag{3.4.17}$$

$$\mathbf{V_0'} \bullet \nabla \mathbf{V_1'} + \mathbf{V_1'} \bullet \nabla \mathbf{V_0'} = -\nabla P_1', \tag{3.4.18}$$

$$V_{r_1}'(1, \theta) = 0, \tag{3.4.19}$$

and

$$|V_{r_1}'| < \infty, |V_{\theta_1}'| < \infty \text{ as } r \to 0. \tag{3.4.20}$$

The solution is

$$V_{r_1}' = \frac{3}{2}c(1 - r^2)\cos\theta, \tag{3.4.21}$$

$$V_{\theta_1}' = 3c\left(r^2 - \frac{1}{2}\right)\sin\theta, \tag{3.4.22}$$

and

$$P_1' = K_1 + \frac{9c}{4}[(r^4 - r^2)\sin^2\theta + (2r^2 - r^4)\cos^2\theta]. \tag{3.4.23}$$

In Equations (3.4.21) to (3.4.23), c is an unknown constant that is to be determined by matching with the inner field and is related to the magnitude of the vorticity within the drop; K_1 is an unknown constant that cannot be determined without considering the normal stress balance at the surface of the drop.

3.4.4 Inner Velocity and Pressure Fields

Next, we turn our attention to the inner fields in the continuous and drop phase fluids. These boundary layer fields are needed to ensure that the tangential stress is continuous across the interface. The inner expansion in the two fluids is written as follows:

$$v_r = \epsilon\left(v_{r_0} + \epsilon v_{r_1} + \cdots\right), \tag{3.4.24}$$

$$v_\theta = v_{\theta_0} + \epsilon v_{\theta_1} + \cdots, \tag{3.4.25}$$

$$p = p_0 + \epsilon p_1 + \cdots, \tag{3.4.26}$$

$$v_r' = \epsilon\left(v_{r_0}' + \epsilon v_{r_1}' + \cdots\right), \tag{3.4.27}$$

$$v_\theta' = v_{\theta_0}' + \epsilon v_{\theta_1}' + \cdots, \tag{3.4.28}$$

and

$$p' = p_0' + \epsilon p_1' + \cdots. \tag{3.4.29}$$

When the governing equations and boundary conditions are rewritten in terms of the inner variables, we can write the following equations for the leading-order inner fields $\mathbf{v}_0(x, \theta)$, $p_0(x, \theta)$, $\mathbf{v}_0'(x', \theta)$, and $p_0'(x', \theta)$.

$$\frac{\partial v_{r_0}}{\partial x} + \frac{1}{\sin\theta}\frac{\partial}{\partial\theta}\left(v_{\theta_0}\sin\theta\right) = 0, \tag{3.4.30}$$

$$v_{r_0}\frac{\partial v_{\theta_0}}{\partial x} + v_{\theta_0}\frac{\partial v_{\theta_0}}{\partial\theta} = -\frac{\partial p_0}{\partial\theta} + \frac{\partial^2 v_{\theta_0}}{\partial x^2}, \tag{3.4.31}$$

$$\frac{\partial p_0}{\partial x} = 0, \tag{3.4.32}$$

$$-\sqrt{\frac{\gamma}{\alpha}}\frac{\partial v_{r_0}'}{\partial x'} + \frac{1}{\sin\theta}\frac{\partial}{\partial\theta}\left(v_{\theta_0}'\sin\theta\right) = 0, \tag{3.4.33}$$

$$-\sqrt{\frac{\gamma}{\alpha}}v_{r_0}'\frac{\partial v_{\theta_0}'}{\partial x'} + v_{\theta_0}'\frac{\partial v_{\theta_0}'}{\partial\theta} = -\frac{\partial p_0'}{\partial\theta} + \frac{\partial^2 v_{\theta_0}'}{\partial x'^2}, \tag{3.4.34}$$

$$\frac{\partial p_0'}{\partial x'} = 0, \tag{3.4.35}$$

$$v_{\theta_0}(0, \theta) = v_{\theta_0}'(0, \theta), \tag{3.4.36}$$

$$v_{r_0}(0, \theta) = v_{r_0}'(0, \theta) = 0, \tag{3.4.37}$$

$$\frac{\partial v_{\theta_0}}{\partial x}(0, \theta) = -\sqrt{\alpha\gamma}\frac{\partial v_{\theta_0}'}{\partial x'}(0, \theta), \tag{3.4.38}$$

$$v_{\theta_0}(x, 0) = v_{\theta_0}'(x', 0) = 0, \tag{3.4.39}$$

$$v_{\theta_0}(x \to \infty, \theta) = V_{\theta_0}(r \to 1, \theta) = \frac{3}{2}\sin\theta, \tag{3.4.40}$$

$$v_{\theta_0}'(x' \to \infty, \theta) = V_{\theta_0}'(r \to 1, \theta) = \frac{3}{2}\sin\theta, \tag{3.4.41}$$

$$p_0(x \to \infty, \theta) = P_0(r \to 1, \theta), \tag{3.4.42}$$

$$p_0'(x' \to \infty, \theta) = P_0'(r \to 1, \theta). \tag{3.4.43}$$

Equation (3.4.36) represents the continuity of tangential velocities across the interface. Equation (3.4.37) is the kinematic condition, and Equation (3.4.38) is obtained from the continuity of tangential stress. Equation (3.4.39) arises from symmetry considerations, and Equations (3.4.40) through (3.4.43) represent matching requirements. As in Section 3.3, it can be shown that the leading-order inner fields in the two fluids are merely the leading-order outer fields evaluated near the surface of the drop. Thus there is no boundary layer correction to the outer fields at leading order.

We next consider the inner fields at $O(\epsilon)$. The radial momentum equations can be simplified as follows:

$$\frac{\partial p_1}{\partial x} = v_{\theta_0}^2 \tag{3.4.44}$$

and

$$\frac{\partial p_1'}{\partial x'} = -\sqrt{\frac{\alpha}{\gamma}} v_{\theta_0}'^2. \tag{3.4.45}$$

The solutions for p_1 and p_1' that match with the respective outer fields are

$$p_1 = \frac{9}{4} x \sin^2 \theta \tag{3.4.46}$$

and

$$p_1' = K_1 + \frac{9}{4}\left(c \cos^2 \theta - \sqrt{\frac{\alpha}{\gamma}} x' \sin^2 \theta\right). \tag{3.4.47}$$

The tangential momentum equations can be simplified to yield the following:

$$\frac{3}{2}\frac{\partial}{\partial \theta}(v_{\theta_1} \sin \theta) - 3x \cos \theta \frac{\partial v_{\theta_1}}{\partial x} = \frac{\partial^2 v_{\theta_1}}{\partial x^2} \tag{3.4.48}$$

and

$$\frac{3}{2}\frac{\partial}{\partial \theta}(v_{\theta_1}' \sin \theta) - 3x' \cos \theta \frac{\partial v_{\theta_1}'}{\partial x'} = \frac{9}{2} c \sin \theta \cos \theta + \frac{\partial^2 v_{\theta_1}'}{\partial x'^2}. \tag{3.4.49}$$

The boundary conditions are given below.

$$v_{\theta_1}(0, \theta) = v_{\theta_1}'(0, \theta), \tag{3.4.50}$$

$$\frac{\partial v_{\theta_1}}{\partial x}(0, \theta) + \sqrt{\alpha \gamma}\frac{\partial v_{\theta_1}'}{\partial x'}(0, \theta) = \frac{3}{2}(1 - \alpha)\sin \theta, \tag{3.4.51}$$

$$v_{\theta_1}(x, 0) = v_{\theta_1}'(x', 0) = 0, \tag{3.4.52}$$

$$v_{\theta_1}(x \to \infty, \theta) \to -\frac{3}{2} x \sin \theta, \tag{3.4.53}$$

$$v_{\theta_1}'(x' \to \infty, \theta) \to \frac{3}{2} c \sin \theta - 6\sqrt{\frac{\alpha}{\gamma}} x' \sin \theta. \tag{3.4.54}$$

Equation (3.4.50) represents continuity of tangential velocities across the interface. Equation (3.4.51) is a consequence of the tangential stress balance, and Equation (3.4.52) arises from symmetry considerations. Equations (3.4.53) and (3.4.54) represent matching requirements with the respective outer fields.

3.4.5 Solution for the Tangential Velocity Fields in the Boundary Layer

To solve the above set of equations for v_{θ_1} and v'_{θ_1}, it is convenient to transform the dependent and independent variables as follows:

$$v_{\theta_1} = -\frac{3}{2}x\sin\theta + \frac{f}{\sin\theta}, \tag{3.4.55}$$

$$v'_{\theta_1} = \frac{3}{2}c\sin\theta + \sqrt{\frac{\alpha}{\gamma}}\left(\frac{f'}{\sin\theta} - 6x'\sin\theta\right), \tag{3.4.56}$$

$$\eta = x\sin^2\theta, \tag{3.4.57}$$

$$\eta' = x'\sin^2\theta, \tag{3.4.58}$$

and

$$\xi = \frac{2}{9}(2 - 3\cos\theta + \cos^3\theta). \tag{3.4.59}$$

The equations for $f(\xi, \eta)$ and $f'(\xi, \eta')$, together with the boundary conditions, can be written as

$$\frac{\partial f}{\partial \xi} = \frac{\partial^2 f}{\partial \eta^2}, \tag{3.4.60}$$

$$\frac{\partial f'}{\partial \xi} = \frac{\partial^2 f'}{\partial \eta'^2}, \tag{3.4.61}$$

$$f(\xi, 0) - \sqrt{\frac{\alpha}{\gamma}}f'(\xi, 0) = \frac{3}{2}\sqrt{\frac{\alpha}{\gamma}}\tilde{c}\sin^2\theta, \tag{3.4.62}$$

$$\frac{\partial f}{\partial \eta}(\xi, 0) + \alpha\frac{\partial f'}{\partial \eta'}(\xi, 0) = 3\left(1 + \frac{3}{2}\alpha\right), \tag{3.4.63}$$

and

$$f(\xi, \eta \to \infty) = f'(\xi, \eta' \to \infty) \to 0, \tag{3.4.64}$$

where $\tilde{c} = \sqrt{\frac{\gamma}{\alpha}}c$. In addition, "starting" conditions on f and f' must be specified at some value of ξ. Conditions at $\theta = 0$ have been specified for v_{θ_1} and v'_{θ_1}, but ξ, η, and η' are zero at $\theta = 0$. Therefore, conditions on $f(0, \eta)$ and $f'(0, \eta')$ must be obtained. At the forward stagnation point, no vorticity enters the boundary layer from outside the drop. This can be shown to yield $f(0, \eta) = 0$. It is not obvious how a condition for $f'(0, \eta')$ can be determined, however. Harper and Moore have shown that, in the limit of large Reynolds number within the drop, a starting condition for f' can be obtained by examining the nature of the internal momentum wake in the drop. The condition is that the distribution for f' near the forward stagnation point is the same as that near the rear stagnation point. The momentum wake merely convects this distribution passively from the rear to the forward stagnation point; viscous effects during this transit cause a negligible change. It will be evident below that this condition leads to an integral equation for $f'(0, \eta')$.

$$f(0, \eta) = 0, \tag{3.4.65}$$

$$f'(0, \eta') = f'(\xi(\pi), \eta') \equiv g(\eta'). \tag{3.4.66}$$

Here $g(\eta')$ is an unknown distribution and $\xi(\pi) = \frac{8}{9}$. The unknown constant $\tilde{c}$ is determined by the following constraint:

$$g(\eta' \to \infty) \to 0. \tag{3.4.67}$$

This constraint ensures that the solution is consistent with the boundary condition on $f'(\xi, \eta' \to \infty)$ in Equation (3.4.64). Harper and Moore have obtained a solution for f and f' in terms of $g(\eta')$:

$$f(\xi, \eta) = \frac{\sqrt{\frac{\alpha}{\gamma}}}{\alpha + \sqrt{\frac{\alpha}{\gamma}}} \left[-3(2 + 3\alpha)\sqrt{\xi}\phi(\zeta) \right.$$
$$\left. + \alpha \left(\frac{3}{2}\tilde{c}N(\xi, \eta) + \frac{1}{\sqrt{\pi\xi}} \int_0^\infty g(\tilde{\eta}) \exp\left(-\frac{(\eta' + \tilde{\eta})^2}{4\xi} \right) d\tilde{\eta} \right) \right], \tag{3.4.68}$$

and

$$f'(\xi, \eta') = -\frac{3}{\alpha + \sqrt{\frac{\alpha}{\gamma}}} \left[(2 + 3\alpha)\sqrt{\xi}\phi(\zeta') + \frac{1}{2}\sqrt{\frac{\alpha}{\gamma}}\tilde{c}N(\xi, \eta') \right]$$
$$+ \frac{1}{2\sqrt{\pi\xi}} \int_0^\infty g(\tilde{\eta}) \left[\exp\left(-\frac{(\eta' + \tilde{\eta})^2}{4\xi} \right) \right.$$
$$\left. + \frac{\alpha - \sqrt{\frac{\alpha}{\gamma}}}{\alpha + \sqrt{\frac{\alpha}{\gamma}}} \exp\left(-\frac{(\eta' - \tilde{\eta})^2}{4\xi} \right) \right] d\tilde{\eta}, \tag{3.4.69}$$

where

$$\zeta = \frac{\eta}{2\sqrt{\xi}}, \quad \zeta' = \frac{\eta'}{2\sqrt{\xi}}, \tag{3.4.70}$$

$$\phi(\zeta) = \frac{1}{\sqrt{\pi}}e^{-\zeta^2} - \zeta \operatorname{erfc} \zeta, \tag{3.4.71}$$

and

$$N(\xi, \eta) = \frac{\eta}{2\sqrt{\pi}} \int_0^\xi \frac{\sin^2\theta(\tau)}{(\xi - \tau)^{\frac{3}{2}}} \exp\left(-\frac{\eta^2}{4(\xi - \tau)} \right) d\tau. \tag{3.4.72}$$

In Equation (3.4.72), θ must be expressed as a function of τ by solving Equation (3.4.59) for θ in terms of ξ and then replacing ξ by τ. Application of the condition in Equation (3.4.66) leads to an integral equation for $g(\eta')$:

$$g(\eta') = -\frac{3}{\alpha + \sqrt{\frac{\alpha}{\gamma}}} \left[(2 + 3\alpha)\sqrt{\xi(\pi)}\,\phi\left(\frac{\eta'}{2\sqrt{\xi(\pi)}} \right) + \frac{1}{2}\sqrt{\frac{\alpha}{\gamma}}\tilde{c}N(\xi(\pi), \eta') \right]$$
$$+ \frac{1}{2\sqrt{\pi\xi(\pi)}} \int_0^\infty g(\tilde{\eta}) \left[\exp\left(-\frac{(\eta' + \tilde{\eta})^2}{4\xi(\pi)} \right) \right.$$
$$\left. + \frac{\alpha - \sqrt{\frac{\alpha}{\gamma}}}{\alpha + \sqrt{\frac{\alpha}{\gamma}}} \exp\left(-\frac{(\eta' - \tilde{\eta})^2}{4\xi(\pi)} \right) \right] d\tilde{\eta}. \tag{3.4.73}$$

Harper and Moore (1968) obtained a numerical solution for $g(\eta')$ that satisfies Equation (3.4.67) for various values of α and γ. The reader may consult this source for further details. They also found that the values for $\tilde{c}$ that are numerically determined are close to those given by the following equation:

$$\tilde{c} \approx -\frac{2.5}{\sqrt{2}} \frac{(2+3\alpha)\left(2\alpha + \sqrt{\frac{\alpha}{\gamma}}\right)}{\sqrt{\frac{\alpha}{\gamma}}\left(2\sqrt{\frac{\alpha}{\gamma}} + 3\alpha\right)}. \tag{3.4.74}$$

Thus, the diffusion of momentum in the boundary layer tends to reduce the magnitude of the vorticity within the drop from that given by the solution at leading order.

By considering the dissipation of energy in the momentum boundary layers, Harper and Moore showed that when α is nonzero, the correction to Equation (3.4.5) for the hydrodynamic force on the drop F_D occurs at $O(\epsilon \log \epsilon)$. As noted in Section 3.3, Moore (1963) determined that for a gas bubble for which $\alpha = 0$, the first correction occurs only at $O(\epsilon)$. In the limit $\alpha \to 0$, the result for F_D given by Harper and Moore (1968) reduces to Equation (3.3.42), which is the result obtained by Moore (1963).

3.5   Motion of Highly Deformed Bubbles and Drops

3.5.1   Introduction

In this section, we discuss the motion of bubbles and drops that are large and are significantly deformed. Davies and Taylor (1950) performed experiments on the motion of air bubbles in nitrobenzene and observed that the rising bubbles possessed a distorted, umbrellalike shape. The shape is actually composed of a curved surface that is close to being spherical at the front of the bubble and a flat, often unsteady rear surface. Such a shape is achieved when the Weber number is approximately 20. These spherical cap bubbles have been observed in other liquids by Bhaga and Weber (1981) and by others. Bhaga and Weber noted that the wake behind the bubble either can be closed and toroidal in shape or open and unsteady; they reported a sharp transition from a closed to an open wake.

Davies and Taylor derived a simple semiempirical result for the rise velocity of the bubble. Moore (1959) improved the theory, using ideas similar to Helmholtz's free-streamline theory for the flow past a blunt body, such as a flat plate oriented normal to the direction of the approaching flow. The subject also has been reviewed by Harper (1972). The main ideas of Moore's theory are reproduced below. Physical quantities with appropriate dimensions are used throughout this section. Let r_c^* be the radius of curvature of the spherical cap of the bubble at its front. The cap extends to a value of the polar angle $\theta = \theta_m$, measured from the forward stagnation streamline, beyond which the shape is a horizontal plane surface that intersects the spherical cap. The Reynolds and Weber numbers in the continuous phase, defined in the same manner as in Section 3.2 by using an equivalent radius as the length scale, are assumed large compared with unity. Further, it is assumed that the flow around the spherical cap can be approximated by a potential flow over the completed sphere. Beyond $\theta = \theta_m$, the flow separates and is assumed to form a free streamline downstream of the bubble. In other words, in a reference frame moving with the bubble, a surface of discontinuity separates regions where there is flow from the wake in which fluid is stagnant. The stagnant wake with a constant hydrodynamic pressure is assumed to extend to infinite distance downstream of the bubble.

3.5.2 Determination of the Rise Velocity

The surface velocity in the spherical cap region is given by potential flow theory to be

$$v_s^*(\theta) = \frac{3}{2} v_\infty^* \sin \theta, \tag{3.5.1}$$

where v_∞^* is the rise velocity of the bubble. The pressure at the bubble surface in this region, including a contribution from the hydrostatic force, is obtained by using Bernoulli's theorem to be

$$p_s^*(\theta) = p_\infty^* - \frac{1}{2}\rho\left(v_s^{*2}(\theta) - v_\infty^{*2}\right) + \rho g r_c^*(1 - \cos\theta), \tag{3.5.2}$$

where p_∞^* is the hydrodynamic pressure far away from the bubble and g is the magnitude of the acceleration due to gravity. The datum for the hydrostatic pressure is located at the horizontal plane passing through the point $\theta = 0$ on the surface of the bubble. Because the pressure inside the bubble is a constant, the balance of normal stress on the surface of the bubble requires that

$$-p_\infty^* + \frac{1}{2}\rho\left(v_s^{*2}(\theta) - v_\infty^{*2}\right) - \rho g r_c^*(1 - \cos\theta) - \frac{2\sigma^*}{r_c^*} = \text{constant}, \tag{3.5.3}$$

where σ^* is the surface tension. Because the Reynolds number is assumed to be large, viscous contributions to the normal stress are ignored in Equation (3.5.3). Clearly, the left side of Equation (3.5.3) cannot be constant everywhere on the bubble surface. Davies and Taylor satisfied Equation (3.5.3) approximately by expanding the θ-dependence in the left side and setting the coefficient of θ^2 to zero. This leaves an error at higher order in θ. The resulting approximation to the bubble velocity is

$$v_\infty^* = \frac{2}{3}(g r_c^*)^{\frac{1}{2}}. \tag{3.5.4}$$

Moore (1959) determined the value of θ_m using the following argument. Because the wake behind the bubble is stagnant, the hydrodynamic pressure everywhere in the wake is constant and is equal to p_∞^*. The hydrodynamic pressure everywhere on the separated streamline is also equal to this value. By Bernoulli's theorem, the velocity must equal v_∞^* everywhere on the separated streamline. Thus,

$$v_s^*(\theta_m) = v_\infty^*. \tag{3.5.5}$$

The value of the constant in the right side of Equation (3.5.3) is determined by evaluating the left side at $\theta = 0$. A subsequent evaluation of Equation (3.5.3) at $\theta = \theta_m$ yields

$$\cos\theta_m = \frac{7}{9}; \quad \theta_m = 38.9°. \tag{3.5.6}$$

Harper (1972) noted that using Equation (3.5.6) in Equation (3.5.1) yields $v_s^*(\theta_m) = 0.9428 v_\infty^*$ and not v_∞^* as required for consistency. This shows the approximate nature of the assumed shape of the bubble and the flow field. Curiously, if Equation (3.5.5) is substituted into Equation (3.5.1), $\theta_m = \sin^{-1}\frac{2}{3} = 41.8°$ is obtained. Then, if Equation (3.5.3) is satisfied at $\theta = 0$ and at $\theta = \theta_m$, the bubble velocity is obtained to be $v_\infty^* = 0.714(g r_c^*)^{1/2}$, in contrast to the result in Equation (3.5.4). Harper has mentioned other approximations to the bubble velocity and the extent of the spherical cap, but the results are not given. He pointed out that the correct solution for the bubble shape is one such that the irrotational flow past it satisfies $v_s^{*^2}(\theta) = 2g r_c^*(1 - \cos\theta)$, $v_s^*(\theta_m) = v_\infty^*$. Such a shape has been numerically calculated by Rippin and Davidson (1967). The true shape is close to a spherical cap, with $\theta_m = 50°$ and $v_\infty^* = 0.787(gr_c^*)^{1/2}$.

Churchill (1988) obtained an expression for the rise velocity of a highly deformed drop by following a path analogous to that used in deriving Equation (3.5.4). The velocity is

$$v_\infty^* = \frac{2}{3}\left(\frac{gr_c^*(\rho - \rho')}{\rho}\right)^{\frac{1}{2}}. \tag{3.5.7}$$

Several investigators have performed experiments to verify the prediction of Davies and Taylor for the bubble velocity in Equation (3.5.4) (see Harper (1972) and Clift et al. (1978)). Bhaga and Weber (1981) stated that Equation (3.5.4) agrees with their data to within 5% for Reynolds numbers larger than five if the radius of the spherical cap is obtained from the experimentally obtained shape of the bubble by fitting a sphere over a region covering $0 \le \theta \le 37.5°$. Equation (3.5.7) for the velocity of a drop does not appear to have been tested experimentally.

CHAPTER FOUR

Thermocapillary Motion

4.1 Governing Equations

4.1.1 Introduction

In this chapter, we consider problems involving the motion of an isolated drop in a continuous phase of infinite extent when that motion is caused by gradients of interfacial tension. Because we primarily consider these gradients when they are caused by a nonuniform temperature distribution, the focus is on thermocapillary migration. Where convenient, we also include the effect of a body force such as gravity. When the gravitational force is accommodated, it will be in the context of calculating the net force on an object such as a drop, and we shall ignore the possibility of buoyant convection, which can be caused by gravity acting in conjunction with density gradients. It is possible to minimize buoyant convection in experiments carried out on Earth by careful design.

4.1.2 Problem Statement, Physical Assumptions, and Scales

We first pose a fairly general problem, and analyze specialized versions of the problem in subsequent sections. With this in mind, consider the unsteady motion of a drop in a continuous phase of infinite extent, in which a constant temperature gradient ∇T_∞ is maintained by external means. This undisturbed temperature gradient is assumed to be independent of time and position. We select the direction of this gradient to be the z-direction, so that $\nabla T_\infty = |\nabla T_\infty| \, \mathbf{k}$. The drop is assumed immiscible with the continuous phase. The drop can be deformed, and the radius of a sphere of the same volume is designated as R_0. Incompressible Newtonian flow is assumed in both the drop phase and the continuous phase, and the variation of physical properties with temperature is ignored, with the exception of the interfacial tension, which is assumed to vary linearly with temperature. For definiteness, the rate of change of the interfacial tension with temperature, σ_T, is assumed negative, even though situations in which it is positive can be accommodated by simple sign changes where needed. The symbols introduced earlier for designating physical properties and their ratios will continue to be used here.

To obtain the velocity of the drop and its shape, it is necessary to solve the governing momentum and energy equations, along with the associated initial and boundary conditions where needed. Therefore, we can expect to obtain the velocity field and the temperature field in the fluids involved when seeking the required answers. The variables are scaled by using R_0 as the reference length and the quantity v_0 defined below as the reference velocity.

$$v_0 = -\frac{\sigma_T |\nabla T_\infty| R_0}{\mu}.$$ (4.1.1)

Here, μ represents the viscosity of the continuous phase. The reference velocity is obtained by assuming that the tangential stress at the interface resulting from the gradient in interfacial tension drives a flow that causes velocity gradients of the order $\frac{v_0}{R_0}$. Stresses including pressure are scaled by using the typical viscous stress $\mu \frac{v_0}{R_0} = -\sigma_T |\nabla T_\infty|$. Temperature is nondimensionalized by first subtracting its value in the undisturbed fluid at infinity in an isothermal plane containing the center of mass of the drop at time zero and then dividing by the scale $R_0 |\nabla T_\infty|$. Time is scaled by using a reference t_R, which is chosen as the viscous relaxation time in the continuous phase over a distance R_0. Therefore, we take $t_R = \frac{R_0^2}{\nu}$.

4.1.3 Governing Equations and Boundary Conditions

With the scalings selected above, the governing equations in the continuous phase take the form below in a reference frame that is attached to the moving drop.

$$\nabla \bullet \mathbf{v} = 0,$$ (4.1.2)

$$\frac{\partial \mathbf{v}}{\partial t} + \frac{d\mathbf{v}_\infty}{dt} + Re[(\mathbf{v} \bullet \nabla)\mathbf{v}] = -\nabla p + \nabla^2 \mathbf{v},$$ (4.1.3)

$$Pr\frac{\partial T}{\partial t} + Ma[\mathbf{v} \bullet \nabla T] = \nabla^2 T.$$ (4.1.4)

The unknown scaled velocity of the drop is designated $\mathbf{v}_\infty(t)$, where we have explicitly shown its possible dependence on scaled time. In the same reference frame, the equations for the drop phase are very similar; they are written below.

$$\nabla \bullet \mathbf{v}' = 0,$$ (4.1.5)

$$\frac{\gamma}{\alpha}\left(\frac{\partial \mathbf{v}'}{\partial t} + \frac{d\mathbf{v}_\infty}{dt}\right) + Re'[(\mathbf{v}' \bullet \nabla)\mathbf{v}'] = -\frac{1}{\alpha}\nabla p' + \nabla^2 \mathbf{v}',$$ (4.1.6)

$$\frac{\gamma}{\alpha}Pr'\frac{\partial T'}{\partial t} + Ma'[\mathbf{v}' \bullet \nabla T'] = \nabla^2 T'.$$ (4.1.7)

The symbols $Re(=\frac{R_0 v_0}{\nu})$ and $Ma(=\frac{R_0 v_0}{\kappa})$, appearing in the above equations, stand for the Reynolds number and the Marangoni number, respectively. Here, ν represents the kinematic viscosity of the continuous phase and κ its thermal diffusivity. Recall that when the reference velocity is chosen to be characteristic of thermocapillary motion, the Péclet number, which measures the relative importance of convective energy transport compared with conduction, is termed the Marangoni number. The primed quantities refer to the drop phase. The symbols Re' and Ma' are defined in the same way as the symbols Re and Ma, respectively, by replacing the physical properties with those corresponding to the drop phase. The Prandtl numbers $Pr = \frac{\nu}{\kappa}$ in the continuous phase and $Pr' = \frac{\nu'}{\kappa'}$ in the drop phase are important parameters encountered in heat transport problems. The symbols γ and α refer to the density and viscosity ratios, respectively, wherein the numerator in the ratio corresponds to the property within the drop phase as mentioned earlier. Under zero gravity conditions, the pressure in the continuous phase will be uniform in the absence of motion. The pressure within the drop, in the absence of motion, also will be uniform and will be given by $\frac{2\sigma^*}{R_0}$ added to the pressure in the continuous phase wherein σ^* is the interfacial tension. There will be no distinction between the hydrodynamic pressure and the actual pressure. For problems in which we introduce a gravitational force for completeness, we need to be aware that the symbols p and p' appearing in the above equations connote the scaled hydrodynamic pressure.

The appearance of new terms on the left hand sides of the two momentum equations is worthy of some comment. If the velocity of the drop is time dependent, the reference frame traveling with the drop is noninertial. In this case, an additional "body force" acts on the fluid. On a unit volume basis, this force is given by $-\rho\frac{d\mathbf{v}_\infty^*}{dt^*}$, where the asterisk is used to designate dimensional counterparts of the scaled quantities used above. When this "force" is brought to the left side and scaled, it leads to the term shown in Equations (4.1.3) and (4.1.6).

We now write the relevant boundary conditions. Far from the drop, conditions must approach those prevalent in the undisturbed continuous phase fluid. Therefore, if $\mathbf{x}$ is the position vector with respect to the center of mass of the drop, as $|\mathbf{x}| \to \infty$,

$$\mathbf{v} \to -\mathbf{v}_\infty(t), \tag{4.1.8}$$

$$p \to 0, \tag{4.1.9}$$

and

$$T \to z + Re \int_0^t v_\infty(\zeta)\, d\zeta. \tag{4.1.10}$$

Here, the symbol z is used to designate scaled distance in the direction of the temperature gradient measured from the center of mass of the moving drop.

At the interface, designated as the surface S at which the unit normal vector pointing into the continuous phase is $\mathbf{n}$, the velocity and temperature fields must be continuous, and the kinematic boundary condition must be satisfied.

On S,

$$\mathbf{v} = \mathbf{v}', \tag{4.1.11}$$

$$\mathbf{v} \bullet \mathbf{n} = \mathbf{v}' \bullet \mathbf{n}, \tag{4.1.12}$$

and

$$T = T'. \tag{4.1.13}$$

If the shape of the drop is steady, the normal velocities in Equation (4.1.12) can be set to zero. Otherwise, one must use the condition given in Equation (2.3.5). In addition, at the interface, the stress and heat flux balances must be written as follows:

On S,

$$\mathbf{n} \bullet (\mathbf{\Pi} - \mathbf{\Pi}') = \frac{1}{Ca}[\sigma\mathbf{n}(\nabla \bullet \mathbf{n}) - \nabla_s\sigma] = \frac{1}{Ca}\sigma\mathbf{n}(\nabla \bullet \mathbf{n}) + \nabla_s T \tag{4.1.14}$$

and

$$\mathbf{n} \bullet (\nabla T - \beta\nabla T') = -\frac{\sigma_T(\sigma\sigma_0 - e_s)}{\mu k}\nabla_s \bullet \mathbf{v}. \tag{4.1.15}$$

In the above, β is the ratio of the thermal conductivity of the fluid within the drop to that of the continuous phase. The Capillary number $Ca = \frac{\mu v_0}{\sigma_0}$ is based on a reference value of the interfacial tension σ_0. Note that the interfacial tension itself varies with position on the interface and with time at each location because the drop continues to move into warmer fluid. The symbol σ represents the interfacial tension scaled by the reference value σ_0. A better scaling scheme would use a characteristic difference in interfacial tension, $\Delta\sigma^*$, over the drop surface and not σ_0, for nondimensionalizing the tangential stress discontinuity. This would lead to a different group, $\frac{1}{Ca'}$, multiplying $\nabla_s\sigma$ where Ca' is defined in a manner similar to the Capillary number, but using $\Delta\sigma^*$ instead of σ_0. For convenience in notation, we have used a simpler scheme. The vector stress balance is commonly broken into a tangential stress balance and a normal stress balance, as shown below:

On S,

$$\mathbf{n} \bullet (\Pi - \Pi') \times \mathbf{n} = \nabla_s T \times \mathbf{n} \tag{4.1.16}$$

and

$$\mathbf{n} \bullet (\Pi - \Pi') \bullet \mathbf{n} = \frac{1}{Ca}\sigma(\nabla \bullet \mathbf{n}). \tag{4.1.17}$$

Harper, Moore, and Pearson (1967) first pointed out the need for the inclusion of the inhomogeneity in the energy flux balance boundary condition given in Equation (4.1.15), which arises from the creation of fresh interface in the forward half of the drop and destruction of interface in the rear. They concluded that the consequences of accommodating the creation and destruction of the interface would be negligible in practical situations. The importance of this effect was reexamined in detail in the recent work of Torres and Herbolzheimer (1993), who pointed out that the dimensionless quantity multiplying the surface divergence of the velocity field in Equation (4.1.15) can take on significant values at certain liquid-gas interfaces. The magnitude of this quantity, which is negative, depends on temperature.

Although the effect of surface creation and destruction is interesting to study, the only case that has been analyzed is that involving negligible convective transport effects. We provide results in this case in Section 4.9. At the present time, no experiments appear to have been performed to verify the predictions in this case. With the exception noted above, we shall set the right-hand side of Equation (4.1.15) to zero from here onward. This yields the simpler energy flux balance given below.

On S,

$$\mathbf{n} \bullet (\nabla T - \beta\nabla T') = 0. \tag{4.1.18}$$

It now remains to complete the set of boundary conditions by writing conditions at the center, which indicate that the velocity, pressure, and temperature fields are bounded. At $|\mathbf{x}| = 0$,

$$|\mathbf{v}'|, |p'|, |T'| < \infty. \tag{4.1.19}$$

When the complete problem within and outside the drop is solved along with the boundary conditions, the velocity of the drop is obtained as part of the solution. When the shape of the drop is prescribed, however, the ability to satisfy the balance of normal stresses given in Equation (4.1.17) is lost. In this case, Newton's law of motion for the drop, given below, is used to obtain its velocity, wherein the effect of gravity is accommodated.

$$\frac{4}{3}\pi\gamma\frac{d\mathbf{v}_\infty}{dt} = \int_S d\mathbf{S}\bullet\mathbf{\Pi} - \frac{4\pi(\rho'-\rho)R_0\mathbf{g}}{3\sigma_T|\nabla T_\infty|} \tag{4.1.20}$$

This completes the posing of the problem. When the interfacial tension depends solely on temperature in a linear fashion, and the energy flux balance is given by Equation (4.1.18), there are eight independent parameters in the problem including the parameter that multiplies $\mathbf{g}$ in the right side of Equation (4.1.20).

Equation (4.1.10) tells us that the temperature in the undisturbed fluid is time dependent. This is a consequence of our choice of reference frame. Riding with the drop, the fluid surrounding it will continue to get warmer with the passage of time. When the physical properties are independent of temperature and the shape is steady, however, it is possible for the temperature gradient field in both fluids to achieve a steady state, leading to migration of the drop at a constant velocity. It is convenient to analyze this situation by using a slightly different choice of datum for the temperature. Recall that we subtracted the value that exists on an isothermal plane passing through the center of mass of the drop at time zero. Instead, if we choose a new datum as the temperature on an isothermal plane in the undisturbed continuous phase fluid passing through the center of mass of the drop at any given instant, we can define scaled temperatures T_1 and T_1' as follows:

$$T_1 = T - Re\int_0^t v_\infty(\zeta)\,d\zeta \tag{4.1.21}$$

and

$$T_1' = T' - Re\int_0^t v_\infty(\zeta)\,d\zeta. \tag{4.1.22}$$

The only equations affected by this transformation are the governing energy equations and the boundary condition as $|\mathbf{x}| \to \infty$. They are provided below.

$$Pr\frac{\partial T_1}{\partial t} + Ma[v_\infty + \mathbf{v}\bullet\nabla T_1] = \nabla^2 T_1, \tag{4.1.23}$$

$$\frac{\gamma}{\alpha}Pr'\frac{\partial T_1'}{\partial t} + Ma'[v_\infty + \mathbf{v}'\bullet\nabla T_1'] = \nabla^2 T_1', \tag{4.1.24}$$

$$T_1 \to z \quad \text{as} \quad |\mathbf{x}| \to \infty. \tag{4.1.25}$$

Beginning with the next section, we drop the subscript 1 on these scaled temperatures for convenience. The scaled temperature at infinity is now steady, and it will be possible to pose steady problems for the temperature and velocity everywhere. Note that the transformation naturally leads to a sink term Mav_∞ in the energy equation for the continuous phase and a similar term in the equation for the drop phase. The physical interpretation of this sink is as follows. Consider the undisturbed fluid far from the drop. This fluid possesses a steady linear variation in temperature in the z-direction and has a velocity $-v_\infty\mathbf{k}$. As warm fluid moves in the negative z-direction, the sink causes it to lose heat at a rate proportional to the velocity $v_\infty(t)$, thereby cooling it to the right temperature corresponding to the location in question.

In subsequent sections, we shall use suitably simplified versions of the governing equations and boundary conditions stated in this section as our starting point. In problems wherein we ignore deformation of shape from a sphere, we drop the subscript 0 on the radius of the drop, designating the radius simply as R.

4.2 Motion When Convective Transport Is Negligible

4.2.1 Introduction

The simplest situation for analysis is one in which convective transport of energy as well as that of momentum are considered negligible compared with molecular transport of these respective entities. The analysis would proceed along the lines followed in Section 3.1 for the gravitational settling of a drop. The requirement here is that the Reynolds and Marangoni numbers both be negligible. From the definition given in the previous section, one can infer that sufficiently small drops moving in gentle temperature gradients in viscous fluids would satisfy the requirement of negligible Reynolds number. For the Marangoni number to be negligible, one would further require that the Prandtl number not be too large. For a fluid of high kinematic viscosity to satisfy this requirement, it must also have a large thermal diffusivity. In practical terms, for an air bubble of radius R mm, migrating in a Dow-Corning DC-200 series silicone oil (a dimethylsiloxane polymer) of nominal kinematic viscosity $v = 10^{-4}$ m^2/s in a temperature gradient of 1 K/mm, the Reynolds and Marangoni numbers are approximately $0.006R^2$ and $6R^2$ respectively. Therefore, for a bubble of radius 100 μm, $Re = 6 \times 10^{-5}$ and $Ma = 0.06$. If the viscosity is halved, the Marangoni number would be doubled, whereas the Reynolds number would be quadrupled. The thermal diffusivity of silicone oils is small, and Marangoni numbers for similar situations in other fluids can be expected to be smaller. In experimental work on air bubbles performed on Earth, we have used silicone oils because of their relative resistance to contamination by surface active agents and the wide range of viscosities available. The latter feature permits one to vary the dimensionless groups involved with relative ease because all the other relevant properties remain nearly constant in a given series of silicone oils. There are a variety of silicone oils, but experiments have been performed most commonly in dimethylsiloxane polymer fluids, beginning with the pioneering work of Young, Goldstein, and Block (1959). A discussion of experimental results can be found in Section 4.16.

4.2.2 The Case of Zero Gravity

We begin by considering the case of zero gravity. Later, we comment on how the motion of a drop under the combined effects of gravity and thermocapillarity can be handled by taking advantage of the linearity of the governing equations.

   The analysis follows the original development of Young et al. We make all the assumptions stated in Section 4.1 and also set the Reynolds and Marangoni numbers equal to zero. We further assume that the drop is moving at a constant velocity in the direction of the temperature gradient, ignoring the initial transient, and that it remains spherical, with a radius R. Figure 3.1.1 shows the spherical polar coordinates used in the analysis. We use the scalings from Section 4.1, replacing R_0 by R, and define $s = \cos\theta$. Because physical properties vary with temperature, the motion can at best be quasi-steady, as noted in Section 2.4. The assumption regarding the spherical shape remains to be verified later. The continuity and Navier–Stokes equations in the continuous phase can be written as follows, in a reference frame translating with the drop:

$$\nabla \bullet \mathbf{v} = 0 \tag{4.2.1}$$

and

$$\nabla^2 \mathbf{v} = \nabla p. \tag{4.2.2}$$

The simplifications from the equations presented in Section 4.1 are a direct consequence of the additional assumptions that have been made. Similar equations can be written in the drop phase fluid. As in Section 3.1, in view of the axial symmetry, a Stokes streamfunction can be defined in this problem. This leads to replacement of Equations (4.2.1) and (4.2.2) and their counterparts in the drop phase by Stokes's equation for the streamfunction:

$$E^4 \psi = 0 \tag{4.2.3}$$

and

$$E^4 \psi' = 0. \tag{4.2.4}$$

The assumptions lead to the temperature fields within and outside the drop satisfying Laplace's equation:

$$\nabla^2 T = 0 \tag{4.2.5}$$

and

$$\nabla^2 T' = 0. \tag{4.2.6}$$

Note that both the momentum and energy equations have become linear in the limit being considered. Solutions of Laplace's equation are known as *harmonic functions*, and there is a substantial literature dealing with such solutions in different geometries under a variety of boundary conditions. Carslaw and Jaeger (1959) and Crank (1975) have provided a good collection of these solutions.

The boundary conditions can be written directly from the results given in Section 4.1.

As $r \to \infty$,

$$\psi \to \frac{1}{2} v_\infty r^2 (1 - s^2) \tag{4.2.7}$$

and

$$T \to rs. \tag{4.2.8}$$

At the drop surface, $r = 1$,

$$v_r(1, s) = v_r'(1, s) = 0, \tag{4.2.9}$$

$$v_\theta(1, s) = v_\theta'(1, s), \tag{4.2.10}$$

$$\tau_{r\theta}(1, s) - \tau_{r\theta}'(1, s) = \frac{\partial T}{\partial \theta}, \tag{4.2.11}$$

$$T(1, s) = T'(1, s), \tag{4.2.12}$$

and

$$\frac{\partial T}{\partial r}(1, s) = \beta \frac{\partial T'}{\partial r}(1, s). \tag{4.2.13}$$

At the center of the drop, $r = 0$,

$$|\psi'|, |T'| < \infty. \tag{4.2.14}$$

It is now clear that the temperature field can be obtained without knowledge of the velocity field. Once the temperature on the surface is known, it can be used in the tangential stress boundary condition given in Equation (4.2.11) as a known inhomogeneity that drives the motion. The reader may wish to review the earlier discussion at the end of Section 2.3.3 regarding the role of such inhomogeneities in producing nontrivial solutions.

The solution of Laplace's equation is straightforward when one uses the following general result applicable in the axisymmetric case.

$$T(r, s) = \sum_{n=0}^{\infty} \left(A_n r^n + \frac{B_n}{r^{n+1}} \right)[P_n(s) + C_n Q_n(s)]. \tag{4.2.15}$$

The result given above can be written for the temperature field in the continuous phase, and a similar result can be used for the temperature field in the drop phase. The symbols A_n, B_n, and C_n stand for sets of arbitrary constants that are specialized by using boundary conditions relevant to a given problem. The symbol $P_n(s)$ represents the Legendre Polynomial, which was introduced in Section 3.1. The symbol $Q_n(s)$ stands for the second linearly independent solution of Legendre's differential equation and is called the *Legendre function of the second kind*. It has the property of being singular when $s = \pm 1$, that is, on the forward and rear stagnation streamlines. Because temperatures have to be finite except at infinite distance away from the drop, the constants C_n can be set to zero immediately. When the other boundary conditions are used, the following solution can be written for the temperature field in each phase:

$$T(r, s) = \left(r + \frac{1 - \beta}{2 + \beta} \frac{1}{r^2} \right) P_1(s) \tag{4.2.16}$$

$$T'(r, s) = \frac{3}{2 + \beta} r P_1(s). \tag{4.2.17}$$

We note that $P_1(s) = s = \cos\theta$. Therefore, the scaled temperature on the surface $T(1, s) = T'(1, s) = \frac{3}{2+\beta} \cos\theta$.

It is straightforward to specialize the constants in the general solution for the streamfunction in spherical polar coordinates given in Equations (3.1.9) and (3.1.10) by the application of the pertinent boundary conditions. As mentioned earlier, when the momentum equations within and outside the drop are solved together, and the shape of the drop is not assumed but determined from the normal stress balance as part of the solution, the velocity of the drop is obtained as part of the solution. We have assumed the shape of the drop to be a sphere and applied the boundary conditions at this spherical surface. Because we ignore the normal stress balance in setting up the problem, we must write Newton's law for the drop, given in Equation (4.1.20). When specialized to steady motion, this requires that the net force on the drop be set equal to zero. In the absence of gravity, this means that the hydrodynamic force on the drop is equal to zero. Examination of the result for the hydrodynamic force in Equation (3.1.30) shows that the constant D_2 in the series solution for the streamfunction ψ in Equation (3.1.9) must be set equal to zero. Use of this force balance, in conjunction with the boundary conditions, leads to the following final results for the fields and the scaled migration velocity of the drop:

$$\psi = A\left(r^2 - \frac{1}{r}\right)\sin^2\theta, \tag{4.2.18}$$

$$v_r = -2A\left(1 - \frac{1}{r^3}\right)\cos\theta, \tag{4.2.19}$$

$$v_\theta = 2A\left(1 + \frac{1}{2r^3}\right)\sin\theta, \tag{4.2.20}$$

$$p = p_\infty, \tag{4.2.21}$$

$$\psi' = \frac{3}{2}A(r^4 - r^2)\sin^2\theta, \tag{4.2.22}$$

$$v'_r = 3A(1 - r^2)\cos\theta, \tag{4.2.23}$$

$$v'_\theta = -3A(1 - 2r^2)\sin\theta, \tag{4.2.24}$$

$$p' = C_1 - 30\alpha Ar\cos\theta, \tag{4.2.25}$$

$$\mathbf{v}_\infty = 2A\mathbf{k}. \tag{4.2.26}$$

Here the constant A is given by

$$A = \frac{1}{(2 + 3\alpha)(2 + \beta)}. \tag{4.2.27}$$

The scaled pressure in the continuous phase is uniform, and its value is that existing at infinity. The constant C_1 appearing in the expression for the scaled pressure within the drop is related to p_∞ through the normal stress balance. We provide the connection in Section 4.3, where the normal stress balance is used to calculate small deformations from the spherical shape caused by inertia.

It is instructive to examine the result for the velocity in dimensional form, which we label $\mathbf{v}^*_\infty$ below:

$$\mathbf{v}^*_\infty = \frac{2(-\sigma_T)Rk}{(2\mu + 3\mu')(2k + k')}\nabla T_\infty. \tag{4.2.28}$$

The following inferences can be made from the expression for the velocity of the drop. Its dependence is linear in the applied temperature gradient and the radius of the drop as well as the rate of change of interfacial tension with temperature. As the viscosity of either the drop phase or that of the continuous phase is increased, the effect is to reduce the velocity of the drop. In the limit of infinite viscosity of the drop phase, the velocity approaches zero, and this corresponds to the case of a rigid sphere. Although a rigid sphere can move in a temperature gradient as discussed in Chapter One, it is not due to the thermocapillary effect. As the interior viscosity approaches zero, we observe that the drop approaches a limiting velocity that approximates that of a gas bubble. When the thermal conductivity of the drop phase becomes large, the effect is to reduce the variation of temperature on the drop surface, and this leads to a decrease in the velocity of the drop. In the limit as the ratio of the thermal conductivities, β, approaches infinity, the velocity of the drop approaches zero inversely as this ratio. It is commonly assumed that β can be set to zero in the gas bubble limit without too much error. Although this is true for some liquids, in the case of air bubbles in silicone oils, a system that has been used extensively in experiments, this ratio is approximately 0.2, and some error is incurred in setting it equal to zero. It is worth noting that the linear dependence on various parameters could have been inferred from the linearity of the governing equations and boundary conditions, as was done by Young et al.

The only limit that appears unrealistic is that when the viscosity of the continuous phase approaches zero. In this case, Equation (4.2.28) predicts a specific velocity of the drop in terms of the remaining parameters, when one would expect no motion of the drop. One might believe this anomaly is caused by the choice of the reference velocity used earlier because it becomes undefined in the limit $\mu \rightarrow 0$. This is not the root cause of the problem however, as is observed by retracing the analysis using an unspecified reference velocity. In reality, the hydrodynamic force on the drop, which includes both viscous and pressure contributions in the continuous phase, is proportional to μ, and when the viscosity approaches zero, that force approaches zero. The drop will experience no force from the continuous phase regardless of the motion that occurs at the interface and within the drop and therefore will not move.

We now make some comments about the fields. The velocity field in the continuous phase consists of a disturbance to the uniform stream that is equivalent to a flow caused by a potential dipole at the drop center. The uniform stream is merely a consequence of our choice of reference frame and will disappear if we return to the laboratory reference frame in which the fluid at infinity is quiescent. The potential dipole decays with distance from the drop as $\frac{1}{r^3}$. Therefore, when one moves away from the drop by approximately five radii, the disturbance velocity would be less than 1% of the velocity prevalent near the drop surface. This can be contrasted with the result obtained in Section 3.1 for gravity driven motion of a drop wherein the disturbance velocity is a Stokeslet decaying only as $\frac{1}{r}$ and therefore reaching approximately 1% of the values prevalent near the drop at 100 radii. The flow within the drop is a Hill's vortex that we have seen in the context of gravity driven motion of a drop in Section 3.1. The fields given in Equations (4.2.18) through (4.2.25) satisfy the normal stress balance condition. Therefore, the spherical shape, assumed initially, may be justified a posteriori. Further discussion of this topic can be found in Section 4.3.

Now, we provide illustrative streamlines in a reference frame attached to the moving drop, as well as in the laboratory reference frame. The viscosity and thermal conductivity ratios appear only in the multiplicative constant A in the results for the streamfunctions in Equations (4.2.18) and (4.2.22). Level curves of ψ/A outside the drop and ψ'/A within the drop are displayed in Figure 4.2.1 in a meridian plane, in a reference frame attached to the moving drop. Equal increments in ψ/A are used for displaying streamlines outside to a distance from the symmetry axis which is twice the radius. In the same way, the range between 0 (at the drop surface) and ψ'_{min} is divided into equal increments that are negative for displaying streamlines within the drop. The streamline structure is qualitatively similar to that observed in the case of body force driven motion. The streamfunctions in a laboratory reference frame can be obtained from Equations (4.2.18) and (4.2.22) by subtracting the quantity $Ar^2 \sin^2 \theta$ from the expressions on the right sides. The resulting streamlines in a meridian plane in the laboratory reference frame are displayed in Figure 4.2.2, using equal increments in the streamfunction. There is some similarity to the flow structure displayed in Figure 3.1.3 in the body force driven motion case in the fore and aft regions. Significant differences are observed in the

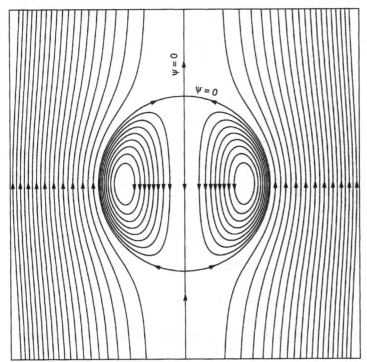

Figure 4.2.1 Streamlines in a reference frame attached to the drop, $\Delta(\frac{\psi}{A}) = \frac{7}{38}$ and $\Delta(\frac{\psi'}{A}) = -\frac{1}{24}$; $\psi = \psi' = 0$ on the surface of the drop, and $\Delta(\frac{\psi}{A})$ and $\Delta(\frac{\psi'}{A})$ are increments measured away from this surface.

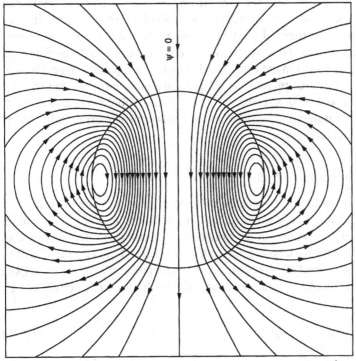

Figure 4.2.2 Streamlines in the laboratory reference frame, $\Delta(\frac{\psi}{A}) = \Delta(\frac{\psi'}{A}) = -\frac{1}{19}$; $\psi = \psi' = 0$ along the symmetry axis, and $\Delta(\frac{\psi}{A})$ and $\Delta(\frac{\psi'}{A})$ are increments measured away from this axis.

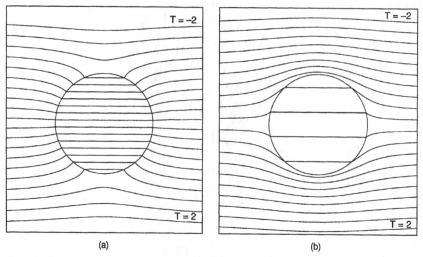

(a)   (b)

Figure 4.2.3   Isotherms (a) $\beta = 0$, (b) $\beta = 5$; $\Delta T = \Delta T' = \frac{4}{19}$.

equatorial region, however. A detailed discussion of these flow structures can be found in Section 4.17.

The temperature field in the continuous phase consists of a disturbance to the applied linear field which decays as $\frac{1}{r^2}$ with distance from the drop. Note that the disturbance temperature gradient falls off as $\frac{1}{r^3}$ just as the disturbance velocity field. These features will be useful when we attempt to understand the influence of neighboring surfaces and objects on the thermocapillary motion of a given drop. Sample isotherms are provided in Figure 4.2.3 for two cases illustrating the role of the ratio of thermal conductivities, β. The figure shows that when the thermal conductivity of the drop is negligible compared with that of the continuous phase, the temperature gradient that exists in the undisturbed fluid is sharpened near the drop. On the other hand, when the drop has a larger thermal conductivity than the continuous phase, this temperature gradient is weakened. The case $\beta = 0$ is commonly taken as an approximation for a gas bubble moving in a liquid. As pointed out earlier, caution must be used in using this approximation in liquids of relatively low thermal conductivity.

4.2.3   Inclusion of the Gravitational Effect

The above results can be extended to include the role of gravity in the limited sense mentioned earlier. Neglecting buoyant convection, in the linear limit considered here, we can simply superpose the solution of the gravitational settling problem analyzed in Section 3.1 with that of the thermocapillary migration problem. If the velocity of a drop when subjected to the combined effects of gravity and thermocapillarity is labeled U^*, using v_0 defined in Equation (4.1.1) as a reference velocity, the scaled velocity $\mathbf{U}$ can be written as

$$\mathbf{U} = \frac{2}{2+3\alpha}\left[\frac{1}{2+\beta}\mathbf{k} - (1+\alpha)G\mathbf{e}\right], \tag{4.2.29}$$

where the dimensionless group G is given by

$$G = \frac{gR(\rho' - \rho)}{3\sigma_T|\nabla T_\infty|}. \tag{4.2.30}$$

Here, the unit vector **e** points in the direction of gravity. One might think of the dimensionless group G as a dynamic version of the Bond number, a term suggested by Ostrach (1982). The Bond number appears in problems involving the static shapes of drops and bubbles when they are distorted by the action of the unequal variation of the hydrostatic pressure on either side of an interface. The group G arises naturally in problems wherein motion in the fluid is driven by interfacial tension gradients and gravity plays a role in affecting this motion. It reflects the relative importance of the gravitational effect to the thermocapillary effect. An analogous group can be coined when the interfacial tension gradients are the consequence of composition variations.

One way to minimize buoyant convection is to arrange for a vertical temperature gradient in experiments. An upward gradient would place warm fluid above cool fluid and lead to a stable situation, assuming the density of the liquid decreases with increasing temperature. On the other hand, a downward temperature gradient will normally lead to unstable stratification of the fluid. In this case, if a group known as the *Rayleigh number* exceeds a certain critical value, buoyant convection will occur spontaneously. But judiciously applied downward temperature gradients can be used in experiments. For example, gas bubbles would normally rise because of buoyancy in a liquid but can be forced downward by the application of such a gradient as demonstrated experimentally by Young et al. (1959). For a downward temperature gradient, the unit vectors **e** and **k** are identical, and we can write

$$\mathbf{U} = \frac{2}{2 + 3\alpha}\left[\frac{1}{2 + \beta} - (1 + \alpha)G\right]\mathbf{k}. \tag{4.2.31}$$

Note that G is positive if the density of the drop phase is smaller than that of the continuous phase because σ_T is negative in most common systems. In this case, the drop would be stationary in Earth's gravitational field if

$$G = \frac{1}{(1 + \alpha)(2 + \beta)} \tag{4.2.32}$$

and would sink in the fluid instead of rising if G is smaller than the right side in Equation (4.2.32). Naturally, even when the drop is stationary, the fluid surrounding it and within it is not at rest. A detailed discussion of the flow patterns in problems involving both gravity and thermocapillarity can be found in Section 4.17.

Now we consider the stability of the stationary position of a drop. The position of a drop is neutrally stable with regard to horizontal displacements because such displacements will cause no change in the balance of forces on the drop. Nonetheless, it is not possible to make a drop stationary if slight vertical displacements from its stationary location would cause it to move away from that location. Let us consider a gas bubble in a downward temperature gradient. Even if conditions are such that $U = 0$ at a certain location in the liquid, a slight downward movement of the bubble will move it into slightly warmer liquid, which would be slightly less dense. Acting alone, this effect would be destabilizing. This is because the reduced density of the continuous phase implies a slightly smaller buoyant force, whereas the thermocapillary contribution stays unchanged. The consequence would be further downward migration of the bubble. Similarly, a slight upward movement of the bubble also would lead to continued movement in that direction. This argument assumes that the radius of the bubble does not change because of its displacement. In fact, it will change both because the hydrostatic pressure in the liquid will be different at the new location and because the temperature of the bubble will change to the value at the new location. First consider a downward displacement. The reduction in size due to the increase in pressure will be virtually immediate. On the other hand, the temperature of the bubble will increase, causing a corresponding increase in its radius over a time scale of the order of thermal diffusion time over the radius of the bubble, assuming conduction is the dominant mechanism for heat transport within the bubble. We can calculate the change in radius assuming the ideal gas law to hold. If the radius of the bubble at the new location is larger than the value at the previous location, the relative importance of buoyancy compared with the thermocapillary effect will be larger at the new location, providing a driving force for moving the bubble back toward its original location. If this contribution is larger than that from the reduced density of the liquid that acts to destabilize, the net result would be to make the position where $U = 0$ locally stable to vertical perturbations. A similar argument can be made for an infinitesimal excursion upward. By considering the force balance on the bubble, for an infinitesimal vertical displacement in either direction, we can show that the quantity

$$\delta F = \zeta + \frac{1}{3}\left(\frac{\rho g}{p^*|\nabla T_\infty|} - \frac{1}{T^*}\right) \tag{4.2.33}$$

needs to be negative to provide local stability. Positive values of this quantity would result in instability. Here, T^*, ρ, and p^* represent the absolute temperature of the liquid, its density, and its pressure, respectively, all at the initial location of the bubble. The symbol ζ is used to designate the coefficient of thermal expansion of the liquid, defined as

$$\zeta = -\frac{1}{\rho}\frac{d\rho}{dT^*}. \tag{4.2.34}$$

Because the coefficient of thermal expansion of liquids is generally positive, the influence of the density change of the continuous phase is to introduce instability. Also, the change in radius due to a change in hydrostatic pressure will have the same effect. Therefore, the instantaneous effect of moving the bubble slightly away from its stationary location is to cause a destabilizing influence. Countering this is the contribution from the change in temperature at the new location to the size of the bubble. If the time taken for the temperature change to occur is small compared with the time it takes for the bubble to move appreciably, this effect must be taken into account in determining the answer to the stability question. Using values of $\rho \approx 1000$ kg/m³, $T^* = 300$ K, $p = 101$ kPa, $g = 9.81$ m/s², and $\zeta \approx 7.5 \times 10^{-4}$ (K)$^{-1}$ for silicone oils, we find that the stationary location will be unstable for $|\nabla T_\infty| < 0.09$ K/mm. If the gradient is larger, stability is possible. It cannot be immediately established whether the position would be locally stable, because one has to consider the kinetic question of how rapidly the temperature of the gas in the bubble will change.

On a larger time scale, dissolution and growth of the bubble will influence its motion. To see this, assume that the entire liquid body is saturated with the gas in question at the temperature prevailing at the stationary location. Then, the liquid will be supersaturated at temperatures above this value and undersaturated at temperatures below it. If we consider a downward displacement from an equilibrium position, the gas from the supersaturated liquid will enter the bubble, making it larger. The opposite holds for upward displacements. The dramatic role of this effect in affecting the motion of a gas bubble in a downward temperature gradient on a relatively long time scale is examined in Section 4.16, where we discuss results from experiments.

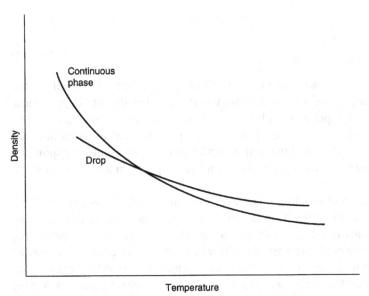

Figure 4.2.4 Sketch showing the qualitative behavior of the densities of the drop and the continuous phase plotted against temperature.

When a drop has the same density as the continuous phase, gravity will not cause it to move because the net hydrostatic force on it is zero. It also is possible for a drop to attain a specific stationary location in a continuous phase when subjected to a temperature gradient. Imagine a drop phase with a density that matches that of the continuous phase at a certain temperature, but decreases with increasing temperature at a lower rate than that of the continuous phase. Figure 4.2.4 qualitatively illustrates the situation.

In this case, one can establish an upward temperature gradient in the continuous phase so that $\mathbf{k} = -\mathbf{e}$. The scaled velocity of the drop is given by

$$\mathbf{U} = \frac{2}{2+3\alpha}\left[\frac{1}{2+\beta} + (1+\alpha)G\right]\mathbf{k}. \tag{4.2.35}$$

If the drop is introduced at a location where the temperature is below the value at which the densities are matched, the drop will move up because the net hydrostatic force on it will be upward ($\rho' < \rho$ leading to a positive value of G) and because the thermocapillary effect also will propel it in that direction. When it reaches the location where its density matches that of the continuous phase $(G = 0)$, the hydrostatic force on it will be zero. The thermocapillary effect will cause the drop to continue to move upward into warmer fluid, however. Now the drop becomes more dense than the continuous phase, resulting in a net hydrostatic force downward because $\rho' > \rho$, implying a negative value of G. Ultimately, if the density difference continues to increase with increasing temperature, one can see that the drop will reach a stationary position. This was experimentally demonstrated first by Delitzsch, Eckelmann, and Wuest (1984). Subsequently, others have investigated the phenomenon. It is worthwhile to obtain a result for the density difference required to hold the drop stationary in this situation. This can be done by setting the magnitude of U equal to zero in Equation (4.2.35), which yields the following result:

$$\rho' - \rho = 3\frac{(-\sigma_T)|\nabla T_\infty|}{(1+\alpha)(2+\beta)gR}.$$ (4.2.36)

Therefore, the smaller the radius of the drop, the larger will be the density difference needed to stop its motion. If we assume a linear variation of density with temperature, given the linear applied temperature field, we see that the distance above the neutral density location occupied by a given drop is inversely proportional to its radius. This presumes that the average density of the drop at any given location is that corresponding to the temperature existing in the undisturbed continuous phase in a horizontal plane that bisects the drop.

The stationary location of the drop is stable to vertical perturbations, provided that the two fluids are immiscible. If the drop moves a bit above this location, the increased density difference will act to pull it down to that location; if it moves a bit below, the hydrostatic force from the decreased density difference will be smaller than that needed to balance the upward force on the drop from the thermocapillary effect, causing the drop to go back up to the stationary location. If the two phases are soluble in each other to some extent, size and density changes that occur because of mass transfer effects between the drop and the continuous phase will need to be accommodated.

As a corollary, when the density of a drop decreases with increasing temperature more rapidly than that of a continuous phase, a downward temperature gradient can be used to position drops based on size at different vertical locations below the neutral density location. This is a stable situation like the one considered above. If the temperature gradient were to be reversed in either case, however, the situation would become unstable because moving the drop slightly away from the stationary location would lead to a change in the density difference that would cause its continued movement away from the stationary location.

Two stable and two unstable cases also can be identified when the interfacial tension increases with increasing temperature, leading to a positive σ_T. When the density of a drop decreases less rapidly with increasing temperature than that of a continuous phase, a drop again would be able to achieve a stable stationary location in an upward temperature gradient. Here, because thermocapillarity will cause drops to migrate into cooler fluid, movement from the neutral density location would be downward into cooler regions wherein the net gravitational force on the drop would now resist this motion until it becomes sufficiently large to arrest it altogether. For the reverse case where the density of a drop decreases more rapidly than that of the continuous phase, a stable stationary location can be found in a downward temperature gradient. The corresponding cases with temperature gradients in the opposite direction in each case would then lead to unstable stationary locations.

Note that the quantitative results given here apply only when the assumptions made earlier are valid. Nonetheless, even in physical situations where these assumptions do not hold (for example, when convective transport effects are not negligible), one can move gas bubbles downward by applying a sufficiently large temperature gradient, or position drops at stable stationary locations. In fact, a substantial amount of the experimental work on drops reaching stationary locations has been performed under conditions wherein convective transport is important, so that the above theoretical results cannot be used in interpreting them. Examples can be found in the work of

Wozniak (1986), Maris, Seidel, and Williams (1987), Hähnel, Delitzsch, and Eckelmann (1989), Rashidnia and Balasubramaniam (1991), Chen et al. (1997), and Ma (1998).

4.3 Motion When Inertia Is Important, but Convective Transport of Energy Is Negligible

4.3.1 Introduction

In the previous section, we assumed all convective transport to be negligible. It so happens that one can easily deal with the problem when inertia is important but with an interfacial tension gradient that is still proportional to $\sin \theta$ on the surface of a sphere. This is the case when we make the assumptions given in Section 4.1 and add the assumption of steady thermocapillary migration when convective transport of energy is negligible, that is, $Ma = Ma' = 0$. As observed in Section 4.2, in this limit, the temperature fields within and outside the drop satisfy Laplace's equation, and the solution is decoupled from that of the momentum equation. When the Reynolds number is not negligible, the assumption that the Marangoni number is negligible implies a fluid of low Prandtl number. Therefore, the analysis presented in this section is useful only when both the drop phase and exterior fluid have relatively large thermal diffusivities compared with their respective kinematic viscosities, if the source of the surface tension gradient is the temperature variation on the interface. Of course, the results can be used where such a gradient arises for other reasons, such as from a surfactant concentration gradient. For example, several problems of that type are considered in the book by Levich (1962) in the context of Stokes motion of a drop. We show here how the results can be used to calculate slight deformations from the spherical shape when they are caused by inertia. Finally, we provide a result from Haj-Hariri, Nadim, and Borhan (1990) for the correction to the drop migration velocity when such shape deformation is accommodated.

4.3.2 Calculation of Small Departures of the Shape from a Sphere Due to Inertia

Inertial effects can be handled with ease under the stated assumptions because the solution for the velocity and streamfunction fields given in Section 4.2 for the purely thermocapillary migration of a spherical drop happens to satisfy the complete Navier–Stokes equation including inertia. This was first noted by Crespo and Manuel (1983) for the solution in the continuous phase, which is the flow induced by a potential dipole located at the center of the drop. Crespo and Manuel made their observations in the context of the gas bubble limit, wherein the solution within the drop is not used. As mentioned earlier, the solution for the interior flow is Hill's spherical vortex, which, although not irrotational, also satisfies the complete Navier–Stokes equation. The fact that the fields both inside and outside a spherical drop, obtained neglecting inertia, satisfy the complete Navier–Stokes equations was independently discovered by Balasubramaniam and Chai (1987). More recently, the analysis has been extended to include a calculation of the effect of the shape deformation on the migration velocity of the drop by Haj-Hariri et al. (1990). These authors confirmed the results reported by Balasubramaniam and Chai and also gave their results in compact invariant form.

The streamfunction and velocity fields both outside the drop and within it are given in Equations (4.2.18) through (4.2.20) and (4.2.22) through (4.2.24). One can use these fields in the Navier–Stokes equations in the drop and continuous phase fluids to calculate the pressure fields. These are reported below. For convenience, we continue to use the symbol A as defined in Equation (4.2.27).

$$p = p_\infty + 2Re\,A^2\left[1 - \left(1 - \frac{1}{r^3}\right)^2 \cos^2\theta - \left(1 + \frac{1}{2r^3}\right)^2 \sin^2\theta\right] \qquad (4.3.1)$$

$$p' = C_1 - 30\alpha Ar\cos\theta + 9A^2\alpha Re'\left[\frac{1}{2}(r^4 - r^2)\sin^2\theta + \left(r^2 - \frac{r^4}{2}\right)\cos^2\theta\right] \quad (4.3.2)$$

The inertial contributions to the pressure fields reported above do not affect the net force on the drop. Therefore, provided the shape of the drop is a sphere, the migration velocity given in Equation (4.2.26) will continue to hold. The shape is indeed not a sphere when inertial corrections to the pressure are included. Later in the present section, we provide a result that permits the correction to the migration velocity to be calculated when inertial effects are small.

With the above pressure fields, one can verify that the balance of normal stress is not satisfied in general at a spherical boundary when the Reynolds number is not negligible. The exception occurs when the densities of the drop and the continuous phase are equal. The normal stress balance in Equation (4.1.17) is then satisfied exactly. This determines the unknown constant in Equation (4.3.2) to be

$$C_1 = p_\infty + \frac{2}{Ca} - \frac{5}{2}A^2\,Re. \qquad (4.3.3)$$

When the Reynolds number is set equal to zero, the spherical boundary is the correct shape for any value of the density ratio $\gamma = \frac{\rho'}{\rho}$. In that case, the above result reduces to

$$C_1 = p_\infty + \frac{2}{Ca}, \qquad (4.3.4)$$

yielding the result for C_1 applicable to the situation considered in Section 4.2. The Capillary number involves σ_0 in its definition. This is the value of the interfacial tension evaluated at the temperature far from the drop on the isothermal plane that passes through the location of the drop center. Because the drop continues to move into fluid of a different temperature, this interfacial tension will change with time. When the drop is a sphere, the curvature is uniform on its surface, and a decrease of the interfacial tension by the same extent everywhere on the surface will lead to a decrease in the pressure within the drop with time. Because this time dependence appears only in C_1, spatial gradients of the pressure are unaffected, and the steady solution for the velocity fields is correct.

The situation is not as simple when the drop is not spherical because the curvature varies over the surface. In that case, a decrease of the interfacial tension with time by the same extent everywhere on the surface of the drop will lead to time-dependent changes in the local curvature. This in turn will cause the shape of the drop to vary with time. A quasi-steady shape can still be calculated for a drop using the local value of the interfacial tension σ_0 corresponding to the instantaneous location of the drop. Such an approximation will become worse with increase in the magnitude of the rate of change of interfacial tension with temperature. We now proceed to obtain a result for small deformations from the spherical shape in this quasi-steady sense.

When the densities of the drop and the continuous phase are unequal, the drop will deform from the spherical shape as a consequence of inertial effects. Because the fields inside and outside have been calculated by using a spherical boundary for applying the boundary conditions, one might wonder if they are correct. The approach we take is that the deviations from a spherical shape are small. Therefore, the shape can be expressed in the form of a small perturbation from that of a sphere. If the radius of a sphere of equivalent volume is R_0, and the actual radius is $R^*(\theta)$, the dimensionless radius $R(\theta) = \frac{R^*}{R_0}$ can be written as

$$R(\theta) = 1 + Ca\, Re\, F(\theta) = 1 + We\, F(\theta) = 1 + f(\theta), \tag{4.3.5}$$

where we have introduced the function $f(\theta) = We\, F(\theta)$ for convenience in the subsequent development. Note the appearance of both the Capillary number and the Reynolds number in front of the correction to the spherical shape. If the Reynolds number is zero, the drop will retain its spherical shape regardless of the value of the Capillary number. Normally, the role of the interfacial tension is to resist deformation while a jump in normal stress (that includes both viscous normal stress and pressure) across the interface, which varies with position on the interface, acts to produce variations in curvature and, therefore, deformation from the spherical shape. In the zero Reynolds number limit, however, regardless of the value of the interfacial tension, there is no tendency for the drop to deform. It is only when inertia is included that the normal stress balance is not satisfied by the spherical shape. Therefore, the variation from a spherical shape can be expected to be small for small Reynolds number. Now, for any Reynolds number, one can expect the deformation to be small for small values of the Capillary number. It is logical that the correction to the spherical shape is proportional to the product of the Reynolds and Capillary numbers, which is the Weber number as mentioned in Chapter Two.

Now, we can apply the normal stress balance given in Equation (4.1.17). The unit normal vector to the surface may be obtained from Equation (4.3.5) as follows:

$$\mathbf{n} = \frac{(1+f)\,\mathbf{i}_r - f'\mathbf{i}_\theta}{\sqrt{1 + 2f + f^2 + (f')^2}}. \tag{4.3.6}$$

In the above, the prime refers to differentiation with respect to θ. Now, we can calculate $\nabla \bullet \mathbf{n}$ in a straightforward manner. Because we assume the deformation to be small, it is appropriate to neglect terms involving products of f and its derivatives and powers of these quantities. The linearized result for the curvature is

$$\nabla \bullet \mathbf{n} \sim 2 - 2f - f'\cot\theta - f''. \tag{4.3.7}$$

The normal stress difference in the left side of Equation (4.1.17), evaluated at the spherical surface $r = 1$, can be written as $-p + \tau_{rr} + p' - \tau'_{rr}$, and this can be worked out from the solution for the fields given in Section 4.2. The scaled interfacial tension $\sigma = 1 - Ca\, T(1, \theta)$, which permits the right-hand side to be expressed in terms of the results for the temperature field on the surface. Using all of the above information, the normal stress balance finally yields the following result:

$$\left[\frac{1}{Ca} - \frac{3}{2+\beta}s\right]\left[\frac{d}{ds}(1-s^2)\frac{df}{ds} + 2f\right] = \left[p_\infty + \frac{2}{Ca} - A^2 Re\left(1 + \frac{3}{2}\gamma\right) - C_1\right]$$

$$+ 6\left[A(2+3\alpha) - \frac{1}{2+\beta}\right]P_1(s)$$

$$+ 3A^2 Re(1 - \gamma)P_2(s). \tag{4.3.8}$$

As before, the symbol $s = \cos\theta$. The viscosity ratio $\alpha = \frac{\mu'}{\mu}$, and the thermal conductivity ratio $\beta = \frac{k'}{k}$. When the Capillary number is very small, the term $\frac{3}{2+\beta}s$ will be negligible compared with $\frac{1}{Ca}$. This is equivalent to the statement that in this case, the variation of the interfacial tension over the drop surface $(\Delta\sigma^*)$ is small compared with its equatorial value (σ_0). The resulting differential equation for the function $f(\theta)$ is a nonhomogeneous version of the governing equation for the first Legendre Polynomial $P_1(s)$. A solution can be constructed in a straightforward manner as shown by Brignell (1973). First, certain constraints must be applied. The following result may be written from the specification of the volume of the drop as $\frac{4}{3}\pi R_0^3$:

$$\int_{-1}^{+1} R^3(s)\,ds = 2. \tag{4.3.9}$$

Inserting the result for $R(\theta)$ in the above and linearizing in f leads to

$$\int_{-1}^{+1} f(s)\,ds = 0, \tag{4.3.10}$$

where, for convenience, we have retained the same symbol f for the function of s that is equivalent to $f(\theta)$. Equation (4.3.10) implies that the solution cannot contain any contribution from $P_0(s)$, meaning that the constant term in the right side of Equation (4.3.8) must be zero. Therefore, the constant C_1 can be written as

$$C_1 = p_\infty + \frac{2}{Ca} - A^2 Re\left(1 + \frac{3}{2}\gamma\right). \tag{4.3.11}$$

Note that this reduces to the result in Equation (4.3.3) when the density ratio $\gamma = 1$.

The second condition is that the solution must be bounded along $\theta = 0$ and π. This eliminates a contribution to the solution, which is a constant times $[P_1(s)Q_0(s) - 1]$, where $Q_0(s)$ is the Legendre function of the second kind of order zero. Furthermore, it also requires that the coefficient of $P_1(s)$ in the right side of Equation (4.3.8) must be zero because such a term causes resonance, and the solution, which is a constant times $[P_1(s)\log(1 - s^2)]$, will be unbounded along $\theta = 0$ and π. This requirement is already met here, which can be verified by substituting for the constant A from Equation (4.2.27). If a force balance had not been used on the drop to infer its velocity as was done in Section 4.2, this new result can be used to obtain A and, therefore, the velocity v_∞. This illustrates that the use of the normal stress balance makes the use of Newton's law for the drop redundant.

The next constraint is that the origin of coordinates is the center of mass of the drop. This leads to the following result:

$$\int_{-1}^{+1} (1 + f)^4 P_1(s)\,ds = 0. \tag{4.3.12}$$

Upon linearization as before, the constraint can be rewritten as

$$\int_{-1}^{+1} f P_1(s)\,ds = 0. \tag{4.3.13}$$

This implies that there can be no contribution to the solution for $f(s)$ of the form $DP_1(s)$ where D is a constant. Now, we can finally obtain the solution for $f(s)$, which only contains a contribution from the $P_2(s)$ mode to the deformation:

$$f(s) = \frac{3}{4} A^2 \, We(\gamma - 1) P_2(s).$$
(4.3.14)

Using the relation $f(s) = We \, F(\theta)$ and the definition of the second Legendre Polynomial, the following result can be written for the function $F(\theta)$, which describes the shape deformation:

$$F(\theta) = \frac{3}{8} A^2 (\gamma - 1)(3\cos^2\theta - 1).$$
(4.3.15)

The shape of the drop is a spheroid. The ratio S of the semiaxes of the spheroid in the direction of migration and normal to it, at this order of approximation, is

$$S = \frac{R(0)}{R(\frac{\pi}{2})} = 1 + \frac{9}{8} A^2 \, We(\gamma - 1).$$
(4.3.16)

A gas bubble moving in a liquid corresponds to a negligible value of γ. Therefore, it will contract in the direction of the temperature gradient, making its shape oblate. In fact, the oblate shape will be achieved by any drop that is less dense than the continuous phase. If the drop is more dense than the continuous phase, however, it should take on a prolate shape. These predictions are in contrast to those made in Section 3.2 regarding deformation due to inertia when a drop moves because of the action of a body force. In that case, in principle, it is possible for the drop to take on either a prolate or an oblate shape. For most physically realizable situations, however, Taylor and Acrivos (1964) showed that the a drop should become an oblate spheroid due to inertial deformation, regardless of whether γ is less than or greater than unity.

4.3.3 Correction to the Migration Velocity Due to Deformation

It is possible to obtain the migration velocity of the deformed drop when inertial effects are small. In this case, the question is one of calculating the migration velocity of a slightly deformed spherical drop in Stokes flow. This can be accomplished without solving for the detailed fields by the use of the reciprocal theorem. This well-known result, which holds for Stokes flows, is attributed to Lorentz (1907); some generalizations have been given by Happel and Brenner (1965) and Leal (1980). It has been applied in the present context by Haj-Hariri et al. (1990), and the reader is encouraged to consult this reference for details.

The original Lorentz reciprocal theorem, as given by Happel and Brenner (1965), may be stated as follows. Let $\mathbf{\Pi}_i$ and $\mathbf{v}_i (i = 1, 2)$ be the stress tensor and velocity vector fields, respectively, corresponding to any two incompressible motions of a given fluid that satisfy Stokes's equation. The following result holds for an arbitrary closed surface S bounding any fluid volume:

$$\int_S \mathbf{n} \bullet \mathbf{\Pi}_2 \bullet \mathbf{v}_1 \, dS = \int_S \mathbf{n} \bullet \mathbf{\Pi}_1 \bullet \mathbf{v}_2 \, dS.$$
(4.3.17)

The surface S can consist of a number of distinct surfaces apart from each other and the important idea is that the set of surfaces completely enclose a fluid volume. The unit vector $\mathbf{n}$ points outward from the volume enclosed by the surface. Here, the surface S is taken to be made up of a spherical boundary S_∞ with a large radius R_∞, which we can later permit to become arbitrarily large, and an internal boundary S_D that coincides with the surface of a drop. Together, S_∞ and S_D enclose a region occupied by the continuous phase, and Equation (4.3.17) can be rewritten as follows:

$$\int_{S_\infty} \mathbf{n} \bullet \mathbf{\Pi}_2 \bullet \mathbf{v}_1 \, dS - \int_{S_D} \mathbf{n} \bullet \mathbf{\Pi}_2 \bullet \mathbf{v}_1 \, dS = \int_{S_\infty} \mathbf{n} \bullet \mathbf{\Pi}_1 \bullet \mathbf{v}_2 \, dS - \int_{S_D} \mathbf{n} \bullet \mathbf{\Pi}_1 \bullet \mathbf{v}_2 \, dS.$$

(4.3.18)

In this result, $\mathbf{n}$ is the unit normal vector pointing outward from the drop on S_D (explaining the origin of the negative sign in front of the integrals over S_D) and similarly outward from the spherical surface S_∞. As $R_\infty \to \infty$, the integrals over S_∞ can be simplified. Assume uniform streaming at infinity in both problems with $\mathbf{v}_1 \to \mathbf{U}_1$, and similarly, $\mathbf{v}_2 \to \mathbf{U}_2$ where $\mathbf{U}_1$ and $\mathbf{U}_2$ are spatially constant vectors. The integral of $(\mathbf{n} \bullet \mathbf{\Pi}_i)$ over S_∞ can be replaced by the integral of the same quantity over the surface S_D for $i = 1, 2$. This is possible because the divergence of the stress tensor $\mathbf{\Pi}$ in Stokes flow is zero everywhere, and therefore its volume integral is zero in any volume occupied by a given fluid. By applying the divergence theorem, this volume integral can be converted to the integral of $(\mathbf{n} \bullet \mathbf{\Pi})$ over the surface bounding this volume. Therefore, one is led to the stated result. Because

$$\int_{S_D} \mathbf{n} \bullet \mathbf{\Pi}_i \, dS = \mathbf{F}_i,$$

(4.3.19)

where $\mathbf{F}$ is the force exerted by the continuous phase on the drop and the subscript corresponds to cases $i = 1, 2$, we obtain the following useful result:

$$\mathbf{U}_2 \bullet \mathbf{F}_1 - \mathbf{U}_1 \bullet \mathbf{F}_2 = \int_S \mathbf{n} \bullet (\mathbf{\Pi}_1 \bullet \mathbf{v}_2 - \mathbf{\Pi}_2 \bullet \mathbf{v}_1) \, dS.$$

(4.3.20)

If subscript 1 corresponds to a model problem in which the fields and the force on the drop when it moves at a velocity $\mathbf{U}_1$ are known, it is possible to use the above equation to obtain a connection between the force and the velocity in the second problem. The velocity and stress fields at the interface that are required in the integrals can be obtained by using the boundary conditions at the interface in conjunction with the reciprocal theorem applied to the fluid in the interior of the drop.

The above development becomes useful only if we know the solution for a related problem with the same boundary. The boundary of the drop in which we are interested is deformed. To get around this difficulty, one would expand the field variables in a Taylor series about the spherical boundary, keeping the first nonzero terms. This would permit boundary conditions to be written at a spherical boundary for the deformed drop case. Now, Equation (4.3.20) can be applied by choosing the known problem to be the one we analyzed in Section 3.1, namely the Hadamard-Rybczynski problem. In this case, the velocity and stress fields, as well as the force on the drop, can be written in terms of its velocity. Because the force on the drop in the thermocapillary problem is zero, Equation (4.3.20) can be used to calculate the velocity of the slightly deformed drop. The actual calculations are lengthy, and we simply report the result provided by Haj-Hariri et al. (1990). If the scaled migration velocity v_∞ is written as

$$v_\infty = v_{\infty_0} + We \, v_{\infty_1},$$

(4.3.21)

the leading order result v_{∞_0} is that given in Equation (4.2.26), and the correction v_{∞_1} is given by the authors as

$$v_{\infty_1} = \frac{3}{5} A^3 (\gamma - 1) \left(1 + \frac{15}{2 + 3\alpha} - \frac{9}{2 + \beta} \right). \tag{4.3.22}$$

The sign of this correction depends on the values of the parameters. For instance, in the case of a gas bubble, γ, α, and β may all be set to zero, leading to a reduction in the migration velocity. In the general case, however, a drop less dense than the continuous phase will move less rapidly than predicted by the leading order theory for a given viscosity ratio if the thermal conductivity ratio is sufficiently large. The opposite is true for drops more dense than the continuous phase.

In closing, we observe that deformation problems normally are difficult to solve because the shape of the boundary needs to be determined as part of the solution, and yet the boundary conditions must be applied at this "unknown" location. One generally resorts to numerical methods because of this difficulty, but even these are not straightforward to implement. It is only in the case of small deformations from a simple shape that the type of perturbational approach that we have illustrated here can be used. Its value lies in the fact that the nature of the shape change (and in this problem, the sign of the correction to the migration velocity) can be predicted without excessive labor.

4.4 The Case of an Arbitrary Temperature Gradient

4.4.1 Introduction

It is possible to obtain some results in the case when the applied temperature field is arbitrary and not necessarily linear in distance. We can only consider this case in the Stokes limit, namely, $Re = 0$. In this situation, we shall write a result for the hydrodynamic force exerted on a spherical drop held fixed at a certain location, regardless of the value of the Marangoni number. This can be done without recourse to the full solution of the governing equations. The result will be in terms of the interfacial tension gradient on the drop surface, however, which must be known. When the Marangoni number is negligible, we can proceed further and write a result for the force in terms of the undisturbed temperature gradient evaluated at the location of the center of the drop. In this section, physical variables will be used throughout for convenience. The undisturbed temperature gradient field in the continuous phase in the absence of the drop is designated by $\nabla T_\infty(\mathbf{x})$.

4.4.2 Analysis

The treatment, detailed by Subramanian (1985), depends on the use of the Lorentz reciprocal theorem, which was introduced in Section 4.3. According to the theorem, if $(\mathbf{v}_1, \Pi_1)$ and $(\mathbf{v}_2, \Pi_2)$ are solutions for the velocity and stress fields for two incompressible flows of the same fluid that satisfy Stokes's equation, and a surface S completely encloses an arbitrary region R occupied by the fluid,

$$\int_S d\mathbf{S} \bullet \Pi_2 \bullet \mathbf{v}_1 = \int_S d\mathbf{S} \bullet \Pi_1 \bullet \mathbf{v}_2. \tag{4.4.1}$$

Here, $d\mathbf{S}$ is the area element oriented in the direction of the normal pointing outward from the region R. We now define the region R to be bounded within by the surface of a drop of arbitrary shape and on the outside by an arbitrarily large spherical boundary. When the fluid at infinity is quiescent, the velocity field $\mathbf{v}$ must decay at least as rapidly as the inverse of the distance from the drop, and the stress will decay as the inverse square of this distance. Therefore, Equation (4.4.1) can be rewritten as

$$\int_{S_D} d\mathbf{S} \bullet \Pi_2 \bullet \mathbf{v}_1 = \int_{S_D} d\mathbf{S} \bullet \Pi_1 \bullet \mathbf{v}_2, \tag{4.4.2}$$

where S_D stands for the surface of the drop and $d\mathbf{S}$ points into the continuous phase fluid. This result is independent of the boundary conditions used at the surface S_D and holds whether the object is a rigid body or a drop, stationary, or in quasi-steady motion. The only condition imposed in addition to Stokes flow is that the fluid at infinity be quiescent.

The fields designated with subscripts 1 and 2 above are assumed to be caused by two identically shaped drops with different interfacial tension gradients $\nabla_s \sigma_1$ and $\nabla_s \sigma_2$ on their surfaces, respectively. The drops are held fixed in space. Use of the boundary conditions pertinent to this situation permits one to obtain the following variant of the reciprocal theorem for problems involving motion driven by capillarity, in which the symbol dS represents the magnitude of the area element $d\mathbf{S}$:

$$\int_{S_D} \nabla_s \sigma_2 \bullet \mathbf{v}_1 \, dS = \int_{S_D} \nabla_s \sigma_1 \bullet \mathbf{v}_2 \, dS. \tag{4.4.3}$$

To proceed further, we need to restrict the analysis to drops of spherical shape. We shall use the following result obtained by Brenner (1964) for the hydrodynamic force on a sphere of radius R, placed in a quiescent fluid of viscosity μ, in terms of the velocity field on the surface, S, of the sphere:

$$\mathbf{F} = -\frac{3\mu}{2R} \int_S \mathbf{v} \, dS. \tag{4.4.4}$$

Consider the flow driven by an interfacial tension gradient, which is the tangential component on the drop surface of an arbitrary but constant spatial vector. We choose this to correspond to case 1 in Equation (4.4.3), and choose case 2 to be that of a drop for which the interfacial tension gradient on the surface is prescribed. With these choices, we can obtain the hydrodynamic force on a spherical drop in terms of the interfacial tension gradient on its surface using Equation (4.4.4):

$$\mathbf{F} = -\frac{1}{2(1+\alpha)} \int_S \nabla_s \sigma \bullet \mathbf{W} \, dS. \tag{4.4.5}$$

Here, α stands for the ratio of the viscosity of the drop phase to that of the continuous phase, and $\mathbf{W}$ can be shown to be a tensor that maps spatial vectors into their projections onto the tangent plane at every point on the drop surface. Because the interfacial tension gradient already lies on the tangent plane, $\mathbf{W}$ leaves it unaffected, thereby leading to the following result:

$$\mathbf{F} = -\frac{1}{2(1+\alpha)} \int_S \nabla_s \sigma \, dS. \tag{4.4.6}$$

This tells us that the hydrodynamic force on a drop held fixed can be calculated if the interfacial tension distribution on its surface is known. Note that the force is independent of the viscosity of either the continuous phase or the drop phase, instead depending only on the ratio of these two viscosities, and can be calculated without solving for the velocity fields.

Because the variation of temperature on the interface is the cause of the gradient in interfacial tension, the above result can be rewritten as follows:

$$\mathbf{F} = -\frac{\sigma_T}{2(1+\alpha)} \int_S \nabla_s T \, dS. \tag{4.4.7}$$

Here, we have assumed that σ_T is constant as mentioned in Section 4.1. If convective heat transfer is assumed negligible, both within the drop and outside, the temperature fields will satisfy Laplace's equation. In this case, it can be shown that

$$\int_S \nabla_s T \, dS = -\frac{8\pi R^2}{2+\beta} (\nabla T_\infty)_0, \tag{4.4.8}$$

where β is the ratio of the thermal conductivity of the drop to that of the continuous phase and the subscript 0 on the undisturbed temperature gradient stands for evaluation at the center of the drop. Using Equation (4.4.8) in Equation (4.4.7) leads to the following expression for the force on the drop:

$$\mathbf{F} = -\frac{4\pi R^2 \sigma_T}{(1+\alpha)(2+\beta)} (\nabla T_\infty)_0. \tag{4.4.9}$$

This is precisely the result one would obtain in the problem solved in Section 4.2 if the drop were immobilized by use of an external force. An example would be to hold it on a wire. When the cause of the interfacial tension gradient is a species concentration field that satisfies Laplace's equation within and outside the drop, the result can be easily modified. In this case, one would multiply β in the above equation by an equilibrium constant K, which is the ratio of the concentration in the drop phase to that in the continuous phase at the interface S, then replace the rate of change of interfacial tension with temperature with a similar rate of change with concentration, and replace the temperature gradient by the concentration gradient. In the linear problem considered in obtaining Equation (4.4.9), a simpler route could have been used. Given the isotropy of the sphere, it can be shown that the force has to depend linearly on $(\nabla T_\infty)_0$, and the constant of proportionality can be obtained from the simpler problem involving a uniform temperature gradient in the undisturbed continuous phase fluid. Of course, the result for the force in Equation (4.4.7) does not require the temperature field to satisfy a linear problem and is broader in its applicability. The following discussion is restricted to the linear case where the temperature field satisfies Laplace's equation.

Because Stokes's equation is linear, by using the principle of superposition, we can write a more general result for the hydrodynamic force when the continuous phase is not quiescent, but executing a Stokes flow with an arbitrary $\mathbf{u}_\infty(\mathbf{x})$, and the drop translates with a velocity $\mathbf{U}$. In the absence of interfacial tension gradients, the force was given by Hetsroni and Haber (1970); one can add the current contribution to yield

$$\mathbf{F} = \frac{2\pi R}{1+\alpha} \left[-\frac{2R\sigma_T}{2+\beta} (\nabla T_\infty)_0 + \mu(2+3\alpha)\{(\mathbf{u}_\infty)_0 - \mathbf{U}\} + \mu' \frac{R^2}{2} (\nabla^2 \mathbf{u}_\infty)_0 \right]. \tag{4.4.10}$$

The instantaneous quasi-steady velocity of a drop can be calculated by adding any hydrostatic contribution to the hydrodynamic force given in Equation (4.4.10) and setting the total force on the drop to zero.

4.4.3 A Useful Special Case

As a special case, consider an undisturbed velocity field, $\mathbf{u}_\infty(\mathbf{x}) = \Lambda \nabla T_\infty(\mathbf{x})$, where Λ is a constant. This field $\mathbf{u}_\infty(\mathbf{x})$ represents a potential flow, which also is a solution of Stokes's equation with a uniform pressure field. Now, we can choose the value of Λ such that the drop is stationary under zero gravity conditions, with zero hydrodynamic force. This value happens to be negative when σ_T is negative, and is given by

$$\Lambda = \frac{2\sigma_T R}{\mu(2 + 3\alpha)(2 + \beta)}. \tag{4.4.11}$$

Physically, this represents a situation where the drop would migrate with a certain velocity in a quiescent continuous phase with the given temperature gradient. The imposed flow, in the opposite direction, would make the drop move with the same velocity in the opposite direction. Therefore, in combination, the two lead to a motionless drop.

In the gas bubble limit $\alpha = \beta = 0$, Equation (4.4.11) reduces to

$$\Lambda = \frac{\sigma_T R}{2\mu}. \tag{4.4.12}$$

As noted by Wang, Mauri, and Acrivos (1994), in this limit, we can immediately write the solution for the disturbance flow $\mathbf{u}(\mathbf{x})$ produced by the bubble, in terms of the disturbance temperature gradient field $\nabla T(\mathbf{x})$. These two fields are related precisely in the same way that the undisturbed velocity and temperature gradient fields are related:

$$\mathbf{u}(\mathbf{x}) = \Lambda \nabla T(\mathbf{x}). \tag{4.4.13}$$

It is straightforward to verify that this solution satisfies the correct boundary conditions and the condition of zero hydrodynamic force on the bubble. Although this appears to be a contrived solution of no physical significance, we show how it can be put to use when we discuss interactions among bubbles of equal size in Chapter Seven. Note that Equation (4.4.13) describes a potential flow that satisfies the full Navier–Stokes equation for any value of the Reynolds number. Of course, the Weber number must be negligible for the bubble to be spherical. Furthermore, for the solution for the temperature field to satisfy Laplace's equation, it is required that the Marangoni number be negligible, even if the Reynolds number is not negligible, implying a sufficiently small Prandtl number.

4.5 The Case of an Axisymmetric Interfacial Tension Gradient

4.5.1 Introduction

It will prove useful in subsequent sections to have a general solution available for the streamfunction and velocity fields for the motion of a drop driven by capillarity, subject to the assumptions made in Section 4.1, and the additional assumptions that the motion is axisymmetric and quasi-steady, and inertial effects as well as deformation from the spherical shape are negligible. We shall use spherical polar coordinates, shown in Figure 3.1.1. All the variables are scaled in the same manner as in Section 4.1. The velocity field satisfies Stokes's equation. The standard boundary conditions presented in Section 4.2 can be used, with the exception of the scaled tangential stress balance, which is stated in slightly more general form below:

$$\tau_{r\theta}(1, s) - \tau'_{r\theta}(1, s) = -\frac{1}{Ca} \frac{d\sigma}{d\theta}. \tag{4.5.1}$$

The Capillary number Ca, defined as $Ca = \frac{\mu v_0}{\sigma_0}$, provides a measure of the relative importance of viscous forces when compared with surface tension. The symbol σ refers to the interfacial tension scaled with the reference value σ_0.

4.5.2 Solution for the Fields

Beginning from the general solution given in Section 3.1 in spherical polar coordinates, the application of all the boundary conditions on the velocity field, with the exception of the tangential stress balance, leads to the following results, where $s = \cos\theta$:

$$\psi = \frac{1}{2}v_\infty\left(r^2 - \frac{1}{r}\right)(1 - s^2) - \sum_{n=2}^{\infty} D_n\left(\frac{1}{r^{n-1}} - \frac{1}{r^{n-3}}\right)C_n(s), \tag{4.5.2}$$

$$v_r = -v_\infty\left(1 - \frac{1}{r^3}\right)s + \sum_{n=2}^{\infty} D_n\left(\frac{1}{r^{n+1}} - \frac{1}{r^{n-1}}\right)P_{n-1}(s), \tag{4.5.3}$$

$$v_\theta = v_\infty\left(1 + \frac{1}{2r^3}\right)\sqrt{1 - s^2} + \sum_{n=2}^{\infty} D_n\left(\frac{n-1}{r^{n+1}} - \frac{n-3}{r^{n-1}}\right)\frac{C_n(s)}{\sqrt{1 - s^2}}, \tag{4.5.4}$$

$$p = p_\infty - \sum_{n=2}^{\infty} \frac{2(2n-3)}{n}\frac{D_n}{r^n}P_{n-1}(s), \tag{4.5.5}$$

$$\psi' = -\frac{3}{4}v_\infty(r^2 - r^4)(1 - s^2) - \sum_{n=2}^{\infty} D_n(r^n - r^{n+2})C_n(s), \tag{4.5.6}$$

$$v_r' = \frac{3}{2}v_\infty(1 - r^2)s + \sum_{n=2}^{\infty} D_n(r^{n-2} - r^n)P_{n-1}(s), \tag{4.5.7}$$

$$v_\theta' = -\frac{3}{2}v_\infty(1 - 2r^2)\sqrt{1 - s^2} - \sum_{n=2}^{\infty} D_n[nr^{n-2} - (n+2)r^n]\frac{C_n(s)}{\sqrt{1 - s^2}}, \tag{4.5.8}$$

and

$$p' = C_1 - \alpha\left[15v_\infty rs + \sum_{n=2}^{\infty} \frac{2(2n+1)}{(n-1)}D_n r^{n-1} P_{n-1}(s)\right]. \tag{4.5.9}$$

In writing the results for the scaled hydrodynamic pressure, the value in the undisturbed fluid at infinity is taken to be p_∞. The constant C_1 in Equation (4.5.9) will need to be determined from the normal stress balance. The constants D_n in the solution are obtained by satisfying the tangential stress balance given in Equation (4.5.1), followed by the use of the orthogonality property of the Gegenbauer Polynomials, $C_n(s)$. The Gegenbauer Polynomials and the Legendre Polynomials, $P_n(s)$, were first introduced in Section 3.1. Their properties can be found in the references mentioned in that section.

$$D_n = -\frac{n(n-1)}{4(1+\alpha)}I_n - \frac{(2+3\alpha)}{2(1+\alpha)}v_\infty\delta_{n2}. \tag{4.5.10}$$

Here, I_n are defined by

$$I_n = \frac{1}{Ca}\int_{-1}^{1} C_n(s)\frac{d\sigma}{ds}ds = \frac{1}{Ca}\int_{-1}^{1} P_{n-1}(s)\sigma(s)\,ds. \tag{4.5.11}$$

When the interfacial tension depends only on the temperature at the interface and this is a linear dependence as assumed earlier, using the connection $\sigma = 1 - Ca\,T$, Equation (4.5.11) can be rewritten as

$$I_n = -\int_{-1}^{1} P_{n-1}(s)T(1,s)\,ds.$$
(4.5.12)

4.5.3 The Hydrodynamic Force and the Quasi-Steady Velocity

The hydrodynamic force exerted on the drop, scaled using $\mu\,Rv_o$, is given by

$$\mathbf{F}_D = -\frac{2\pi}{(1+\alpha)}\left[(2+3\alpha)v_\infty + I_2\right]\mathbf{k}.$$
(4.5.13)

This expression for the force also could have been obtained from the general result given in Equation (4.4.6) by specializing to the axisymmetric case and adding the hydrodynamic force due to motion of the drop at a scaled velocity $v_\infty\mathbf{k}$.

Quasi-steady motion at zero hydrodynamic force will occur at a scaled velocity $v_\infty\mathbf{k}$, the magnitude of which can be obtained by setting $D_2 = 0$ as

$$v_\infty = -\frac{1}{(2+3\alpha)}I_2 = -\frac{1}{Ca(2+3\alpha)}\int_{-1}^{1} C_2(s)\frac{d\sigma}{ds}\,ds$$

$$= -\frac{1}{Ca(2+3\alpha)}\int_{-1}^{1} P_1(s)\sigma(s)\,ds.$$
(4.5.14)

Again, if the interfacial tension depends linearly on temperature and there are no surface active agents present, the last quantity can be written in terms of an integral involving the interfacial temperature field. This yields the following useful result:

$$v_\infty = \frac{1}{(2+3\alpha)}\int_{-1}^{1} P_1(s)T(1,s)\,ds.$$
(4.5.15)

If desired, a hydrostatic force can be included in the force balance to yield the quasi-steady scaled velocity at zero net force. In this case, using notation introduced in Section 4.2 and assuming the gravity vector to be given by $g\mathbf{k}$, we set $D_2 = G$. This yields the following result for the velocity of the drop under the combined influence of gravity and an arbitrary interfacial tension gradient:

$$v_\infty = -\frac{1}{2+3\alpha}\left[I_2 + 2(1+\alpha)G\right].$$
(4.5.16)

The extension to the more general, nonaxisymmetric case in spherical polar coordinates is straightforward but requires us to abandon the streamfunction formulation. It is given in Section 9.8.

A spherical shape, as assumed, is only compatible with a special distribution of interfacial tension of the form $\sigma(s) = a + bP_1(s)$, where a and b are constants. For other distributions of the interfacial tension, the shape will have to be determined along with the solution for the fields. In the asymptotic limit $Ca \to 0$, a perturbation approach can be used. In this case, the first estimate of the small deviation from a spherical shape can be calculated from the solution given above by using the normal stress balance.

4.6 Motion in the Presence of Surfactants

4.6.1 Introduction

So far, we have implicitly assumed the interface to be free of surface active agents, known as surfactants. In practice, it is difficult to avoid contamination by trace amounts of chemicals that can be surface active. In this section, we provide a model for estimating the effect of surfactants, which is used in the next two sections to make predictions of thermocapillary migration velocities in relatively simple situations.

4.6.2 The Way in which Surfactants Affect the Motion of Drops

Consider an isothermal case where the surfactant is present in some uniform concentration in the continuous phase. When a drop is placed in this fluid, the surfactant will adsorb onto the interface. If it is soluble in the drop phase, it will desorb into that fluid. If the drop is stationary, an equilibrium state ultimately will be reached wherein the concentration of surfactant within the drop will be uniform. The adsorbed concentration on the interface also will be uniform, leading to a uniform reduction of the interfacial tension. Because the interfacial tension will not vary with position in this instance, there will be no interfacial tension gradient. This situation is depicted pictorially in Figure 4.6.1(a).

Now imagine that the same drop moves in the continuous phase fluid. This will significantly alter the situation. The adsorbed surfactant near the front of the drop will be swept by the flow along the interface to the rear of the drop where it will build up as pictorially illustrated in Figure 4.6.1(b). This immediately leads to a concentration gradient of surfactant on the interface and therefore an interfacial tension gradient. Because the interfacial tension will be lower near the rear of the drop where the surfactant concentration is larger, the tangential stress resulting from the surfactant composition

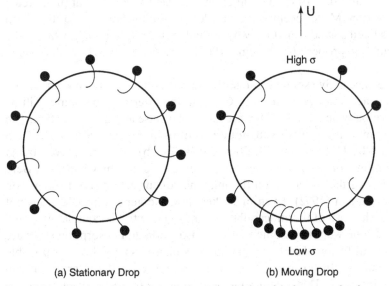

(a) Stationary Drop   (b) Moving Drop

Figure 4.6.1 Sketch depicting (a) the uniform distribution of surfactant molecules on the surface of a stationary drop and (b) the nonuniformity that develops when the drop is moving with a velocity U in the direction shown.

gradient will act to oppose the motion of the neighboring fluid, leading to more drag. In a situation where the diffusion of surfactant along the interface is slow and surfactant only desorbs very slowly from the interface, it is possible to build up enough surfactant in the rear to lead to stagnant conditions over a portion of the surface. Such a region is known as a *stagnant cap*.

Bond (1927) and Bond and Newton (1928) observed that small bubbles and drops moved because of the action of gravity in a second fluid at speeds that were close to those predicted by Stokes for rigid spheres rather than those corresponding to a fluid sphere with a mobile interface. These authors observed that as the radius increased through a relatively narrow range of values, the speed of a drop exhibited a transition from the result for a rigid sphere to that for a fluid sphere; larger drops moved at speeds consistent with the prediction for a fluid sphere. It also was found that this transition region in the radius was different for different pairs of fluids. Bond made a reference to surface contamination of air bubbles as a source of the discrepancy from theory but offered no specific mechanism. In fact, Bond and Newton suggested that the "critical radius" for the transition depends on the surface tension itself and equated it to the square root of the ratio of the surface tension to the product of the density difference between the two fluids and the acceleration due to gravity. The correct explanation for the observations of Bond and Newton, in terms of the action of adsorbed surfactant molecules, was offered by Frumkin and Levich (1947) and is discussed in detail in Levich's book (1962), in which he considered a variety of models. The existence of a stagnant region over the rear of a drop, mentioned in the preceding paragraph, was experimentally demonstrated by Savic (1953) who also presented an analysis of the Stokes problem in that case. The boundary presented by the cap region is assumed rigid, whereas the rest of the surface is assumed free of surfactant. This analysis was developed further by Davis and Acrivos (1966), who postulated that conditions at the rear pole of the drop are determined by the assumption that the surfactant monolayer collapses at that point. This allowed them to predict the qualitative feature of the transition from a rigid to a fluid interface that occurs over a small range of radius values, in contrast to models that predicted a more gradual transition. More recently, the problem of Stokes flow over a drop with a stagnant cap has been solved analytically by Sadhal and Johnson (1983), who used a mathematical technique proposed by Collins (1961) for solving mixed boundary value problems.

In general, a surface active solute can dissolve in both fluids in the bulk phase and adsorb on the interface and desorb from it. Once on the interface, this solute will be convected by the velocity field on the interface and will diffuse along the interface. The influence of surfactants on gravity driven motion of drops in this general case has been considered in detail by Holbrook and LeVan (1983a, 1983b). The role played by an electric field in conjunction with surfactants has been considered by Chang and Berg (1985), and a review of the literature on the influence of surfactants on drop motion can be found in Quintana (1992). Even in Stokes flow, the general problem has not been tackled yet in the case of thermocapillary migration due to the complexity of the requisite analysis. If one were to assume that the adsorption and desorption steps are very slow or that the diffusion of surfactant in the bulk phases is very slow, it is possible to use an approximation. In this case, if surfactant is present at the interface, it will mostly stay on the interface. This is commonly known as the *insoluble surfactant limit*. As mentioned in an earlier section, McLaughlin (1996) has calculated the gravitational rise velocity and the shape of air bubbles rising in water in this limiting situation. In the case of thermocapillary migration, the insoluble surfactant limit has been analyzed in some detail by Kim and Subramanian (1989a, 1989b) and Nadim and Borhan (1989). In most work, including ours, when it becomes necessary to relate the interfacial tension to the surfactant concentration at the interface, a linear connection is assumed. This is acceptable only when the surfactant concentration on the interface is small. At larger surfactant concentrations, one must be concerned with saturation of the monolayer that is adsorbed, as well as nonideal interactions among the surfactant molecules. This is discussed in the context of gravity-driven drop motion by He, Maldarelli, and Dagan (1991) and Chen and Stebe (1996). More recently, Chen and Stebe (1997) have analyzed the thermocapillary migration of drops, with careful accommodation of the features that arise when the adsorbed surfactant concentration is not small. The reader interested in extending the analyses given here should consult these sources.

The nonuniform distribution of surfactant along the interface, brought about by flow, will have a stronger effect on thermocapillary migration than on body force driven motion. This is because the driving force for motion resides on the interface in the former case. The interfacial tension gradient caused by surfactant will act to oppose the interfacial tension gradient from the temperature gradient. Under the right conditions, the driving force can be nearly annihilated. In body force driven motion, even if the entire interface were to appear rigid, the drop would continue to move, albeit at a reduced velocity when compared with the situation with no surfactant.

4.6.3 Modeling the Effect of Surfactants on Thermocapillary Migration

We now proceed to model the influence of an adsorbed insoluble surfactant layer at the interface between a drop and a continuous phase on the motion of the drop due to a temperature gradient. The assumptions made earlier in Section 4.2 apply here so that the analysis is restricted to negligible values of the Reynolds and Marangoni number, both within and outside the drop. Also, it is assumed that the surfactant has had adequate time to distribute itself suitably on the interface so that the drop motion can be considered steady. In writing the force balance on the drop, we shall include a force due to gravity as well, so that the results can be used for drop motion caused by a vertical temperature gradient on Earth. As before, the drop is assumed to be spherical in shape, which means that the normal stress balance cannot be satisfied in general. Instead, it can be used in the limit $Ca \to 0$ to calculate slight departures from the assumed spherical shape.

The governing equations, as well as the boundary conditions on the temperature fields, are identical to those in Section 4.2 and need not be reproduced here. The temperature field is given by Equations (4.2.16) and (4.2.17). The symbol s will continue to represent $\cos \theta$. The boundary conditions on the velocity field are identical to those used in Section 4.2 with the exception of the stress balance that is provided in a more general form in Equation (4.5.1). The reference value σ_o, used for making the interfacial tension dimensionless, can be taken conveniently to be the value corresponding to the surfactant-free interface, at a temperature T_0 prevailing on the isothermal plane in the undisturbed continuous phase passing through the center of the drop. We can use the solution given in Equations (4.5.2) to (4.5.9) for an arbitrary axisymmetric interfacial tension gradient but will need to establish a connection between the interfacial tension and the temperature and surfactant concentration at the interface.

Although it is possible to develop models at different levels of complexity for the dependence of the interfacial tension on surfactant concentration, we use a linearized model that invokes the Gibbs equation and an ideal film of surfactant as mentioned in

Chapter One. The dependence on temperature also is assumed linear.

$$\sigma^* = \sigma_0 + \sigma_T (T^* - T_0) - R^* T^* \Gamma^*. \tag{4.6.1}$$

Here σ^* refers to the interfacial tension, and T^* is the absolute temperature. The symbol Γ^* represents the surfactant concentration at the interface measured in moles of surfactant per unit area of the interface, and R^* is the gas constant. When Equation (4.6.1) is nondimensionalized, we get the following result:

$$\sigma = 1 - Ca(T + E\Gamma + \Lambda T\Gamma). \tag{4.6.2}$$

The symbol Γ represents the surfactant concentration scaled using a reference value Γ_0, which stands for the mean value of the concentration at the interface. Three dimensionless groups appear in Equation (4.6.2). The first is the Capillary number, Ca, which has been defined and discussed earlier. The second group is called the Elasticity number. It is defined below.

$$E = \frac{R^* T_0 \Gamma_0}{\mu v_0}. \tag{4.6.3}$$

Inserting the definition of the reference velocity v_0 in the denominator in the above equation provides a physical interpretation of the Elasticity number in the present context. It represents a measure of the relative role of the surfactant in lowering the interfacial tension, when compared with the role of the varying temperature in altering the interfacial tension. The third group is a nondimensional gas constant and is given by

$$\Lambda = \frac{R^* \Gamma_0}{(-\sigma_T)}. \tag{4.6.4}$$

When the surfactant concentration at the interface is relatively small, a representative value of Γ_0 can be taken to be 10^{-7} mol/m^2. Using a typical value of σ_T equal to -0.05 mN/(m•K), the value of Λ works out to approximately 0.017. Thus, we can neglect the nonlinear term in the right-hand side of Equation (4.6.2) and write the following linearized approximation of the scaled thermodynamic relationship between the interfacial tension, temperature, and surfactant concentration:

$$\sigma = 1 - Ca(T + E\Gamma). \tag{4.6.5}$$

This result will be used as needed in subsequent sections. To complete the statement, we need an equation governing the surfactant concentration distribution on the interface. Equation (2.3.17) describes the conservation of surfactant in the insoluble limit. When written in dimensionless form, at steady state, it reads as follows:

$$Pe_s \nabla_s \bullet (\mathbf{v}_s \Gamma) = \nabla_s^2 \Gamma. \tag{4.6.6}$$

The surface Péclet number Pe_s is introduced in the above equation. It is defined in a manner very similar to the Péclet number in the bulk.

$$Pe_s = \frac{R v_0}{D_s}. \tag{4.6.7}$$

Note the appearance of the surface diffusivity of the surfactant, D_s, which has the same units as the standard diffusivity in the bulk but a different magnitude. The velocity on the surface $\mathbf{v}_s$ is equal to $v_\theta \mathbf{i}_\theta$. The surface gradient operator is defined in Section 2.3.3.

It is evident that in the absence of surfactant, the problem reduces to the one already solved in Section 4.2. Inclusion of surfactant effects drastically alters the nature of the problem because we need to know the distribution of surfactant concentration for solving the fluid mechanical problem and, in turn, the velocity distribution on the surface plays a role in determining the surfactant concentration. This coupling is a forerunner of the feature we shall encounter in later sections in which we consider the surfactant-free case when the Marangoni number is not negligible. Then, the surface temperature field will not be known from the conduction solution but rather will be influenced by the motion itself.

We present solutions of the governing equations in two situations in the next two sections. In Section 4.7, we analyze the case when convective transport of surfactant along the interface dominates diffusion. In Section 4.8, the more general situation, where both convection and diffusion of surfactant on the interface play significant roles, is analyzed.

4.7 Effect of Insoluble Surfactant When Convective Transport Dominates – Stagnant Cap Limit

4.7.1 Solution for the Drag on a Drop and Its Migration Velocity

The first situation we consider is one in which the surfactant concentration distribution along the interface can be decoupled from the fluid mechanical problem. It is possible to do this when convective transport of surfactant overwhelms diffusive transport along the interface. This leads to the *stagnant cap model* mentioned in Section 4.6. Motivation for considering this model is provided by experimental observations dating back to Savic (1953), who noted that a portion of the rear of a moving drop appeared to be stagnant. When the surface Péclet number, $Pe_s \to \infty$, Equation (4.6.6) reduces to

$$\frac{\partial}{\partial \theta}[v_\theta(1, s)\Gamma \sin \theta] = 0 \tag{4.7.1}$$

at leading order, where $s = \cos \theta$. The solution of this equation, which must remain bounded at the poles, $\theta = 0$ and π, is $v_\theta(1, s)\Gamma(s) = 0$. Therefore, at every point on the interface, either the surface velocity must be zero or the surfactant concentration must be zero. Over the rear of the drop, the surfactant concentration is not zero, and therefore the velocity, $v_\theta(1, s)$, must vanish. This is assumed to occur over the region $(\pi - \phi) < \theta \leq \pi$. This region is called the stagnant cap, and ϕ is termed the cap angle. Over the remaining portion of the interface, $0 \leq \theta \leq (\pi - \phi)$, the velocity is permitted to be nonzero, which means that the surfactant concentration must be zero; that is, the remaining portion of the interface is clean and free of surfactant. The distribution of surfactant over the stagnant cap is not uniform. Rather, it is determined by the distribution of the tangential stress corresponding to the boundary condition being imposed on that surface. The stress balance condition given in Equation (4.5.1) is replaced by the following set:

$$\tau_{r\theta}(1, s) - \tau'_{r\theta}(1, s) = \frac{\partial T}{\partial \theta}(1, s), \quad 0 \leq \theta \leq (\pi - \phi) \tag{4.7.2}$$

and

$$v_\theta(1, s) = v'_\theta(1, s) = 0, \quad (\pi - \phi) < \theta \leq \pi. \tag{4.7.3}$$

The solution given in Equations (4.5.2) to (4.5.9) can be used here because it satisfies all the boundary conditions except for the stress balance. Therefore, we can proceed to evaluate the constants D_n in the solution by applying the boundary conditions in

Equations (4.7.2) and (4.7.3). Given the nature of these conditions, it is not possible to directly use the orthogonality of the Gegenbauer Polynomials. When the results for the velocity and temperature fields are substituted in the above boundary conditions, we get the following set of dual series equations:

$$\sum_{n=2}^{\infty} (2n-1) D_n C_n(s) = \frac{1}{2(1+\alpha)} \left[\frac{6}{2+\beta} - 3v_\infty(2+3\alpha) \right] C_2(s),$$

$$0 \leq \theta \leq (\pi - \phi) \quad (4.7.4)$$

and

$$\sum_{n=2}^{\infty} D_n C_n(s) = -\frac{3}{2} v_\infty C_2(s), \quad (\pi - \phi) < \theta \leq \pi. \quad (4.7.5)$$

The procedure used for solving these equations comes from Collins (1961). First the equations are rewritten in terms of Ferrer functions, which are associated Legendre functions, and then a straightforward but lengthy process is used for obtaining the constants. The final result, obtained by Kim and Subramanian (1989a), is reported below:

$$D_n = \frac{(-1)^{n-1}}{4\pi(1+\alpha)} \left[\frac{2}{2+\beta} + v_\infty \right] \Phi_n, \quad n \geq 3 \quad (4.7.6)$$

and

$$D_2 = \frac{1}{4\pi(1+\alpha)} \left[\frac{2}{2+\beta} (2\pi - \Phi_2) - v_\infty(\Phi_2 + 2\pi\{2+3\alpha\}) \right] \quad (4.7.7)$$

where

$$\Phi_n(\phi) = \frac{n}{n-2} \sin(n-2)\phi + \sin(n-1)\phi - \sin n\phi - \frac{n-1}{n+1} \sin(n+1)\phi. \quad (4.7.8)$$

Note that $\Phi_n(\phi) \to 0$ as $\phi \to 0$, and $\Phi_n(\phi) \to 2\pi \delta_{n2}$ as $\phi \to \pi$.

The drag on the drop, scaled by $\mu R v_o$, can be written in terms of the coefficient D_2 as

$$\mathbf{F}_D = 4\pi D_2 \mathbf{k}. \quad (4.7.9)$$

Now, the quasi-steady velocity of the drop can be obtained by setting the net force on it to be zero. If we include a hydrostatic force for accommodating the effect of gravity, which is assumed to point in the z-direction in Figure 3.1.1, we obtain the following result:

$$\mathbf{v}_\infty = \frac{2}{[2\pi(2+3\alpha) + \Phi_2]} \left[\frac{2\pi - \Phi_2}{2+\beta} - 2\pi(1+\alpha)G \right] \mathbf{k}. \quad (4.7.10)$$

If desired, one can set this velocity to zero to infer the temperature gradient needed to hold a drop stationary in the Earth's gravitational field when a surfactant cap is present. The result in Equation (4.7.10) reduces to the proper limit given in Section 4.2 when there is no surfactant present, that is, when $\phi = 0$. Also, it approaches the result provided by Sadhal and Johnson (1983) when the gravitational force dominates over the thermocapillary effect, that is, as $G \to \infty$. Of course, in this limit, it would be sensible to rescale the velocity by using $v_0 G$ as a characteristic velocity.

The effect of a surfactant cap on the thermocapillary migration of a drop is more severe than that on gravity driven motion of the same drop. The drop offers more resistance to motion in both cases because of the cap region that appears rigid, but in the case of thermocapillary migration, the "engine" that drives the motion is the temperature difference over the clean surface of the drop. This is necessarily smaller than the temperature difference between the forward and rear stagnation points. It can be seen from Equation (4.7.10) that regardless of the value of G, the thermocapillary contribution to the velocity decreases as the cap angle increases, approaching zero when the cap covers the entire surface. Of course, the motion of the drop is necessary to provide a nonuniform surfactant coverage in the first place. Therefore, the fact that Equation (4.7.10) predicts a zero velocity in the absence of a body force, when the cap completely covers the surface, is an artifact resulting from the manner in which the problem has been handled. The surface Péclet number, Pe_s, was permitted to approach infinity, and it was defined by using a reference velocity based on a clean interface. This reference velocity is not appropriate in the limit as the cap angle ϕ approaches π in the purely thermocapillary migration problem, and surfactant diffusion along the interface will have to be included in the treatment to avoid the self-contradictory limit predicted by Equation (4.7.10).

The result for the scaled migration velocity depends on the viscosity and thermal conductivity ratios, the cap angle, and the parameter G. A normalized migration velocity can be defined, however, as $U = \frac{v_\infty(\phi) - v_\infty(\pi)}{v_\infty(0) - v_\infty(\pi)}$, where only the dependence on the cap angle is explicitly stated. It can be shown that U is given by

$$U = \frac{(2\pi - \Phi_2)(2 + 3\alpha)}{\Phi_2 + 2\pi(2 + 3\alpha)}. \tag{4.7.11}$$

We see that U only depends on the cap angle and the viscosity ratio, α. It is insensitive to the driving force causing drop motion. Numerical evaluation shows that U also is relatively insensitive to variation in the viscosity ratio over a good range of values.

4.7.2 Surfactant Distribution along the Cap

Using the solution obtained here, it is possible to infer the distribution of surfactant concentration along the interface over the stagnant cap region. This distribution must be such as to provide an interfacial tension gradient that balances the discontinuity in tangential stress over the cap. We can calculate this gradient by evaluating the jump in tangential stress by using Equation (4.5.1). This leads to the following result:

$$\frac{1}{Ca}\frac{d\sigma}{d\theta} = \frac{3}{2}v_\infty(2 + 3\alpha)\sin\theta + 2(1 + \alpha)\sum_{n=2}^{\infty}(2n - 1)D_n\frac{C_n(s)}{\sin\theta}. \tag{4.7.12}$$

This interfacial tension gradient can be integrated over the extent of the cap to obtain the following closed form result for the scaled interfacial tension difference $\Delta\sigma = \sigma(\pi - \phi) - \sigma(\pi)$:

$$\Delta\sigma = Ca\,\Omega[3\phi + 3\sin\phi - \phi(1 + \cos\phi)] - \frac{3Ca}{(2 + \beta)}(1 - \cos\phi), \tag{4.7.13}$$

where

$$\Omega = \frac{1}{\pi}\left[\frac{2}{2 + \beta} + v_\infty\right] = \frac{4(1 + \alpha)}{[2\pi(2 + 3\alpha) + \Phi_2]}\left[\frac{3}{2 + \beta} - G\right]. \tag{4.7.14}$$

The first term in the right side of Equation (4.7.13) represents the increase in interfacial tension due to the decrease in surfactant concentration, as one moves from the rear stagnation pole to the edge of the surfactant cap. The second term corresponds to the interfacial tension decrease due to the increase in temperature and represents the lost driving force for thermocapillary migration.

Integration of Equation (4.7.12), from the edge of the surfactant cap where the surfactant concentration is zero ($\Gamma(\phi) = 0$), to an arbitrary value of θ within the cap, permits the calculation of the surfactant concentration distribution along the cap. To obtain this, we need to make use of the known relationship between the scaled interfacial tension and the scaled temperature and surfactant concentration given in Equation (4.6.5) and use the known temperature distribution along the interface. The result is given below.

$$\Gamma(\theta) = \frac{\Omega}{E} f(\theta), \tag{4.7.15}$$

where

$$f(\theta) = -3 \cos \theta \arcsin\left(\frac{\xi}{\sqrt{1 - \cos \theta}}\right) + (1 - 2 \cos \phi) \arctan\left(\frac{\xi}{\sqrt{1 + \cos \phi}}\right)$$

$$+ 3\xi \sqrt{1 + \cos \phi}, \quad (\pi - \phi) < \theta \leq \pi \tag{4.7.16}$$

and

$$\xi^2 = -(\cos \theta + \cos \phi). \tag{4.7.17}$$

In Equation (4.7.17), we are assured that ξ is real because $\cos \theta$ over the cap region is negative and larger in magnitude than $\cos \phi$. The definition of the reference surfactant concentration as the mean value Γ_m requires that

$$\int_0^\pi \Gamma(\theta) \sin \theta \, d\theta = 2. \tag{4.7.18}$$

It follows that the Elasticity number is related to the cap angle and Ω as follows:

$$E = \frac{\Omega}{2}\left(\phi - 2\phi \cos \phi - \frac{1}{2} \sin 2\phi + 2 \sin \phi\right). \tag{4.7.19}$$

The range of ϕ values is limited to the interval $[0, \pi]$. The Elasticity number corresponding to the maximum extent of the cap, attained when surfactant covers the entire drop surface ($\phi = \pi$), is designated E_{max} and is reported below.

$$E_{max} = \frac{3}{2 + \beta} - G. \tag{4.7.20}$$

The minimum value of E, which occurs when there is no surfactant ($\phi = 0$), is zero. Therefore, we can expect an admissible solution for the cap angle ϕ when the value of E lies in the interval $[0, E_{max}]$. We shall comment on the situation when $E > E_{max}$ a little later. Using the relation between E and Ω given in Equation (4.7.19), we can write the surfactant concentration distribution along the cap from Equation (4.7.15) as follows:

$$\Gamma(\theta) = \frac{2 f(\theta)}{\phi - 2\phi \cos \phi - \frac{1}{2} \sin 2\phi + 2 \sin \phi}, \quad (\pi - \phi) < \theta \leq \pi. \tag{4.7.21}$$

It is seen that the scaled surfactant distribution over the cap depends on only the cap angle as a parameter. It is insensitive to the ratios of physical properties or the driving force for the motion. This behavior is analogous to that of the normalized migration velocity U noted earlier.

It is worthwhile to discuss the logical set of independent parameters. We have treated the cap angle as a specified parameter in the model, along with the ratios of the viscosities and thermal conductivities and the parameter G. This permits one to explicitly calculate the scaled velocity of the drop, and as seen above, the parameters Ω and E. If one were to specify the total amount of surfactant and therefore its mean concentration on the interface, however, the Elasticity number can be calculated from its definition given in Section 4.6. Then, one may infer the cap angle, as well as the scaled velocity of the drop.

The case where E is greater than E_{max} is worthy of comment. Physically, when $E = E_{max}$, the amount of surfactant present on the interface is just sufficient to cause the entire surface of the drop to become stagnant. Larger values of E correspond to the presence of additional surfactant on the interface, but the entire surface will continue to appear to be rigid. Therefore, thermocapillarity can have no influence on the motion, which will be driven purely by gravity in this situation. The surfactant distribution on the interface can be obtained in the same manner as before. It is especially simple and is reported below.

$$\Gamma(\theta) = 1 - \frac{E_{max}}{E}\cos\theta. \tag{4.7.22}$$

The above distribution is achieved at $E = E_{max}$. Beyond that point, the presence of additional surfactant merely serves to increase the concentrations everywhere by a suitable constant. Of course, one should recognize that the linearized model used here is good only for small surfactant concentrations and will need to be replaced by a more realistic description as the concentration is increased.

In Figure 4.7.1, the surfactant concentration, normalized using the value at the rear stagnation point, is plotted against a polar angle $\theta^* = \pi - \theta$ for different values of the cap angle ϕ. The polar angle was measured from the rear stagnation streamline by Kim

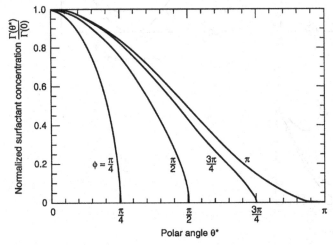

Figure 4.7.1 Normalized surfactant concentration on the drop surface plotted against the polar angle $\theta^* = (\pi - \theta)$ (Courtesy of Academic Press).

(1988), who prepared the original drawings. This is the reason for the use of this polar angle in the drawings presented in this section. The inflection in the curve for the case $\phi = \pi$ results from symmetry at $\theta^* = \pi$, which requires that $d\Gamma/d\theta^*$ be zero at that point when the cap covers the entire surface.

4.7.3 Surface Velocity Distribution over the Clean Surface

The surface velocity distribution $v_s(\theta)$ can be written from the result for v_θ given in Equation (4.5.4) evaluated at $r = 1$. This is given below.

$$v_s(\theta) = \frac{\Omega}{2(1+\alpha)}\left[(2\pi - \Phi_2)\frac{C_2(s)}{\sin\theta} + \sum_{n=3}^{\infty}(-1)^{n-1}\Phi_n\frac{C_n(s)}{\sin\theta}\right]. \tag{4.7.23}$$

It is evident from the above result that the scale for the velocity depends on all the relevant parameters including the cap angle but that the profile itself is only dependent on the cap angle. Working out the algebra in the premultiplying factor in the right side of Equation (4.7.23) shows that the quantity $v_s/(\frac{3}{2+\beta} - G)$ depends only on the viscosity ratio α and the cap angle ϕ, in addition to its dependence on the polar angle θ. In Figure 4.7.2, we have displayed the behavior of this quantity for selected values of the viscosity ratio and cap angle. Once again, we use the polar angle θ^* from Kim (1988). The figure shows that the velocity is zero over the stagnant cap. The velocity achieves a peak value at a location closer to the edge of the stagnant cap than to the clean pole.

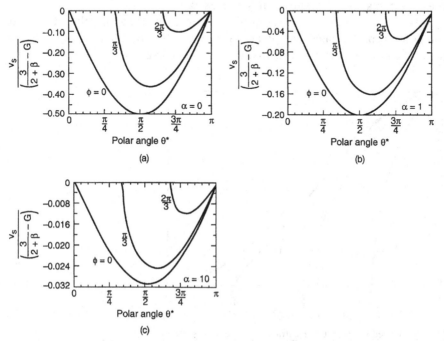

Figure 4.7.2 Scaled interfacial velocity profile. Plotted is $v_s/(\frac{3}{2+\beta} - G)$ against the polar angle $\theta^* = (\pi - \theta)$ for a series of values of the viscosity ratio (Courtesy of Academic Press).

4.8 Effect of Insoluble Surfactant When Both Surface Convection and Diffusion Are Important

4.8.1 Introduction

When a drop of a sufficiently small size moves slowly, it is possible for surface diffusion of surfactant to play a role comparable to that of surface convection. The objective of this section is to consider the solution of the problem posed in Section 4.6 in the general situation when the surface Péclet number Pe_s is finite. In this case, diffusion of surfactant on the interface plays a role that cannot be ignored. We begin with Equation (4.6.6), which represents conservation of insoluble surfactant. Axial symmetry permits it to be rewritten in the following simple form:

$$\frac{d}{d\theta}\left[\sin\theta\left(\Gamma v_s(s) - \frac{1}{Pe_s}\frac{d\Gamma}{d\theta}\right)\right] = 0. \tag{4.8.1}$$

Here $s = \cos\theta$ as usual. Use of the average surfactant concentration at the interface as the reference concentration again yields Equation (4.7.18), as in the stagnant cap model problem. We can formally integrate the surfactant conservation equation, along with the requirement that Γ and $\frac{d\Gamma}{d\theta}$ be bounded, to yield the following solution:

$$\Gamma(s) = K\exp\left[Pe_s\int_s^1 \frac{v_s(x)}{(1-x^2)^{\frac{1}{2}}}\,dx\right]. \tag{4.8.2}$$

The constant K is obtained by using Equation (4.7.18).

$$K = 2\left[\int_{-1}^{+1}\exp\left\{Pe_s\int_s^1 \frac{v_s(x)}{(1-x^2)^{\frac{1}{2}}}\,dx\right\}ds\right]^{-1} \tag{4.8.3}$$

If the interfacial velocity distribution $v_s(s)$ is known, the above results can be used to calculate the surfactant distribution. To obtain the velocity of the drop, we can use a force balance, setting the sum of the hydrodynamic and hydrostatic forces on it to zero. The solution of the problem in the case of a general axisymmetric interfacial tension gradient, given in Section 4.5, can be used for this purpose. The scaled velocity of the drop, including the effect of gravity, is given by the following result:

$$\mathbf{v}_\infty = -\frac{1}{2+3\alpha}[I_2 + 2(1+\alpha)G]\mathbf{k}. \tag{4.8.4}$$

Let us explore the question of whether we know $v_s(s)$. We can use the solution for v_θ reported in Equation (4.5.4) evaluated at the interface $r = 1$.

$$v_s(s) = \frac{3}{2}v_\infty \sin\theta + 2\sum_{n=2}^{\infty} D_n \frac{C_n(s)}{\sin\theta}. \tag{4.8.5}$$

The set of constants D_n depend on the integrals I_n and on the velocity of the drop v_∞, as displayed in Equation (4.5.10). We can see that the velocity v_∞ depends on I_2 from Equation (4.8.4) above. Therefore, knowledge of the integrals I_n is necessary to complete the solution. By substituting the expression in Equation (4.6.5) for the interfacial tension in terms of scaled temperature and surfactant concentration, Equation (4.5.11) for I_n can be rewritten as follows:

$$I_n = -\int_{-1}^{+1} [T(1,s) + E\Gamma(s)] P_{n-1}(s)\, ds. \tag{4.8.6}$$

As noted earlier, the temperature field is given by Equations (4.2.16) and (4.2.17). Therefore, the contribution to the integrals from it can be evaluated immediately. This leads to the following result:

$$I_n = -\left[\frac{2}{2+\beta}\delta_{n2} + E\int_{-1}^{+1} \Gamma(s) P_{n-1}(s)\, ds \right]. \tag{4.8.7}$$

The nature of the difficulty is evident. The interfacial velocity field $v_s(s)$ depends on the interfacial surfactant concentration distribution $\Gamma(s)$. This is the coupling mentioned in Section 4.6. This makes the problem for the surfactant distribution nonlinear. Kim and Subramanian (1989b) obtained a solution by numerical means, and we shall provide some results from their solution shortly. First, we show how perturbation methods can be used to obtain analytical solutions for small values of the surface Péclet number, and similarly for small values of the Elasticity number. These solutions can be found in Kim and Subramanian (1989b), and additional details pertaining to them are given in Kim (1988). Before proceeding with the analysis, we provide the following information for subsequent use in this section.

Consider the expansion of the surfactant distribution on the interface in Legendre Polynomials, which form a natural basis set in the axisymmetric case.

$$\Gamma(s) = \sum_{n=0}^{\infty} A_n P_n(s). \tag{4.8.8}$$

The orthogonality property of the Legendre Polynomials leads to the following result for the coefficients in the above expansion:

$$A_n = \frac{2n+1}{2} \int_{-1}^{+1} \Gamma(s) P_n(s)\, ds. \tag{4.8.9}$$

It is now possible to rewrite the result for I_n in terms of these expansion coefficients.

$$I_n = -\left[\frac{2}{2+\beta}\delta_{n2} + \frac{2}{2n-1} E A_{n-1} \right]. \tag{4.8.10}$$

Similarly, the surface velocity distribution can be written as follows:

$$v_s(s) = \frac{H + E A_1}{2 + 3\alpha}\sin\theta + \frac{E}{1+\alpha}\sum_{n=3}^{\infty} \frac{n(n-1)}{2n-1} A_{n-1}\frac{C_n(s)}{\sin\theta}. \tag{4.8.11}$$

The symbol H is defined below.

$$H = \frac{3}{2+\beta} - G. \tag{4.8.12}$$

It is useful to think of H as a combined "driving force" parameter. The first term corresponds to the thermocapillary effect, and the second term stands for the gravitational contribution. The migration velocity can be written as

$$v_\infty = \frac{2}{2+3\alpha}\left[\frac{1}{2+\beta} - (1+\alpha)G + \frac{1}{3}E A_1 \right]. \tag{4.8.13}$$

4.8.2 Expansion for Small Surface Péclet Number

The solution for the surfactant distribution in Equation (4.8.2) can be expanded for small Pe_s as shown below.

$$\Gamma(s; Pe_s) \sim \sum_{j=0}^{N} Pe_s^j \Gamma_j(s). \tag{4.8.14}$$

The surfactant concentration depends on several parameters, in addition to its dependence on the variable s. We have displayed the dependence on Pe_s explicitly on the left side to indicate that this is the parameter in which the perturbation series is being developed. The expansion is assumed to be carried out to N terms. The choice of N is usually dictated by the difficulty of the problem.

From Equation (4.8.7), it can be seen that the use of the above expansion for the surfactant concentration distribution will, in turn, lead to a corresponding expansion of the integrals I_n. Therefore, the migration velocity v_∞ and the interfacial velocity distribution $v_s(s)$ will have similar expansions. The expansion for the interfacial velocity can be written as

$$v_s(s; Pe_s) \sim \sum_{j=0}^{N} Pe_s^j v_{s,j}(s). \tag{4.8.15}$$

Now, we can proceed with the solution. The above expansions for $\Gamma(s)$ and $v_s(s)$ are substituted into Equation (4.8.1) for the surfactant concentration. The result is rearranged and the coefficients of like powers of Pe_s are equated to yield a set of ordinary differential equations for the coefficient functions $\Gamma_j(s)$ given below.

$$\frac{d\Gamma_0}{d\theta} = 0, \tag{4.8.16}$$

$$\frac{d\Gamma_j}{d\theta} = \sum_{i=0}^{j-1} \Gamma_i v_{s,j-i-1}, \quad j = 1, 2, 3 \dots. \tag{4.8.17}$$

We also need to make use of Equation (4.7.18) by substitution of the expansion for the surfactant concentration into it. This leads to the following result:

$$\int_{-1}^{+1} \Gamma_j(s)\, ds = 2\delta_{j0}. \tag{4.8.18}$$

We can solve the above set of equations to any desired order j. The procedure is tedious but straightforward. The final results, good to $O(Pe_s^3)$, are given below.

$$\Gamma(s; Pe_s) = 1 - \frac{H}{2 + 3\alpha} P_1(s) Pe_s + \frac{H}{(2 + 3\alpha)^2} \left[EP_1(s) + \frac{H}{3} P_2(s) \right] Pe_s^2$$

$$- \frac{H}{15(2 + 3\alpha)^3} \left[(15E^2 - H^2) P_1(s) \right.$$

$$+ \left. \frac{12 + 13\alpha}{1 + \alpha} EHP_2(s) + H^2 P_3(s) \right] Pe_s^3 + O(Pe_s^4), \tag{4.8.19}$$

$$
\begin{aligned}
v_s(s; Pe_s) = \frac{2H}{2+3\alpha} &\left[1 - \frac{E}{2+3\alpha}Pe_s + \frac{E^2}{(2+3\alpha)^2}Pe_s^2 \right.\\
&\left. - \frac{E(15E^2 - H^2)}{15(2+3\alpha)^3}Pe_s^3 \right]\frac{C_2(s)}{\sin\theta}\\
&+ \frac{2H^2E}{5(1+\alpha)(2+3\alpha)^2}\left[Pe_s^2 - \frac{E(12+13\alpha)}{5(1+\alpha)(2+3\alpha)}Pe_s^3 \right]\frac{C_3(s)}{\sin\theta}\\
&- \frac{4H^3E}{35(1+\alpha)(2+3\alpha)^3}Pe_s^3\frac{C_4(s)}{\sin\theta} + O(Pe_s^4),
\end{aligned}
\tag{4.8.20}
$$

$$
\begin{aligned}
v_\infty = \frac{2}{2+3\alpha}&\left[\frac{1}{2+\beta} - G(1+\alpha) \right] - \frac{2HE}{3(2+3\alpha)^2}Pe_s + \frac{2HE^2}{3(2+3\alpha)^3}Pe_s^2\\
&- \frac{2HE(15E^2 - H^2)}{45(2+3\alpha)^4}Pe_s^3 + O(Pe_s^4).
\end{aligned}
\tag{4.8.21}
$$

The first correction from convection is seen to decrease the interfacial velocity and the migration velocity. The reason for this is clear. Weak surface convection will produce principally a P_1-mode surfactant concentration profile. The resulting compositional contribution to the surface stress is proportional to EPe_s and opposes the stress due to thermocapillarity. The corrections obtained above are not sensitive to the specific driving force for motion and depend only on the total intensity of this driving force as measured by the value of the parameter H.

4.8.3 Expansion for Small Elasticity Number

When the Elasticity number E is small, implying relatively small contributions from the surfactant distribution on the interface, it is possible to write a perturbation expansion in this parameter. Setting $E = 0$ leads to a relatively simple surface velocity distribution:

$$
v_s(s) = \frac{H}{2+3\alpha}\sin\theta.
\tag{4.8.22}
$$

By using the above velocity distribution in Equations (4.8.2) and (4.8.3), a leading order result can be obtained for the surfactant concentration:

$$
\Gamma(s) = \frac{Q}{\sinh Q}\exp(-Qs).
\tag{4.8.23}
$$

Here, Q is given by

$$
Q = \frac{H}{2+3\alpha}Pe_s.
\tag{4.8.24}
$$

We can use this to obtain results for the migration velocity of the drop as well as the interfacial velocity distribution, both good to $O(E)$.

$$
v_\infty = \frac{2}{2+3\alpha}\left[\frac{1}{2+\beta} - (1+\alpha)G + E\left(\frac{1}{Q} - \coth Q \right) \right] + O(E^2)
\tag{4.8.25}
$$

$$
\begin{aligned}
v_s(s) = \frac{2H}{2+3\alpha}\frac{C_2(s)}{\sin\theta} &+ E\left[\frac{6}{2+3\alpha}\left(\frac{1}{Q} - \coth Q \right)\frac{C_2(s)}{\sin\theta} \right.\\
&\left. + \frac{1}{2(1+\alpha)}\sum_{n=3}^{\infty}n(n-1)\mathcal{J}_n\frac{C_n(s)}{\sin\theta} \right] + O(E^2).
\end{aligned}
\tag{4.8.26}
$$

Here,

$$\mathcal{J}_n = \frac{Q}{\sinh Q} \int_{-1}^{+1} \exp(-Qs) P_{n-1}(s) \, ds \qquad (4.8.27)$$

so that

$$\mathcal{J}_2 = 2\left(\frac{1}{Q} - \coth Q\right), \qquad (4.8.28)$$

a result used in writing Equations (4.8.25) and (4.8.26).

Nadim and Borhan (1989) also solved this problem by expanding in a small parameter that is related to E. To calculate the correction to the migration velocity from the presence of surfactant, they used a subsequent expansion of the tangential stress balance in Pe_s, yielding a result for the migration velocity at $O(E)$ that is correct to $O(Pe_s^2)$. Then they proceeded to calculate slight deformations from the spherical shape, finding that when moving in the presence of surfactants, the drop assumes a prolate spheroidal shape. Later, Nadim, Haj-Hariri, and Borhan (1990) obtained the correction to the migration velocity caused by this deformation. These authors pointed out that the additional correction to the migration velocity due to deformation can be positive or negative, depending on the values assumed by the parameters.

4.8.4 Series Improvement Using Padé Approximants

The above expansions in the surface Péclet number, Pe_s, and in the Elasticity number, E, are perturbation series. As noted in Section 3.2, such series are not necessarily convergent but asymptotic in nature. Shortly, we shall see, from a comparison with numerical solutions, that their performance is not as good as one might hope. However, one can extract information from a power series and use it to obtain improved approximations to the function involved. Here, we introduce a technique for improving series that requires little labor, yet provides useful information. It is possible to do this with the series in Pe_s in this problem because a sufficient number of terms have been calculated. We cannot do much with the series in the Elasticity number since only the leading corrections are obtained because of the complexity of the algebra involved.

The properties of power series for a function are best understood on the complex plane. Here, the perturbation parameter, Pe_s, is clearly a real positive number. Nonetheless, when considering a function of this quantity such as the migration velocity, expressed as a power series in Pe_s, the series will converge only within the largest circle centered on the origin on the complex plane which just touches the first singularity of the function. Therefore, convergence can be lost because of the presence of a singularity of the function anywhere on the complex plane for the variable Pe_s, even if it is not on the positive real axis. Several techniques can be used to improve the performance of power series in this situation, and they are discussed by Van Dyke (1975) and Hinch (1991). If the singularity is known, one attempts to move it away to the point at infinity by a transformation. If the series is known to a sufficient number of terms, one can use the method of Domb and Sykes (1957) to estimate the location of the singularity. In cases such as the present where only the first few terms are known, however, that cannot be done with confidence. Therefore, we can illustrate only one method that appears to work in this and similar problems.

Padé approximants are rational fraction approximations to a power series and sometimes can provide analytic continuation of the function approximated by the power series. The $[M/N]$ Padé approximant of a power series of $f(\epsilon)$ of degree $M + N$ in the variable ϵ is defined as follows:

$$f(\epsilon) = \frac{P_M(\epsilon)}{Q_N(\epsilon)}. \qquad (4.8.29)$$

Here, $P_M(\epsilon)$ and $Q_N(\epsilon)$ are polynomials in ϵ of degree M and N, respectively. The coefficients in the polynomials can be determined by solving the set of linear algebraic equations obtained by equating coefficients of like powers of ϵ in the following re-arranged version of the above equation:

$$Q_N(\epsilon) f(\epsilon) = P_M(\epsilon). \qquad (4.8.30)$$

Although the sum $M + N$ cannot exceed the degree of the polynomial, it can be smaller. Therefore, one usually constructs all the available approximants, varying M and N systematically. For a third degree polynomial $f(\epsilon)$, ten such approximants can be constructed, one of which is the power series for $f(\epsilon)$ itself. Only a few are useful. Here the best performers, when compared with the solution obtained by numerical means, were the [1/2] and [2/1] approximants:

$$U_p[1/2] = \frac{U_0(U_0 U_2 - U_1^2) - (U_0^2 U_3 - 2U_0 U_1 U_2 + U_1^3) Pe_s}{(U_0 U_2 - U_1^2) - (U_0 U_3 - U_1 U_2) Pe_s + (U_1 U_3 - U_2^2) Pe_s^2} \qquad (4.8.31)$$

and

$$U_p[2/1] = \frac{U_0 U_2 + (U_1 U_2 - U_0 U_3) Pe_s + (U_2^2 - U_1 U_3) Pe_s^2}{U_2 - U_3 Pe_s}. \qquad (4.8.32)$$

Here, the symbols U_0, U_1, U_2, and U_3 represent the coefficients of the power series for the migration velocity in the parameter Pe_s in Equation (4.8.21).

When a singularity of a function lies on the negative real line, it can be mapped away to the point at infinity by the Euler transformation. Consider a function $f(\epsilon)$ that is singular at a negative value of ϵ given by ϵ_1. We can define a new variable $\zeta = \frac{\epsilon}{\epsilon - \epsilon_1}$. The power series in the original variable ϵ can be recast as a new power series in the variable ζ, which can be obtained by matching coefficients of like powers of ϵ. Because Padé approximants can sometimes closely mimic the singularities of power series they represent, it is common practice to find the zeros of the denominator in a Padé approximant and use each as a guess for the singularity of the function. If the singularity lies on the positive real line, this procedure will not work, so we must look for negative real roots of the denominator.

Kim (1988) used this procedure to obtain improved series from the best Padé approximants $U_p[1/2]$ and $U_p[2/1]$ and termed them $U_E[1/2]$ and $U_E[2/1]$, respectively. The results for the improved series are given below.

$$U_E[1/2] = g(\epsilon_1, \zeta_1) \qquad (4.8.33)$$

and

$$U_E[2/1] = g(\epsilon_2, \zeta_2), \qquad (4.8.34)$$

where

$$g(\epsilon_i, \zeta_i) = U_0 - U_1 \epsilon_i \zeta_i - (U_1 \epsilon_i - U_2 \epsilon_i^2) \zeta_i^2 - (U_1 \epsilon_i - 2U_2 \epsilon_i^2 + U_3 \epsilon_i^3) \zeta_i^3$$
$$(i = 1, 2), \qquad (4.8.35)$$

and

$$\zeta_i = \frac{Pe_s}{Pe_s - \epsilon_i} \qquad (i = 1, 2).$$ (4.8.36)

Here, ϵ_1 and ϵ_2 are the zeros of the denominators of the Padé approximants and are given by

$$\epsilon_1 = \frac{1}{2(U_1 U_3 - U_2^2)}$$
$$\times \left[U_0 U_3 - U_1 U_2 \pm \sqrt{(U_0 U_3 - U_1 U_2)^2 - 4(U_1 U_3 - U_2^2)(U_0 U_2 - U_1^2)} \right]$$ (4.8.37)

and

$$\epsilon_2 = \frac{U_2}{U_3}.$$ (4.8.38)

The two values of ϵ_1 in Equation (4.8.37) can be complex conjugate or real, depending on the values of the parameters. For the ranges considered here, it was found that the roots were always real and of opposite sign to each other. In the improved series reported above in Equation (4.8.33), the negative root was employed.

4.8.5 Numerical Solution

As a general rule, we do not provide extensive numerical solutions in this book, focusing instead on analysis and physical interpretation. In a few cases where a numerical solution provides insight, however, we include it. This is one such case where the numerical result is used principally to test the range of validity of the asymptotic approximations, and to demonstrate the utility of improved versions of the series for small Pe_s.

Equation (4.8.2) can be solved for the surfactant distribution numerically by guessing an initial surfactant distribution on the interface. This can be used to calculate the interfacial velocity distribution $v_s(s)$ from Equation (4.8.11), which can be used now in Equation (4.8.2) to calculate a new surfactant distribution. One might then iterate with this new distribution and proceed to do this until convergence is obtained, but in fact, this scheme converges poorly. Therefore, Kim (1988) solved the equation in his doctoral thesis by using a successive underrelaxation method. For details regarding such methods, we refer the reader to Press et al. (1992). The recipe is as follows:

$$\Gamma^{(n+1)}(s_j) = \lambda \tilde{\Gamma}^{(n+1)}(s_j) + (1 - \lambda) \Gamma^{(n)}(s_j).$$ (4.8.39)

Here, the superscript n designates the iteration number and $\tilde{\Gamma}^{(n+1)}$ is the $(n+1)$st guess of the surfactant concentration distribution obtained using the straightforward iteration procedure described above. The parameter λ was adjusted by trial and error for best convergence. The domain from -1 to $+1$ in the surface coordinate s has to be divided into a finite number of intervals, with s_j designating the j'th mesh point. For the required surface integrations, a 32-point Gaussian quadrature scheme was used. The calculations were begun at small values of the surface Péclet number, Pe_s, using the perturbation solution as a guide for the initial guess. At larger values of this parameter, the solution for the previous value was used as an initial guess. More details on the numerical scheme may be found in Kim (1988). Kim performed calculations for a range of values of the parameters α, β, G, and E for values of Pe_s ranging from 0 to 200. He only presented results for $G = 0$, however, which corresponds to zero gravity, and two representative sets of values of α and β. The first set $\alpha = \beta = 0$ approximates the case of a gas bubble. The second set is $\alpha = \beta = 1$ and is taken to be representative of the case of a liquid drop. In the figures in this section, the terms *bubble* and *drop* will carry this connotation. Numerical results for the drop velocity are normalized in the following manner so that they can be compared on the same drawing:

$$\bar{U}_1 = \frac{(v_\infty - v_\infty|_{Pe_s \to \infty})}{(v_\infty|_{Pe_s=0} - v_\infty|_{Pe_s \to \infty})}. \tag{4.8.40}$$

The value corresponding to zero surface Péclet number is the result obtained earlier in Section 4.2. For the case when $Pe_s \to \infty$, the magnitude is obtained from Equation (4.7.10) from the stagnant cap model provided $E < E_{max}$. When $E \geq E_{max}$, the migration velocity is zero in this limit in the absence of gravity. Note that $E_{max} = 1$ for the drop case considered here and $\frac{3}{2}$ for the bubble case. We note that the quantity defined as H in the general case considered in this section is the same as E_{max} defined in the stagnant cap problem in Section 4.7.

The results for the normalized velocity of the drop, $\bar{U}_1$, plotted as a function of the surface Péclet number, Pe_s, are displayed in Figure 4.8.1. Note that a value of Pe_s between 100 and 1,000 appears sufficient for the conditions to approximate the stagnant cap case. The bubble appears to be more easily affected by surfactant than the drop. Of course, this cannot be taken as a generalization because only one set of parameters for the drop case has been examined.

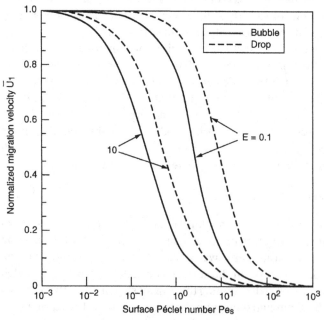

Figure 4.8.1 Normalized migration velocity $\bar{U}_1$ plotted against the surface Péclet number when $G = 0$ (Courtesy of Academic Press).

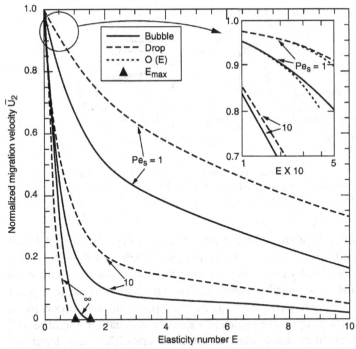

Figure 4.8.2 Normalized migration velocity $\bar{U}_2$ plotted against the Elasticity number E when $G = 0$; in the inset, the region $0 \leq E \leq 0.5$ is magnified, and the $O(E)$ approximation is included for comparison (Courtesy of Academic Press).

In Figure 4.8.2, we show the effect of the Elasticity number on the migration velocity. Here, the velocity is normalized differently to bring out the pertinent features. We define

$$\bar{U}_2 = \frac{v_\infty}{v_\infty|_{E=0}}. \tag{4.8.41}$$

Note that whether $E = 0$ or $Pe_s = 0$, we recover the limiting case of Section 4.2. For both the drop and the bubble, three values of the surface Péclet number are used, namely 1, 10, and ∞. The last value corresponds to the stagnant cap model, in which case results are obtained from Equation (4.7.10) as before. The values of E_{max} for the bubble and drop cases are marked by triangles on the abscissa. Also included for comparison in the inset are results from Equation (4.8.25), which is the asymptotic result good to $O(E)$. As one might observe from this comparison, the asymptotic result is useful only for values of E significantly less than unity and for values of Pe_s approximately of the order of unity. The stagnant cap model predicts that when $E \geq E_{max}$, the migration velocity is zero. We have pointed out the inadequacy of this model in that situation. In any case, for finite values of the surface Péclet number, it is quite possible for the bubble or drop to migrate when E exceeds E_{max}, as can be observed from the figure. Note that the trend observed regarding the impact of surfactant on a bubble versus a drop is reversed in the stagnant cap case.

Now, we move on to the testing of the asymptotic result for small values of the surface Péclet number. This is done for a representative value of $E = 1$ in Figure 4.8.3, for the bubble and drop cases. The drawing also includes two Padé approximants for comparison purposes. First, let us focus on the two approximations shown in each case

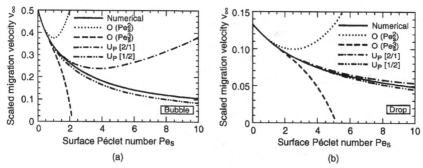

Figure 4.8.3 Scaled migration velocity from the numerical solution plotted against the surface Péclet number and compared with the various approximations for $G = 0$, $E = 1$: (a) bubble case, (b) drop case (Courtesy of Academic Press).

corresponding to truncating Equation (4.8.21) after the term involving Pe_s^2, labeled $O(Pe_s^2)$, and truncating after the next term involving Pe_s^3, labeled $O(Pe_s^3)$. It is seen that the approximations do slightly better in the drop case than in the bubble case. In each case, we see that the series result follows the numerical solution up to a point and then diverges rapidly away from it. The higher-order approximation stays close to the numerical solution for slightly larger values of Pe_s. As noted earlier, the observed behavior is typical of the performance of many asymptotic series. The sharp departure from the correct result, as Pe_s increases, is usually indicative of divergence of the series. Whereas the original series in Equation (4.8.21) is only useful for values of Pe_s up to about unity, it can be observed from the figure that the two Padé approximants given here perform well up to $Pe_s \leq 10$ in the drop case. In the bubble case, the [2/1] approximant does not do as well, beginning to diverge away around $Pe_s \approx 2$. The [1/2] approximant performs well for up to $Pe_s \approx 10$, however.

The performance of the improved series in Equations (4.8.33) and (4.8.34) is illustrated in Figure 4.8.4. In the bubble case, $U_E[2/1]$ offers significant improvement over $U_p[2/1]$, which exhibits divergence, whereas the performance of $U_E[1/2]$ is comparable to that of $U_p[1/2]$ over the same range. In the drop case, the improved series are useful up to $Pe_s \approx 30$. Also, in the drop case, the two results for $U_E[1/2]$ and $U_E[2/1]$ are indistinguishable from each other on the scale of the drawing and are close to the result from the numerical solution.

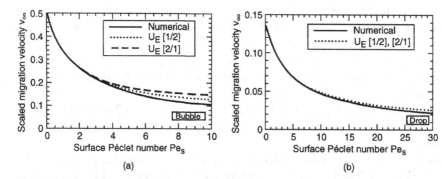

Figure 4.8.4 Scaled migration velocity from the numerical solution plotted against the surface Péclet number and compared with the improved series for $G = 0$, $E = 1$: (a) bubble case, (b) drop case (Courtesy of Academic Press).

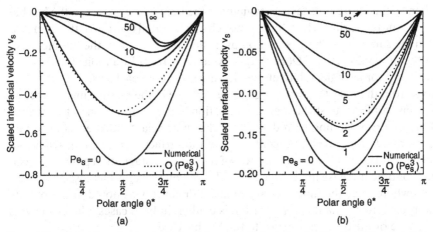

Figure 4.8.5 Scaled interfacial velocity from the numerical solution for $G = 0$ and $E = 1$ plotted against the polar angle $\theta^* = (\pi - \theta)$ and compared with the $O(Pe_s^3)$ approximation: (a) bubble case, (b) drop case (Courtesy of Academic Press).

It is not our intent to provide a detailed discussion of series improvement here. Rather, we have used this example as a vehicle to illustrate certain key ideas. The results are not always as good as those demonstrated here, especially when one has so few terms. The interested reader should consult the references cited earlier to learn more about series improvement techniques.

We close this section with some plots of interfacial velocity and surfactant concentration distributions. The former are shown in Figure 4.8.5, and the latter in Figure 4.8.6, as functions of the polar angle $\theta^* = \pi - \theta$. The reader will recall that this change to θ^* in the plots is necessitated by the fact that Kim (1988) measured the polar angle from the rear stagnation point. In each drawing, results for the bubble and drop cases are displayed for several values of the surface Péclet number Pe_s. Also included for comparison is the case when $Pe_s = \infty$, wherein results are calculated from the stagnant cap model. Note that the Elasticity number $E = 1$. This value corresponds to E_{max} for the

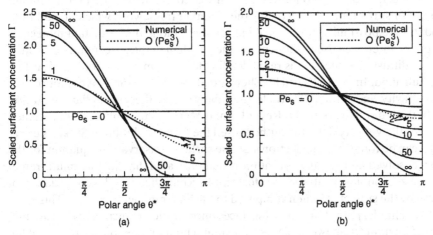

Figure 4.8.6 Scaled surfactant concentration from the numerical solution for $G = 0$ and $E = 1$ plotted against the polar angle $\theta^* = (\pi - \theta)$ and compared with the $O(Pe_s^3)$ approximation: (a) bubble case, (b) drop case (Courtesy of Academic Press).

drop case; hence, the entire interface is stagnant in the limit as $Pe_s \to \infty$. In the bubble case, E_{max} is $\frac{3}{2}$ as noted earlier. Therefore, although a substantial portion of the interface is stagnant in the same limit, there is motion over the front surface of the bubble. Also included in the drawings are the $O(Pe_s^3)$ approximations from Equations (4.8.19) and (4.8.20) calculated for a single, relatively small, value of Pe_s in each case.

Beginning from the case $Pe_s = 0$, progressive increase in the relative importance of convective transport leads to surfactant accumulation in the rear portion of the surface of the bubble or drop, leading to reduction in the interfacial velocities as observed from the drawings. For large Pe_s, concentration of the surfactant in the rear leads to a relatively clean interface in the neighborhood of the forward stagnation point. This can actually lead to some increase in the velocity in this region as Pe_s is increased, whereas velocities elsewhere over the interface are reduced drastically as noted in the bubble case. The approach to stagnant cap formation is evident in both cases. The cap covers the entire surface of the drop in contrast to the bubble case.

The $O(Pe_s^3)$ approximations for the surface velocity and surfactant concentration distribution perform as anticipated from the results displayed earlier for the migration velocity. The approximation is only shown at that Pe_s value for which it still remains reasonably accurate. It does better in the drop case, and the results for $Pe_s = 1$ for the drop are indistinguishable from the numerical solution. So, we display the results instead for $Pe_s = 2$, where slight differences from the numerical solution become discernible.

In Sections 4.6 to 4.8, we have chosen to discuss the case of insoluble surfactant in some detail to illustrate the role a surfactant can play in affecting the motion of drops. Although the gravitational effect is retained, where possible in the analysis, through a hydrostatic force, the emphasis has been on thermocapillary migration. We have not considered the situation where the surfactant diffuses in the bulk. In the thermocapillary migration problem, this is discussed at some length in the doctoral thesis of Kim (1988). It is hoped that the reader would have noted that trace quantities of contaminants, when adsorbed on the interface, can have a large influence on drop motion in reduced gravity.

Chen and Stebe (1996, 1997) recently analyzed a more general problem using two nonlinear adsorption frameworks. The first is the Langmuir framework, which takes into account the fact that a monolayer can only accommodate a certain maximum concentration of surfactant. The second is the Frumkin framework which, in addition, accommodates the possibility of nonideal interactions among the surfactant molecules adsorbed on the interface. The authors posed the problem in the adsorption-controlled limit wherein diffusion in the bulk is rapid, leading to uniform concentration of surfactant in the bulk fluid. In the 1996 article, Chen and Stebe treated the Stokes motion of a drop due to gravity, and in the 1997 article, they analyzed the thermocapillary migration problem. In the latter case, both the Reynolds and Marangoni numbers were assumed negligible. For the velocity field, the authors used the formal solution of Stokes's equations, which can be found here in Section 4.5. The surface conservation equation for the surfactant was solved numerically. The authors demonstrated that the Langmuir framework, that takes monolayer saturation effects into account, leads to less retardation of drop motion by the surfactant when compared to the linear model used here. This is because the linear model permits adsorbed surface concentration to increase without limit, whereas the Langmuir framework places a maximum limit on this concentration. In the Frumkin framework, which also takes interactions among the surfactant molecules into account, cohesion of the molecules increases surface concentration gradients, which leads to strong retardation effects from the surfactant. In contrast, for repulsion among the surfactant molecules, the effects of the surfactant are relatively weaker. These comparisons were made for a fixed amount of surfactant adsorbed on the interface. Chen and Stebe (1997) also reported on flow structures similar to those noted here in Section 4.17. The interested reader should consult their articles for more details.

4.9 Effects of Newtonian Surface Rheology and Temperature Gradients Induced by Motion

4.9.1 Introduction

In this section, we treat certain effects that are important only in special situations. The first effect we consider is the occurrence of dissipative processes in the interfacial region. This is possible when surfactant molecules are adsorbed on the interface in a sufficient concentration that they interact with each other when interfacial motion occurs. The concept of a viscosity that can be ascribed to an interface is traced back to Plateau in 1869 by Edwards, Brenner, and Wasan (1991). According to these authors, in 1913, Boussinesq invoked surface viscosity to explain the additional resistance experienced by small drops settling in another fluid. This is a different mechanism from that proposed by Frumkin and Levich (1947), mentioned in Section 4.6, which is based on interfacial tension gradients arising from surfactant composition gradients. Both explanations are based on the presence of surfactants. The added resistance in the interface leads to an additional discontinuity between the stress vectors on either side of the interface besides that already noted in Equation (2.3.8). This new contribution to the right side of that equation is the negative of the surface divergence of the interfacial excess stress tensor. To relate this quantity to the rate of deformation of interfacial elements, one constructs a rheological model of the interface in much the same manner as one does in a fluid in the bulk. Because the two-dimensional space defined by the interface is curved in general and therefore not Euclidean, one must work with general tensors. Scriven (1960) has provided the framework for a Newtonian model of the interface, in which the surface excess stress is linear in deformation in the surface. We use this model for our illustrative purposes. Unlike in the bulk, both the surface shear viscosity (μ_s^*) and surface dilatational viscosity (λ_s^*) are important in resisting deformation. The reader interested in a detailed discussion of surface rheological models and the manner in which they are incorporated into the stress discontinuity boundary condition should consult Edwards et al. (1991).

The second effect considered here is the inducement of temperature variations on the surface of an otherwise isothermal drop or bubble due to its motion. The physical basis is as follows. Over the forward half of the drop, elements of area of the interface are growing as they move toward the equatorial region. This requires that the needed interfacial internal energy be provided by the neighboring fluid, which is cooled as a consequence. In the rear half, the opposite process occurs, warming the neighboring fluid. This effect was first identified as possibly being of some consequence by Harper et al. (1967), and the results were subsequently extended by Kenning (1969), LeVan (1981), and more recently by Torres and Herbolzheimer (1993), who gave a careful derivation of the boundary condition that we already have reported in Equation (2.3.14). Torres and Herbolzheimer went on to point out that in an otherwise isothermal liquid, a rising swarm of gas bubbles can cause a macroscopic temperature gradient to arise.

Harper et al. were motivated by the possibility that the temperature gradient resulting from this mechanism provides a thermocapillary stress that leads to the bubble experiencing some additional resistance to its motion. This is an alternative hypothesis to that put forth by Frumkin and Levich (1947). After making estimates of the effect, however, Harper et al. dismissed it as being of little consequence, concluding that the explanation of Frumkin and Levich must be correct. Torres and Herbolzheimer have explored the induced temperature gradient question in more detail. They found that in certain fluids, the resulting temperature gradients and their effect can be significant. In the less common situation where the interfacial tension increases with increasing temperature, the thermocapillary stress from this effect would be directed so as to aid the motion of the drop instead of hindering it. Because an increase in the velocity of the drop will increase the surface velocity, a positive feedback situation can arise that will ultimately be limited by nonlinear effects in this problem, such as inertia and convective transport of energy. The motionless state of a drop in an isothermal fluid, in the absence of gravity, appears to be unstable in this case.

LeVan (1981) considered both of the above effects and obtained results for the velocity fields and the drop migration velocity when the two effects act in conjunction. The problems were solved in the limit when convective transport effects are negligible, and the results constitute an extension of those presented here in Section 4.5. As it happens, the fluid mechanical problem including surface rheology can be solved separately for a situation involving an arbitrary interfacial tension gradient. We first provide results for this problem. Then we consider the general problem including both surface rheology effects and the effect of stretching and shrinkage of elements of interfacial area. It will be seen that the contribution from the internal energy considerations associated with stretching and shrinking interfacial area elements affects the heat transport problem directly, thereby influencing the interfacial temperature gradient. Therefore, this will have an impact on the velocity of a drop, whether it moves because of a body force such as gravity or because of an applied temperature gradient. We analyze the situation when a drop moves under the combined action of these two mechanisms.

4.9.2 Motion of a Drop with a Newtonian Interface Subjected to an Arbitrary Interfacial Tension Gradient

Here we summarize results for the motion of a drop when a Newtonian model describes interfacial rheology. All the assumptions made in Section 4.5 apply here. We consider the same physical problem, and Figure 3.1.1 shows the spherical polar coordinates used. The only change from Section 4.5 is in the tangential stress balance boundary condition, which is modified from that given in Equation (4.5.1) as given below. The normal stress balance also must be modified to account for surface rheology, but we do not consider it here, instead assuming a spherical shape for the purpose of this analysis.

$$\tau_{r\theta}(1, s) - \tau'_{r\theta}(1, s) = -\frac{1}{Ca}\frac{d\sigma}{d\theta} - 2\mu_s v_\theta(1, s)$$
$$- (\lambda_s + \mu_s)\frac{\partial}{\partial\theta}\left[\frac{1}{\sin\theta}\frac{\partial}{\partial\theta}\{v_\theta(1, s)\sin\theta\}\right]. \tag{4.9.1}$$

In the above, $s = \cos\theta$, and the dimensionless counterparts of the surface shear and dilatational viscosities are described by the same symbols used earlier but without the asterisk. They are defined as follows:

$$\mu_s = \frac{\mu_s^*}{\mu R_0} \tag{4.9.2}$$

and

$$\lambda_s = \frac{\lambda_s^*}{\mu R_0}. \tag{4.9.3}$$

The solutions for the fields given in Equations (4.5.2) through (4.5.9) continue to hold because they are obtained before the application of the tangential stress balance. Only the result for the constants D_n needs to be modified slightly as given below.

$$D_n = -\frac{n(n-1)(2n-1)}{4[(2n-1)(1+\alpha) + n(n-1)\lambda_s + (n-2)(n+1)\mu_s]} I_n$$
$$-\frac{3(2+3\alpha+2\lambda_s)}{2(3+3\alpha+2\lambda_s)} v_\infty \delta_{n2}. \tag{4.9.4}$$

The constants I_n are defined precisely in the same way as in Section 4.5, being given by Equation (4.5.11). The modified result for the scaled migration velocity of the drop, including a contribution from the gravitational force, as in Section 4.5, is given below.

$$v_\infty = -\frac{1}{3(2+3\alpha+2\lambda_s)} [3I_2 + 2(3+3\alpha+2\lambda_s)G]. \tag{4.9.5}$$

We now proceed to evaluate the integral I_2, accommodating the contribution of stretching and shrinking of surface elements in the energy flux balance at the drop surface, while solving the energy equation.

4.9.3 Contribution Due to Stretching and Shrinkage of Interfacial Area Elements

In this segment, we consider the motion of a drop under the combined action of a uniform temperature gradient and a gravitational field subject to all the assumptions made in Section 4.5, the only changes being in the boundary condition on the flux of energy across the interface, and the inclusion of a Newtonian model of the interfacial rheology. The effect of stretching of interfacial area elements or their shrinkage is accommodated by the boundary condition given in Equation (4.1.15). We make the assumption that the difference $(e_s - \sigma\sigma_0)$ in that equation is approximately constant over the drop surface. Here, σ_0 is a reference value of the interfacial tension used to make it dimensionless, and σ is the dimensionless interfacial tension. In the case of a pure liquid surface, as seen from Equation (2.3.15), the difference $(e_s - \sigma\sigma_0)$ reduces to $-T^*\sigma_T$, where T^* is the absolute temperature. The variation of the absolute temperature over the interface is negligible in comparison with its value in most practical cases. The assumption that σ_T is constant already has been made in the analysis. For a mixture of components, a simple result for $(e_s - \sigma\sigma_0)$ does not seem to be available in the case of a liquid-liquid interface, or even at a gas-liquid interface. Nonetheless, we proceed with the approximation stated above and modify the scaled boundary condition at the surface of the drop given in Equation (4.2.13) as follows:

$$\frac{\partial T}{\partial r}(1,s) - \beta\frac{\partial T'}{\partial r}(1,s) = \frac{E_s}{\sin\theta}\frac{\partial}{\partial\theta}[v_\theta(1,s)\sin\theta]. \tag{4.9.6}$$

Here, E_s is defined by

$$E_s = -\frac{(e_s - \sigma\sigma_0)\sigma_T}{\mu k}.$$ (4.9.7)

Note that the expression multiplying E_s in the right side of Equation (4.9.6) is the surface divergence of the surface velocity field. This is the manner in which the stretching and shrinking of surface elements enters the energy flux balance at the interface.

As before, in the limit when convective transport of energy is negligible, Laplace's equation needs to be solved within and outside the drop. All the other boundary conditions on the temperature fields stated in Section 4.2 continue to apply. For the velocity field $v_\theta(1, s)$ at the interface, we can use the result from Equation (4.5.4), or equivalently Equation (4.5.8), evaluated at the drop surface. This result is good for an arbitrary interfacial tension gradient because the tangential stress balance has not yet been applied. That is where the temperature field would enter the fluid mechanical problem. Here, we only allow for interfacial tension gradients arising from temperature variations at the drop surface. In this case, as noted in Section 4.6, we can directly use Equation (4.5.12). After working through some algebra, it is possible to establish the following simple results for the scaled temperatures in the two fluids:

$$T = \left(r + \frac{B}{r^2}\right)\cos\theta,$$ (4.9.8)

$$T' = (1 + B)r\cos\theta,$$ (4.9.9)

where

$$B = \frac{1 - \beta + (G - 1)\Omega}{2 + \beta + \Omega}$$ (4.9.10)

and

$$\Omega = \frac{2E_s}{2 + 3\alpha + 2\lambda_s}.$$ (4.9.11)

The scaled migration velocity of the drop can be calculated from Equation (4.9.5) and is given below.

$$v_\infty = \frac{1}{3(2 + 3\alpha + 2\lambda_s)}\left[\frac{6}{2 + \beta + \Omega} + 2G\left(\frac{\Omega}{2 + \beta + \Omega} - (3 + 3\alpha + 2\lambda_s)\right)\right]$$ (4.9.12)

It is not surprising that the surface shear viscosity has no influence on the result for the migration velocity. As noted by Edwards et al. (1991), the reason is that the interfacial motion, being axisymmetric, is purely dilatational. As expected, the surface dilatational viscosity plays a direct role in reducing the velocity of the drop. It can be verified easily that the above results reduce to the correct limits when the parameter $E_s = 0$ or when $\lambda_s = 0$. In the case of a highly conducting drop, the temperature of the drop tends to be nearly uniform, and the effect of stretching and shrinking of interfacial elements will be negligible. Otherwise, when the parameter Ω is of the order unity, a significant impact on the velocity of the drop can be expected when the motion is driven either by gravity or by thermocapillarity. If the denominator in the right side of Equation (4.9.11) is of the order unity, a value of E_s of the order unity is sufficient to make an impact as noted by Torres and Herbolzheimer (1993). These authors cited several liquids for which this is the case, such as nitric oxide, methane, and ammonia at low temperatures, as well as cyclopentanone and dimethylphenyl carbinol at intermediate temperatures and abietic acid and water at relatively high temperatures. This is only a partial list, and the reader should consult the reference for information on other liquids. According to Torres and Herbolzheimer, the magnitude of the parameter E_s can be as large as 3.5 in the case of dimethylphenyl carbinol and is often of the order of unity. It generally increases with increasing temperature, mainly due to the decrease of viscosity with temperature. In some cases, a maximum value is reached at some temperature, and the magnitude begins to decrease with further increase in temperature.

4.10 Motion Due to a Source of Energy at the Drop Surface

4.10.1 Introduction

It is possible for a drop to move in an otherwise isothermal fluid, if a source of energy is present at its surface and is distributed in a nonuniform way. As an example, Oliver and DeWitt (1988) suggested a source of electromagnetic radiation, which is impressed on the drop through the continuous phase. If the continuous phase is transparent to this radiation and the drop is opaque to it, absorbing it at the surface, a nonuniform source can result at the drop surface. The drop would seek out the source of heat, moving toward it. Nuclear radiation can be used in a similar manner. A source also can arise at the drop surface from chemical reactions. Usually this would lead to a symmetric situation with no gradients of temperature on the surface, but the symmetry can be broken as noted by Kurdyumov, Rednikov, and Ryazantsev (1994). We are not concerned with the specific mechanism used to produce the nonuniform source. Instead, we assume such a source of energy exists at the drop surface and provide results that can be used in that problem.

4.10.2 Analysis

The assumptions made in analyzing the problem of thermocapillary migration of a spherical drop in Section 4.2 will continue to hold here. The continuous phase fluid is assumed to be isothermal in the undisturbed state far from the drop. We consider only the case when convective transport of energy and that of momentum are both negligible. The effect of a uniform temperature gradient in the continuous phase, and that of a gravitational force, can be accommodated by superposing the results from Section 4.2 with those given here because the problem is linear. Figure 3.1.1 shows the spherical polar coordinates used in the analysis. The scaled radial coordinate r is defined by dividing the radial coordinate by the radius R of the drop. As before, we use $s = \cos\theta$ as needed.

We need to solve first for the temperature fields. The only differences from the problem considered earlier in Section 4.2 are that the continuous phase is isothermal far from the drop and that an axially symmetric nonuniform source of energy, with a source density $q^*(s)$, with dimensions of energy per unit time per unit surface area, is present on the drop surface. The rest of the solution then can be written immediately from results given in Section 4.5. Therefore, we can predict the migration velocity of the drop and the velocity and pressure fields, if desired.

Because the undisturbed continuous phase fluid is isothermal, temperature is scaled by subtracting the value T_0 in the undisturbed fluid and then dividing by a characteristic temperature difference $\Delta T = \frac{q_0^* R_0}{k}$ where q_0^* is the average value of the energy source density over the entire drop surface. The scaled temperatures within the drop and outside it satisfy Laplace's equation. The boundary conditions are given below.

As $r \to \infty$,

$$T \to 0. \tag{4.10.1}$$

At the drop surface, $r = 1$,

$$T(1, s) = T'(1, s) \tag{4.10.2}$$

and

$$\frac{\partial T}{\partial r}(1, s) = \beta \frac{\partial T'}{\partial r}(1, s) - q(s). \tag{4.10.3}$$

Here, the scaled energy source is defined as $q = \frac{q^*}{q_0^*}$. At the center of the drop, $r = 0$,

$$|T'| < \infty. \tag{4.10.4}$$

The solutions for the scaled temperatures can be written as follows:

$$T(r, s) = \sum_{n=0}^{\infty} \frac{A_n}{r^{n+1}} P_n(s), \tag{4.10.5}$$

$$T'(r, s) = \sum_{n=0}^{\infty} A_n r^n P_n(s). \tag{4.10.6}$$

Here, $P_n(s)$ represent the Legendre Polynomials, mentioned first in Section 3.1. The coefficients A_n can be written as

$$A_n = \frac{q_n}{1 + n(1 + \beta)}, \tag{4.10.7}$$

where q_n are the coefficients in an expansion of the scaled energy source in Legendre Polynomials. Using the orthogonality property of Legendre Polynomials, we can obtain q_n as

$$q_n = \frac{2n + 1}{2} \int_{-1}^{+1} q(s) P_n(s) \, ds. \tag{4.10.8}$$

Therefore, once the functional dependence of the energy source at the surface of the drop is given, the temperature field at the surface can be evaluated as

$$T(1, s) = \sum_{n=0}^{\infty} A_n P_n(s). \tag{4.10.9}$$

This result can be used in Equation (4.5.12) to determine the constants I_n for use with the solution presented in Section 4.5:

$$I_n = -\frac{2}{2n - 1} A_{n-1}, \quad n \geq 2. \tag{4.10.10}$$

In particular, the scaled migration velocity of the drop can be written as follows:

$$v_\infty = \frac{2}{3(2 + 3\alpha)} A_1 = \frac{1}{(2 + 3\alpha)(2 + \beta)} \int_{-1}^{+1} sq(s)\,ds. \tag{4.10.11}$$

Here, the reference velocity v_0 is slightly different from that defined in Equation (4.1.1). Because the undisturbed continuous phase fluid is isothermal, the temperature difference used in defining the reference velocity cannot be set to $|\nabla T_\infty| R_0$. Instead, we must use the definition of ΔT given earlier in this section. We define the reference velocity below; it should be used in converting the results from Section 4.5 for the scaled velocity, streamfunction, and pressure fields to their respective physical counterparts.

$$v_0 = \frac{(-\sigma_T)\, q_0^* R_0}{\mu k} \tag{4.10.12}$$

As an example, consider the case treated by Oliver and DeWitt (1988) in which a thermal radiation flux in the form of a uniform parallel beam of intensity Q (energy per area) is incident on a drop with absorptivity ζ. All the energy that is absorbed is assumed to be deposited on the surface. We can continue to use Figure 3.1.1 with the incident photons traveling from large values of z toward the drop in the negative z-direction. The hemisphere of the drop that faces the radiation will then experience a source that is proportional to $\cos\theta$. On the hemisphere that faces away from the radiation, the energy source will be zero. The average energy density of this source can be shown to be $\frac{Q\zeta}{4}$. The dimensionless energy source at the surface then is given by

$$q(s) = 4s, \quad 0 \le s \le 1$$
$$= 0, \quad -1 \le s < 0. \tag{4.10.13}$$

In this case, the constants $q_0 = 1$ and $q_1 = 2$. Of the remaining, those with odd values of the subscript n $(3, 5, 7, \ldots)$ are zero. When n takes on even values $(2, 4, 6, \ldots)$, q_n is given by

$$q_n = (-1)^{\frac{n}{2}+1} \frac{(2n+1)}{2^{\frac{n}{2}}} \frac{1 \cdot 3 \cdot 5 \cdots (n-3)}{\left(\frac{n}{2}+1\right)!}. \tag{4.10.14}$$

The scaled velocity of the drop toward the source of radiation then becomes

$$v_\infty = \frac{4}{3(2 + 3\alpha)(2 + \beta)}. \tag{4.10.15}$$

4.11 Unsteady Motion

4.11.1 Introduction

So far, we have analyzed steady state problems involving the motion of a drop in another fluid in which a uniform temperature gradient is imposed. Even with the idealized assumption that a drop begins its motion from a state of rest when it is first introduced into the continuous phase, it would be useful to know the transient solution, if possible. Also one might like to know approximately how long it will take to achieve a state of quasi-steady motion. The absence of a true steady state in thermocapillary migration has been discussed earlier in Section 2.4. Therefore, one must be satisfied with quasi-steadiness by assuming constancy of relevant physical properties.

In this section, we provide results from a treatment of the unsteady problem given by Dill and Balasubramaniam (1992), who assumed that the motion occurs under conditions of negligible convective transport of energy and momentum. This makes the problem linear so that the method of Laplace transforms can be used. The assumptions made earlier in Section 4.2 continue to apply, including that of a spherical shape, the only change being that the motion is permitted to develop from a state of rest. The spherical shape satisfies the normal stress balance in the unsteady problem as well, so long as convective transport effects are negligible. Figure 3.1.1 shows the spherical coordinate system used in the analysis. We designate R as the radius of the drop.

The same problem, including the effect of a gravitational field, was independently analyzed by Galindo et al. (1994). They obtain the temperature fields in the same manner but provide results for the Laplace transforms of the velocity components as well as the streamfunction. Both articles provide sample calculations of the migration velocity of drops.

4.11.2 Governing Equations and Initial and Boundary Conditions

In the present unsteady problem, a fully analytical solution is not available. Rather, the problem is solved analytically only after taking Laplace transforms in time. The procedure is similar to that of Chisnell (1987), who considered the unsteady motion of a drop in a gravitational field. Full inversion can only be accomplished numerically, but small and large time asymptotic behavior of the migration velocity can be inferred analytically. The necessary reference scales have been given in Section 4.1, and the governing partial differential equations for the scaled velocity and temperature fields are provided as Equations (4.1.2) through (4.1.7). When we set the Reynolds and Marangoni numbers equal to zero in the continuous phase, the Stokes streamfunction $\psi(t, r, \theta)$ in this axisymmetric problem can be shown to satisfy

$$\frac{\partial}{\partial t}(E^2\psi) = E^4\psi. \tag{4.11.1}$$

A similar process leads to the following equation for the Stokes streamfunction $\psi'(t, r, \theta)$ in the drop phase.

$$\frac{\gamma}{\alpha}\frac{\partial}{\partial t}(E^2\psi') = E^4\psi' \tag{4.11.2}$$

The operators E^2 and E^4 already have been defined in Section 3.1. We note that the first step in obtaining the above equations is to take the curl of the Navier–Stokes equation, which eliminates the pressure gradient as well as the inertia term that contains $\frac{dv_\infty}{dt}$. Recall that this inertia term arises because of the transformation to an accelerating reference frame that is attached to the moving drop. The contribution of this term can be included by modifying the pressure gradient in each fluid by an additive function of time because $\frac{dv_\infty}{dt}$ is a spatially uniform vector field. The net effect of this term is to produce a time-dependent hydrostatic force on the drop similar to the steady hydrostatic force exerted by gravity.

The boundary conditions on the velocity field are precisely the same as those used in Section 4.2. They are given in Equations (4.2.7), (4.2.9) through (4.2.11), and (4.2.14). In the tangential stress balance, the appropriate unsteady solution must be used for the temperature field when evaluating the temperature gradient at the interface. It is assumed that the fluids within the drop phase and in the continuous phase are initially at rest and also that the drop is at rest. This requires that any disturbances arising from the process of introducing the drop into the continuous phase be negligible. The initial conditions on the streamfunction fields, and that on the velocity of the drop, are given below.

$$\psi(0, r, \theta) = 0, \tag{4.11.3}$$

$$\psi'(0, r, \theta) = 0, \tag{4.11.4}$$

$$v_\infty(0) = 0. \tag{4.11.5}$$

The scaled temperature fields in the continuous phase and in the drop phase satisfy the following equations:

$$Pr \frac{\partial T}{\partial t} = \nabla^2 T \tag{4.11.6}$$

and

$$\frac{\gamma}{\alpha} Pr' \frac{\partial T'}{\partial t} = \nabla^2 T'. \tag{4.11.7}$$

Here, we define the scaled temperature by first subtracting the temperature in the undisturbed continuous phase fluid at infinity, in the isothermal plane passing through the center of mass of the drop; and then dividing by a reference temperature difference, which is $R|\nabla T_\infty|$.

The boundary conditions on the temperature field are given in Equations (4.1.13), (4.1.18), (4.1.19), and (4.1.25). These conditions are already stated in posing the corresponding steady-state problem in Equations (4.2.8) and (4.2.12) through (4.2.14). We also need to provide initial conditions for the temperature fields. In the continuous phase, we assume that the temperature gradient at time zero is uniform. The drop is initially assumed to be at a uniform temperature equal to that existing in the undisturbed continuous phase in the isothermal plane passing through the center of the drop at time zero. In scaled variables, these initial conditions can be written as follows:

$$T(0, r, \theta) = z \tag{4.11.8}$$

and

$$T'(0, r, \theta) = 0. \tag{4.11.9}$$

4.11.3 Solution by Laplace Transforms

The above equations for the temperature fields can be solved in a straightforward manner by using the method of Laplace transforms. The Laplace transform $\overline{f}(q)$ of a function of time $f(t)$ is defined as follows:

$$\overline{f}(q) = \int_0^\infty e^{-qt} f(t)\, dt. \tag{4.11.10}$$

The results for the transforms of the scaled temperature fields that satisfy the initial and boundary conditions are reported below. We have explicitly included q in the list of variables on which the transforms depend but separated it by a semicolon to indicate that this is a parametric dependence:

$$\overline{T}(r, \theta; q) = \left[\frac{r}{q} + \frac{B}{\sqrt{r}} \frac{K_{\frac{3}{2}}(a_1 r)}{K_{\frac{3}{2}}(a_1)} \right] \cos\theta \tag{4.11.11}$$

and

$$\overline{T}'(r, \theta; q) = \frac{B'}{\sqrt{r}} \frac{I_{\frac{3}{2}}(a_2 r)}{I_{\frac{3}{2}}(a_2)} \cos \theta. \tag{4.11.12}$$

Here, $I_x(r)$ and $K_x(r)$ are modified Bessel functions of order x. Their definitions and properties can be found in Abramowitz and Stegun (1965). In particular, when $x = n + \frac{1}{2}$ where n is an integer, it is possible to represent these Bessel functions in terms of elementary functions. The constants appearing in the above solutions are given below:

$$a_1 = \sqrt{q Pr}, \tag{4.11.13}$$

$$a_2 = \sqrt{q \frac{\gamma}{\alpha} Pr'}, \tag{4.11.14}$$

$$B = \frac{1 - \beta(1 + R_2)}{q[2 + R_1 + \beta(1 + R_2)]}, \tag{4.11.15}$$

$$B' = \frac{3 + R_1}{q[2 + R_1 + \beta(1 + R_2)]}, \tag{4.11.16}$$

where

$$R_1 = \frac{a_1 K_{\frac{1}{2}}(a_1)}{K_{\frac{3}{2}}(a_1)}, \tag{4.11.17}$$

$$R_2 = \frac{a_2 I_{\frac{5}{2}}(a_2)}{I_{\frac{3}{2}}(a_2)}. \tag{4.11.18}$$

The constants R_1 and R_2 also can be rewritten in terms of elementary functions as follows:

$$R_1 = \frac{a_1^2}{1 + a_1} \tag{4.11.19}$$

and

$$R_2 = \frac{(3 + a_2^2) \tanh a_2 - 3a_2}{a_2 - \tanh a_2}. \tag{4.11.20}$$

The most important quantity is the distribution of the transform of the scaled interfacial temperature. This is given below.

$$\overline{T}(1, \theta; q) = \overline{T}'(1, \theta; q) = B' \cos \theta. \tag{4.11.21}$$

It does not seem feasible to provide an analytical inverse of the transform in Equation (4.11.21) valid for all values of time. By using well-known properties of the Laplace transform, however, it is straightforward to infer the asymptotic behavior of the interfacial temperature distribution for small and large t. From Equation (4.11.21), it is clear that we can always write $T(t, 1, \theta) = F(t) \cos \theta$. Asymptotic results for $F(t)$ are reported below.

$$F(t) \sim \frac{1}{1 + \beta \sqrt{\frac{\kappa}{\kappa'}}} + O(\sqrt{t}) \quad \text{as} \quad t \to 0, \tag{4.11.22}$$

$$F(t) \sim \frac{3}{2 + \beta} \left[1 - \frac{1}{6\sqrt{\pi}} \frac{1 - \beta}{2 + \beta} \left(\frac{Pr}{t} \right)^{\frac{3}{2}} + O(t^{-\frac{5}{2}}) \right] \quad \text{as} \quad t \to \infty. \tag{4.11.23}$$

Now, we can solve the equations for the streamfunction fields in a similar manner. After taking Laplace transforms of the governing equations and boundary conditions for the scaled streamfunctions and using the initial conditions, the solutions for the transforms $\overline{\psi}$ and $\overline{\psi}'$ may be written as follows:

$$\overline{\psi}(r, \theta; q) = H(r; q) \sin^2 \theta \tag{4.11.24}$$

and

$$\overline{\psi}'(r, \theta; q) = H'(r; q) \sin^2 \theta. \tag{4.11.25}$$

Here, the functions H and H' are given by

$$H(r; q) = \frac{1}{2} \overline{v}_\infty \left[r^2 - \frac{1}{r} + \frac{1+2\gamma}{3} \left(\frac{1}{r} - \frac{\sqrt{r} K_{\frac{3}{2}}(a_3 r)}{K_{\frac{3}{2}}(a_3)} \right) \right] \tag{4.11.26}$$

and

$$H'(r; q) = \overline{v}_\infty \left[\frac{R_3(1+2\gamma)+9}{6R_4} \right] \left[\frac{\sqrt{r} I_{\frac{3}{2}}(a_4 r)}{I_{\frac{3}{2}}(a_4)} - r^2 \right]. \tag{4.11.27}$$

The constants appearing in the solutions are defined as follows:

$$a_3 = \sqrt{q}, \tag{4.11.28}$$

$$a_4 = \sqrt{q \frac{\gamma}{\alpha}}, \tag{4.11.29}$$

$$R_3 = a_3 \frac{K_{\frac{1}{2}}(a_3)}{K_{\frac{3}{2}}(a_3)}, \tag{4.11.30}$$

and

$$R_4 = \frac{a_4 I_{\frac{5}{2}}(a_4)}{I_{\frac{3}{2}}(a_4)}. \tag{4.11.31}$$

Note that the transform of the scaled velocity of the drop $\overline{v}_\infty$ appears in the above solution. It is obtained by using the transform of Newton's law applied to the drop:

$$\overline{v}_\infty = \frac{6\overline{T}(1, \theta; q)}{q(1+2\gamma) + [2 + \alpha(3 + R_5)][9 + R_3(1+2\gamma)]}. \tag{4.11.32}$$

The constant R_5 is defined as

$$R_5 = \frac{a_4 I_{\frac{7}{2}}(a_4)}{I_{\frac{5}{2}}(a_4)}. \tag{4.11.33}$$

Inversion of the above transform of the scaled migration velocity can only be accomplished numerically in general. Nonetheless, it is possible to provide asymptotic analytical results for small and large values of time:

$$v_\infty(t) \sim \frac{6t}{(1 + \beta\sqrt{\frac{\kappa}{\kappa'}})(1+2\gamma)(1 + \sqrt{\alpha\gamma})} + O(t^{\frac{3}{2}}) \quad \text{as} \quad t \to 0, \tag{4.11.34}$$

$$v_\infty(t) \sim \frac{2}{(2+3\alpha)(2+\beta)} \left[1 - \frac{1}{6\sqrt{\pi}} \left(\frac{1-\beta}{2+\beta} Pr^{\frac{3}{2}} + \frac{1+2\gamma}{3} \right) \frac{1}{t^{\frac{1}{2}}} + O(t^{-\frac{3}{2}}) \right]$$

$$\text{as} \quad t \to \infty. \tag{4.11.35}$$

Note the linear growth of the velocity of the drop as it begins its motion from rest. Asymptotically, as time grows large, it can be seen that the result approaches the steady value obtained in Section 4.2. It is possible for the velocity to overshoot the steady value if the second term within the square brackets that is proportional to $\frac{1}{t^{3/2}}$ works out to be positive. This can happen for a given density ratio γ when the continuous phase has a sufficiently large Prandtl number and the thermal conductivity of the drop is larger than that of the continuous phase.

4.11.4 Discussion

The unsteady motion of a drop settling because of the action of gravity bears some similarity to the present situation. This problem was analyzed by Chisnell (1987) as mentioned earlier. In particular, the asymptotic behavior of the velocity of the drop at small values of time is linear in that problem as well, with a correction that occurs at $O(t^{\frac{3}{2}})$. The behavior at large time in the gravitational settling problem is a bit more involved. The velocity approaches the steady value with a correction that is proportional to $\frac{1}{\sqrt{t}}$ at leading order. This correction does not involve the density ratio; however, Chisnell pointed out that the next correction term, which is proportional to $t^{-\frac{3}{2}}$, dominates when the density ratio γ is large. In the thermocapillary migration problem, the correction to the steady velocity is proportional to $t^{-\frac{3}{2}}$ and to $(1 + 2\gamma)$. Also included at this order is a term that is proportional to $(\frac{Pr}{t})^{\frac{3}{2}}$, which is physical time scaled with the time scale for energy transport by molecular means in the continuous phase fluid over a distance R. The correction proportional to $\frac{1}{\sqrt{t}}$, found in the gravitational settling problem, is absent here. In this context, the principal difference between the two problems is that the hydrodynamic force on the drop asymptotically approaches zero in thermocapillary migration, whereas this force approaches the negative of the hydrostatic force on a drop settling because of the action of gravity.

Dill and Balasubramaniam (1992) calculated results in representative situations by inverting the transform of the scaled migration velocity in Equation (4.11.32) numerically. Their principal observations concern the magnitude of the time required to achieve steady motion. Examination of the problem reveals several time scales. There are two viscous time scales, $\frac{R^2}{\nu}$ and $\frac{R^2}{\nu'}$, and two time scales for molecular transport of energy, $\frac{R^2}{\kappa}$ and $\frac{R^2}{\kappa'}$. Also, an additional time scale, $\gamma \frac{R^2}{\nu}$, emerges from Newton's law applied to the drop. At first glance, it might appear that the slowest of the various relaxation processes will control the time to steady state. This would be true except when a certain molecular transport process only weakly influences the scaled velocity. For example, consider an air bubble in a liquid. The tangential stress communicated to the bubble at the interface is negligible, and virtually all the thermocapillary stress is transmitted to the liquid. In this case, the approach to a steady-state velocity field within the bubble does not have much of an influence on the migration velocity. Also, diffusion of energy within the drop will not affect the results significantly if the heat capacity of the drop phase is negligible compared with that of the continuous phase.

In most cases, the drop can be expected to attain a steady velocity by the time it moves a distance of the order of one radius. This is because the time scale for such movement is $t_c = \frac{R}{v_0}$ and the assumption of negligible Reynolds numbers in both phases provides assurance that this time is large compared with the time for the molecular transport of momentum in either the drop phase or the continuous phase. Similarly, the assumption of negligible Marangoni number guarantees that the transport of energy within the drop and in the continuous phase will occur on a time scale much smaller than the above time scale for appreciable motion. The exception occurs when a drop of large density moves through a fluid of small density. In this situation, the time $\gamma \frac{R^2}{\nu}$ can be much larger than the viscous relaxation time and t_c.

When convective transport becomes important, analytical solution does not seem feasible. A solution by numerical means was obtained by Oliver and De Witt (1994) in the case of Stokes flow when convective transport of energy is important. The authors only dealt with the case of a gas bubble, setting the viscosity and the thermal conductivity of the drop phase to zero. This makes it necessary to solve only the transport problem in the continuous phase. The solution of the momentum equation that applies for steady motion is used, with a parametric dependence on time through the time dependence of the interfacial temperature field. The unsteady energy equation is solved by the method of finite differences using the alternating direction implicit technique, and calculations are carried out for two different initial conditions and up to a value of Marangoni number equal to 200. Also, although not explicitly stated, the velocity of the bubble appears to be calculated by assuming the instantaneous force on the bubble at any given time to be zero. The results indicate that the bubble velocity approaches its asymptotic value by the time the bubble has moved approximately two to five times its own radius. In the case of one of the initial conditions used, at the larger values of the Marangoni number for which results are reported, the velocity is found to approach its asymptotic value through an overshoot that increases with increasing Ma. This must be a direct consequence of an overshoot in the temperature gradient driving force on the bubble surface. The explanation for this is not immediately obvious. Although the Prandtl number is infinite here because the Reynolds number is zero and the Marangoni number is nonzero, the thermal conductivity ratio β is zero. Therefore, the comment made earlier in the context of the results of Dill and Balasubramaniam cannot be used to explain the overshoot in the results of Oliver and De Witt (1994). Welch (1998) has recently obtained numerical solutions in the transient case, permitting inertial effects, as well as convective transport of energy, and accommodated deformation of shape as well. We postpone discussion of Welch's results to Section 4.15.4 but note here that Welch also observed an overshoot in the transient velocity as the Prandtl number was increased for a fixed Reynolds number.

4.12 Influence of Convective Transport of Energy – Asymptotic Analysis for Small Marangoni Number

4.12.1 Introduction

So far, with one exception in Section 4.3, we have neglected convective transport effects altogether. This permitted us to analyze linear problems, which are relatively straightforward. When the problem is nonlinear, a superposition of solutions cannot be used as a general approach, eliminating a powerful technique available for solving linear problems. Although there are several analytical approaches for solving specific nonlinear problems, few general methods exist. In this and the next two sections, we use perturbation methods to treat nonlinear problems in which convective transport influences the fields and the migration velocity of a drop. In the present section, our object is to consider again the physical case analyzed in Section 4.2; however, we permit a small influence to be exerted by convective energy transport by treating the Marangoni number Ma as a small parameter. In perturbation theory, it is conventional to use the symbol ϵ for a small quantity. Therefore, we define $\epsilon = Ma$ but note that this definition holds only in this section. We shall continue to ignore convective transport of momentum by setting the Reynolds number to zero inside the drop and in the continuous phase. Physically, this would mean that the Prandtl numbers of the continuous and drop phase fluids are large so that even for negligible Reynolds number within each phase, the Marangoni number is not entirely negligible. In addition, we require that the ratio of thermal diffusivities $\lambda = \frac{\kappa'}{\kappa}$ not be asymptotically small. In this case, one can anticipate that convective transport of energy within the drop also will play only a small role in influencing the results.

We note here that a similar problem was considered by Bratukhin (1975), who used a regular perturbation expansion to analyze the case wherein convective transport of momentum and that of energy have a small influence. He also permitted deformation of the shape of the drop in the treatment. Bratukhin used the Reynolds number, Re, defined in the usual manner, as a perturbation parameter, and expanded the fields in power series in this quantity. He determined that the velocity correction at $O(Re)$ is zero and obtained a result for the deformation in the shape of a drop. Thompson, De Witt, and Labus (1980) extended this expansion to the next higher order term and determined a nonzero correction to the velocity of the drop at $O(Re^2)$. Correction fields in such an expansion at some order will fail to meet the boundary condition at infinity, as noted by Subramanian (1981), who provided an analysis of the problem considered in the present section.

Figure 3.1.1 can be used as a sketch of the system showing the spherical polar coordinates that are used. We use the scalings and notation introduced in Sections 4.1 and 4.2. All the assumptions of Section 4.2 continue to hold, with the exception of that neglecting convective transport of energy, but the analysis is limited to small values of the Marangoni number. We have discussed a similar situation in Section 3.2 regarding the Navier–Stokes equation. When the Reynolds number is small, a perturbation expansion in Re can be used for the field variables. It was pointed out, however, that the inertia terms become comparable with the viscous terms at a scaled radial distance $r = \frac{1}{Re}$ and therefore cannot be considered small for values of $r \sim O(\frac{1}{Re})$. This makes the asymptotic expansion in Re not valid at large distances from the drop. A separate asymptotic expansion termed the *outer expansion* can be developed for the *outer region* in which $r \geq O(\frac{1}{Re})$. The standard expansion is then termed the *inner expansion* and applies to the *inner region* where $r \sim O(1)$. It is commonly assumed that an overlap region exists in which both expansions have validity, so that they can be matched. This topic is discussed in detail by Van Dyke (1975) and Kevorkian and Cole (1981), who also noted difficulties that can arise sometimes when using this technique known as the method of matched asymptotic expansions. The original work in the context of the Navier–Stokes equation was developed by Kaplun and Lagerstrom (1957), and detailed solutions for the case of a sphere and a cylinder were worked out by Proudman and Pearson (1957). A few years later, the technique was used in the analogous heat transfer problem for a sphere by Acrivos and Taylor (1962). In this section, we apply the method of matched asymptotic expansions to the present problem. The same technique was used in Sections 3.3 and 3.4 in the case of large Reynolds number with $\frac{1}{\sqrt{Re}}$ serving as a small parameter. We begin by showing why a regular perturbation method will lead to an incorrect solution.

4.12.2 The Regular Perturbation Approach

The momentum equations within and outside the drop reduce to Stokes's equations. These already have been solved in Section 4.5 for an arbitrary interfacial tension gradient, subject to all the applicable boundary conditions, and we shall use these solutions for the fields where needed. Note that the velocity field is known to within a set of constants I_n, which can be obtained from Equation (4.5.12) if we know the temperature field at the interface. This result shows that I_n is simply related to the P_{n-1}-mode of the interfacial temperature field. That is, if we expand the interfacial temperature field $T(1, s)$ in a series of Legendre Polynomials as we did in Section 4.10 in Equation (4.10.9), I_n is $-\frac{2}{2n-1}$ times the coefficient A_{n-1} multiplying $P_{n-1}(s)$ in the expansion, as shown in Equation (4.10.10). Here, as usual, $s = \cos\theta$. The migration velocity of the drop, given in Equation (4.5.14), is seen to be proportional to the constant I_2.

The equations satisfied by the scaled temperature fields in the continuous phase and in the drop phase are written below.

$$\epsilon(v_\infty + \mathbf{v} \bullet \nabla t) = \nabla^2 t, \tag{4.12.1}$$

$$\epsilon(v_\infty + \mathbf{v}' \bullet \nabla t') = \lambda \nabla^2 t'. \tag{4.12.2}$$

Lower case symbols are used for the scaled temperature. These will be seen to be inner fields a bit later, and we reserve the upper case symbols for use in the outer solution. Shortly, we shall define the terms *inner* and *outer* in the present context. For convenience, we have chosen to work with the ratio of thermal diffusivities, $\lambda(=\frac{\kappa'}{\kappa})$, as a parameter, instead of defining a separate Ma' for the drop phase fluid. The origin of the sink term in the energy equations already has been discussed in Section 4.1. The boundary conditions on the temperature fields are included here for completeness.

$$t(r, s) \to rs \quad \text{as} \quad r \to \infty, \tag{4.12.3}$$

$$t(1, s) = t'(1, s), \tag{4.12.4}$$

$$\frac{\partial t}{\partial r}(1, s) = \beta \frac{\partial t'}{\partial r}(1, s), \tag{4.12.5}$$

$$|t'(0, s)| < \infty. \tag{4.12.6}$$

We attempt to solve this problem with a straightforward asymptotic expansion of the scaled temperature fields in the small parameter ϵ. This is known as a regular perturbation approach. Assuming that $f_j(\epsilon)$, $j = 0, 1, 2, \ldots$ form an asymptotic sequence as discussed in Section 3.2, we write

$$t \sim \sum_j f_j(\epsilon) t_j(r, s) \tag{4.12.7}$$

and

$$t' \sim \sum_j f_j(\epsilon) t_j'(r, s) \tag{4.12.8}$$

and recognize from the boundary conditions that $f_0(\epsilon) = 1$. If we substitute these expansions in the governing equations, it can be seen that the leading order fields $t_0(r, s)$ and $t_0'(r, s)$ satisfy Laplace's equation. These fields also satisfy the boundary conditions given in Equations (4.12.3) through (4.12.6), and therefore the solutions are already known. We have given them in Equations (4.2.16) and (4.2.17). When the expansion for the temperature field is used in the definitions of the constants I_n and in the result for the scaled migration velocity v_∞, we see that they too possess expansions similar to those written above:

$$I_n \sim \sum_j f_j(\epsilon) I_{n,j} \tag{4.12.9}$$

and

$$v_\infty \sim \sum_j f_j(\epsilon) v_{\infty,j}. \tag{4.12.10}$$

Use of the leading order temperature fields in Equations (4.5.12) and (4.5.15) provides the following leading order solutions for these quantities:

$$I_{n,0} = -\frac{2}{2+\beta} \delta_{n2} \tag{4.12.11}$$

and

$$v_{\infty,0} = 2A_0, \tag{4.12.12}$$

where

$$A_0 = \frac{1}{(2+3\alpha)(2+\beta)}. \tag{4.12.13}$$

The results for the leading order velocity and temperature fields and the migration velocity (identified by the subscript 0) are therefore seen to be precisely those obtained in Section 4.2 when we neglected convective energy transport altogether. Postulating the asymptotic sequence to be a simple one of increasing powers of ϵ subject to later verification, we can write an expansion of the form

$$t \sim t_0(r,s) + \epsilon t_1(r,s) + \epsilon^2 t_2(r,s) + \cdots \tag{4.12.14}$$

for $t(r,s;\epsilon)$ and similar expansions for the other fields. These expansions are substituted in the governing equations and boundary conditions for the temperature fields. The result is a set of equations for the corrections to the temperature fields within and outside the drop. These equations are linear, unlike the governing equations for the original fields. The general governing equations can be written as follows where the symbol v_i is used to designate the coefficient of ϵ^i in the expansion of the velocity field in the continuous phase resulting from using Equation (4.12.14); a similar notation is used within the drop phase.

$$\nabla^2 t_j = v_{\infty,j-1} + \sum_{i=0}^{j-1} v_i \bullet \nabla t_{j-1-i}, \quad j \geq 1, \tag{4.12.15}$$

$$\lambda \nabla^2 t'_j = v_{\infty,j-1} + \sum_{i=0}^{j-1} v'_i \bullet \nabla t'_{j-1-i}, \quad j \geq 1. \tag{4.12.16}$$

The fields $t_j(r,s)$ and $t'_j(r,s)$ for $j = 0,1,2,\ldots$ must each satisfy the conditions given below:

$$t_j(1,s) = t'_j(1,s), \tag{4.12.17}$$

$$\frac{\partial t_j}{\partial r}(1,s) = \beta \frac{\partial t'_j}{\partial r}(1,s), \tag{4.12.18}$$

and

$$|t'_j(0, s)| < \infty. \tag{4.12.19}$$

In addition, we can see from Equation (4.12.3) that

$$t_j(r, s) \to \delta_{j0}\, rs \quad \text{as} \quad r \to \infty. \tag{4.12.20}$$

This reduction of the problem to one of solving a sequence of linear problems is the reason for using a perturbation approach. It is possible to solve for as many of the temperature correction fields as desired. The procedure is straightforward, but the details get laborious with increasing order of the correction. The second correction field in the continuous phase displays a relatively mild problem analogous to that experienced by Whitehead (1889). The result for t_2 is a superposition of a homogeneous solution of Laplace's equation and a particular solution to accommodate the inhomogeneity. It is this particular solution that causes the difficulty. It contains nonadjustable constants multiplying Legendre Polynomials of orders 0, 1, and 3 and does not approach zero as $r \to \infty$. The rest of the solution approaches zero as $r \to \infty$. From Equation (4.12.20), we see that the correction fields for $j \geq 1$ should approach zero in that limit. Of course one might argue that these corrections need not approach zero so long as they are asymptotically smaller than r as $r \to \infty$. In this sense, the second correction to the temperature field is acceptable. As Merritt (1988) has shown by calculating higher order corrections, however, the third correction field contains similar terms that are proportional to r and the fourth has terms proportional to r^2, and it is clear that this problem gets worse with increasing order of the correction field. Therefore, a solution for the temperature field in the continuous phase by a straightforward perturbation expansion is not correct.

4.12.3 Solution by the Method of Matched Asymptotic Expansions

As one might suspect, the origin of the difficulty is the same as that in flow past a sphere at small Reynolds number. Because a uniform temperature gradient is imposed in the undisturbed continuous phase, one might expect conduction to be the only operative mechanism far from the drop, but it is important to think in terms of a disturbance to the background linear temperature field, which is caused by the presence of the moving drop. For this disturbance temperature field, we can approximately estimate the order of the conduction and convection terms in the energy equation at some value of the radial coordinate r^* as follows. The undisturbed temperature field at infinity varies by a magnitude $|\nabla T_\infty|R$ over a distance R. In the pure conduction problem in the presence of a drop, the variation over the same distance in the vicinity of the drop is of the same order of magnitude, but different by a factor that is $O(1)$. For example, for a gas bubble of negligible thermal conductivity, the variation at the surface is $\frac{3}{2}$ that in the undisturbed fluid. Only when the conductivities of the drop and continuous phase are exactly the same will the variation be the same at all locations in the pure conduction problem. Otherwise, we can anticipate that the order of magnitude of the temperature variation of the disturbance field in the continuous phase over a distance R will be given by $\Delta T = O(|\nabla T_\infty|R)$ in the vicinity of the drop. The main idea is that this characteristic temperature difference for the disturbance fields is of $O(1)$. Using this ΔT, the convective transport term will be of the order $\frac{\rho C_p v_0 \Delta T}{r^*}$, and the conduction term will be of the order $k\frac{\Delta T}{r^{*2}}$. Here, C_p is the specific heat at constant pressure, and the other symbols have been introduced in earlier sections. The ratio of the convective transport term to the conduction term is $O(\epsilon r)$, where we have nondimensionalized the result and substituted ϵ for the Marangoni number in the continuous phase. This shows that when $r \sim O(1)$, the convective transport term is indeed asymptotically small compared with the conduction term. At distances $r \sim O(\frac{1}{\epsilon})$, however, the two terms are of comparable order, and the convective transport terms cannot be considered small.

The remedy for the difficulty has been known for some time. As mentioned in the introduction, it involves developing a separate expansion in the region where the standard expansion in Equation (4.12.14) fails to be valid, namely, $r \sim O(\frac{1}{\epsilon})$. From here on, we use the term *inner expansion* for that given in Equation (4.12.14) and analogous expansions of the other field variables. For the outer expansion, we must define a new scaled radial coordinate that is held fixed while the field variables are expanded in a suitable asymptotic sequence of functions of ϵ. Such a coordinate can be written generally as $\rho = \epsilon^\zeta r$, and the value of ζ can be established as unity by transforming the energy equation for the continuous phase to (ρ, s) variables and requiring that the convection and conduction terms should be of equal importance. Therefore, the outer radial coordinate is defined as

$$\rho = \epsilon r. \tag{4.12.21}$$

The use of the symbol ρ for the radial variable in the outer equations should not cause any confusion with the use of the same symbol for the density of the continuous phase in other places in this book. We have adopted this notation because it is customary in this class of problems. To clearly distinguish the temperature field expressed as a function of (ρ, s) from that expressed as a function of (r, s), we use the symbol $T(\rho, s)$ for this field. Note that the solution in the interior of the drop does not pose any difficulties, and therefore we only need the single symbol $t'(r, s)$. The boundary conditions at the surface of the drop given in Equations (4.12.17) and (4.12.18) are applied between the inner field $t(r, s)$ and the field within the drop $t'(r, s)$. We no longer require the inner field $t(r, s)$ to satisfy the boundary condition given in Equation (4.12.20) as $r \to \infty$. Instead, it is required to match the outer solution to the specified order in ϵ. This matching requirement will be discussed in more detail shortly. The outer field $T(\rho, s)$ will not be expected to satisfy the boundary conditions at the drop surface. Instead it will be required to match the inner solution appropriately and satisfy the boundary condition as $\rho \to \infty$.

We define an *outer expansion* as one in which the variables (ρ, s) are fixed and the fields are expanded in a suitable asymptotic sequence in ϵ. The outer expansion for $T(\rho, s)$ can be written as follows:

$$T \sim \frac{\rho s}{\epsilon} + F_0(\epsilon) T_0(\rho, s) + F_1(\epsilon) T_1(\rho, s) + \cdots. \tag{4.12.22}$$

The expansion begins with the known result rs which the scaled temperature field must approach as either r or $\rho \to \infty$. Another way of interpreting this particular form of the expansion is to recognize that it is the disturbance to the established linear temperature field in the undisturbed fluid that is sought. Note that we have assumed the asymptotic sequence to be unknown and, in particular, do not specify $F_0(\epsilon)$. This is important because this leading term in the outer expansion is not necessarily of $O(1)$, and its order must be established during the analysis. This can be done in the present problem by examining the energy equation rewritten in outer variables:

$$\mathbf{V} \bullet \nabla_\rho T = \nabla_\rho^2 T - \frac{1}{\epsilon} v_\infty. \tag{4.12.23}$$

The symbol ∇_ρ is used to designate the scaled gradient operator in the (ρ, s) variables and $\mathbf{V}$ stands for the scaled velocity field in the continuous phase expressed in the (ρ, s) variables. By substituting the expansion for T and taking the limit as $\epsilon \to 0$ in the above equation, we can obtain the problem for $T_0(\rho, s)$:

$$-v_{\infty,0}\mathbf{k} \bullet \nabla_\rho T_0 = \nabla_\rho^2 T_0 + W_0(\rho, s), \tag{4.12.24}$$

where

$$W_0(\rho, s) = -\lim_{\epsilon \to 0} \frac{1}{\epsilon F_0(\epsilon)} (v_\infty + \mathbf{V} \bullet \mathbf{k}). \tag{4.12.25}$$

The quantity $v_\infty + \mathbf{V} \bullet \mathbf{k}$ is $O(\epsilon^3)$. The leading order outer problem for the disturbance field will turn out to be trivial if the governing equation is homogeneous. Therefore, we are led to the choice

$$F_0(\epsilon) = \epsilon^2, \tag{4.12.26}$$

and this yields the result given below for $W_0(\rho, s)$.

$$W_0(\rho, s) = \frac{I_{2,0}}{(2 + 3\alpha)\rho^3} P_2(s) - \frac{3 I_{3,1}}{10(1 + \alpha)\rho^2} [2 P_1(s) + 3 P_3(s)]. \tag{4.12.27}$$

In writing the above result for the inhomogeneity, we have assumed that the correction to the leading order inner field is $O(\epsilon)$. It can be established that the result for $F_0(\epsilon)$ given above is good as long as this correction is of $O(\epsilon)$ or higher. If it is of a higher order than ϵ, the term multiplied by $I_{3,1}$ in Equation (4.12.27) would be absent. In this problem, this is not the case, and the inhomogeneity is correctly given by Equation (4.12.27). In other problems, it is not always possible to find $F_0(\epsilon)$ right away. Rather, one must wait to match the leading order outer and inner fields and determine this quantity from the matching requirement.

The solutions for T_0 and higher-order corrections must satisfy conditions of boundedness and matching requirements with the inner field. Also, they must satisfy the condition

$$\frac{T_j}{\rho} \to 0 \quad \text{as} \quad \rho \to \infty. \tag{4.12.28}$$

The governing equation for T_0 is the well-known Oseen equation containing an inhomogeneity that is specific to the present problem. Physically, it represents a balance between convective transport of energy (with the uniform velocity that exists far from the drop) and conduction, both exerting equal influence. In addition, there is a source term, which can be positive or negative, depending on location. A general homogeneous solution of this equation in spherical polar coordinates may be obtained by separation of variables following a simple transformation. It is reported below.

$$T_{0h} = \left(\frac{\pi}{\rho v_{\infty,0}}\right)^{1/2} \exp\left(-\frac{\rho s}{2} v_{\infty,0}\right) \sum_{n=0}^{\infty} C_n P_n(s) K_{n+\frac{1}{2}}\left(\frac{\rho}{2} v_{\infty,0}\right). \tag{4.12.29}$$

Here, the function $K_{n+\frac{1}{2}}(x)$ is a modified Bessel function of the second kind, which has been introduced in Section 4.11. It can be written as

$$K_{n+\frac{1}{2}}(x) = \left(\frac{\pi}{2x}\right)^{\frac{1}{2}} e^{-x} \sum_{i=0}^{n} \frac{(n+i)!}{(n-i)!i!(2x)^i}, \tag{4.12.30}$$

which permits us to rewrite the homogeneous solution in the following form that is useful when matching is performed:

$$T_{0h} = \frac{\pi}{\rho v_{\infty,0}} \exp\left\{-\frac{\rho}{2}v_{\infty,0}(1+s)\right\} \sum_{n=0}^{\infty} C_n P_n(s) \sum_{i=0}^{n} \frac{(n+i)!}{(n-i)!i!(\rho v_{\infty,0})^i}. \tag{4.12.31}$$

A particular solution for T_0 needs to be written as well, but we postpone doing so. We need to know the solution for t_1 to specialize the constants in the complete solution for T_0 by matching. Therefore, we now consider the inner field. As mentioned earlier, examination of the inner field equations shows that the first correction can be expected to be at $O(\epsilon)$, leading to the choice we made earlier that $f_1(\epsilon) = \epsilon$. This is not an absolute choice, but a reasonable guess based on the fact that with this assumption a nontrivial problem arises for $t_1(r, s)$. If our choice is not correct, this will be revealed by the matching process, and we can go back and redefine the asymptotic sequence. The solution for t_1 can be calculated immediately. The only boundary condition that would have been used in a straightforward perturbation approach, which is no longer applicable, is that as $r \to \infty$. Instead of requiring that $t_1(r, s) \to 0$ in this limit, we match the inner field to this order to the outer field. This brings us to an explanation of the matching principle. Assuming an overlap region exists, the asymptotic matching principle requires that when the inner solution good to $O(f_m(\epsilon))$ is rewritten in outer variables and expanded for small ϵ while keeping the outer variables fixed and truncated now at $O(F_n(\epsilon))$, it must be equal to the result we get when we take the outer solution good to $O(F_n(\epsilon))$, rewrite in inner variables, expand for small ϵ while keeping the inner variables fixed and truncate at $O(f_m(\epsilon))$. Of course, to put this principle to use, one of these solutions must be written back in its original variables.

At this stage, the result for $t_1(r, s)$ contains one set of arbitrary constants that arise in the solution of Laplace's equation in the same way that the result for t_0 contains a set of arbitrary constants. We then take $t_0 + \epsilon t_1$ and rewrite it in (ρ, s). We find that it is sufficient to match this with $\frac{\rho s}{\epsilon}$ from the outer solution to determine all the arbitrary constants. The results reported earlier in the straightforward perturbation approach for the leading order quantities are found to be correct. The results for t_1 and t_1' are given below, along with the contributions they make at this order to the constants I_n.

$$t_1(r, s) = \frac{A_0}{3\lambda(2+\beta)}\left[\frac{k_1}{r} + \frac{k_2}{r^4} + P_2(s)\left(\frac{k_3}{r} + \frac{k_4}{r^3} + \frac{2k_2}{r^4}\right)\right], \tag{4.12.32}$$

$$t_1'(r, s) = \frac{A_0}{\lambda(2+\beta)}\left[l_1 + l_2 r^2 - \frac{3}{4}r^4 + P_2(s)\left(l_3 r^2 + \frac{3}{7}r^4\right)\right], \tag{4.12.33}$$

$$I_{n,1} = \frac{A_0}{2+\beta}\left[\left(\frac{4\beta}{3\lambda}(2+\beta) - (1-\beta)\right)\delta_{n1}\right.$$

$$\left. + \frac{2A_0}{105\lambda(2+\beta)(3+2\beta)}(49\lambda - 7\lambda\beta + 18\beta)\delta_{n3}\right], \tag{4.12.34}$$

$$k_1 = 2[\lambda(1-\beta) - \beta(2+\beta)], \tag{4.12.35}$$

$$k_2 = -\frac{\lambda}{2}(1 - \beta), \tag{4.12.36}$$

$$k_3 = -\lambda(4 - \beta), \tag{4.12.37}$$

$$k_4 = \frac{1}{7(3 + 2\beta)}[7\lambda(8 + 5\beta - 4\beta^2) - 18\beta], \tag{4.12.38}$$

$$l_1 = \frac{1}{12}[6\lambda(1 - \beta) - (8\beta^2 + 20\beta + 17)], \tag{4.12.39}$$

$$l_2 = \frac{1}{6}(2\beta + 13), \tag{4.12.40}$$

$$l_3 = -\frac{1}{21(3 + 2\beta)}[7\lambda(7 - \beta) + 9(3 + 4\beta)]. \tag{4.12.41}$$

Because we find that $I_{2,1}$ is zero, the correction $v_{\infty,1} = 0$. We must go to a higher order to determine the influence of convective energy transport on the migration velocity.

Now, we return to the outer field. It is possible to solve the governing equation, along with the boundary conditions, for the function $T_0(\rho, s)$. The general homogeneous solution has been given, and a particular solution suitable for our purposes can be constructed by considering the effect of the Oseen operator on a simple model function. If we define

$$\mathcal{L} \equiv v_{\infty,0}\mathbf{k} \bullet \nabla_\rho + \nabla_\rho^2, \tag{4.12.42}$$

then it follows that

$$\mathcal{L}\left(\frac{1}{\rho^m} P_n(s)\right) = [m(m - 1) - n(n + 1)]\frac{1}{\rho^{m+2}} P_n(s)$$
$$+ \frac{v_{\infty,0}}{(2n + 1)\rho^{m+1}}[n(n - m + 1)P_{n-1}(s)$$
$$- (n + 1)(m + n)P_{n+1}(s)]. \tag{4.12.43}$$

Using this result, the particular solution corresponding to the inhomogeneity in Equation (4.12.24) that is given in Equation (4.12.27) can be constructed. The arbitrary constants appearing in the homogeneous solution after application of the boundary conditions are determined by matching requirements. This particular matching is accomplished between $t_0 + \epsilon t_1$ and $\frac{\rho s}{\epsilon} + \epsilon^2 T_0$. The inner solution, after rewriting in outer variables, is expanded for small ϵ and terms up to and including those of $O(\epsilon^2)$ are retained. Similarly, the outer solution, after rewriting in inner variables, is expanded for small ϵ, and terms up to and including those of $O(\epsilon)$ are retained. Matching of these two results yields all the arbitrary constants in the solution for T_0. The final result is reported below.

$$T_0(\rho, s) = \frac{\pi}{\rho v_{\infty,0}} \exp\left\{-\frac{\rho}{2}v_{\infty,0}(1 + s)\right\}\left[C_0 + C_1 P_1(s)\left(1 + \frac{2}{\rho v_{\infty,0}}\right)\right]$$
$$- A_1\left[\frac{49\lambda - 7\beta\lambda + 18\beta}{105\lambda}\left(1 + \frac{P_2(s)}{2}\right)\frac{1}{\rho}\right.$$
$$+ \frac{(2 + \beta)[14\lambda(8 + 11\beta) + 7\alpha\lambda(9 + 23\beta) - 18\beta(2 + 3\alpha)]}{140\lambda}\left.\frac{P_1(s)}{\rho^2}\right], \tag{4.12.44}$$

where

$$C_0 = \frac{A_1 v_{\infty,0}^2}{840\lambda\pi}[14\lambda(162 + 20\beta - 53\beta^2) + 7\alpha\lambda(366 + 55\beta - 109\beta^2)$$
$$- 4\beta(366 + 463\beta + 140\beta^2) - 2\alpha\beta(678 + 899\beta + 280\beta^2)], \qquad (4.12.45)$$

$$C_1 = \frac{A_1 v_{\infty,0}^2}{280\lambda\pi}\left[14\lambda(46 + 20\beta - 9\beta^2) + 7\alpha\lambda(78 + 35\beta - 17\beta^2) - \frac{18\beta}{A_0}\right],$$
$$\qquad (4.12.46)$$

and

$$A_1 = \frac{1}{(1+\alpha)(2+\beta)(3+2\beta)}. \qquad (4.12.47)$$

Now, we can proceed to calculate the inner correction fields at $O(\epsilon^2)$. As with t_1, we apply all the boundary conditions except the one as $r \to \infty$. The results, although lengthy, are simple to obtain. The arbitrary constants in the solution are determined by matching the inner solution $t_0 + \epsilon t_1 + \epsilon^2 t_2$ with the outer solution $\frac{ps}{\epsilon} + \epsilon^2 T_0$. The final results for t_2 and t'_2 are lengthy and therefore not reproduced here. The interested reader will find them in Subramanian (1983). Knowledge of these fields permits one to finally calculate the correction to the migration velocity. The principal result of this section is that for the migration velocity of the drop, including a contribution from convective transport of energy. This is given below.

$$v_\infty \sim 2A_0(1 + \epsilon^2\Omega + \cdots), \qquad (4.12.48)$$

where

$$\Omega(\alpha, \beta, \lambda) = -\frac{A_0^2 A_1}{22,050}[7(12,642 + 3,619\beta - 3,070\beta^2)$$
$$+ 7\alpha(13,818 + 4,431\beta - 3,210\beta^2)$$
$$- 2\frac{\beta}{\lambda}(11,916 + 17,925\beta + 6,510\beta^2)$$
$$- 2\frac{\alpha\beta}{\lambda}(10,404 + 16,665\beta + 6,510\beta^2)]. \qquad (4.12.49)$$

Ma (1998) has shown that the predictions from Equation (4.12.48) are adequate for small values of the Marangoni number in a few test cases by comparing them with those obtained from a numerical solution of the governing equations. Typically, the asymptotic prediction begins to deviate from the numerical results when the Marangoni number exceeds a value of approximately 5.

The result in the gas bubble limit $\alpha \to 0$, $\beta \to 0$ reduces to

$$v_\infty \sim \frac{1}{2} - \frac{301}{14,400}\epsilon^2 + \cdots. \qquad (4.12.50)$$

This prediction does not do as well as that in the more general case. Shankar and Subramanian (1988) tested it against the value obtained from a numerical solution of the governing equations. They found that it was not useful beyond a value of $Ma \approx 0.5$. Working with Padé approximants, the authors constructed an improved series that was useful up to $Ma \le 25$. We provide that result in Section 4.15, where results from numerical solutions are presented and discussed.

Subramanian (1981) made the statement that the correction to the scaled migration velocity at $O(\epsilon^3)$ will be zero because the corresponding inner temperature correction field will not have a $P_1(s)$ component. This assertion is based on the observation that the inhomogeneity in the equation for t_3 will consist only of even order harmonics. As Merritt (1988) noted, however, there will be a $P_1(s)$ component in the solution for t_3, which arises from the homogeneous solution. Therefore, it happens that the correction at $O(\epsilon^3)$ is not zero. In the bubble limit, Merritt provided the following result calculated with the help of the computer algebra program MACSYMA:

$$v_\infty \sim \frac{1}{2} - \frac{301}{14,400}\epsilon^2 + \frac{13}{1,440}\epsilon^3 - \frac{31}{56,448}\epsilon^4 + \cdots. \tag{4.12.51}$$

It is common to encounter terms of $O(\epsilon^n \log \epsilon)$ in problems of this type. We have not noticed them in the above results. In solving the problem of flow past a sphere at small values of the Reynolds number, Proudman and Pearson (1957) observed that a correction field in the inner solution at $O(Re^2)$ contains a term proportional to $r^2 \log r$ which, when rewritten in outer variables, produces a contribution that cannot be matched. Therefore, they concluded that the inner asymptotic sequence must contain a term of $O(Re^2 \log Re)$ to cancel the contribution that cannot be matched. Note that this is of lower order in the sequence than $O(Re^2)$. This term in turn leads to a corresponding logarithmic term in the outer asymptotic sequence. A similar situation is noted by Acrivos and Taylor (1962) in the analogous heat transfer problem. Merritt observed that in the thermocapillary migration problem, the correction field t_3 contains a term $-\frac{\log r}{3600 r^5} P_4(s)$. The origin of this term is not clear, however, and the algebra is too formidable to attempt to check the results by hand. If correct, this term would lead to a contribution at $O(\epsilon^8 \log \epsilon)$ when rewritten in outer variables. If this is not matched, its cancellation will require a term $O(\epsilon^3 \log \epsilon)$ in the inner field, which would precede the $O(\epsilon^3)$ term in the asymptotic sequence. A problem with this line of reasoning is that the corresponding contribution to the inner field at $O(\epsilon^3 \log \epsilon)$ that is proportional to $r^4 P_4(s)$ cannot be matched. As a practical matter, it would be difficult to carry out the calculation involved to settle the question, which must therefore be left open.

A corresponding result in the gas bubble limit has been obtained recently by Crespo, Migoya, and Manuel (1998), when the velocity field in the continuous phase is given by potential flow. The velocity components in that case are given in Equations (3.3.1) and (3.3.2). The procedure is similar, and the final result is

$$v_\infty \sim \frac{1}{2} - \frac{49}{2,880}\epsilon^2 + \cdots. \tag{4.12.52}$$

The Reynolds number must be large for potential flow to be a good approximation. Therefore, the Prandtl number must be small to yield a sufficiently small Marangoni number for the expansion to be useful. According to Crespo et al. (1998), Equation (4.12.52) predicts results within approximately 10% of the value obtained from a numerical solution of the governing equations when $Ma < 2$.

4.13   Influence of Convective Transport of Energy – Asymptotic Analysis for Large Marangoni Number in the Gas Bubble Limit

4.13.1   Introduction

In the previous section, we analyzed the situation when convective energy transport effects are relatively small compared with conduction. Here we consider the opposite extreme wherein convective transport of energy dominates over conduction. This corresponds to large values of the Marangoni number. It is possible to treat this problem analytically, using perturbation theory, only in two asymptotic limits in the context of momentum transport. The case of large values of the Reynolds number can be analyzed by using a potential flow velocity field in the continuous phase as a first approximation. The case of Stokes flow, which corresponds to negligible Reynolds number, also is amenable to analysis. The mathematical framework for both problems is the same, although physically these are two opposite limits. As usual, the Reynolds and Marangoni numbers are defined based on the properties of the continuous phase. The analysis in this section is confined to the gas bubble limit, wherein the physical property ratios $\alpha \to 0$, $\beta \to 0$. Detailed analyses of these problems have been provided by Crespo and Jiménez-Fernández (1992a, 1992b) and by Balasubramaniam and Subramanian (1996). In this section, we follow the development presented in Balasubramaniam and Subramanian (1996), giving the essential steps and results.

We already have encountered the assumption of large values of the Reynolds number in the context of gravity driven motion of bubbles and drops in Sections 3.3 and 3.4. The ideas developed in those sections will prove useful in the present discussion. This is the limit when inertia dominates over viscous forces. As noted in Section 3.3, although there is a momentum boundary layer at the surface of a bubble or drop in which viscous forces become important, only the stress field undergoes a change of $O(1)$ in this boundary layer, which has a thickness of $O(\frac{1}{\sqrt{Re}})$. The velocity field calculated from potential flow theory changes only by a quantity of $O(\frac{1}{\sqrt{Re}})$ in the boundary layer. Therefore, we can use the potential flow velocity field in the continuous phase as a good first approximation to the velocity everywhere in the energy equation, regardless of whether the Prandtl number is small or large. In the opposite extreme when we assume Stokes flow, the Reynolds number is negligible. To achieve a large value of the Marangoni number, the Prandtl number must be large. The practical utility of the solutions obtained here will be seen in subsequent sections, wherein comparisons are made with results from a numerical solution and with data obtained from reduced gravity experiments.

In this section, we consider the gas bubble limit which is analytically simpler and will serve to bring out the important ideas. In this limit, we must solve the transport problem only in the continuous phase. A significant question that needs to be answered concerns the behavior of the scaled migration velocity v_∞ in the limit as $Ma \to \infty$; does it approach a value of zero or a constant, or does it grow with increasing Ma? Interestingly, we find that it approaches a constant in the gas bubble limit but grows linearly with increasing Ma in other cases, as will be seen in the next section. In the development given here, we make the assumption that v_∞ is of $O(1)$ and find that this leads to a self-consistent analytical framework.

4.13.2 Governing Equations

Consider the steady motion of a gas bubble under the action of an applied temperature gradient under conditions wherein convective transport of energy dominates over conduction. A sketch of the system is given in Figure 3.1.1. We neglect deformation from the spherical shape, assuming the Capillary number to be sufficiently small. The scalings used in Section 4.2, and the physical assumptions made there, continue to apply, with the exception of that regarding the Reynolds and Marangoni numbers being negligible. Here, we assume that the Marangoni number is a large parameter and examine the problem in the asymptotic limit $Ma \to \infty$. Therefore, we introduce $\epsilon = \frac{1}{\sqrt{Ma}}$, and perform an asymptotic analysis in the limit $\epsilon \to 0$. The Reynolds number is permitted to be either negligible, in which case the solution for the streamfunction and velocity fields given in Equations (4.5.2) through (4.5.4) can be used, or it is assumed to be very large, in which case the potential flow solution in Equations (3.3.1) and (3.3.2) can be employed. Recall that these velocity distributions are in a reference frame attached to the moving bubble. In the case of Stokes motion, determination of the constants appearing in the infinite series for the velocity field components, and the scaled steady migration velocity of the bubble v_∞, requires a knowledge of the temperature distribution on the surface of the bubble. In potential flow, the only unspecified constant in the velocity distribution is the scaled migration velocity of the bubble, and we shall see that its determination also requires knowledge of the temperature distribution on the bubble surface. Therefore, we proceed to the solution of the energy equation.

As in Section 4.2, we use spherical polar coordinates (r, θ) with r being the radial coordinate scaled by the radius of the bubble and θ the polar angle measured from the forward stagnation streamline. As usual, we define $s = \cos\theta$ and use (r, θ) and (r, s) interchangeably as the set of independent variables. The energy equation for the scaled temperature field $T(r, s)$ is given below.

$$v_\infty + \mathbf{v} \bullet \nabla T = \epsilon^2 \nabla^2 T. \tag{4.13.1}$$

The symbol $\mathbf{v}$ represents the scaled velocity vector. Far from the bubble, the temperature field in the continuous phase must approach that in the undisturbed fluid.

$$T(r \to \infty, s) \to rs. \tag{4.13.2}$$

The assumption of negligible thermal conductivity of the gas inside the bubble leads to an adiabatic boundary condition at the surface of the bubble.

$$\frac{\partial T}{\partial r}(1, s) = 0. \tag{4.13.3}$$

4.13.3 Solution by the Method of Matched Asymptotic Expansions

A straightforward expansion, using the asymptotic sequence $F_j(\epsilon)$, can be written for the solution of the energy equation as follows:

$$T \sim F_0(\epsilon) T_0(r, s) + F_1(\epsilon) T_1(r, s) + \cdots. \tag{4.13.4}$$

We establish $F_0(\epsilon) = 1$ from the boundary condition given in Equation (4.13.2). We then substitute the expansion in Equation (4.13.4) into the governing energy equation and take the limit as $\epsilon \to 0$. This process yields the following governing equation for T_0:

$$v_{\infty,0} + \mathbf{v}_0 \bullet \nabla T_0 = 0. \tag{4.13.5}$$

The symbol $\mathbf{v}_0$ is used to designate the leading order result in the asymptotic expansion for the velocity field, and the symbol $v_{\infty,0}$ represents the leading order result for the scaled migration velocity. Note that the order of the energy equation has been reduced by one because the conduction term is absent. This reduces the number of boundary conditions that can be satisfied by the solution. It can be shown that we have lost the ability to satisfy the boundary condition at the bubble surface given in Equation (4.13.3).

The solution for T_0 can be expected to satisfy only the boundary condition far from the bubble.

$$T_0(r \to \infty, s) \to rs. \tag{4.13.6}$$

The inability to satisfy the entire set of boundary conditions, because of a reduction in the order of the differential equation, leads to the appearance of a boundary layer at the surface of the bubble in which conduction must be important. Therefore, for an entirely different reason from that in the case of small values of the Marangoni number considered in Section 4.12, the present problem also poses an exercise in matched asymptotic expansions. By requiring that the conduction term be as important as the convective transport term in the boundary layer, the thickness of the boundary layer can be established to be of $O(\epsilon)$. Therefore, we define an inner coordinate x as follows:

$$x = \frac{r - 1}{\epsilon}. \tag{4.13.7}$$

The temperature field in the boundary layer is designated $t(x, s)$. The outer field $T(r, s)$ will be expected to satisfy only the condition at $r \to \infty$ and match the inner field as $\epsilon \to 0$ while keeping the outer variables fixed. Similarly, the inner field $t(x, s)$ should satisfy the adiabatic boundary condition at the bubble surface given in Equation (4.13.3), and match the outer solution as $\epsilon \to 0$ while keeping the inner variables fixed. When a transformation is made to the inner variables (x, s), the following governing differential equation is obtained for the inner temperature field $t(x, s)$:

$$v_\infty + \frac{v_r}{\epsilon} \frac{\partial t}{\partial x} - \sqrt{1 - s^2} \frac{v_\theta}{1 + \epsilon x} \frac{\partial t}{\partial s}$$
$$= \frac{1}{(1 + \epsilon x)^2} \left[\frac{\partial}{\partial x} (1 + \epsilon x)^2 \frac{\partial t}{\partial x} + \epsilon^2 \frac{\partial}{\partial s} \left((1 - s^2) \frac{\partial t}{\partial s} \right) \right]. \tag{4.13.8}$$

As noted, this inner field will satisfy the adiabatic boundary condition at the surface of the bubble:

$$\frac{\partial t}{\partial x}(0, s) = 0. \tag{4.13.9}$$

It is evident from Equation (4.13.8) that conduction in the tangential direction is negligible at leading order. Superficially, the radial convective transport term appears to be asymptotically large, but the radial velocity vanishes at the bubble surface due to the kinematic condition. Therefore, this term is of the same order as the convective transport term in the tangential direction in the boundary layer. This can be seen by expanding the velocity field in a Taylor series about the interface in the inner variable x.

We now write an inner asymptotic expansion as follows:

$$t(x, s) \sim f_0(\epsilon) t_0(x, s) + f_1(\epsilon) t_1(x, s) + \cdots. \tag{4.13.10}$$

When this is substituted into the governing equation, and the limit as $\epsilon \to 0$ is taken, the leading order temperature field in the thermal boundary layer $t_0(x, s)$ can be shown to satisfy the following equation:

$$v_{\infty,0} + h_1(s) x \frac{\partial t_0}{\partial x} - h_2(s) \frac{\partial t_0}{\partial s} = \frac{\partial^2 t_0}{\partial x^2}. \tag{4.13.11}$$

Here,

$$h_1(s) = \frac{\partial v_{r,0}}{\partial r}(1, s) \tag{4.13.12}$$

and

$$h_2(s) = \sqrt{1 - s^2} v_{\theta,0}(1, s), \tag{4.13.13}$$

and the symbols $v_{r,0}$ and $v_{\theta,0}$ are used to designate the leading order components of the velocity field, when expanded for small ϵ while keeping r fixed. In writing the above, we have assumed that $f_0(\epsilon) = 1$. This assumption will be reexamined when matching the inner and outer fields. The field $t_0(x, s)$ must satisfy the adiabatic boundary condition at the bubble surface.

$$\frac{\partial t_0}{\partial x}(0, s) = 0. \tag{4.13.14}$$

Also, from symmetry, we can write $\frac{\partial t_0}{\partial \theta} = 0$ at $\theta = 0$. This implies that

$$\left| \frac{\partial t_0}{\partial s}(x, 1) \right| < \infty. \tag{4.13.15}$$

A similar symmetry condition also can be written on the outer field where needed. To proceed further, we must work with a specific velocity field. First, we consider the situation when the Reynolds number is large.

4.13.4 The Case of Large Reynolds Number

In the limit $Re \to \infty$, the potential flow velocity field is given in Equations (3.3.1) and (3.3.2). It is reproduced below, along with a result for the streamfunction, for convenience.

$$v_r(r, s) = -v_\infty \left(1 - \frac{1}{r^3} \right) s, \tag{4.13.16}$$

$$v_\theta(r, s) = v_\infty \left(1 + \frac{1}{2r^3} \right) \sqrt{1 - s^2}, \tag{4.13.17}$$

$$\psi(r, s) = \frac{1}{2} v_\infty \left(r^2 - \frac{1}{r} \right)(1 - s^2). \tag{4.13.18}$$

In steady potential flow, the hydrodynamic force exerted by the continuous phase on the bubble is always zero, regardless of the velocity of the bubble. Therefore, we cannot use the condition of zero hydrodynamic force to infer the velocity of the bubble unless we properly account for the momentum boundary layer. Instead, we use a dissipation argument along the lines presented in Section 3.3, which is valid at steady state. Over the surface of the bubble, any differential element of fluid experiences an unbalanced force caused by the interfacial tension gradient, which acts in the tangential direction. The rate at which work is done by the interfacial tension force on the element is given by the product of this force and the tangential velocity of the element. Integrated over the surface of the bubble, this leads to the total rate at which work is done on the continuous phase fluid, which must equal the rate at which energy is dissipated by viscosity. There is no contribution to the work done from the hydrodynamic force because that is identically zero. The dissipation integral is written in precisely the same manner as in Section 3.3. The result for the steady scaled velocity of the bubble, obtained by this route, is given below.

$$v_\infty = \frac{1}{2} \int_{-1}^{1} \frac{\partial t}{\partial s}(0, s) C_2(s) \, ds = \frac{1}{2} \int_{-1}^{1} t(0, s) P_1(s) \, ds. \tag{4.13.19}$$

The Gegenbauer Polynomials $C_n(s)$, and the Legendre Polynomials $P_n(s)$, have been encountered earlier in Section 3.1. Whereas we have derived Equation (4.13.19) using energy dissipation arguments, Crespo and Jiménez-Fernández (1992a) obtained the same result from the requirement that the mass flux in the θ-direction in the momentum boundary layer should be bounded at $\theta = \pi$.

First, consider the outer problem for the temperature field. Because there is no conduction in the outer equation, the temperature field of a fluid element moving along each streamline is independent of that of its neighbors and is solely determined by the balance between convective transport and the sink term. A formal solution can be written for the leading order outer field $T_0(r, s)$ by using the method of characteristics, after transforming to (ψ_0, r) as the independent variables.

$$T_0(r, s) = rs + \int_{r}^{\infty} \frac{\dfrac{3\psi_0 \tilde{r}}{v_{\infty,0}(\tilde{r}^3 - 1)} - 1}{(\tilde{r}^3 - 1)\sqrt{1 - \dfrac{2\psi_0 \tilde{r}}{v_{\infty,0}(\tilde{r}^3 - 1)}}} d\tilde{r}. \tag{4.13.20}$$

Here, ψ_0 is the leading order result for the streamfunction when it is expanded in the same way as the temperature field is expanded in Equation (4.13.4). While Equation (4.13.20) is formally correct, we cannot perform the integration analytically to obtain an explicit solution at all spatial locations. Therefore, we pursue a result useful in the vicinity of the surface of the bubble. For this, we first write an alternative result for T_0, also obtained from the method of characteristics, after transforming to (ψ_0, s) as the independent variables.

$$T_0(r, s) = G(\psi_0) + \int^{s} \frac{2r^4}{1 + 2r^3} \frac{d\tilde{s}}{1 - \tilde{s}^2}. \tag{4.13.21}$$

On the right side, the integrand must be expressed as a function of ψ_0 and $\tilde{s}$. Because ψ_0 appears as a parameter in Equations (4.13.20) and (4.13.21), we can expand T_0 about a given value of ψ_0, specifically $\psi_0 = 0$. This is tantamount to obtaining an expansion for the thin bundle of streamlines surrounding the axis of symmetry. This is the region where we need the outer solution for matching purposes. Equation (4.13.20) is used to expand T_0 in the vicinity of $\theta = 0$ for arbitrary r. This provides a result for the solution from the point at infinity to a location close to the surface of the bubble, for small values of θ. Equation (4.13.21) is used to expand T_0 in the vicinity of $r = 1$ for $0 \le \theta \le \pi$. These two expressions for T_0 are then patched in the neighborhood of $r = 1$ and $\theta = 0$ so that they are equivalent representations of the same function. This process yields

$$G(\psi_0) \sim a_0 + a_1 \log \psi_0 + a_2 \psi_0 + a_3 \psi_0 \log \psi_0, \tag{4.13.22}$$

and the constants are determined by the patching procedure outlined above. In this way, it is possible to obtain the following result for T_0 near $r = 1$ that is good to $O(\psi_0)$

from Equation (4.13.21):

$$T_0(r,s) = \left(1 + \frac{\pi}{6\sqrt{3}} - \frac{1}{6}\log 432\right) - \frac{1}{18}\left(\frac{\pi}{\sqrt{3}} + \log 432\right)\left(r^2 - \frac{1}{r}\right)(1 - s^2)$$

$$+ \frac{1}{3}\log\left(r^2 - \frac{1}{r}\right) + \frac{2}{3}\log(1+s) + \frac{2}{9}\left(r^2 - \frac{1}{r}\right)s$$

$$+ \frac{1}{9}\left(r^2 - \frac{1}{r}\right)\log\left(r^2 - \frac{1}{r}\right)(1 - s^2) + \frac{2}{9}\left(r^2 - \frac{1}{r}\right)(1 - s^2)\log(1+s).$$

(4.13.23)

This contains sufficient information for matching purposes because it is only the be-havior of the outer solution as $r \to 1$ that is needed. In particular, T_0 is logarithmically singular as the bubble surface is approached. This behavior must be rectified by the in-ner solution. Also, note the appearance of a logarithmic singularity in the outer solution as $s \to -1$, which corresponds to the rear stagnation line, $\theta = \pi$. This is a consequence of the absence of conduction in the θ-direction in the governing equation for T_0. This singularity will be seen to persist in the inner solution as well.

It is possible to demonstrate that the correction to T_0 occurs at $o(\epsilon)$, but we do not pursue higher-order terms. Instead, we proceed to obtain the inner solution. The solution of Equation (4.13.11) for the leading order inner field, $t_0(x, s)$, along with the boundary conditions given in Equations (4.13.14) and (4.13.15), can be written as follows:

$$t_0 = c_1 + \log(f(s)) - \frac{1}{3}\log(h_2(s)) + \frac{1}{2K}\log(\xi(s)) + \frac{2}{K}F(\zeta).$$

(4.13.24)

Here, c_1 and K are constants whose values are to be established by matching, and

$$f(s) = (1+s)^{\frac{2}{3}},$$

(4.13.25)

$$\frac{d\xi}{ds} = -h_2(s),$$

(4.13.26)

$$\xi(s) = \frac{1}{2}v_{\infty,0}(1-s)^2(2+s),$$

(4.13.27)

$$\zeta = \frac{h_2(s)x}{2\sqrt{\xi(s)}},$$

(4.13.28)

and

$$F(\zeta) = \int_0^\zeta D(w)\,dw = \int_0^\zeta e^{-w^2}\left(\int_0^w e^{z^2}\,dz\right)dw.$$

(4.13.29)

When integrating Equation (4.13.26) to obtain Equation (4.13.27), the condition $\xi(1) = 0$ has been used. The function $D(w)$, appearing in Equation (4.13.29), is known as *Dawson's function*, and its values are tabulated in references provided in Abramowitz and Stegun (1965). As mentioned earlier, the inner solution also contains a logarithmic singularity as $s \to -1$, which corresponds to the rear stagnation line, $\theta = \pi$, because of the absence of conduction in the θ-direction in the inner equation for the temperature field.

Now, we can proceed to match the inner and outer solutions at leading order. When the outer solution T_0 is rewritten in inner variables, expanded for small ϵ, truncated at

$O(1)$ and rewritten in outer variables, we obtain

$$T_0 \sim 1 + \frac{\pi}{6\sqrt{3}} - \frac{1}{6}\log 48 + \frac{1}{3}\log(r-1) + \log\left(f(s)\right). \tag{4.13.30}$$

Likewise, the inner solution t_0 is rewritten in outer variables, expanded for small ϵ and truncated at $O(1)$. It is necessary to use the asymptotic result that as $\zeta \to \infty$, $F(\zeta) \to \frac{1}{2}\log\zeta + 0.4909$. The final result is

$$t_0 \sim c_1 + \frac{1}{K}(0.9818 - \log 2) + \log(f(s)) + \left(\frac{1}{K} - \frac{1}{3}\right)\log(h_2(s))$$

$$+ \frac{1}{K}\log(r-1) - \frac{1}{K}\log\epsilon. \tag{4.13.31}$$

It is evident that matching requires us to set the constant $K = 3$. Notice, however, the appearance of the term $\frac{1}{K}\log\epsilon$ that cannot be matched. This implies that the inner field must begin at $O(\log\epsilon)$ instead of at $O(1)$ as initially assumed. It is straightforward to establish that the only solution at that order, which is consistent with the outer solution, is precisely the constant $\frac{1}{3}\log\epsilon$, which will cancel the corresponding $\log\epsilon$ term in Equation (4.13.31). After obtaining the constant c_1, the inner solution at $O(1)$, incorporating the constant $\frac{1}{3}\log\epsilon$, which appears at leading order, can be written as follows:

$$t_0 = \frac{1}{3}\log\epsilon + 0.6727 + \frac{\pi}{6\sqrt{3}} - \frac{1}{6}\log(27v_{\infty,0}^2) + \frac{2}{3}\log(1+s) - \frac{1}{3}\log(1-s^2)$$

$$+ \frac{1}{6}\log(\xi(s)) + \frac{2}{3}F(\zeta). \tag{4.13.32}$$

The appearance of a leading term in the inner expansion that is proportional to $\log\epsilon$ deserves comment. This implies that the surface of the bubble will become progressively cooler as the Marangoni number is increased. This behavior can be understood by examining the temperature field in a bundle of streamlines surrounding the forward stagnation streamline. We use Figure 3.1.1 and discuss the physical picture from the perspective of a reference frame attached to the bubble. As fluid elements move toward the bubble in the negative z-direction, Equation (4.13.5) predicts that they lose heat because of the sink term and become cooler. The temperature change of a fluid element from an upstream location to one near the bubble is proportional to the transit time of the fluid element. Far from the symmetry axis, and similarly, far from the bubble along the forward stagnation streamline, the elements are moving at a constant velocity, $-v_\infty\mathbf{k}$, and losing energy at a constant rate. This naturally yields the linear variation of temperature with distance that is imposed as a boundary condition in the undisturbed fluid. In the bundle of streamlines of interest, however, as the fluid elements approach the bubble, they move at a smaller velocity. This causes a greater cooling of these elements of fluid for a given traverse distance, when compared with elements that are located far from the symmetry axis. In fact, the velocity near the forward stagnation point, along the stagnation streamline, decreases to a value of zero with distance from the bubble surface, leading to a logarithmic divergence of the transit times of the elements along that line. This is the origin of the leading term in the inner expansion, which is proportional to $\log\epsilon$. As a corollary, we can conclude that the time taken to achieve a steady temperature field and therefore to approach a steady migration velocity also will scale as $\log\frac{1}{\epsilon}$. Examining the situation from the perspective of an observer situated in the laboratory, of course, there is no sink term. As the bubble moves along, it carries pockets of relatively cool fluid with it along the forward and rear stagnation regions, which leads to the same physical picture.

When the inner solution given in Equation (4.13.32) is used in Equation (4.13.19), we obtain the leading order result for the scaled migration velocity of the bubble as

$$v_{\infty,0} = \frac{1}{3} - \frac{1}{8} \log 3 \approx 0.1960. \tag{4.13.33}$$

It is possible to proceed further with the analysis. Rewriting the outer solution to $O(1)$ in inner variables shows that the inner expansion must proceed as follows:

$$t(x,s) \sim \frac{1}{3} \log \epsilon + t_0(x,s) + \epsilon \log \epsilon \, t_{\ell 1}(x,s) + \epsilon \, t_1(x,s) + \cdots. \tag{4.13.34}$$

Note the appearance of a term of $O(\epsilon \log \epsilon)$, which is forced by the matching requirement with the outer solution. As noted in Section 3.2, we must also retain the next higher order term of $O(\epsilon)$ in the analysis. The corresponding expansion for v_∞ must be of the form

$$v_\infty \sim v_{\infty,0} + \epsilon \log \epsilon \, v_{\infty,\ell 1} + \epsilon v_{\infty,1} + \cdots. \tag{4.13.35}$$

The governing equation and boundary conditions for $t_{\ell 1}(x,s)$ are given below.

$$-v_{\infty,0}\left(3xs\frac{\partial t_{\ell 1}}{\partial x} + \frac{3}{2}(1-s^2)\frac{\partial t_{\ell 1}}{\partial s}\right) = \frac{\partial^2 t_{\ell 1}}{\partial x^2} - \frac{v_{\infty,\ell 1}}{v_{\infty,0}}\frac{\partial^2 t_0}{\partial x^2}, \tag{4.13.36}$$

$$\frac{\partial t_{\ell 1}}{\partial x}(0,s) = 0, \tag{4.13.37}$$

$$\left|\frac{\partial t_{\ell 1}}{\partial s}(x,1)\right| < \infty, \tag{4.13.38}$$

$$t_{\ell 1}(x \to \infty, s) \to \frac{1}{3}x(1-s^2). \tag{4.13.39}$$

Equation (4.13.37) is the result of substituting the inner expansion into the adiabatic boundary condition given in Equation (4.13.9). Equation (4.13.38) is obtained from symmetry at $\theta = 0$, and Equation (4.13.39) is obtained from matching requirements with the outer solution. The solution for $t_{\ell 1}$ is

$$t_{\ell 1}(x,s) = \frac{1}{3}x(1-s^2) + \frac{4\sqrt{\xi}}{9v_{\infty,0}}\left[\frac{1}{\sqrt{\pi}}e^{-\zeta^2} - \zeta \operatorname{erfc}\zeta\right] - \frac{1}{6}\frac{v_{\infty,\ell 1}}{v_{\infty,0}}D'(\zeta). \tag{4.13.40}$$

When this solution is substituted into Equation (4.13.19), we obtain the corresponding contribution to the scaled migration velocity $v_{\infty,\ell 1}$ as

$$v_{\infty,\ell 1} = \sqrt{\frac{1}{72\pi v_{\infty,0}}}\left(\frac{64}{21} - \frac{16}{7}\sqrt{3}\right) \approx -0.1369. \tag{4.13.41}$$

In a like manner, the governing equation and boundary conditions for $t_1(x,s)$ can be written as

$$v_{\infty,0}\left(2x - 3xs\frac{\partial t_1}{\partial x} - \frac{3}{2}(1-s^2)\frac{\partial t_1}{\partial s}\right) = \frac{\partial^2 t_1}{\partial x^2} + 2\frac{\partial}{\partial x}\left(x\frac{\partial t_0}{\partial x}\right) - \frac{v_{\infty,1}}{v_{\infty,0}}\frac{\partial^2 t_0}{\partial x^2}, \tag{4.13.42}$$

$$\frac{\partial t_1}{\partial x}(0,s) = 0, \tag{4.13.43}$$

and

$$\left|\frac{\partial t_1}{\partial s}(x, 1)\right| < \infty. \tag{4.13.44}$$

Equation (4.13.43) arises from the adiabatic boundary condition at the surface of the bubble, and Equation (4.13.44) is a consequence of symmetry at $\theta = 0$. In addition, we must require the solution for t_1 to satisfy matching requirements. The details of the solution are lengthy and are given in Balasubramaniam and Subramanian (1996). Here, we report only the final contribution to v_∞, which is obtained after performing a double integration numerically:

$$v_{\infty,1} \approx 0.6578. \tag{4.13.45}$$

This permits us to write a result for v_∞ good to $O(\epsilon)$:

$$v_\infty \sim \left(\frac{1}{3} - \frac{1}{8}\log 3\right) - 0.1369\epsilon \log \epsilon + 0.6578\epsilon + \cdots. \tag{4.13.46}$$

Balasubramaniam (1995) obtained a numerical solution of the problem considered here for values of the Marangoni number ranging from 0 to 2000, using a potential flow velocity profile. He found the result for v_∞ approaching a value of 0.2 for large Ma. More recently, Crespo et al. (1998) have performed numerical computations for the case of potential flow, extending the maximum value of the Marangoni number to 50,000. These authors reported that their numerical solution for v_∞ is in agreement with the prediction from Equation (4.13.46) for $Ma > 15$, with an error of the order of ten percent. Judging from a figure presented in their article, the agreement is much better for $Ma \geq 500$.

4.13.5 Temperature Field in the Thermal Wake

Now, we briefly discuss the thermal wake behind the bubble. As noted earlier, the singularity along the rear stagnation streamline $\theta = \pi$, in both the inner and outer solutions, is caused by the neglect of conduction in the θ-direction. This can be relieved by obtaining a separate inner solution valid near that streamline. To distinguish this solution from the boundary layer solution over the entire surface of the bubble, we shall designate the new solution as the wake solution. We work only with the leading order fields obtained above, dropping the subscript 0 for convenience. In the wake region, an angular coordinate ϕ is defined as follows:

$$\phi = \pi - \theta. \tag{4.13.47}$$

We need an outer solution in the wake region in the angular coordinate, which we designate as the ϕ-outer solution, that is good for small values of ϕ. For this purpose, it is convenient to write a composite solution, which is obtained by adding appropriate forms of the r-outer and r-inner solutions at leading order and subtracting the common part. There are some difficulties in doing this in the vicinity of the rear stagnation point, as discussed in Balasubramaniam and Subramanian (1996). To obtain a valid composite solution in that region, it becomes necessary to define a new variable η as follows:

$$\eta = \sqrt{\frac{v_{\infty,0}}{32}} \frac{r^2 - \frac{1}{r}}{\epsilon} \phi^2. \tag{4.13.48}$$

The variable η reduces to ζ defined in Equation (4.13.28) as $(r-1) \to 0$ and $\phi \to 0$. The r-composite solution in the wake, $T_{wake}(r, \phi)$, which serves as an outer wake solution, can be written as follows:

$$T_{wake} = \frac{1}{3} \log \epsilon + g(r) + \frac{2}{3} \log \phi + \frac{2}{3} F(\eta). \tag{4.13.49}$$

Here, the function $g(r)$ is given by

$$g(r) = 1.6727 - \frac{\pi}{6\sqrt{3}} - \frac{1}{6} \log(5832 v_{\infty,0}) - r - \frac{1}{3} \log r$$
$$+ \frac{1}{2} \log(r^2 + r + 1) + \frac{1}{\sqrt{3}} \arctan\left(\frac{2r+1}{\sqrt{3}}\right). \tag{4.13.50}$$

The function F already has been defined in Equation (4.13.29).

It is useful to provide an estimate of the thickness of the thermal wake region behind the bubble. We already know that the scaled thickness of the thermal boundary layer around the bubble is $O(\epsilon)$. Therefore, let us set the physical thickness of this boundary layer equal to $cR\epsilon$, where c is a numerical constant of $O(1)$, and R is the radius of the bubble. Using this estimate around the equatorial region of the bubble, we can evaluate the order of magnitude of the mass flow rate of fluid in the boundary layer as $2\pi c\rho R^2 v_0 \epsilon$, where ρ is the density of the liquid. In obtaining this result, we have used the fact that the order of magnitude of the velocity in the boundary layer is v_0, which is the reference velocity, defined in Equation (4.1.1). The fluid in the thermal boundary layer must exit from behind the bubble in the thermal wake, which is approximately a cylindrical region of radius R_{wake}. Because the velocity of this fluid is of the same order as v_0, we conclude that $R_{wake} \sim R\sqrt{2c\epsilon}$. Therefore, the scaled thickness of the wake is $O(\sqrt{\epsilon}) = O(\frac{1}{Ma^{\frac{1}{4}}})$.

We now define an inner variable for the wake as

$$y = \frac{\phi}{\epsilon}. \tag{4.13.51}$$

This permits the accommodation of conduction in the tangential direction and therefore relieves the singularity in T_{wake} as $\phi \to 0$. The temperature field in the inner wake region, $t_{wake}(r, y)$, satisfies the following governing equation and boundary conditions:

$$v_{\infty,0}\left[1 + \left(1 - \frac{1}{r^3}\right)\frac{\partial t_{wake}}{\partial r} - \frac{y}{r}\left(1 + \frac{1}{2r^3}\right)\frac{\partial t_{wake}}{\partial y}\right] = \frac{1}{r^2 y}\frac{\partial}{\partial y}\left(y\frac{\partial t_{wake}}{\partial y}\right), \tag{4.13.52}$$

$$\frac{\partial t_{wake}}{\partial y}(r, 0) = 0, \tag{4.13.53}$$

and

$$\left|\frac{\partial t_{wake}}{\partial r}(1, \phi)\right| < \infty. \tag{4.13.54}$$

Equation (4.13.53) represents a symmetry condition at the rear stagnation point. Equation (4.13.54) precludes the occurrence of a term of the form $\log(r-1)$ in the solution for t_{wake}. It is not possible to impose the adiabatic boundary condition in the r-outer variable. Equivalently, however, we can impose the condition that the r-inner temperature field in the inner wake region should satisfy the adiabatic condition

$$\frac{\partial t_{wake}}{\partial x}(x = 0, y) = 0, \tag{4.13.55}$$

where we have explicitly mentioned the x-dependence. The inner solution in the wake region can be obtained using a similarity transformation as

$$t_{wake} = \log \epsilon + g(r) + \frac{2}{3} \log y + \frac{1}{3} E_1 \left[\frac{1}{4} v_{\infty,0} y^2 \left(\frac{r^2 + r + 1}{r} \right) \right], \tag{4.13.56}$$

where

$$E_1(w) = \int_w^\infty \frac{e^{-p}}{p} \, dp \tag{4.13.57}$$

is an exponential integral. The properties of such integrals can be found in Abramowitz and Stegun (1965). It can be seen from Equation (4.13.51) that the scaled thickness of the inner wake is $O(\epsilon)$.

4.13.6 The Case of Negligible Reynolds Number

Here, we proceed along lines similar to those adopted in the case of large Reynolds number. The streamfunction and velocity fields in Stokes flow in the continuous phase are given by the solution presented in Equations (4.5.2) through (4.5.4), after setting $\alpha = 0$ to accommodate the gas bubble limit. For convenience, the results for the velocity components and the streamfunction are reproduced here after substituting for D_n in terms of I_n using Equation (4.5.10) and setting $D_2 = 0$ because the hydrodynamic force on the bubble is zero.

$$v_r(r, s) = \frac{I_2}{2} \left(1 - \frac{1}{r^3} \right) s + \frac{1}{4} \sum_{n=3}^\infty n(n-1) I_n \left(\frac{1}{r^{n-1}} - \frac{1}{r^{n+1}} \right) P_{n-1}(s),$$
$$\tag{4.13.58}$$

$$v_\theta(r, s) = -\frac{I_2}{2} \left(1 + \frac{1}{2r^3} \right) \sqrt{1 - s^2} + \frac{1}{4} \sum_{n=3}^\infty n(n-1) I_n \left(\frac{n-3}{r^{n-1}} - \frac{n-1}{r^{n+1}} \right) \frac{C_n(s)}{\sqrt{1 - s^2}},$$
$$\tag{4.13.59}$$

$$\psi(r, s) = -\frac{I_2}{4} \left(r^2 - \frac{1}{r} \right) (1 - s^2) + \frac{1}{4} \sum_{n=3}^\infty n(n-1) I_n \left(\frac{1}{r^{n-1}} - \frac{1}{r^{n-3}} \right) C_n(s).$$
$$\tag{4.13.60}$$

In the above, I_n is given by

$$I_n = -\int_{-1}^1 C_n(s) \frac{\partial t}{\partial s} (0, s) \, ds = -\int_{-1}^1 P_{n-1}(s) t(0, s) \, ds. \tag{4.13.61}$$

The scaled velocity of the bubble v_∞ is obtained from Equation (4.5.15) as

$$v_\infty = -\frac{1}{2} I_2 = \frac{1}{2} \int_{-1}^1 t(0, s) P_1(s) \, ds. \tag{4.13.62}$$

This result is identical to that obtained in the case of large Reynolds number through different arguments and given in Equation (4.13.19).

Because the velocity field is given as an infinite series, algebraic complexity is a significant factor in our ability to make progress with the analysis. Therefore, we have only obtained the leading order solutions for the outer and inner fields and a leading order result for the scaled migration velocity of the bubble. Even though the leading order outer equation can be solved in principle by the method of characteristics, this becomes a practical approach only if we restrict the analysis to a bundle of streamlines that are adjacent to the bubble surface. As noted earlier, this proves sufficient for matching purposes. Therefore, we expand the velocity field, appearing in Equation (4.13.5), in Taylor series in $(r - 1)$ and retain the leading order term for each component. This makes it possible to write the solution for $T_0(r, s)$ as

$$T_0(r, s) = c_2 + \frac{1}{A}\log(r - 1) + \log(f(s)), \tag{4.13.63}$$

where c_2 is a constant. In writing this result, we have used a symmetry condition similar to that on the inner field, which is given in Equation (4.13.15). Note the similarity between Equation (4.13.63) and Equation (4.13.30), which is the solution obtained for the same region in the limit as $Re \to \infty$. In Equation (4.13.63), the constant A is given by

$$A = 3 - \frac{1}{2v_{\infty,0}}\sum_{n=3}^{\infty} n(n - 1)I_{n,0}, \tag{4.13.64}$$

where $I_{n,0}$ and $v_{\infty,0}$ represent leading order results obtained from Equations (4.13.61) and (4.13.62), respectively, when the inner expansion is substituted into them. It can be shown that the constant A is proportional to the viscous normal stress on the bubble surface at the forward stagnation point. The function $f(s)$ appearing in Equation (4.13.63) satisfies the following differential equation:

$$\frac{1}{f}\frac{df}{ds} = \frac{v_{\infty,0}(1 - \frac{3}{4}s) + \frac{1}{2A}\sum_{n=3}^{\infty} n(n - 1)I_{n,0}P_{n-1}(s)}{\frac{3}{2}v_{\infty,0}(1 - s^2) - \frac{1}{2}\sum_{n=3}^{\infty} n(n - 1)I_{n,0}C_n(s)}. \tag{4.13.65}$$

Now, we can proceed to Equations (4.13.11) through (4.13.15) for the leading order inner field $t_0(x, s)$. They can be solved in the same manner as in the case $Re \to \infty$. We obtain Equation (4.13.24) again, but the definitions of some of the symbols appearing in it are different. The function $f(s)$ is the solution of Equation (4.13.65), along with the requirement that it be bounded. The definitions of $h_1(s)$ and $h_2(s)$ in Equations (4.13.12) and (4.13.13), respectively, continue to hold as long as the velocity components in the Stokes flow problem, given in Equations (4.13.58) and (4.13.59), are used. The function $\xi(s)$ satisfies Equation (4.13.26) with this proviso. Integration, along with $\xi(1) = 0$, yields

$$\xi(s) = \frac{3}{4}I_{2,0}\left(s - \frac{s^3}{3} - \frac{2}{3}\right) + \frac{1}{2}\sum_{n=3}^{\infty}\frac{n(n - 1)}{2n - 1}I_{n,0}(C_{n+1}(s) - C_{n-1}(s)). \tag{4.13.66}$$

The variable ζ in Equation (4.13.28) continues to be defined in the same way with the understanding that one must use ξ from Equation (4.13.66).

Matching yields the same consequences. The constant K must be equal to the constant A, but the value of A is only known in terms of the coefficients $I_{n,0}$ as seen from Equation (4.13.64). A logarithmic term, of the form $\frac{1}{A}\log\epsilon$, must be added to the inner field for the same reason as in the earlier problem. The constant term appearing in the inner solution cannot be easily established, but this is not a concern if one is interested in obtaining the migration velocity at leading order. Because Equation (4.13.62) shows the migration velocity to be $-\frac{1}{2}I_2$, the result at leading order is

$$v_{\infty,0} = -\frac{1}{2}I_{2,0}. \tag{4.13.67}$$

Unfortunately, $I_{2,0}$ cannot be evaluated in isolation. It is necessary to solve for the entire set of constants $I_{n,0}$ after truncating a nonlinear set of algebraic equations for them, which is obtained by substituting the solution for $t_0(x, s)$ in Equation (4.13.61). The techniques used are described in detail by Balasubramaniam and Subramanian (1996). The solution is obtained by an iterative procedure employing underrelaxation. A straightforward iteration fails if one uses Equation (4.13.64) to calculate intermediate values of the constant A. If the set is truncated at N, odd values of N do not yield a converged set of $I_{n,0}$ by a fixed point iteration method. When even values of N between 10 and 150 are used, the numerically determined values of both A and the migration velocity $v_{\infty,0}$ are found to increase monotonically with increasing N. The problem is resolved by recognizing that the series in Equation (4.13.64) for A is poorly behaved. Balasubramaniam and Subramanian (1996) found that the terms alternate in sign and used an Euler transformation to improve its convergence. They provided a table of values of $I_{2,0}$ to $I_{25,0}$ and obtained

$$A = 2.406 \tag{4.13.68}$$

and

$$v_{\infty,0} = 0.1538. \tag{4.13.69}$$

The difficulty in making the above calculation is caused in part by a singularity, which is present at $\theta = \pi$, in the temperature gradient in the tangential direction on the bubble surface. The origin of this singularity is the absence of conduction in the θ-direction in the inner equation at leading order. A scaling argument, balancing convective transport in the θ-direction against the sink term in the vicinity of the rear stagnation point, suggests that $\frac{\partial t_0}{\partial s} \sim (1 + s)^{-\frac{3}{4}}$ in that region. This singularity in the gradient is integrable when calculating the migration velocity and the coefficients $I_{n,0}$. Balasubramaniam and Subramanian (1996) also found that their numerical results, used in conjunction with a Domb–Sykes plot of the coefficients in the expansion of the viscous normal stress, yield a value close to the result estimated by scaling arguments. The use of the Domb–Sykes plot to investigate singularities of functions represented by infinite series is discussed by van Dyke (1975).

From Equations (4.13.33) and (4.13.69), which give leading order results in the limit $Ma \to \infty$, we see that the scaled migration velocity in the case $Re \to 0$ is not very different from that in the case $Re \to \infty$. The value obtained using potential flow is larger than the value in the Stokes flow limit by only approximately 27.4%. The tendency of the scaled migration velocity to increase gently with increasing values of Reynolds number, at fixed Marangoni number, is exhibited over the entire range of values of Ma. This will be seen in Section 4.15, where we present results from a numerical solution. We discuss the physical reason for this behavior at that point.

We conclude with a discussion of the behavior of the scaled temperature difference over the surface of the bubble. This can be obtained by integrating the tangential stress balance at the surface as

$$\Delta t = -\sum_{n=1}^{\infty} (4n - 1) I_{2n}. \tag{4.13.70}$$

In a numerical procedure, only partial sums Δt_N can be evaluated by truncating the infinite series at some finite value of the index $n = N$. By plotting the values of these partial sums as a function of N, Balasubramaniam and Subramanian (1996) found that these sums are well fitted by the relationship

$$\Delta t_N = \frac{a}{\sqrt{N}} + b. \tag{4.13.71}$$

The constant b is found to be approximately 1.5715 and represents the extrapolated value of the scaled temperature difference over the surface of the bubble as $N \to \infty$. In contrast, when convective transport of energy is negligible, from the results in Section 4.2, we can write $\Delta t = 3$.

It is not possible to obtain a similar result for Δt in the case when $Re \to \infty$ because the tangential stress balance cannot be satisfied without including the momentum boundary layer, which we have ignored in the analysis. Also, unlike in the case $Re \to 0$, the singularity in the temperature gradient along the surface, which occurs at the rear stagnation point, is not integrable. From the analysis of the thermal wake region given here, we see that Δt over the surface of the bubble scales as $-\frac{2}{3}\log\epsilon$. Therefore, this quantity would become unbounded as $\epsilon \to 0$. From a numerical solution of this problem, Balasubramaniam (1995) found that Δt first decreases with increasing Ma for small values of Ma but increases subsequently with increasing Ma as noted above. A numerical solution of the governing Navier–Stokes and energy equations, obtained by Balasubramaniam and Lavery (1989), does not display such behavior. Balasubramaniam and Subramanian (1996) speculated that velocity corrections in the momentum wake, which are not included in the asymptotic analysis performed here, might be important in influencing Δt over the surface of the bubble. The issue of calculating the correct Δt over the surface of the bubble, in the case of large values of the Reynolds number, is unresolved at this time.

4.14 Influence of Convective Transport of Energy – Asymptotic Analysis for Large Marangoni Number for a Drop

4.14.1 Introduction

In the previous section, we analyzed the asymptotic problem, wherein convective transport of energy dominates over conduction, in the gas bubble limit. Here, we consider the more general case of a drop, so that the relevant physical property ratios are not negligible. For a gas bubble, we found that the scaled migration velocity approaches a constant value as the Marangoni number approaches infinity. This is true both when the Reynolds number approaches zero and when it approaches infinity. The constant values are only slightly different from each other. Therefore, we might speculate that the same will be true in the generalization to a drop, with the constant depending on the physical property ratios. In this section, we show that this is not the case. When we consider the transport problem within the drop, coupled to that outside, the scaled migration velocity increases with increase in the Marangoni number. In physical terms, we find that the velocity of a drop is proportional to the square of the temperature gradient and the cube of the radius of the drop in this limiting situation. This observation is in contrast to that in the gas bubble limit, in which the linear dependence on the radius and the applied temperature gradient, which occurs when convective heat transport effects are negligible, is preserved in the opposite situation when convective heat transport dominates. We show how this remarkable feature is a consequence of the demand for energy made by a drop as it moves into warmer surroundings.

We restrict the present analysis to the case of $Re \to \infty$, which is tractable. In this limit, we showed in Section 3.4 that the leading order velocity field outside the drop is given by that in potential flow and that within the drop is Hill's spherical vortex. When the Reynolds number is large, at leading order, we can neglect the contributions from the corrections to these fields caused by viscous effects, which occur in thin momentum boundary layers within and outside the drop. Therefore, we proceed directly to obtain the solution of the coupled energy equations in the two phases, along with suitable boundary conditions, using the known velocity distribution. From this solution, the temperature field on the surface of the drop can be obtained, which permits us to calculate the migration velocity of the drop.

In the limit when the thermal conductivity of the drop phase becomes negligibly small, compared with that of the continuous phase, the energy transport problem in the continuous phase is uncoupled from that within the drop. In this limit, we should expect to recover the results from Section 4.13. We show that this is indeed the case. The leading order contribution to the scaled migration velocity of a drop, which grows with increasing values of the Marangoni number, vanishes. The next higher-order problem is precisely that posed in Section 4.13. Therefore, we recover the correct leading order result in the gas bubble limit.

The following analysis is based on the development given in Balasubramaniam and Subramanian (2000).

4.14.2 Governing Equations and Asymptotic Scalings

Consider the steady migration of a drop in a fluid of infinite extent. Figure 3.1.1 can be used as a sketch of the system, showing the spherical polar coordinates that are used in the analysis. All the assumptions made in Section 4.2 apply, with the exception that convective transport of momentum and energy both play a dominant role here, whereas in that section, we considered the opposite limit when both were negligible. Deformation from the spherical shape is neglected, assuming that a suitably defined Weber number is sufficiently small. We use the scalings introduced in Section 4.1. The radial coordinate, scaled by the drop radius R, is r and the polar angle, measured from the forward stagnation streamline, is θ. As usual, we use $s = \cos\theta$. The ratios of the viscosity, thermal conductivity, density, and thermal diffusivity of the drop phase to the corresponding properties of the continuous phase are designated by the symbols α, β, γ, and λ, respectively. The Reynolds and Marangoni numbers in the continuous phase, designated by Re and Ma, respectively, are defined in Section 4.1.

The drop moves at a steady scaled migration velocity v_∞, and we use a reference frame attached to the drop in the analysis. In a natural manner, we find that the analysis must be performed in the asymptotic limit when $\epsilon \to 0$, where $\epsilon = \frac{1}{\sqrt{Ma v_\infty}}$. Note the inclusion of the scaled migration velocity of the drop in the definition of ϵ. This is an important difference from the analysis performed in Section 4.13. Of course, we do not know the dependence of v_∞ on ϵ at this point. We only assume that $\epsilon \to 0$ as $Ma \to \infty$ and verify this assumption a posteriori.

The energy equations for the scaled temperature fields $T(r, s)$ in the continuous phase, and $T'(r, s)$ within the drop, are given below.

$$1 + \mathbf{v} \bullet \nabla T = \epsilon^2 \nabla^2 T, \tag{4.14.1}$$

$$1 + \mathbf{v}' \bullet \nabla T' = \lambda \epsilon^2 \nabla^2 T'. \tag{4.14.2}$$

When $Re \to \infty$, the leading order results for the velocity fields given in Section 3.4, which are good to $O(Re^{-\frac{1}{2}})$, can be used. The fields $\mathbf{v}$ and $\mathbf{v}'$, which appear in Equations (4.14.1) and (4.14.2), respectively, have been normalized by division by v_∞. Their components are given below.

$$v_r(r, s) = -\left(1 - \frac{1}{r^3}\right)s, \tag{4.14.3}$$

$$v_\theta(r, s) = \left(1 + \frac{1}{2r^3}\right)\sqrt{1 - s^2}, \tag{4.14.4}$$

$$v_r'(r, s) = \frac{3}{2}(1 - r^2)s, \tag{4.14.5}$$

$$v_\theta'(r, s) = 3\left(r^2 - \frac{1}{2}\right)\sqrt{1 - s^2}. \tag{4.14.6}$$

Note that the unknown drop velocity v_∞ appears in the problem only implicitly, through the quantity ϵ. Because in potential flow we are unable to satisfy the tangential stress balance at the drop surface, we use an energy dissipation argument to determine the migration velocity. This approach is similar to that used in Sections 3.4 and 4.13. The rate at which work is done by the thermocapillary stress on the surface elements must equal the rate of viscous dissipation in the two fluids. The final result is

$$v_\infty = \frac{1}{(2 + 3\alpha)} \int_{-1}^{1} \frac{\partial T}{\partial s}(1, s) C_2(s)\, ds = \frac{1}{(2 + 3\alpha)} \int_{-1}^{1} T(1, s) P_1(s)\, ds. \tag{4.14.7}$$

Now, we write the boundary conditions that must be satisfied by the temperature fields in the two fluids. The temperature in the continuous phase must approach the undisturbed field far from the drop.

$$T \to rs \quad \text{as} \quad r \to \infty. \tag{4.14.8}$$

The temperature and the heat flux are continuous across its interface.

$$T(1, s) = T'(1, s), \tag{4.14.9}$$

$$\frac{\partial T}{\partial r}(1, s) = \beta \frac{\partial T'}{\partial r}(1, s). \tag{4.14.10}$$

In addition, we require that the temperature within the drop remain bounded.

In the limit $\epsilon \to 0$, a perturbation expansion in ϵ will be singular both within the drop and outside it. The reason is the inability of the solution, obtained by using this expansion, to satisfy the conditions imposed at the drop surface. This is because the loss of the conduction terms leads to a reduction in the order of the governing equations. Therefore, it will be necessary to use the method of matched asymptotic expansions in solving this problem. The straightforward perturbation, in which we expand for small ϵ keeping (r, s) fixed, will yield outer solutions $T(r, s)$ and $T'(r, s)$. To accommodate radial conduction in the crucial region near the interface, where it must be of the same order as convective transport, we magnify the distance coordinate from the interface as follows:

$$x = \frac{r - 1}{\epsilon}, \quad r \geq 1; \tag{4.14.11}$$

$$x' = \frac{1 - r}{\epsilon\sqrt{\lambda}}, \quad r \leq 1. \tag{4.14.12}$$

In the inner (boundary layer) variables, we designate the temperature fields as $t(x, s)$ and $t'(x, s)$. These are the symbols that should be used in Equation (4.14.7) for the migration velocity.

Next, we determine the asymptotic scalings for the leading order temperature fields in the outer and inner regions. The outer temperature field in the continuous phase is of $O(1)$, because this scaling is controlled by the temperature field far away from the drop.

$$T = T_1 + \cdots. \tag{4.14.13}$$

The reason for the choice of the subscript 1 on the leading order outer field in Equation (4.14.13) will become clear as we pursue the analysis. To determine the scaling for the inner field in the continuous phase, consider the following result obtained from an overall energy balance for the drop:

$$\int_{-1}^{1} \frac{\partial t}{\partial x}(0, s)\, ds = -\frac{\beta}{\lambda} \int_{-1}^{1} \frac{\partial t'}{\partial x'}(0, s)\, ds = \frac{2\beta}{3\lambda} \frac{1}{\epsilon}. \tag{4.14.14}$$

Because $\frac{\beta}{\lambda}$ is independent of ϵ, it is evident that the inner field $t(r, s)$ must be of the form

$$t = \frac{1}{\epsilon} t_0 + \cdots. \tag{4.14.15}$$

The inner field within the drop must be expanded as

$$t' = \frac{1}{\epsilon} t_0' + \cdots. \tag{4.14.16}$$

To determine the scaling of the outer temperature field within the drop, let us examine the consequence of substituting a straightforward expansion in ϵ of the form $T' = T_0' + o(1)$ into Equation (4.14.2). This yields a problem for the leading order temperature field in which convective transport of energy balances the sink. Thus, as a fluid element goes around a closed circulation loop in the Hill's vortex within the drop, it would lose energy. This is not acceptable because the steady temperature field cannot be multivalued within the drop. Therefore, we must begin the expansion at an order other than $O(1)$. We can rule out the possibility that T' is of $o(1)$ because it is not possible to balance the sink term using such an expansion. Therefore, we must conclude that T' is of an order that is lower than $O(1)$. A solution that is of the form $T' = \frac{1}{\epsilon} T_0' + o(\frac{1}{\epsilon})$ leads to the same difficulty as that encountered before. In this case, there is no problem with the leading order outer temperature field, which is constant along each closed streamline. The next higher-order field, however, which must be of $O(1)$ to balance the sink term, violates the requirement that the temperature field be single-valued within the drop. Therefore, we are led to the following outer expansion for the scaled temperature field within the drop:

$$T' = \frac{1}{\epsilon^2} T_0' + T_1' + \cdots. \tag{4.14.17}$$

When the expansion in Equation (4.14.17) is substituted into Equation (4.14.2), it can be seen that the sink is balanced against convective transport of energy in the problem for T_1, but in addition, there is a source-sink term in the governing equation for T_1 of the form $\lambda \nabla^2 T_0'$, that balances the sink when integrated around a closed streamline. This leads to a self-consistent formulation, wherein the leading order temperature field is

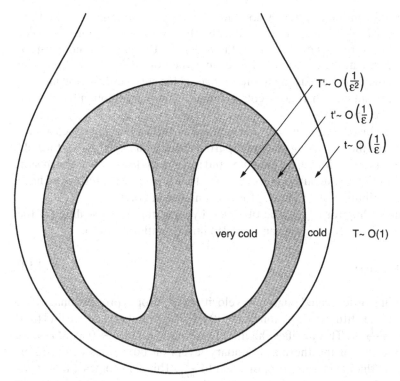

Figure 4.14.1 Sketch showing the asymptotic scalings in the outer and inner regions.

determined by conduction across the streamlines balancing the sink term. Higher-order fields satisfy the condition that they be single valued within the drop.

From Equation (4.14.17), we observe that the outer temperature field within the drop scales as $O(\frac{1}{\epsilon^2})$. Figure 4.14.1 shows a sketch of the various regions near the interface of the drop and the order of magnitude of the scaled temperature in these regions.

The order of the scaled temperature in Equation (4.14.17) implies that the magnitude of the temperature within the drop is large compared with that in the continuous phase. Because it is thermodynamically impossible for the drop to be warmer than the continuous phase, the interior of the drop must be very cold, relative to the surrounding fluid. The extent of the temperature difference is characterized by a factor $O(\frac{1}{\epsilon^2})$. The reason for these cold conditions is the need to sustain a sensible heat flux into the drop in the limit when convective transport dominates over conduction. Physically, this can be regarded as the limit when the thermal conductivities are small, so that it takes a large difference in temperature to drive the requisite heat flux at the interface. One might wonder about the mechanism by which the interior of the drop can become so cold compared with the exterior fluid. We can explain it by considering the initial period, during which the drop achieves its quasi-steady physical velocity v_∞^*. In the following discussion, we assume that this is the order of magnitude of the velocity in the transient period.

The physical temperature of the interior of the drop can change only when the changes at the surface are communicated to it. Because the streamlines in the Hill's vortex within the drop are closed, and convective transport is rapid compared with conduction, the temperature changes at the surface are communicated to the stagnation

ring of the Hill's vortex in the interior by conduction across the streamlines. The time
scale for this process is $\frac{R^2}{\lambda \kappa}$. During this period, the drop moves a distance of order
$\frac{v_\infty^* R^2}{\lambda \kappa} = \frac{R}{\lambda} Ma v_\infty$. Therefore, this distance is of the order $\frac{R}{\epsilon^2}$. The physical temperature
of the continuous phase at the location of the drop is then of $O\left(\frac{|\nabla T_\infty| R}{\epsilon^2}\right)$, whereas the
temperature within the drop is adjusting to the changes at the surface. Thus, the interior
of the drop, near the stagnation ring, is cold relative to the continuous phase when
conduction effects are small.

We have not introduced a term at $O(\frac{1}{\epsilon})$ in Equation (4.14.17). Formally, such a term
should be included, but the governing first order partial differential equation for the field
is homogeneous at this order. Matching the outer and inner fields leads to the conclusion
that the only contribution to the outer solution at $O(\frac{1}{\epsilon})$ is a constant. Therefore, we have
labeled the next contribution at $O(1)$ as T_1 for convenience in notation.

The asymptotic scaling for v_∞ can be obtained by substituting the scaling for the
inner temperature field given in Equation (4.14.15) into Equation (4.14.7).

$$v_\infty = Ma\, v_{\infty_0} + \cdots \tag{4.14.18}$$

Therefore, the leading order scaled migration velocity of the drop is proportional to the
Marangoni number. Substituting Equation (4.14.18) into the definition of ϵ, we find that,
at this order, $\epsilon = \frac{1}{Ma\sqrt{v_{\infty_0}}}$. This justifies the initial assumption that $\epsilon \to 0$ as $Ma \to \infty$.
Because the thicknesses of the thermal boundary layers on both sides of the inter-
face scale as ϵ, these thicknesses are proportional to $\frac{1}{Ma}$. This is in contrast to the gas
bubble limit in which we found the thickness of the thermal boundary layer in the
continuous phase to be proportional to $\frac{1}{\sqrt{Ma}}$. There is no inconsistency here, however.
In that problem, $v_\infty \sim O(1)$ as $Ma \to \infty$. Therefore, we recover the correct scaling of
the thermal boundary layer thickness when using the more general definition of ϵ given
here. Later, we establish that in the gas bubble limit, when $\beta \to 0$, v_{∞_0} in Equation
(4.14.18) is zero, and the correct leading order result for v_∞ is recovered. Finally, it
is worth noting that when higher-order terms are included in the expansion, the relationship between ϵ and Ma, and therefore that between the thickness of the thermal
boundary layers and Ma, will need to be corrected appropriately.

4.14.3 Leading Order Outer Temperature Fields

Now, we proceed to obtain the solution of the governing equations, along with the
associated boundary conditions and matching requirements. For this, we begin with the
equation for the outer temperature field T_0 at leading order, which can be obtained
from Equations (4.14.1) and (4.14.13) as

$$1 + \mathbf{v} \bullet \nabla T_1 = 0. \tag{4.14.19}$$

The boundary condition to be satisfied by this field is

$$T_1 \to rs \quad \text{as} \quad r \to \infty. \tag{4.14.20}$$

In addition, $T_1(r, s)$ must be matched with the inner field. The solution is the same as
that in the case of a gas bubble, which is given in Equation (4.13.20). It is reproduced

below in the notation used in the present section.

$$T_1 = rs + \int_r^\infty \frac{\frac{3\psi\tilde{r}}{(\tilde{r}^3-1)} - 1}{(\tilde{r}^3 - 1)\sqrt{1 - \frac{2\psi\tilde{r}}{(\tilde{r}^3-1)}}} \, d\tilde{r}. \tag{4.14.21}$$

Here, $\psi = \frac{1}{2}(1 - s^2)(r^2 - \frac{1}{r})$ is the normalized streamfunction in the continuous phase.

The equation satisfied by the leading order temperature field T_0' within the drop, obtained from Equations (4.14.2) and (4.14.17), is given below.

$$\mathbf{v}' \bullet \nabla T_0' = 0. \tag{4.14.22}$$

The solution is

$$T_0' = \tilde{F}(\psi'), \tag{4.14.23}$$

where ψ' is the normalized streamfunction within the drop, given by

$$\psi' = \frac{3}{4}(1 - s^2)(r^4 - r^2) \tag{4.14.24}$$

and $\tilde{F}(\psi')$ is an unknown function yet to be determined. The physical implication of Equations (4.14.23) and (4.14.24) is that the temperature is constant along any of the closed streamlines within the drop. To determine the unknown field $\tilde{F}(\psi')$, we must consider the governing equation for the next higher-order temperature field T_1'.

$$1 + \mathbf{v}' \bullet \nabla T_1' = \lambda \nabla^2 T_0'. \tag{4.14.25}$$

To solve Equation (4.14.25), we follow the procedure used by Kronig and Brink (1950) and Brignell (1975) and transform to coordinates along a streamline, and orthogonal to it, in a meridian plane. The transformed coordinates (m, q) are related to the coordinates (r, s) as shown below.

$$m = -\frac{16}{3}\psi'(r, s), \tag{4.14.26}$$

$$q = \frac{r^4 s^4}{2r^2 - 1}. \tag{4.14.27}$$

The metrical coefficients for the transformation are

$$h_m = \frac{1}{8r\Delta\sqrt{1 - s^2}}, \tag{4.14.28}$$

$$h_q = \frac{(2r^2 - 1)^2}{4r^3 s^3 \Delta}, \tag{4.14.29}$$

and

$$h_\phi = r\sqrt{1 - s^2}, \tag{4.14.30}$$

where

$$\Delta^2 = (1 - 2r^2)^2(1 - s^2) + (1 - r^2)^2 s^2 = (1 - 2r^2)^2 + r^2 s^2(2 - 3r^2). \tag{4.14.31}$$

Writing Equation (4.14.25) in (m, q) coordinates and multiplying both sides of the resulting equation by $h_m h_q h_\phi$, followed by integration with respect to q for one circuit along a streamline, yields the following equation for $F(m) = \tilde{F}(\psi')$:

$$\lambda \frac{d}{dm}\left(J(m)\frac{dF}{dm}\right) = H(m),$$ (4.14.32)

where

$$J(m) = \oint \frac{h_q h_\phi}{h_m} dq$$

$$= \frac{2\sqrt{2}}{3}(1 + \sqrt{1-m})^{\frac{1}{2}}(4 - 3m)E\left(\frac{2\sqrt{1-m}}{1+\sqrt{1-m}}\right) - \frac{16}{3}mH(m)$$

$$= \frac{4}{3}(1 + \sqrt{m})^{\frac{1}{2}}\left[(4 - 3m)E\left(\frac{1 - \sqrt{m}}{1 + \sqrt{m}}\right) - (4\sqrt{m} - 3m)K\left(\frac{1 - \sqrt{m}}{1 + \sqrt{m}}\right)\right],$$

(4.14.33)

and

$$H(m) = \oint h_m h_q h_\phi dq = \frac{\sqrt{2}}{8(1 + \sqrt{1-m})^{\frac{1}{2}}}K\left(\frac{2\sqrt{1-m}}{1+\sqrt{1-m}}\right)$$

$$= \frac{1}{4(1 + \sqrt{m})^{\frac{1}{2}}}K\left(\frac{1 - \sqrt{m}}{1 + \sqrt{m}}\right),$$ (4.14.34)

and $K(x)$ and $E(x)$ are the complete elliptic integrals of the first and second kinds, respectively. The definitions and properties of these elliptic integrals can be found in Abramowitz and Stegun (1965). Kronig and Brink (1950) obtained the above results for $H(m)$ and $J(m)$, and we have rewritten them in the notation used by Abramowitz and Stegun.

The solution for $F(m)$ is given below.

$$F = \frac{1}{\lambda}\left[B_0' + \int_0^m \frac{dx}{J(x)}\int_1^x H(\tilde{m})\,d\tilde{m}\right].$$ (4.14.35)

Here, B_0' is an unknown constant. Later, for the purpose of matching with the inner solution, we shall need the outer temperature field T' near the surface of the drop. Hence, we obtain an expansion of the function $F(m)$ near $m = 0$. It can be shown that

$$\int_0^1 H(m)\,dm = \frac{1}{3}.$$ (4.14.36)

Also, $J(1) = 0$, and $H(1) = \frac{\pi}{8\sqrt{2}}$. We can expand $J(m)$ and $H(m)$ near $m = 0$ as

$$J(m) = \frac{16}{3} - 5m + O(m^2 \log m)$$ (4.14.37)

and

$$H(m) = \frac{3}{8}\log 2 - \frac{1}{16}\log m + O(m\log m).$$ (4.14.38)

Using the above results, the solution for the leading order outer temperature field within the drop can be written in a form that is applicable near the surface of the drop, as follows:

$$T_0' = \frac{1}{\lambda}\left[B_0' - \frac{1}{16}m + \frac{3}{16}\left(\frac{3}{16}\log 2 - \frac{7}{64}\right)m^2 - \frac{3}{512}m^2 \log m \right] + O(m^3 \log m).$$

(4.14.39)

Note from Equation (4.14.39) that $\frac{\partial T_0'}{\partial r}$ is positive as $r \to 1$. The drop gains energy at a constant rate from the continuous phase to increase its physical temperature at a constant rate, as it moves into the warmer surroundings. At $O(\frac{1}{\epsilon^2})$, however, the outer temperature field in the continuous phase is zero and changes in the scaled temperature T occur only at $O(1)$. Therefore, the applied temperature gradient far away from the drop, which is responsible for its migration, enters into the problem at leading order indirectly, via the demand for energy within the drop that is represented by the sink in Equation (4.14.25).

Because the leading order temperature fields in the two fluids, in the absence of conduction, are unable to satisfy the boundary conditions of continuity of temperature and heat flux at the interface, inner (boundary) layers are present on both sides of the interface, as noted earlier. Matching with the inner field, to be obtained shortly, leads to the result that $B_0' = 0$. It will become necessary to introduce a constant temperature field, labeled B', at $O(\epsilon)$ to satisfy matching requirements at leading order. Therefore, the physical picture is as follows. The interior of the drop displays scaled temperature variations of $O(\frac{1}{\epsilon^2})$, overlaid on a constant scaled temperature that is of $O(\frac{1}{\epsilon})$. We now proceed to the analysis of the inner temperature fields.

4.14.4 Leading Order Inner Temperature Fields

We first transform Equations (4.14.1) and (4.14.2) to the inner variables defined in Equations (4.14.11) and (4.14.12), respectively. Substitution of the expansions in Equations (4.14.15) and (4.14.16) then yields the following governing equations for the leading order inner temperature fields:

$$-\left(3xs\frac{\partial t_0}{\partial x} + \frac{3}{2}(1-s^2)\frac{\partial t_0}{\partial s}\right) = \frac{\partial^2 t_0}{\partial x^2}$$

(4.14.40)

and

$$-\left(3x's\frac{\partial t_0'}{\partial x'} + \frac{3}{2}(1-s^2)\frac{\partial t_0'}{\partial s}\right) = \frac{\partial^2 t_0'}{\partial x'^2}.$$

(4.14.41)

The boundary conditions on these fields are obtained by substitution of the expansions into the conditions at the drop surface given in Equations (4.14.9) and (4.14.10), after transformation to the inner variables.

$$t_0(0, s) = t_0'(0, s)$$

(4.14.42)

$$\delta \frac{\partial t_0}{\partial x}(0, s) = -\frac{\partial t_0'}{\partial x'}(0, s)$$

(4.14.43)

Here, we have introduced the physical property ratio $\delta = \frac{\sqrt{\lambda}}{\beta} = \sqrt{\frac{k\rho C_p}{k'\rho'C_p'}}$. The requirement that the inner field in the continuous phase must match with the outer field leads to the condition that

$$t_0(x \to \infty, s) \to 0.$$

(4.14.44)

When we rewrite the leading order outer field in Equation (4.14.39) in the inner variables (x', s) and expand for small ϵ, it is seen that only the terms up to $O(m)$ need to be matched at leading order in the inner field. This matching requirement yields

$$t_0'(x' \to \infty, s) \to B' - \frac{1}{2\sqrt{\lambda}} x'(1 - s^2). \tag{4.14.45}$$

Here, B' is the constant outer temperature field at $O(\frac{1}{\epsilon})$, as noted earlier. In the same manner as in Section 3.4, we must also provide "starting" conditions at $\theta = 0$, or equivalently, at $s = 1$. The starting condition for t_0 may simply be written as $t_0 \to 0$ as $s \to 1$. In other words, near the forward stagnation point, the fluid in the continuous phase is isothermal in the outer region. Of course, the outer temperature field demands variations of $O(1)$, but at $O(\frac{1}{\epsilon})$, there is no change in the temperature with distance along the forward stagnation streamline and in a narrow bundle of streamlines surrounding it. The starting condition for t_0' is complicated because the temperature distribution near the forward stagnation point within the drop is influenced by the inner thermal wake that convects energy from the rear stagnation region to the forward stagnation region. We follow Harper and Moore (1968), as in Section 3.4, and assume that the temperature distributions at leading order near the forward and rear stagnation points are identical and that the internal thermal wake within the drop merely convects the temperature field passively along streamlines. A simple argument, originally given by Brignell (1975), can be used to justify this assumption. The thermal boundary layer at the drop surface is of thickness ϵ, and the leading order temperature variation across it is of $O(\frac{1}{\epsilon})$. The internal wake comprises the boundary layer fluid that turns around near the rear stagnation region. Therefore, the temperature change across the internal wake is $O(\frac{1}{\epsilon})$ as well. As in Section 4.13, where we established the order of the thickness of the thermal wake behind a bubble, we can determine the order of the thickness of the thermal wake within the drop. We use the idea that the rate at which mass enters the wake is the rate at which it leaves the thermal boundary layer. This permits us to determine that the thermal wake within the drop is of thickness $O(\sqrt{\epsilon})$. The balance of convection along streamlines in the wake and conduction across the wake yields the result that the temperature change along the streamlines is $O(1)$ and can be neglected at leading order. The expressions for the starting conditions are provided a bit later after transforming to new variables.

It is convenient to transform the dependent and independent variables as follows:

$$f(\xi, \eta) = \sqrt{\lambda}\, t_0(x, s), \tag{4.14.46}$$

$$f'(\xi, \eta') = \sqrt{\lambda}\, [t_0'(x', s) - B'] + \frac{2\sqrt{2}}{3} \eta', \tag{4.14.47}$$

$$\eta = \frac{3}{4\sqrt{2}} x(1 - s^2), \tag{4.14.48}$$

$$\eta' = \frac{3}{4\sqrt{2}} x'(1 - s^2), \tag{4.14.49}$$

and

$$\xi(s) = \frac{1}{16}(2 - 3s + s^3). \tag{4.14.50}$$

The governing equations for $f(\xi, \eta)$ and $f'(\xi, \eta')$, along with the associated boundary conditions, are written below.

$$\frac{\partial f}{\partial \xi} = \frac{\partial^2 f}{\partial \eta^2}, \tag{4.14.51}$$

$$\frac{\partial f'}{\partial \xi} = \frac{\partial^2 f'}{\partial \eta'^2}, \tag{4.14.52}$$

$$f(\xi, 0) - f'(\xi, 0) = \frac{B}{\delta}, \tag{4.14.53}$$

$$\delta \frac{\partial f}{\partial \eta}(\xi, 0) + \frac{\partial f'}{\partial \eta'}(\xi, 0) = \frac{2\sqrt{2}}{3}, \tag{4.14.54}$$

$$f(\xi, \eta \to \infty) \to 0, \tag{4.14.55}$$

$$f'(\xi, \eta' \to \infty) \to 0, \tag{4.14.56}$$

$$f(0, \eta) = 0, \tag{4.14.57}$$

$$f'(0, \eta') = f'(\xi(-1), \eta') \equiv g(\eta'), \tag{4.14.58}$$

$$g(\eta' \to \infty) \to 0. \tag{4.14.59}$$

The constant B in Equation (4.14.53) is $B = \frac{\lambda}{\beta} B'$. Equations (4.14.53) and (4.14.54) represent the continuity of temperature and heat flux at the surface of the drop, respectively. Equations (4.14.55) and (4.14.56) are obtained from matching the inner fields with the respective outer fields. Equations (4.14.57) and (4.14.58) represent the starting conditions for the inner fields discussed earlier. Equation (4.14.59) is a constraint that ensures that the solution is consistent with the boundary condition on $f'(\xi, \eta' \to \infty)$ in Equation (4.14.56). Equations (4.14.51) through (4.14.56) can be solved in the same way as Equations (3.4.60) through (3.4.64), using methods suggested by Harper and Moore (1968). The solutions are

$$f(\xi, \eta) = \frac{1}{1+\delta} \left[R_+(\xi, \eta) - R_-(\xi, \eta) + \left(\frac{2\sqrt{2}}{3} \eta + \frac{B}{\delta} \right) \operatorname{erfc}\left(\frac{\eta}{2\sqrt{\xi}} \right) \right.$$
$$\left. - \frac{4}{3} \sqrt{\frac{2\xi}{\pi}} \exp\left(-\frac{\eta^2}{4\xi} \right) \right] \tag{4.14.60}$$

and

$$f'(\xi, \eta') = \frac{1}{1+\delta} \left[R_+(\xi, \eta') + \delta R_-(\xi, \eta') + \left(\frac{2\sqrt{2}}{3} \eta' - B \right) \operatorname{erfc}\left(\frac{\eta'}{2\sqrt{\xi}} \right) \right.$$
$$\left. - \frac{4}{3} \sqrt{\frac{2\xi}{\pi}} \exp\left(-\frac{\eta'^2}{4\xi} \right) \right], \tag{4.14.61}$$

where the function $R_\pm(\xi, \eta)$ is defined as follows:

$$R_\pm(\xi, \eta) = \frac{1}{2\sqrt{\pi\xi}} \int_0^\infty g(\tilde{\eta}) \left[\exp\left(-\frac{(\eta - \tilde{\eta})^2}{4\xi} \right) \pm \exp\left(-\frac{(\eta + \tilde{\eta})^2}{4\xi} \right) \right] d\tilde{\eta}. \tag{4.14.62}$$

The starting condition given in Equation (4.14.58) yields a Fredholm integral equation of the second kind for $G(\eta') = (1 + \delta)g(\eta')$.

$$G(\eta') = \left(\frac{2\sqrt{2}}{3}\eta' - B\right)\text{erfc } \eta' - \frac{2}{3}\sqrt{\frac{2}{\pi}}e^{-\eta'^2}$$
$$+ \frac{1}{\sqrt{\pi}}\int_0^\infty G(\tilde{\eta})\left(e^{-(\eta'-\tilde{\eta})^2} + \frac{1-\delta}{1+\delta}e^{-(\eta'+\tilde{\eta})^2}\right)d\tilde{\eta}. \tag{4.14.63}$$

The unknown constant B is determined by applying the condition given in Equation (4.14.59). The inner temperature field on the surface of the drop is now used in Equation (4.14.7) to determine the migration velocity of the drop at leading order.

$$v_{\infty_0} = \frac{Ma}{\lambda(2+3\alpha)^2(1+\delta)^2}\left[\int_{-1}^1\left(\frac{1}{(1+\delta)\sqrt{\pi\xi}}\int_0^\infty G(\tilde{\eta})e^{-\frac{\tilde{\eta}^2}{4\xi}}d\tilde{\eta}\right.\right.$$
$$\left.\left. - \frac{1}{3}(1-s)\sqrt{\frac{2}{\pi}(2+s)}\right)s\, ds\right]^2. \tag{4.14.64}$$

The above result for v_{∞_0} can be written as

$$v_{\infty_0} = \frac{4h(\delta)Ma}{\lambda(2+3\alpha)^2(1+\delta)^2}. \tag{4.14.65}$$

For a chosen value of δ, B, $g(\eta')$, and $h(\delta)$ are determined numerically. The details of the numerical procedure used to solve the integral equation and obtain the scaled migration velocity are provided in Balasubramaniam and Subramanian (2000). Some values of $h(\delta)$ and $B(\delta)$ obtained from the numerical calculations are given in Table 4.14.1 It is seen that $h(\delta)$ is a monotonically increasing function, which varies only by approximately 45% over the entire range of values of δ. The value of $h(\infty)$ is $\frac{1}{18\pi}(\frac{16}{7} - \frac{12\sqrt{3}}{7})^2 = 0.0082618$. This is obtained by analytically evaluating the integral in the right side of Equation (4.14.64) in the limit $\delta \to \infty$. We have not performed the computations needed to determine $B(\infty)$.

Upon rewriting the result from Equation (4.14.39) for T_0' in inner variables, we see that the first correction beyond the leading order is of $O(\log \epsilon)$. Therefore, the next terms in the inner expansions in Equations (4.14.15) and (4.14.16) must be of $O(\log \epsilon)$. Substituting an inner expansion of the form $\frac{1}{\epsilon}t_0 + \log \epsilon\, t_{\ell 1}$ into Equation (4.14.7), and

Table 4.14.1　Numerically Determined Values of the Functions $h(\delta)$ and $B(\delta)$

δ	$h(\delta)$	$B(\delta)$
0	0.00568	−0.418
0.25	0.00611	−0.407
0.5	0.00642	−0.399
1	0.00683	−0.389
5	0.00775	−0.366
10	0.00798	−0.361
∞	0.00826	—

using the definition of ϵ, we obtain the following expansion for v_∞:

$$v_\infty = v_{\infty_0} + a_{\ell,1} \log Ma + a_1 + a_{\ell 2} \frac{(\log Ma)^2}{Ma} + b_{\ell 2} \frac{\log Ma}{Ma} + \cdots. \qquad (4.14.66)$$

The calculations involved in obtaining the unknown constants in this expansion are involved, and we therefore have not performed them. For a test case in which $Pr = \alpha = \beta = \lambda = 1$, Balasubramaniam and Subramanian (2000) found that Equation (4.14.66) fits the numerical results of Ma, Balasubramaniam, and Subramanian (1999) to within 0.2%, in the range $200 \leq Ma \leq 1000$, provided the leading term is written as $a_0 Ma$, wherein the coefficient a_0 is obtained from the best fit. Even at such large values of the Marangoni number, the higher order terms included in Equation (4.14.66) make a significant contribution and are needed to obtain a good fit to the numerical results. In addition, we note that the result for the fitted coefficient a_0 is 0.00236, whereas the coefficient of Ma in Equation (4.14.65) for this set of parameters is 0.00205. The reason for this discrepancy is not evident, and we leave its resolution to a future study of this problem.

It is useful to write a result for the physical migration velocity of the drop, at leading order, in the asymptotic situation when convective transport of both momentum and energy dominate over molecular transport:

$$v_{\infty_0}^* = \frac{4|\sigma_T|^2 |\nabla T_\infty|^2 R^3 h(\delta)}{\lambda \kappa (2\mu + 3\mu')^2 (1+\delta)^2}. \qquad (4.14.67)$$

We see that the velocity of the drop depends nonlinearly on its radius, the magnitude of the applied temperature gradient, and the physical properties. This is in sharp contrast to the corresponding result given in Equation (4.2.28), which is applicable when convective transport of both momentum and energy are negligible. Also, the result is remarkably different from that obtained in Section 4.13 for a gas bubble when the Marangoni number is large. We now proceed to demonstrate that the result for a gas bubble can be obtained from the more general analysis given in this section.

4.14.5 The Gas Bubble Limit

The gas bubble limit is obtained when the viscosity and the thermal conductivity of the drop are negligibly small compared with the corresponding properties in the continuous phase. This occurs in the limit $\alpha \to 0$, $\beta \to 0$. In the limit $Re \to \infty$, the velocity field around a gas bubble is given by the potential flow solution in Equations (4.14.3) and (4.14.4). Equation (4.14.7) can be used to relate the migration velocity of the bubble to the temperature distribution on its surface.

From the definition of δ, we note that $\delta \to \infty$ when $\beta \to 0$. In the limit $\delta \to \infty$, we see from Equations (4.14.60) and (4.14.65) that the inner temperature field at $O(\frac{1}{\epsilon})$ in the continuous phase vanishes, as does the migration velocity at $O(Ma)$. This is not surprising because the leading order inner temperature fields for a drop are driven by the demand for energy within the drop and in the gas bubble limit, no heat enters the bubble. The leading order inner temperature field in the continuous phase must therefore be of $O(1)$ to match with the outer temperature field given in Equation (4.14.21). The governing equation for this temperature field $t_1(x, s)$ can be written as follows:

$$1 - 3xs \frac{\partial t_1}{\partial x} - \frac{3}{2}(1 - s^2) \frac{\partial t_1}{\partial s} = \frac{\partial^2 t_1}{\partial x^2} + (2 - 6x^2 s) \frac{\partial t_0}{\partial x} - 3x(1 - s^2) \frac{\partial t_0}{\partial s}. \qquad (4.14.68)$$

The boundary conditions that must be satisfied by $t_1(x, s)$ are

$$\frac{\partial t_1}{\partial x}(0, s) = 0 \tag{4.14.69}$$

and

$$\left| \frac{\partial t_1}{\partial s}(x, 1) \right| < \infty. \tag{4.14.70}$$

Equation (4.14.69) represents the adiabatic condition on the surface of the gas bubble. Equation (4.14.70) follows from symmetry. Also, t_1 must be matched with the $O(1)$ outer temperature field given in Equation (4.14.21); the expansion of this solution near $r = 1$ is given in Equation (4.13.23). Because $t_0 \to 0$ as $\delta \to \infty$, the inhomogeneity on the right side of Equation (4.14.68) vanishes in this limit. The governing equation and boundary conditions for t_1 become identical to Equations (4.13.11), (4.13.14), and (4.13.15) for the leading order temperature field given in Section 4.13, in the limit $Re \to \infty$. Therefore, the analysis of Section 4.13 is correctly recovered in the limit $\alpha \to 0$, $\beta \to 0$.

4.15 Migration When Convective Transport Is Important – Results from Numerical Solution of the Governing Equations

4.15.1 Introduction

In the last three sections, we have presented analytical results obtained from perturbation theory for handling the case when convective transport of energy affects the migration velocity of a drop. These are only useful in asymptotic situations when the Marangoni number is very small or very large. Also, in treating the momentum transport problem, we were forced to assume either Stokes flow, neglecting convective transport altogether, or potential flow, which ignores the boundary layer in the vicinity of the drop and the wake. In the case where the Reynolds and Marangoni numbers take on intermediate values, it does not appear possible to make progress via analysis. This has prompted several investigators to obtain numerical solutions of the governing equations along with the associated boundary conditions. As computers have become more powerful over the years, it has become possible to extend the calculations to larger values of the Reynolds and Marangoni numbers. The calculations, with a few exceptions, are limited to a spherical shape, in which case the assumption of negligible Capillary number is implicit. In such computations, the normal stress balance is not used.

There are seven independent parameters in this problem. One set might be the Reynolds, Marangoni, and Capillary numbers based on the properties of the continuous phase, and the viscosity, thermal conductivity, thermal diffusivity, and density ratios. Other combinations are possible. Because the number of parameters is large, typically calculations are carried out for a few sets of values of the parameters spanning the range. The gas bubble limit is simpler to handle because of the need to consider only the transport problems in the continuous phase. This reduces the number of independent parameters to three, namely the Reynolds, Marangoni, and Capillary numbers based on the continuous phase properties. For the reader's convenience, in Table 4.15.1, we have summarized the ranges of the Reynolds, Marangoni, and Prandtl numbers in each phase, along with the range of Capillary numbers, where applicable, in each of the available studies. This is not to imply that these are independent parameters because

Table 4.15.1 A Summary of the Parameter Ranges Used in Numerical Studies

Author(s)	Gas Bubble Limit Only	Assumed Spherical Shape	Re	Pr	Ma	Ca	Re'	Pr'	Ma'
Shankar and Subramanian (1988)	Yes	Yes	0	∞	0-200				
Szymczyk and Siekmann (1988)	Yes	Yes	0-100	0.01-10	0-1,000				
			0-50	100	0-5,000				
Balasubramaniam and Lavery (1989)	Yes	Yes	0-2,000	0.01-1000	0-1,000				
Chen and Lee (1992)	Yes	Yes	1-100	1	1-100	0.01-0.5			
Ehmann, Wozniak, and Siekmann (1992)		Yes	0-34	23	0-780		0-0.025	1140	0-29
			0-0.025	1,140	0-29		0-34	23	0-780
Nas (1995)			1-10	1-60	1-60	0.01-0.1333	1-10	1-30	0.5-30
Treuner et al. (1996)	Yes	Yes	0-1,000	1-100	0-2,500				
Haj-Hariri, Shi, and Borhan (1997)			0-50	0.02-1,000	0-1,000	0-0.5	0-250	0.025-1.25	0-50
Welch (1998)	Yes		1-100	0.01-100	1-100	0.001-0.1			
Ma, Balasubramaniam, and Subramanian (1999)		Yes	0-12	83-3,469	0-1,000		0-155	17.9	0-3,175
								22.3	

the product of the Reynolds and Prandtl number yields the Marangoni number. The object is mainly to show the ranges of these physically meaningful dimensionless groups covered in each study.

4.15.2   Results in the Gas Bubble Limit

Results from a numerical solution of the governing momentum and energy equations in the gas bubble problem were first reported by Szymczyk and Siekmann (1988). Shortly thereafter, Balasubramaniam and Lavery (1989) published results from their numerical solution of the same problem, covering a larger range of values of the Reynolds number. Here, we present results obtained in the gas bubble limit using the code developed by Balasubramaniam and Lavery (1989). In this limit, the viscosity ratio α and the thermal conductivity ratio β are both set equal to zero. Therefore, only the governing equations in the continuous phase need to be considered, along with the associated boundary conditions and a force balance on the bubble. The predictions for the migration velocity are good to within 2 to 5%, and this is adequate for our graphical presentation purposes. These results were calculated by discretizing the governing Navier–Stokes and energy equations and the associated boundary conditions, using the method of finite differences, for a gas bubble undergoing steady thermocapillary migration. These equations already have been given in Section 4.1. The bubble is assumed to be spherical, and the normal stress balance is ignored. The boundary conditions at infinity have to be applied at a finite value of the scaled radial coordinate $r = R_\infty$, and this location is chosen as $R_\infty = 5$. A uniform grid is formed by dividing the range $r = 1$ to 5 into 64 equal intervals, and the range $\theta = 0$ to π into 32 equal intervals. All derivatives, except those of the pressure, are replaced by three-point second-order finite difference approximations at the vertices of the grid. The pressure derivatives are replaced by the same order approximations formulated slightly differently. The continuity equation is discretized to second-order accuracy at the centers of the cells formed by the grid. The resulting system of nonlinear algebraic equations is solved by an iterative procedure described by the authors who also outline precautions taken to assure that their results are sufficiently accurate. The results displayed in Figures 4.15.1, 4.15.4, and 4.15.5 were obtained by using the above code.

In Figure 4.15.1, we show the predictions for the migration velocity, scaled by its value when $Re \to 0$, $Ma \to 0$, namely, the result in Equation (4.2.28). We refer to the latter as v_{YGB} to signify that it was originally obtained by Young et al. (1959). It is possible to choose either the Marangoni number or the Reynolds number for the abscissa. Using the former will show the strong sensitivity of the scaled velocity to variation in the Marangoni number for any given Prandtl number. The data are somewhat clustered in such a drawing, however, and the appearance is improved by using the Reynolds number as the variable along the abscissa. Each symbol in the figure corresponds to a single Prandtl number. Results are displayed for five values of the Prandtl number in the range 0.01 to 100. The dashed curves show the behavior of the scaled velocity for a fixed value of the Marangoni number as the Reynolds number is varied. To complete these curves, some calculations were performed at Prandtl numbers other than the five listed in the legend. We also have included, on the right border of the drawing, the asymptotic predictions from Equation (4.13.46) for the case $Re \to \infty$, $Ma \to \infty$ for each Prandtl number. The numerical results approach the asymptotic value in each case. The tendency of the migration velocity to increase with increasing values of Reynolds

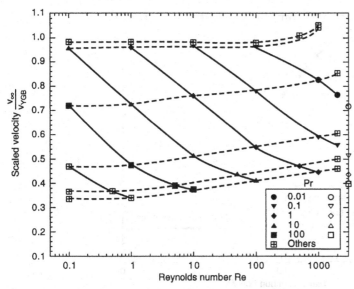

Figure 4.15.1 Scaled velocity of the bubble plotted against the Reynolds number for a series of values of the Prandtl number; the symbols on the right border represent the asymptotic prediction in each case for $Re \to \infty$, $Ma \to \infty$, from Equation (4.13.46). Each solid curve corresponds to a fixed Prandtl number, and each dashed curve corresponds to a fixed Marangoni number. (Reproduced with permission from Balasubramaniam, R., and Lavery, J. E. Numerical simulation of thermocapillary bubble migration under microgravity for large Reynolds and Marangoni numbers. *Num. Heat Transfer A* 16, 175–187. Copyright 1989, Taylor and Francis, Inc.)

number, for fixed Marangoni number, appears counter to intuition, but there is a simple explanation. Increasing the role of inertia will, in fact, lead to more resistance to the motion of the bubble; however, this is mitigated by the effect of decreasing the Prandtl number as the Reynolds number is increased at fixed Marangoni number. The consequence of the latter is to increase the temperature difference over the surface of the bubble.

In Figure 4.15.2, the velocity of the bubble, scaled in the same manner as in Figure 4.15.1, is plotted against the Marangoni number for the case of Stokes flow. These numerical results are from Shankar and Subramanian (1988), who employed the method of finite differences to solve the energy equation, but used analytical results for the velocity field of the form given in Equations (4.5.3) and (4.5.4). Details regarding the numerical method can be found in Shankar (1984). The asymptotic result for $Re \to 0$ and $Ma \to \infty$ is shown in the figure, along with an approximate result provided by Shankar and Subramanian (1988). These authors found the asymptotic expansion for small values of Ma given in Equation (4.12.51) to do well only for Ma up to a value of ≈ 0.5. But, they were able to use series improvement techniques, similar to those discussed in Section 4.8, to obtain a new result that is indistinguishable from the numerical result for values of the Marangoni number up to about 25. This result, included in Figure 4.15.2, is given below.

$$v_\infty = \frac{1}{2} - 0.1125\chi^2 - 0.1123\chi^3. \tag{4.15.1}$$

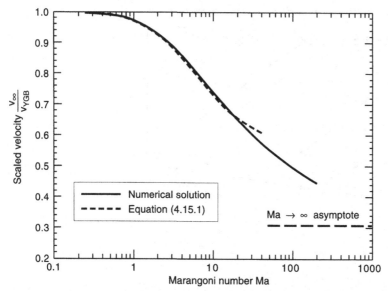

Figure 4.15.2   Scaled velocity of the bubble plotted against the Marangoni number when the Reynolds number is negligible; also shown are an approximation from Equation (4.15.1) and the asymptotic prediction for $Re \to 0$, $Ma \to \infty$, from Equation (4.13.69).

Here,

$$\chi = \frac{Ma}{(Ma + 2.32)}. \tag{4.15.2}$$

The numerical results obtained by Shankar and Subramanian (1988) were fitted, in the range $75 \le Ma \le 200$, by

$$v_\infty = \frac{1.59}{1.84 + \log Ma}. \tag{4.15.3}$$

The simple form of the fit, which, surprisingly, is within 0.6% of the numerically obtained results for $25 \le Ma \le 200$, led the authors to suggest that it hints at the asymptotic behavior of v_∞ as $Ma \to \infty$. This implies that the migration velocity approaches a value of 0 as $Ma \to \infty$ when $Re = 0$. As shown in Section 4.13, we have found that this is not the case, and the correct asymptotic result is a nonzero constant.

In Figure 4.15.3, we show representative isotherms and level curves of the scaled hydrodynamic pressure. These were obtained by using the code developed by Ma (1998) for liquid drops, by using values of the parameters appropriate for a gas bubble. The results are presented for the case $Re = 50$ and $Ma = 500$ and are typical of the behavior of these fields in the calculations performed. The main feature observed from the isotherms is that there is a long thermal wake behind the bubble that extends to several radii and that the temperature field ahead of the bubble quickly adjusts to the undisturbed state by $r \approx 2$. In plotting the pressure distribution, we have used the arbitrary value of 0 as the scaled pressure at large distances from the bubble. Note the sharp pressure gradient in the vicinity of the rear stagnation point.

In Figure 4.15.4, the scaled velocity at the interface $v_\theta(1, \theta)$ is displayed as a function of the polar angle θ for some representative values of the Marangoni number for a fluid of Prandtl Number $Pr = 10$. The solid curve represents the potential flow velocity profile, which is included for comparison. The results corresponding to the case $Ma = 0$

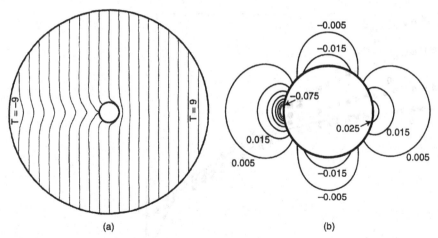

(a)   (b)

Figure 4.15.3 (a) Isotherms, $\Delta T = 1$, and (b) level curves of the hydrodynamic pressure, $\Delta p = 0.01$; $Re = 50$, $Ma = 500$. The applied temperature gradient is directed from left to right.

were calculated from Equation (4.2.20) by using a value of $A = \frac{1}{4}$ corresponding to the gas bubble limit. It is clear from the figure that as the relative importance of convective transport increases, the maximum in the scaled interfacial velocity decreases, the velocities tend to change less with position near the maximum leading to broader peaks, and the location of the maximum shifts slightly toward the rear of the bubble. The velocity profile for $Ma = 500$ is closely approximated by the potential flow profile in the forward half of the bubble. Substantial deviations are noted in the rear half, however. This is because the Reynolds number is only 50 at this value of Ma. We have made calculations for $Re = 1000$ and $Ma = 1000$ and verified that indeed the velocity profile approaches that from potential flow more closely over the entire surface of the bubble.

Scaled temperature profiles along the interface, also for a fluid of Prandtl Number $Pr = 10$, are displayed in Figure 4.15.5. The results corresponding to the case $Ma = 0$ are from Equation (4.2.16), with $\beta = 0$. The solid curve represents the prediction from the asymptotic theory for $Re \to \infty$, $Ma \to \infty$. We have calculated a composite solution

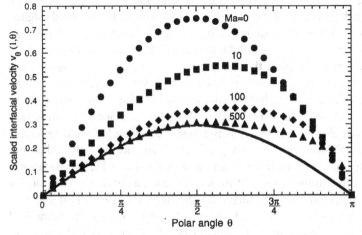

Figure 4.15.4 Scaled interfacial velocity plotted against the polar angle for $Pr = 10$ for a set of values of Ma, including the potential flow velocity profile, which can be regarded as the asymptote for $Re \to \infty$.

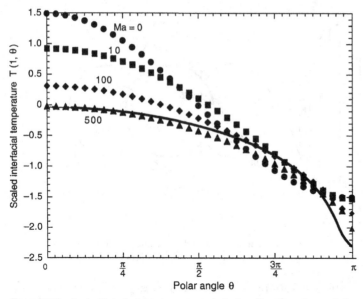

Figure 4.15.5 Scaled interfacial temperature plotted against the polar angle for $Pr = 10$ for a set of values of Ma, including the asymptote for $Re \to \infty$ and $Ma \to \infty$ from Equation (4.15.4); in the asymptotic result, the value of Ma has been set equal to 500.

from Equations (4.13.32) and (4.13.56) for this purpose:

$$t_0 = \frac{1}{3}\log \epsilon + 0.6727 + \frac{\pi}{6\sqrt{3}} - \frac{1}{6}\log\left(27v_{\infty,0}^2\right) + \frac{2}{3}\log\left(\frac{1+\cos\theta}{\sin\theta}\right)$$
$$+ \frac{1}{6}\log[\xi(s)] + \frac{2}{3}F(\zeta) + \frac{1}{3}E_1\left[\frac{3}{4}v_{\infty,0}y^2\right]. \tag{4.15.4}$$

All the symbols used in Equation (4.15.4) have been defined in Section 4.13. Because it is necessary to use a specific value of the Marangoni number to include the contribution of the leading order constant in Equation (4.15.4), we have calculated this solution for $Ma = 500$, so that the solid curve can be compared with the numerical results for that case. It is evident that, in this range of Ma values, the temperature at the forward stagnation point is decreasing more rapidly with increasing values of Ma than that at the rear stagnation point. The net consequence is a reduction in the overall temperature difference over the surface of the bubble. Note that the asymptotic result is in good agreement with that from the numerical solution in the forward half but deviates from it for $\theta \geq \frac{\pi}{2}$. In this region, the temperature field on the bubble surface will be affected by the separated reverse flow wake that will be discussed in Section 4.17, which is not accommodated in the asymptotic analysis. We suspect that this is a possible reason for the differences between the numerical solution and the asymptotic solution and offer the following argument in support of this conjecture. As mentioned in Section 4.13, Balasubramaniam (1995), who performed numerical computations by using a potential flow velocity profile, noted that the temperature difference over the surface of the bubble appears to increase with increasing values of Ma for large Ma. On the contrary, the numerical calculations made by Balasubramaniam and Lavery (1989) do not display such behavior.

Calculations similar to those of Balasubramaniam and Lavery have been performed by Treuner et al. (1996) who sought to obtain theoretical results corresponding to their experimental conditions. These authors used a nonuniform grid to adequately resolve the sharp gradients near the bubble surface and repeated the calculations with different values of R_∞ to obtain assurance that the results did not depend on its value. The authors performed the calculations up to $Ma = 2500$. Numerical calculations in the gas bubble limit, permitting shape deformation, have been made by Chen and Lee (1992) and, more recently, by Welch (1998). We discuss shape deformation effects later in this section.

In Section 4.11, we discussed the numerical calculations reported by Oliver and DeWitt (1994) for the transient thermocapillary migration of a bubble for values of Ma up to 200. The authors assumed Stokes flow and used the corresponding quasi-steady analytical result for the velocity distribution, solving the energy equation by the method of finite differences. Their principal results are that a steady state can be expected after a bubble has moved a distance approximately equal to 1 to 5 radii and that, for sufficiently large values of the Marangoni number, the velocity overshoots the terminal value before achieving that value asymptotically. Oliver and DeWitt, as well as all the other investigators, assumed constant physical properties, assuring the existence of an asymptotic steady migration velocity.

4.15.3   Results in the Case of Drops

The first detailed numerical computations, in the case of drops for which the ratios α and β assume nonzero values, appear to have been reported by Ehmann, Wozniak, and Siekmann (1992), who assumed a spherical shape. The authors chose to make calculations for their experimental system, which consisted of paraffin oil drops in an aqueous ethanol solution and the opposite case of ethanol solution drops in paraffin oil. More recently, Nas (1995), as well as Haj-Hariri et al. (1997), have made computations for liquid drops permitting deformation of shape. Ma (1998), assuming a spherical shape, has carried out numerical computations for two pairs of fluids that were used in our reduced gravity experiments, discussed in the next section. The principal results from his computations have been reported in Ma et al. (1999). We now display and discuss some of these results.

The scaled migration velocity, v_∞, normalized using v_{YGB}, is plotted against the Marangoni number in Figure 4.15.6 for two specific pairs of fluids used in our flight

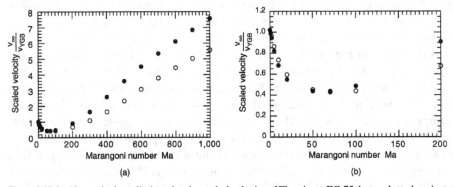

Figure 4.15.6   Numerical predictions for the scaled velocity of Fluorinert FC-75 drops plotted against the Marangoni number in two DC-200 silicone oils: (a) full range of Marangoni number values from 0 to 1000, (b) magnified view of the range $0 \leq Ma \leq 200$; the filled circles correspond to 50 cs silicone oil and the unfilled circles to 10 cs silicone oil. (Reproduced with permission from Ma, X., Balasubramaniam, R., and Subramanian, R. S. Numerical simulation of thermocapillary drop motion with internal circulation. *Num. Heat Transfer A* 35, 291–309. Copyright 1999, Taylor and Francis, Inc.)

experiments. In both cases, the drop phase is Fluorinert FC-75, which is a perfluori-
nated liquid, and the continuous phase is a Dow-Corning silicone oil. In making these
numerical computations, the values of the Prandtl number for each fluid were chosen
to be representative of the experimental conditions in each case. For the FC-75/10 cs
silicone oil pair, the Prandtl numbers are 17.9 and 83.3, respectively. For the FC-75/50 cs
silicone oil pair, the values are 22.3 and 469, respectively. The difference in the values of
the Prandtl number for the FC-75 fluid between the two cases arises from a difference
in the temperature at which the physical properties were evaluated.

Figure 4.15.6(a) shows the results in the range $Ma = 0$ to 1000, whereas Figure
4.15.6(b) displays a magnified view of the region $Ma = 0$ to 200. At small values of Ma,
the behavior is similar to that noted in the gas bubble limit. That is, the scaled velocity
decreases as the Marangoni number is increased. For small values of the Marangoni
number, Ma (1998) has confirmed the correctness of the asymptotic result given in
Equation (4.12.48). As noted in Section 4.12, Ma showed that these results appear
to be good for values of the Marangoni number up to about 5 for the two pairs of fluids
considered in the calculations. The most noteworthy feature displayed in the figures is
the increase of the scaled velocity with increasing values of the Marangoni number that
occurs at sufficiently large values of Ma. This behavior is consistent with the prediction
from the asymptotic theory presented in Section 4.14 and is in stark contrast with the
behavior in the gas bubble limit. The predictions from the asymptotic theory are not
included in Figure 4.15.6 because the Reynolds numbers used in the numerical computations are too small for the asymptotic theory to be valid. We have discussed the
performance of the leading order prediction given in Equation (4.14.65), when compared with numerical results for a more appropriate set of parameters, in Section 4.14.

In Figure 4.15.7, we display the evolution of the isotherm structure within the drop
from the conduction limit to that dominated by convective transport for the FC-75/50
cs silicone oil pair. It can be observed from Ma et al. (1999) that when $Ma = 400$, the
isotherm structure inside the drop is similar to the streamline structure, which we have
not included here. Physically, convective transport is rapid in this situation, so that the
temperature remains the same along a streamline in a first approximation, except in
a thin thermal boundary layer near the surface of the drop. Energy is transferred by
conduction across the streamlines within the drop.

Figure 4.15.8 shows representative isotherms in the continuous phase for $Ma = 400$.
These isotherms are striking in their contrast with those displayed in Figure 4.15.3(a),
for the case of a gas bubble. The Prandtl numbers of the continuous phase for the
isotherms displayed in Figures 4.15.3(a) and 4.15.8 are 10 and 469, respectively; the
corresponding Reynolds numbers are 50 and 0.853. Although the difference in the values
of the Reynolds number will alter the flow structure to some extent, the substantial
contrast in the isotherm structure in the thermal wake is due to qualitative differences
in the temperature distributions in the thermal boundary layers between the gas bubble
limit, and that of a liquid drop, discussed in Section 4.14. Specifically, we see from the
asymptotic solution obtained in that section that the scaled temperature near the drop
surface is negative, and proportional to Ma at leading order. This implies not only very
cold conditions in the thermal boundary layer, but also large changes in temperature
because the temperature must approach values of $O(1)$ at the edge of the boundary
layer. On the other hand, in the gas bubble limit, Equation (4.13.32) shows that the
temperature near the bubble surface is $-\frac{1}{6} \log Ma + O(1)$. Even when $Ma = 500$, the
logarithmic term leads to only a contribution of order unity in lowering the temperature
in the thermal boundary layer. Therefore, in the gas bubble limit, moderately cold

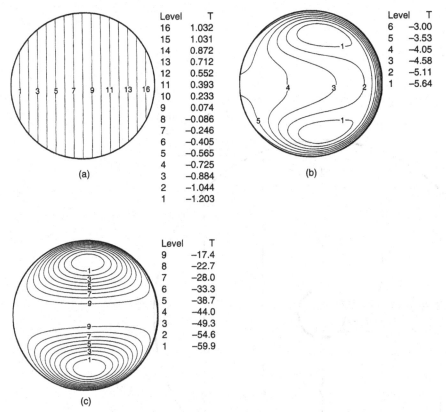

Figure 4.15.7 Predicted isotherms within a Fluorinert FC-75 drop migrating in a 50 cs DC-200 silicone oil: (a) $Ma = 0$, (b) $Ma = 100$, and (c) $Ma = 400$. The applied temperature gradient is directed from left to right.

conditions prevail in the thermal boundary layer near the bubble surface, with variations of $O(1)$. The asymptotic temperature scalings in the thermal wakes behind a gas bubble and a drop are similar to the scalings in the thermal boundary layers in the respective cases, discussed above. Therefore, for a comparable value of the Marangoni number, in the case of a drop, we see relatively colder conditions, as well as larger variations of temperature in the thermal wake.

In Figure 4.15.9, the scaled temperature difference over the interface is displayed as a function of the Marangoni number for the FC-75/50 cs silicone oil pair. The variation of the scaled interfacial temperature with the polar angle is shown in Figure 4.15.10. We see that the total scaled temperature difference over the surface of the drop first decreases as the Marangoni number increases from a value of zero. When Ma increases past about 100, however, this temperature difference begins to increase with further increase of the Marangoni number, whereas the drop surface cools relative to the undistorted continuous phase fluid. This is consistent with the asymptotic picture presented in Section 4.14. The scaled interfacial velocity distribution tracks this behavior as is evident from Figure 4.15.11.

4.15.4 Influence of Shape Deformation

Now, we discuss available results that are calculated by taking shape deformation into account. As noted earlier, Chen and Lee (1992) were probably the first investigators

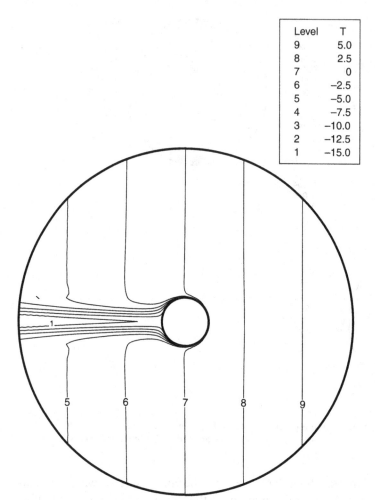

Level	T
9	5.0
8	2.5
7	0
6	−2.5
5	−5.0
4	−7.5
3	−10.0
2	−12.5
1	−15.0

Figure 4.15.8   Predicted isotherms in a 50 cs DC-200 silicone oil, for $Ma = 400$, when a drop of Fluorinert FC-75 migrates through it. The applied temperature gradient is directed from left to right. (Reproduced with permission from Ma, X., Balasubramaniam, R., and Subramanian, R. S. Numerical simulation of thermocapillary drop motion with internal circulation. *Num. Heat Transfer A* 35, 291–309. Copyright 1999, Taylor and Francis, Inc.)

to permit shape deformation in obtaining a numerical solution. Chen and Lee solved the problem for a gas bubble. An important observation that these authors made is that a small deformation from the spherical shape has a significant impact on the migration velocity of the bubble. Later, Welch (1998) performed computations permitting the motion to be unsteady, using a finite volume method on an unstructured moving grid to obtain the solution. The unsteadiness arises from the fact that the deformation changes with the passage of time because of the variation of surface tension with temperature. As the bubble moves into warmer regions, the average surface tension decreases, leading to increased deformation and a concomitant reduction in velocity. Other physical properties, such as the viscosity, density, and thermal conductivity, are assumed to be constant. Welch also reported that the computed migration velocities for a small value of the Capillary number, $Ca = 0.001$, for $Re = 1$, and $Ma = 1, 10$, and

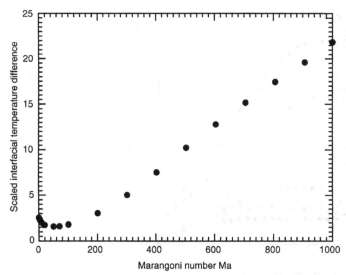

Figure 4.15.9 Predicted scaled interfacial temperature difference $[T(1,0) - T(1,\pi)]$ plotted against the Marangoni number for Fluorinert FC-75 drops in 50 cs DC-200 silicone oil. (Reproduced with permission from Ma, X., Balasubramaniam, R., and Subramanian, R. S. Numerical simulation of thermocapillary drop motion with internal circulation. *Num. Heat Transfer A* 35, 291–309. Copyright 1999, Taylor and Francis, Inc.)

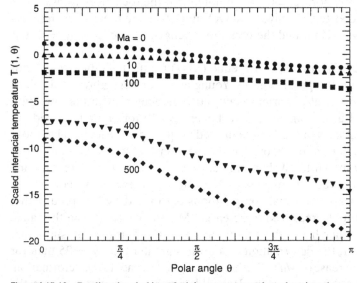

Figure 4.15.10 Predicted scaled interfacial temperature plotted against the polar angle for Fluorinert FC-75 drops in 50 cs DC-200 silicone oil. (Reproduced with permission from Ma, X., Balasubramaniam, R., and Subramanian, R. S. Numerical simulation of thermocapillary drop motion with internal circulation. *Num. Heat Transfer A* 35, 291–309. Copyright 1999, Taylor and Francis, Inc.)

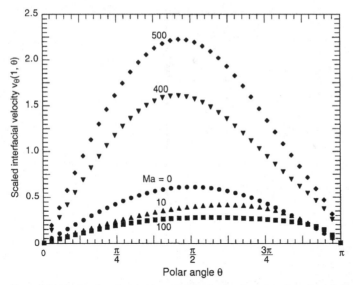

Figure 4.15.11 Predicted scaled interfacial velocity plotted against the polar angle for Fluorinert FC-75 drops in 50 cs DC-200 silicone oil. (Reproduced with permission from Ma, X., Balasubramaniam, R., and Subramanian, R. S. Numerical simulation of thermocapillary drop motion with internal circulation. *Num. Heat Transfer A* 35, 291–309, Copyright 1999, Taylor and Francis, Inc.)

100, were in agreement with those obtained by Balasubramaniam and Lavery (1989) and Haj-Hariri et al. (1997), for $Ca = 0$. Welch noted that the oscillatory transients in the migration velocity, reported by Oliver and DeWitt (1994), are present when deformation of shape is allowed. He found the oscillatory behavior to occur at $Re = 1$, but not at $Re = 100$.

The results of Nas (1995) and Haj-Hariri et al. (1997) for the migration velocity of a drop are consistent with the predictions of Young et al. for negligibly small values of Re and Ma. Haj-Hariri et al. reported quantitative agreement with the migration velocities computed by Balasubramaniam and Lavery (1989) for an undeformed gas bubble with $Re = 1$ and Ma up to 100. The computed drop shapes are identical to those predicted by the small deformation theory given in Section 4.3 for a Weber number of 0.1 and are qualitatively similar for Weber numbers equal to 2.5 and 5. There are some qualitative differences between the results of Nas and Haj-Hariri et al. For $Re = 1$ and $Ca = 0.0666$, the two-dimensional computations performed by Nas predict that the deformation of a drop decreases with increasing Marangoni number in the range $10 \le Ma \le 60$. On the other hand, the computations of Haj-Hariri et al. show that for $Re = 50$ and $Ca = 0.1$, the deformation of a drop is greater for $Ma = 25$ than for $Ma = 1$. With a further increase of Ma to 50, there is a slight decrease in the deformation. It is unclear whether the difference in the Reynolds number is playing a role in the dependence of the deformation of a drop on the Marangoni number. It also is unclear whether the results of Nas can be qualitatively extrapolated to the three-dimensional case. The property ratios in the two simulations are comparable. Another difference between the two sets of predictions is in the behavior of the scaled migration velocity of a drop as the Reynolds number is increased for a fixed value of the Marangoni number. Nas obtained a decrease in the scaled migration velocity with increasing Reynolds number, in the range $1 \le Re \le 10$, for $Ma = 1$ and $Ca = 0.0666$. Haj-Hariri et al. showed that the scaled migration velocity of a drop is a weakly increasing function of Re, in the range $0 \le Re \le 100$, for $Ma = 50$ and $Ca = 0.1$. When the shape of a drop is forced to be a sphere, Ma et al. (1999) found more complex behavior. In the region where the scaled velocity decreases with increasing Marangoni number at a fixed Prandtl number, increasing the value of Re while keeping the value of Ma fixed leads to an increase in the scaled migration velocity. This trend is similar to that noted in the gas bubble limit. On the other hand, at larger values of Ma, wherein the scaled migration velocity increases with increase in Ma at a fixed value of Pr, increasing Re at a fixed value of Ma results in a decrease in the scaled migration velocity. These results illustrate the need for a more thorough numerical study of the thermocapillary motion of an isolated drop to permit one to draw conclusions about general trends.

It is not our purpose to provide an exhaustive account of the numerical investigations on this problem in this section. Instead, the highlights of available results are detailed here. The reader can see that the full range of parameters relevant to this problem has not yet been explored. Also, none of the studies has accommodated the role of the variation of physical properties with temperature. In particular, the change in viscosity with temperature will cause the velocity of a drop to be inherently unsteady. Using a potential flow approximation, Balasubramaniam (1998a) has shown, in the asymptotic case $Ma \to \infty$, that the effect of viscosity variation with temperature is not fully accommodated by using a value of the viscosity corresponding to the temperature in the plane $z = 0$ (in Figure 3.1.1) in the undisturbed fluid. It will be seen in the next section, in which experimental results are presented, that there is a need for computations that fully account for the time dependence introduced by the variation of physical properties with temperature.

4.16 Experimental Results on Isolated Bubbles and Drops

4.16.1 Introduction

So far in this chapter, we have focused on theoretical development. Available results from experiments will be discussed now, confining attention to bubbles and drops unaffected by boundaries or neighboring objects. Experiments have been carried out in Earth-based laboratories, as well as in reduced gravity conditions. At the outset, it is useful to mention the limitations imposed by Earth's gravitational field on such experiments, so that one can see the rationale for carrying out experiments in low gravity.

Two problems are introduced by Earth's gravitational field. The gravitational force causes buoyant convection, also known as *free* or *natural* convection, and, in addition, imposes a net hydrostatic force on a drop, which must be balanced by the hydrodynamic force for quasi-steady motion. Buoyant convection always arises when density gradients are present in a direction normal to the gravitational acceleration vector. Although it will occur both within the drop and in the continuous phase, for relatively small drops the more critical concern is with buoyant motion in the continuous phase. Because drops and bubbles will be affected by it, and because it is difficult to quantitatively characterize the extent of the interference from it, this situation must be avoided in Earth-based experiments. Therefore, experiments designed to observe thermocapillary migration have been carried out in vertical temperature gradients. In experiments on gas bubbles, downward temperature gradients have been used to attempt to hold a bubble still or make it move downward. The resulting density stratification in the liquid is unfavorable, and a sufficiently large density gradient will cause buoyant convection. Such motion is termed *cellular* because fluid rises and sinks within cells. The length scale of the cells depends on the parameters. The onset of this instability is characterized by a critical value of the Rayleigh number, which depends on the geometry and boundary conditions. A definition of the Rayleigh number and a discussion of the instability problem can be found in Chandrasekhar (1961). This instability limits the magnitude of the temperature gradient that can be used in a given liquid. In the case of liquid drops moving in a second liquid, upward temperature gradients have been used to achieve a stable density stratification, and the drop phase has been chosen as the more dense phase. Of course, the mere presence of the drop will lead to lateral temperature gradients, and therefore lateral density variations, leading to some contribution from buoyant convection. If the radius of the drop is sufficiently small, such effects can at least be minimized because the magnitude of the buoyant convection velocities decreases rapidly as the length scale is reduced. Even when using a vertical temperature gradient, precautions must be taken to insulate the side walls of the container holding the continuous phase, so that heat loss through them would be negligible. This is necessary to keep lateral temperature variations as small as possible. In early experiments, no attempt was made to monitor the background convection in the cell to assure that it is indeed at a negligible level, but more recent experiments have always included such a check.

The second aspect of the interference from Earth's gravitational field is even more troublesome. In the linear case when convective transport is negligible, one can conduct experiments on Earth and deduce the thermocapillary contribution to the velocity of a drop by subtracting out the gravitational contribution. When convective transport of momentum, energy, or both becomes important, this is no longer possible. When inertia is important, it is easy to see that the problem is nonlinear; therefore, the contributions from the two driving forces are not simply additive. Even if inertia is negligible, the inclusion of convective transport of energy makes the problem nonlinear. This is because the unknown velocity field is coupled to the temperature distribution on the surface of the drop. Hence, to study the purely thermocapillary migration of a drop or bubble under conditions outside of the linear regime, a reduced gravity environment is necessary. This has been the primary motivating force for the conduct of experiments on this subject in drop towers, sounding rockets, and the space shuttle.

Of course, it is not straightforward to conduct an experiment on drops and bubbles in reduced gravity. Virtually all of the experimental work has been performed without the direct interaction of a scientist with the experiment in the manner that is possible on Earth. It is necessary to have automated facilities to reproducibly form drops of a given size at a given location within a second fluid and release them with minimal disturbance. Furthermore, when the drop has completed its motion in the test cell, it must be removed by a reliable mechanism. Another complication, which becomes important in large test cells, is the effect of the background gravitational field in causing some buoyant convection and distorting the temperature field. Nevertheless, these obstacles have been overcome, and results have been obtained recently that we report in the present section.

First, experimental work performed on Earth will be summarized, and then the information available from experiments performed in reduced gravity aboard spacecraft will be discussed.

4.16.2 Experiments on Earth

The first reported experiments were performed by Young et al. (1959). These authors suspended a small column of liquid between the anvils of a machinist's micrometer. Both anvils were passed through copper blocks in which mercury thermometers were inserted for the measurement of temperature. A nichrome wire was wound around the lower copper block so that it could be heated. Three Dow-Corning silicone oils of the DC-200 series, of nominal viscosity 20, 200, and 1000 centistokes, were used for obtaining the reported data. Also, control experiments were carried out in n-hexadecane. As an aside, silicone oils have commonly been used in studies involving surface tension driven motion because they appear unsusceptible to surface contamination. The DC-200 series fluids are dimethylsiloxane polymers that, at any given nominal viscosity, contain a collection of polymer molecules of varying chain length. They no doubt contain other chemical substances that are used during the polymerization process. These are dissolved in the bulk and do not appear to prefer the interface, however.

The liquid column contained a collection of air bubbles. At the beginning, when the lower anvil was heated, the bubbles were found to sink toward it and collect at the anvil surface. The authors then reduced the temperature gradient gradually and found that the larger bubbles began to rise slowly. Another adjustment of the temperature gradient was found to be sufficient to keep them nearly motionless near the middle of the liquid column, this condition being characterized by the authors as one in which the bubbles moved with a velocity less than 10 μm/s. As we know from the arguments presented in Section 4.2, it is not always possible to hold gas bubbles fixed by the use of a downward temperature gradient in a liquid in which the density decreases with increasing temperature. This is because the equilibrium position can become unstable to small vertical disturbances from it under certain conditions. Nonetheless, a very slowly moving bubble, or one reversing direction, can be imagined to be nearly stationary. In any case, even with considerable scatter in their data, Young et al. found that the temperature gradient needed to hold a bubble nearly fixed increased with the diameter of the bubble, consistent with their theoretical prediction. When they plotted the diameter against the temperature gradient, they also observed that the results were insensitive to the viscosity of the silicone oil that was varied by a factor of 50, also consistent with their prediction. The authors found that bubbles did not move downward in hexadecane contaminated with a small amount of silicone oil when a downward temperature gradient was applied. They took this to be confirmation of the correctness of their hypothesis, namely that the phenomenon being observed in the silicone oils was driven by surface tension gradients caused by temperature variations. They suggested that the failure to observe the same phenomenon in hexadecane occurred because the silicone oil served as a surface active contaminant in this liquid.

It is remarkable that the pioneering experiments of Young et al. provided such a clear trend because the observations were subject to several experimental uncertainties. The liquid column had a free liquid surface at which the temperature gradients would certainly have led to interfacial tension gradients. In silicone oils, this would cause upward motion at the free surface (because the lower anvil was heated) and sinking of the fluid in the center of the liquid column. Also, because of heat loss to the surrounding air, one would expect radial temperature gradients to arise with warm fluid in the middle and relatively cooler fluid at the surface. The resulting buoyant convection would have opposed the motion driven by thermocapillarity at the free liquid surface and may have compensated somewhat for it. Furthermore, there were several bubbles present at a given time, leading to potential problems with interactions. Finally, observations of size were made through a cylindrical surface that must have caused some distortion. The authors must have been aware of this because they mentioned a substantial uncertainty in the measurement of the diameter.

Young et al. contented themselves with drawing three straight lines through their plot of the diameter of the nearly stationary bubbles versus the applied temperature gradient, representing three values of the rate of change of interfacial tension with temperature σ_T. Interestingly, the values of σ_T, reported by the authors and attributed to an earlier article by Young, are nearly the same for all three silicone oils to within the experimental uncertainty quoted. When this correct value is used, it can be seen that most of the reported bubble diameters are smaller than those expected from the prediction by the authors.

Young et al. were well aware of the possibility of the onset of buoyant convection due to the Rayleigh instability in a liquid column in which the density increases upward. They used sufficiently gentle temperature gradients, in the given fluids and column geometry, to keep the Rayleigh number below the critical value.

Many of the pitfalls in the experiments of Young et al. were eliminated by Hardy (1979) who conducted similar studies on one air bubble at a time in an enclosed rectangular cavity filled with a Dow-Corning DC-200 silicone oil of nominal viscosity 1000 centistokes. Hardy used horizontal brass plates that were soldered to tubes heated with nichrome wire wound around them and implanted miniature thermocouples in the plates so that he could directly measure the temperatures at the top and bottom surfaces of the cell. By using a third thermocouple inserted into the cell, Hardy confirmed that the vertical variation of temperature in the liquid was linear, as assumed in the theoretical model. He used flat double-pane glass windows to provide insulation while permitting observation through a microscope. Hardy found that a bubble was never stationary in the cell. Furthermore, the radius of a bubble also changed continuously. He noted that a typical bubble, upon insertion into the cell, was sufficiently large that it rose to the upper surface and came to rest there. This is the location of the coolest fluid in the cell, however, and the bubble was observed to shrink in size because of dissolution of the components of air into the liquid which, at that location, was presumably undersaturated. When the bubble became sufficiently small, it started moving downward because of the thermocapillary effect. This is because the relative influence of gravity, when compared with thermocapillarity, is proportional to the radius, as we saw in Section 4.2.3. Therefore, when the bubble shrank in size, thermocapillarity began assuming an increasingly dominant role over gravity. When the bubble began moving down, it continued to shrink in size for a brief period, but then began to grow again. This was because the bubble was moving into warmer liquid with the same amount of the components of air dissolved in it, but the liquid was supersaturated because the temperature was higher. Soon, the growing bubble came to a stop and started rising because the relative effect of the buoyant force became too large. After growing for some more time until it passed the temperature level at which the liquid was just saturated, the bubble once again began to shrink, and the cycle repeated itself. Therefore, the bubbles in Hardy's experiments could not be held motionless because of the consequences of dissolution and growth.

Hardy still was able to gather useful data from his experiments. He noted that when the bubble just reversed direction, its velocity was zero; he concluded that he could use the theory of Young et al. to interpret the data at that point. Of course, the fact that the bubble is slowing down and changing direction implies that it is constantly accelerating and therefore under the action of a nonzero force. Nonetheless, one can argue that for very small bubbles moving at low Reynolds number, the motion is nearly quasi-steady and occurs at negligible net force at every instant. In any case, Hardy found that for bubbles ranging in diameter from about 50 μm to about 250 μm, the data were consistent with the prediction that the temperature gradient necessary to hold a bubble nearly stationary be proportional to the diameter. From the slope of the straight line that best fits the data plotted in suitable form, the value of σ_T can be inferred for this silicone oil, and Hardy obtained a value of -0.055 mN/(m·K). From independent measurements of the surface tension of the silicone oil at different temperatures, Hardy found $\sigma_T = -0.058$ mN/(m·K), which was sufficiently close to the value inferred from the bubble experiments. Hardy also reported some data on bubble velocities that were somewhat smaller in magnitude than those predicted by Young et al. Neither Young et al. nor Hardy reported any values of the Reynolds and Marangoni numbers in their experiments, but from the conditions of these experiments, one can conclude that these groups must have been very small.

Hardy's design of the cell was adapted by Merritt and Subramanian (1988), who performed experiments on the motion of air bubbles in three different Dow-Corning DC-200 series silicone oils of nominal viscosity 200, 500, and 1000 centistokes. Just like Hardy, these authors also observed that the bubbles never achieved a stationary location in the cell. Their objective was to measure the velocities of the bubbles and to test the predictions of Young et al. To do this, they arranged Equation (4.2.31), after setting $\alpha = \beta = 0$, in the following form for the physical velocity of the bubble v_∞^*:

$$\frac{\mu v_\infty^*}{\rho g R^2} = \frac{\sigma_T |\nabla T_\infty|}{2\rho g R} + \frac{1}{3}. \tag{4.16.1}$$

All the symbols appearing here have been defined earlier in Section 4.2. The authors plotted the quantity appearing on the left side against $\frac{|\nabla T_\infty|}{2\rho g R}$ for a given fluid. They found the data fell on a straight line confirming the theoretical prediction. The slope of a straight line, fitted using the method of least squares, gave the coefficient σ_T. The values were in the range -0.060 to -0.063 mN/(m·K) for the three silicone oils, consistent with those reported by Young et al., but a bit higher than the values reported by Hardy.

Merritt (1988) indicated that the Reynolds number in his experiments did not exceed 10^{-5} and that the Péclet number was less than 0.003 using the bubble velocity as the reference in defining these groups. From the relatively small values of these dimensionless groups, one can safely assume that convective transport effects were indeed small in the experiments discussed above. Merritt confirmed that the temperature profile in the silicone oils used in the experiments was linear in the absence of bubbles, and he made careful measurements of the background buoyant convection velocities in the silicone oils arising from lateral temperature variations caused by heat loss to the surroundings. He found these velocities were less than about 0.35 μm/s, whereas the measured velocities ranged up to about 25 μm/s, making the contribution from buoyant convection in the silicone oils a negligible factor. Morick and Woermann (1993) reported experimental results on the motion of gas bubbles in a vertical temperature gradient that are consistent with the observations made by Hardy (1979) and Merritt and Subramanian (1988).

In a different type of experiment, McGrew, Rehm, and Griskey (1974) used a device constructed from metallic wire to hold bubbles at the end of a cantilever in a liquid, in which an upward temperature gradient was introduced. The device kept a bubble in place with a minimum of contact, so that disturbances introduced by it could be regarded as being small. The authors determined the force exerted on the bubbles from the deflection of the cantilever. From this, by subtracting the hydrostatic force, they could obtain the hydrodynamic force, which they labeled as the *thermophoretic force* and which we shall term the *thermocapillary force*. In the absence of specific information, one must presume the bubbles contained air. The liquids used were methanol and ethanol. Temperature gradients ranged from approximately 1 to 6.5 K/mm. The authors do not provide the range of bubble radii, indicating only that experiments were conducted over a variety of bubble sizes and temperature gradients in each liquid. From a calibration drawing included in the article, in which the hydrostatic force is plotted against the bubble volume, we infer that the range of bubble radii was from approximately 0.52 mm to 0.93 mm. The authors plotted the ratio of the thermocapillary force to the buoyant force, multiplied by the radius of the bubble, against the applied temperature gradient. Although the data show scatter, in the case of butyl alcohol, they lie on either side of a straight line based on the linear theory, which neglects convective transport altogether. This is puzzling because the conditions of the experiments were such that convective transport effects must have played a substantial role. In the case of methanol, the force resulting from thermocapillarity appears too large compared with similar predictions. The authors advance an explanation based on the fact that methanol has a large vapor pressure at these temperatures. The influence of thermocapillarity on bubbles containing a significant amount of condensable vapor does not appear to have been explored experimentally since the appearance of the work of McGrew et al. (1974).

Experiments on the thermocapillary migration of liquid drops in another liquid have taken two directions. One type of experiment is similar to that on gas bubbles performed by Merritt (1988). In the other type, drops are held stationary against Earth's gravitational force by the application of a suitable temperature gradient. Unlike in the case of gas bubbles, it is possible to achieve stable stationary positioning of liquid drops under suitable conditions. We provide a discussion of both types of experiments below.

The first reported experiments were performed by B. Facemire and are described in a National Aeronautics and Space Administration (NASA) report by Lacy et al. (1982). The authors were studying the behavior of a model system consisting of a binary mixture of ethyl salicylate (ES) and diethylene glycol (DEG), that exhibits a miscibility gap. When mixed and heated to a sufficiently high temperature, DEG and ES form a single liquid phase. When this liquid is cooled, at a specific temperature that depends on the composition, drops of one phase are formed spontaneously. In the experiments reported in Lacy et al., a liquid rich in DEG was placed in a spectrophotometer cell that was cooled from below and heated from above. As soon as cooling was initiated, a cloud of drops rich in ES was formed near the cold bottom plate and began to move upward. Because the drops are slightly more dense than the continuous phase, they should sink in the absence of other mechanisms which influence their motion. The authors observed that the drops always rose and took this to be conclusive evidence that the motion was influenced by thermocapillarity. From a knowledge of the physical properties and the experimental conditions, the authors also estimated that the contribution of thermocapillarity to the motion was an order of magnitude larger than that of gravity. As the drops moved toward the upper (hot) surface of the cell, they were expected to dissolve slowly. This was confirmed by the observation of a wake in which the refractive index was sufficiently different from that of the continuous phase that it could be seen. All the measurements were made when the drops were within 1 to 2 mm of the upper surface. The authors reported measured velocities that are lower than those predicted, with the ratio varying from about 0.5 to 0.92. This is remarkable, considering the nature of the uncertainties involved. The cell was not well insulated, which no doubt led to buoyant convection, a fact the authors mention. Also, the temperature distribution was not directly measured, and it is unclear whether the drops can be regarded as being isolated. The temperature distribution was unsteady during the experiment, even though the authors reported reaching steady temperatures at the top and bottom surfaces. Lacy et al. speculated that the observed discrepancies also may be caused by the effects of concentration gradients on interfacial tension.

Barton and Subramanian (1989) used the same pair of liquids in an improved version of the apparatus used by Merritt (1988). The authors filled the cavity in the cell with pure DEG and injected drops of pure ES one at a time. Observations of the behavior of a large number of isolated drops yielded data that established that the drops normally settle downward in the absence of a temperature gradient and that they rise in an upward temperature gradient. The upward velocity of the drops was found to increase with increasing drop size and with increasing temperature gradient. Control experiments, performed with a surfactant that suppressed the upward motion in a temperature gradient, were offered as evidence that the observed upward motion was indeed driven by the thermocapillary stress. The authors rearranged Equation (4.2.29) as shown below and plotted the quantity appearing on the left side against the radius of the drops.

$$|\nabla T_\infty|^{-1}\left[\frac{3\mu(2+3\alpha)}{2}v_\infty^* + (\rho'-\rho)g(1+\alpha)R^2\right] = -\left(\frac{3\sigma_T}{2+\beta}\right)R. \qquad (4.16.2)$$

Their objective was to verify that, after subtracting the gravitational contribution to the motion, the resulting thermocapillary migration velocity displayed the correct linear dependence on the temperature gradient and the first power of the drop radius. The temperature gradient was varied from 2.4 to 9.5 K/mm, and the radius of the drops from approximately 10 to 250 μm. The Reynolds number was no greater than 0.006, and the maximum Marangoni number was approximately unity. Their drawing, which verifies the scalings predicted by the linear theory, is reproduced as Figure 4.16.1. The straight

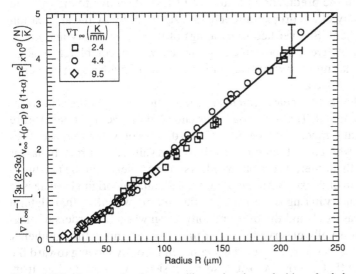

Figure 4.16.1 Data on the thermocapillary migration velocities of ethylsalicylate drops in diethylene glycol; the left side of Equation (4.16.2) is plotted against the radius of the drop demonstrating the linear scaling predicted by theory (Courtesy of Academic Press).

line included in the drawing was fitted to the data by the method of least squares. Equation (4.16.2) also predicts that the intercept in the figure should be zero. Nonetheless, a small nonzero intercept can be observed, which is likely due to experimental uncertainties. Barton and Subramanian (1990) noted that the coefficient σ_T estimated from the slope of the straight line in Figure 4.16.1 is substantially smaller in magnitude than the same coefficient evaluated from an expression given by Lacy et al. The authors also provided some possible explanations for this observation and presented similar results for two other average temperatures at which the thermocapillary migration measurements were made. More details are given in Barton (1990).

Later, Nallani and Subramanian (1993) set out to identify a pair of liquids suitable for reduced gravity experiments on thermocapillary migration. Using an apparatus modeled after that of Barton, the authors measured the velocities of methanol drops in a Dow-Corning DC-200 silicone oil of nominal viscosity 50 centistokes, when subjected to a temperature gradient. Under isothermal conditions, methanol drops rise in this continuous phase. When an upward temperature gradient was imposed, the drops rose at higher velocities, and the difference in velocities between the two cases increased with increasing drop radius. The drops ranged in radius from 50 to 146 μm, and the temperature gradients used were 8.4, 11.9, and 13.0 K/mm. The maximum Reynolds number in the experiments was 0.005, and the maximum Marangoni number was 1.5. The rise velocities in the isothermal experiments followed the predictions of Hadamard (1911) and Rybczyński (1911) for the larger drops but approached those of Stokes in the case of small drops. The authors surmised that this was due to contamination by trace impurities that must have been surface active. They fitted the data to the predictions for a drop with a stagnant cap, given in Section 4.7, but added a note of caution that this should not necessarily be considered as proof of the existence of such a cap. Because of the problem with impurities in this system, Srividya (1993) chose to investigate Fluorinert FC-75 drops settling in a DC-200 series silicone oil of nominal viscosity 50 centistokes when subjected to an upward temperature gradient. Fluorinert is more dense than the silicone oil used, and the drops were found to sink under isothermal conditions in accord with the prediction in Equation (3.1.33). Srividya found that the downward settling velocity of the drops was decreased significantly by the application of the temperature gradient. She verified the scalings of the linear theory and noted some influence from convective transport of energy, which was not completely negligible. Although the Reynolds number was no larger than 0.015 in Srividya's experiments, the Péclet number was as large as 5.

Bratukhin et al. (1984) made measurements of the motion of air bubbles in water contained in a rectangular cell. The water was maintained at an average temperature of 4°C. It is known that at temperatures below 4°C, the density of water increases with increasing temperature, whereas at temperatures above this value, the normal behavior of decreasing density with temperature is observed. As a consequence, in the neighborhood of 4°C, the variation of density with temperature is small. The original article is in Russian, and from a summary in English provided by the authors, we gather that a lateral temperature gradient was used, and the buoyant convection was a few orders of magnitude weaker than that usually observed in water under such temperature gradients. The authors found the bubbles to rise upward while simultaneously moving toward the heated wall. They report that the rise velocities were consistent with predictions from Equation (3.1.33), whereas the lateral velocities attributed to thermocapillarity were consistent with predictions from Equation (4.2.28).

Both Merritt (1988) and Barton (1990) made the assumption that the velocity of a drop or bubble is quasi-steady in interpreting their results. During any given experiment, the velocity of a bubble or drop varied with time. Even if the physical properties remain constant throughout the interior of the test cell, it would take a certain amount of time for an object released from the needle to achieve a steady thermocapillary migration velocity, as noted in Section 4.11. When convective transport effects are negligible, as was the case in the above experiments, the longest time scale of the form $\frac{R^2}{\zeta}$, where ζ is the smallest value among the kinematic viscosities and thermal diffusivities of the two fluids involved in a given experiment, determines the time to steady state in all cases except when a liquid drop moves in a gas. An additional complication in the experiments on air bubbles arises from the change of radius with time, but Dill (1991) has pointed out that in Stokes flow, quasi-steady conditions can be expected to prevail, provided the dimensionless group $\Upsilon = |(1 - \gamma)\frac{dR}{dt^*}|/v_0$ is small compared with unity. Here v_0 is the reference velocity defined in Equation (4.1.1). A typical value of Υ in Merritt's experiments is 0.02 and that in Barton's experiments is much smaller. Later, when presenting results from reduced gravity experiments, we discuss the appropriateness of making the quasi-steady state assumption when convective transport effects are important.

The first experiments, in which drops were held fixed against Earth's gravitational force by the use of a temperature gradient, were reported by Delitzsch et al. (1984). These authors established steady upward temperature gradients in an organic liquid contained in a test cell, with the help of Peltier heating and cooling elements. Butyl benzoate was used in some experiments for this purpose and fluorobenzene in some others. Water was employed as the drop phase. When the densities of water and either organic liquid are plotted against temperature, the curves look similar to those shown in Figure 4.2.4. The densities are equal at a certain temperature, above which the density of water decreases more gently with temperature than that of the organic liquid. The reader may find it useful to review the discussion given in Section 4.2.3 regarding this type of experiment. One can expect a drop to reach a stationary location where the forces on it are balanced. The density difference at this location can be inferred from Equation (4.2.36), provided convective transport of momentum and energy can be considered negligible and the drop is away from other drops and boundaries. The density difference necessary to hold a drop stationary is then predicted to be directly proportional to the ratio of the applied temperature gradient to the drop radius. Delitzsch et al. stated that the data support the prediction. They only used a single temperature gradient in each liquid, however. Therefore, only the scaling with respect to the radius could have been verified by the authors. Later Hähnel et al. (1989) presented a substantial amount of data gathered from a similar apparatus on water drops that reached stationary locations in butyl benzoate. Data were obtained at different temperature gradients and for drops of different diameters. The authors plotted the inverse of the diameter of a drop against the height reached by the drop above the stationary location. Assuming a linear dependence of the difference in densities on temperature and a linear variation of temperature with height, Equation (4.2.36) shows that a straight line should result when data are plotted this way. This, of course, presumes that convective transport effects are negligible. Indeed, the data of Hähnel et al. lie approximately on a straight line when plotted in this manner, but data obtained by using different temperature gradients fall on straight lines of different slopes. The slopes are larger than those predicted from the linear theory, and the authors collapsed the results using an empirical effective temperature gradient. They suggested

that the Marangoni number may not have been negligible in their experiments, being
of the order of unity.

Ma (1998) recently performed similar experiments on the same pair of liquids. He
used water from different sources because of a concern about contamination by surface
active chemicals. The water used by Hähnel et al. was doubly distilled and then stored
in a 15-year-old glass bottle for use in the experiments. Ma conducted experiments
by using drops of tap water, deionized water, water used for high-performance liquid
chromatography (HPLC), and freshly made doubly distilled water. Among these, he
found that water drops from the first two sources did not rise significantly above the level
of equal densities, suggesting that the thermocapillary stress must have been balanced
by a compensating tangential stress, which is likely to have been caused by surface-active
contaminants. For a given pair of fluids, one can plot the data in dimensionless form by
using the parameter G, defined in Equation (4.2.30), against the Marangoni number Ma
in the continuous phase. Ma plotted his data for the HPLC and freshly doubly distilled
water in this form. His drawing is displayed in Figure 4.16.2. Also shown is a prediction
from a numerical solution that Ma obtained by using the method of finite differences.
The ordinate is normalized by dividing G by the result that holds in the linear case
(labeled G_{YGB}). This result is given in Equation (4.2.32). Therefore, as the Marangoni
number approaches zero, the ordinate should approach a value of unity. Data obtained
from the HPLC-grade water show little sensitivity to change in the Marangoni number

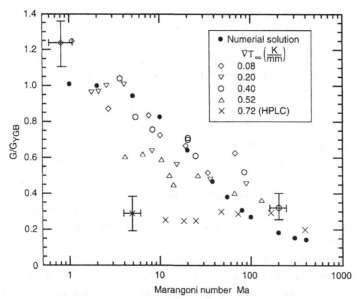

Figure 4.16.2   Data on the stationary positions reached by water drops in butyl-
benzoate in four temperature gradients; the dynamic Bond number G, normalized by G_{YGB}, is plotted against the Marangoni number, and compared with
predictions from a numerical solution. The crosses are for water used for high-
performance liquid chromatography, whereas the rest of the experimental data
were obtained by using doubly distilled water. (The plot is reproduced with
permission from Ma, 1998. "Numerical Simulation and Experiments on Liquid Drops in a Vertical Temperature Gradient in a Liquid of Nearly the Same
Density." Ph. D. diss. in Chemical Engineering, Clarkson University, Potsdam,
New York.)

and lie away from the prediction. Ma (1998) has noted the same behavior when the data of Hähnel et al. are replotted in this form. In that case, the data for each temperature gradient fall on a different line, and for two of the gradients, these lines are nearly horizontal. A simple explanation cannot be offered for this behavior. It is possible that surface active contaminants may play some role, but it is unlikely that their role would lead to such a special effect.

During the period in which Delitzsch et al. were performing their experiments, Wozniak (1986) was studying the same phenomenon, using paraffin drops moving in an upward temperature gradient in a solution of ethanol and water. The apparatus was equipped with suitable windows and an interferometry system, which permitted Wozniak to measure temperature distributions in the presence of the drop using this noninvasive technique. Wozniak reported data on the stationary positions attained by the paraffin drops, as well as the trajectories of these drops. The experiments were performed under conditions when convective transport effects were important. Therefore, he was unable to make comparisons with predictions from a theoretical model.

The behavior of vegetable oil drops, ranging from 1 mm to 10 mm in diameter, in a DC-200 silicone oil of nominal kinematic viscosity 5 centistokes, was studied by Rashidnia and Balasubramaniam (1991). These authors made measurements of the final stationary position reached by the drops when subjected to different temperature gradients and also reported the maximum velocities occurring at the interface. The latter was accomplished by illuminating fine tracer particles that were present in the drops with a laser light sheet, which also permitted the authors to visualize the flow patterns. The authors indicated that the Reynolds numbers were low, but convective transport of energy was important in the experiments. An interesting observation reported by Rashidnia and Balasubramaniam (1991) is that drops rose to greater heights above the matched density location than predicted by linear theory. This is opposite to the trend in other systems, and the authors offered several possible reasons for this behavior. The most plausible is based on the fact that the drops were located between 3 and 10 radii from the cold bottom wall. As will be seen later in Section 7.1, interactions with boundaries can serve to enhance the relative contribution made by thermocapillarity, in contrast to that of gravity, on the motion of drops near a surface. This is because of the difference in the rates at which the disturbance flows decay away from the drop in low Reynolds number flows.

4.16.3 Experiments in Reduced Gravity

It was noted by Yee et al. (1975) and other investigators that gas bubbles were incorporated in materials solidified aboard the United States Skylab in the early 1970s. This prompted Papazian and Wilcox (1978) to study the behavior of bubbles at a solidification interface in carbon tetrabromide aboard a NASA sounding rocket flight, which provided 5 min of reduced gravity. The authors found that the bubbles did not move, even though they were certain that a temperature gradient should have existed in the liquid near the interface. They advanced several explanations, the most likely being that there may have been surface active impurities present in the melt. In experiments conducted aboard similar sounding rocket flights, Wilcox and coworkers (Smith et al. (1982), Meyyappan et al. (1982)) observed the motion of gas bubbles in a temperature gradient in a sodium borate melt contained in a long narrow channel of rectangular cross-section. A series of photographs, taken during the flight, revealed that bubbles indeed moved toward warmer regions in the melt, and large bubbles moved more rapidly than smaller ones. Also, acceleration of the bubbles was noted as they moved into less viscous liquid. While Meyyappan et al. suggested that the velocities were of the correct order of magnitude, quantitative comparisons with theory were not attempted because of experimental complications, such as interactions among the bubbles and with neighboring surfaces.

The first experiments involving a systematic study of the thermocapillary motion of bubbles were those of Thompson (1979), who used the drop tower at the NASA Lewis Research Center for this purpose. The results from this work are described in Thompson, DeWitt, and Labus (1980). The drop tower provided a free-fall period of approximately 5 s. Thompson used a cylindrical container of diameter 120 mm and length 120 mm. Ethylene glycol, a Dow-Corning silicone oil, ethanol, and water were used as the continuous phase in separate experiments. Nitrogen was used as the gas in the bubbles, which ranged from 3 to 4 mm in radius. Photographic records were made of the motion executed by the bubbles. Before each experiment, the liquid in the container was heated from above for a period ranging from 210 to 285 min. The temperature profile in the container was nonlinear, but was approaching linearity in the upper half of the container. After the start of the free-fall period, bubbles were injected from the bottom of the container, at the center, and appear to have had a significant residual velocity when released. Also, Thompson mentioned that the bubbles oscillated when introduced and that the oscillations decayed rapidly, leading to a spherical shape. Frequently, multiple bubbles were introduced by accident. Also, several experiments were performed under isothermal conditions. In these cases, the bubbles moved throughout the run but decelerated during the course of the run. Thompson et al. reported that the bubbles did not exhibit thermocapillary migration in water, but that such migration was observed in the other three fluids. Bubbles in ethylene glycol, up to a Reynolds number of 5.66 and Marangoni number as large as 713, were found to move at approximately the velocity given in Equation (4.2.28). Because this result is only expected to hold good in the linear limit as $Re \to 0$ and $Ma \to 0$, this is a puzzling observation. In the silicone oil and in ethanol, however, bubbles were observed to move at smaller velocities than those predicted by Equation (4.2.28). Given the conditions of Thompson's experiments, one must conclude that the nonlinearity of the temperature profile, coupled with the initial transients, must have had significant impact on the velocities of the bubbles.

Langbein and Heide (1984) performed experiments on sounding rockets using a binary liquid mixture of cyclohexane and methanol. The system exhibits a miscibility gap. The experiments showed drops forming when the liquid was cooled, and the drops moved in the direction of the temperature gradient. No quantitative comparison with theory was attempted.

Siekmann and coworkers carried out experiments on bubble motion in a temperature gradient aboard the D-1 mission of the space shuttle in November 1985. Results from these experiments have been reported in Nähle et al. (1987) and Szymczyk, Wozniak, and Siekmann (1987). A silicone oil of Prandtl number 687 was used as the continuous phase fluid. Air bubbles were introduced in one cell and water drops were injected in a second cell. The water drops were found not to move. A collection of bubbles was employed, and the temperature field was evolving during the experiment. From the measured temperatures at the hot and cold ends, using a transient solution of the conduction equation, the temperature profiles were estimated and used in interpreting the results. Data from six bubbles, with Ma values ranging from nearly 0 to 288, were presented in Szymczyk et al. and shown to be consistent with predictions from a numerical solution of the governing equations.

Aboard the same D-1 mission of the space shuttle, another experiment on thermocapillary migration was carried out by Neuhaus and Feuerbacher (1987). These authors used three different silicone oils as the continuous phase. According to the authors, silicone oil AK100 is a methylsiloxane, with 6% phenyl groups on the free valences of the siloxane chain, whereas in silicone oil AP100, the amount of phenyl groups is reported to be 28%. Silicone oil AS100 contains an intermediate fraction of phenyl groups. Air bubbles were introduced into each silicone oil under isothermal conditions, at which point heating of one side of the cell was initiated. Later, the heating was shut down, but the experiment was continued. The velocity of the bubbles appears to have been estimated using holograms taken at regular intervals. The authors mentioned that the velocities of air bubbles in the AK100 silicone oil were consistent with the predictions from Equation (4.2.28), whereas bubbles moved at velocities smaller than those predicted in AS100 and did not move at all in AP100. No information is available on the Reynolds or Marangoni numbers corresponding to these experiments. Gravitational rise experiments, conducted on the ground, yielded rise velocities in AK100 and AS100 silicone oils that were consistent with the predictions for a mobile interface given here in Section 3.1, whereas the results for AP100 fell on a curve corresponding to the motion of a rigid sphere. The authors concluded from these observations that one must account for dissipation processes at the interface through an interfacial dilational viscosity, but it is not obvious why this particular hypothesis is put forward. Interpretation of the observations of Neuhaus and Feuerbacher (1987) remains an open question.

Wozniak (1991) conducted experiments aboard sounding rockets on drops of paraffin oil moving in a solution of ethanol and water, the same pair on which he had conducted studies on the ground earlier. A period of approximately seven minutes of reduced gravity time was available. The drops ranged in radius from 0.7 to 2.4 mm, and a temperature gradient of 0.8 – 0.9 K/mm was employed. The maximum Reynolds and Marangoni numbers in the continuous phase were approximately 25 and 588, respectively. Predictions were made from a numerical solution of the governing equations. Wozniak found that the measured velocities ranged from approximately 3.6% to 31.7% of the predicted values, with the worst agreement being displayed for the smallest drop. He attributed the discrepancy to possible interactions with the boundaries and to the influence of surface active contaminants, even though every effort was made to work with pure liquids.

Braun et al. (1993) performed experiments aboard a sounding rocket that provided approximately 5.5 min. of experiment time in reduced gravity conditions, using a mixture of 2-butoxyethanol and water, which exhibits an inverted miscibility gap. Prior to launch, the sample was kept in a rectangular cell at a temperature 0.2 K below the phase separation temperature, so that it was a homogeneous single phase. Two seconds after launch, the sample was elevated to a temperature 0.45 K above the phase separation temperature. Drops rich in 2-butoxyethanol were nucleated, and these drops grew to a mean diameter of 11 μm. A temperature gradient of 0.37 K/mm was imposed at this time, and the authors indicated that a steady temperature distribution was reached in 20 s. The drops were found to migrate toward cooler fluid. This was consistent with expectations because σ_T is positive in this system. The Marangoni numbers ranged from 10^{-5} to 10^{-6}, and the Reynolds numbers must have been smaller because the Prandtl numbers are larger than unity for both liquids. Results on migration velocities were found to be consistent with predictions from Equation (4.2.28).

Recently, Treuner et al. (1996) reported results from experiments carried out in the drop tower in Bremen, Germany, which provides approximately 4.7 s. of low gravity conditions. An upward temperature gradient was established over a two-hour period in the liquid, and the first 0.5 second of the low gravity period was used to form bubbles of air of diameters from 0.5 to 2 mm. Experiments were performed in three liquids, n-octane, n-decane, and n-tetradecane. Treuner et al. provided several plots of data on the velocities of the bubbles, as well as an interferometry image revealing refractive index distributions around the moving bubbles. In addition, the authors presented a numerical solution of the governing equations, which is used for comparison with the data. According to the authors, the data were consistent with the theoretical predictions for quasi-steady migration to within approximately 20%. The authors also provided comments regarding the time dependence of the migration velocity. Generally, the observed velocities were smaller than the predicted quasi-steady velocities. Treuner et al. attributed this to the fact that the bubbles were accelerating to achieve the local quasi-steady velocity as they moved into warmer fluid, but never quite reached it. We make additional comments regarding the issue of transient migration later in the context of liquid drops.

We conclude this section with a discussion of our own results from reduced gravity experiments conducted aboard the NASA space shuttle *Columbia* in two series, the first in summer 1994 and the second in summer 1996. The experiments in 1994 were included in the International Microgravity Laboratory-2 (IML-2) mission, and those in 1996 were part of the Life and Microgravity Spacelab (LMS) mission. We have described the apparatus, procedure, and results in detail in Balasubramaniam et al. (1996) and Hadland et al. (1999). Only a summary is provided here.

Typically the gravitational acceleration levels in orbit were of the order of $10^{-6}g$, where g is the magnitude of the acceleration due to gravity on the surface of the Earth. The apparatus – known as the Bubble, Drop, Particle Unit (BDPU) – was designed and built under the auspices of the European Space Agency and made available for our use through a cooperative arrangement with NASA. The BDPU consisted of a facility that provided power, optical diagnostics and illumination, imaging facilities including a video camera and a motion picture camera, and other support services such as heating and cooling. Point Diffraction Interferometry was available in the IML-2 mission; in addition, shearing interferometry using a Wollaston prism was available in the LMS experiments. The interferometry images proved useful in inferring temperature distributions in gentle temperature gradients. Within the BDPU facility, different test cells could be inserted by the astronauts when needed. We used two identical test cells differing only in the fluid that was injected into a silicone oil contained in the cell. The cell was of rectangular cross-section, 60 mm long, and 45×45 mm square in cross-section. In the long dimension, the cell was bounded by aluminum flanges which were heated or cooled using Peltier elements; these end temperatures could be held constant to within $\pm0.1°C$. The other four walls were made of fused silica that was coated with a thin layer of indium tin oxide to minimize heat loss due to radiation while maintaining transparency to visible light. The continuous phase used in the IML-2 mission was a DC-200 silicone oil of nominal viscosity 50 centistokes, whereas that used in the LMS mission was a DC-200 silicone oil of nominal viscosity 10 centistokes. One test cell permitted the injection of air bubbles, whereas the second test cell was equipped to inject drops of Fluorinert FC-75. During injection or extraction of a bubble or drop, the test cell was connected to mechanical systems that ensured the compensation of the volume of the bubble or drop.

After the astronaut inserted the test cell within the BDPU facility and turned the power on, the experiment was controlled from the ground. First, a temperature difference was set up between the aluminum flanges, and a period of approximately 2 h was allowed to ensure that a steady temperature gradient had been established. The experiments were initiated at that point. In each case, a bubble or drop was injected at a location close to the cold wall and released by retracting the injector rapidly. The bubbles and drops ranged in radius from approximately 0.5 mm to 8 mm. Their traverses, captured on videotape, and on film in selected cases, were analyzed later on the ground to obtain information on their position and velocity, as well as size, as a function of time. When a bubble reached the hot wall, it was extracted through a tube of small diameter. After the elapse of an adequate amount of time for the disturbance to decay, the experiment was repeated with a different bubble. When a sufficient number of runs had been made, the temperature gradient was reset to a new value, and the experiment was repeated. It was found that the bubbles and drops traversed the cell rapidly enough for size change to be negligible. In evaluating the velocity of an object from the video or film, data on position versus time near the middle of the cell, away from the end walls, were used.

A brief digression is necessary to discuss whether the velocity of a bubble in these experiments is quasi-steady. When physical properties are constant, Oliver and DeWitt (1994) showed that for negligible inertia, steady state can be expected when a bubble has moved a distance approximately equal to 1 to 5 radii, at values of the Marangoni number up to 200. One can arrive at this result from the idea that at sufficiently large values of the Marangoni number, a thermal boundary layer is formed around the bubble. The time to achieve a steady temperature gradient field in this boundary layer, and therefore on the bubble surface, is of the order of the time taken by fluid to move from the vicinity of the forward stagnation point to the neighborhood of the rear stagnation point. The situation is more complicated than implied by this picture because the temperature field in the thermal boundary layer is influenced by the outer temperature field, to which the boundary layer solution must be matched. The time required for the gradient in this outer temperature field to reach a steady distribution increases with increasing Marangoni number. As noted in Section 4.13, this time to steady state asymptotically grows proportionally to log Ma. There are additional considerations in the case of a liquid drop. Even when convective transport is rapid within a drop, conduction across the streamlines determines the time to achieve a steady temperature gradient field. The thermocapillary migration problem in this unsteady situation is difficult to solve. However, an approximate answer can be obtained by considering an analogous problem in which a drop, at a given uniform temperature, is moving steadily through a fluid. The temperature at the surface of the drop is suddenly altered to a new value and held fixed at that value, and we need to estimate the time it takes for the entire drop to achieve this new temperature. A mass transfer analog of this problem was solved by Kronig and Brink (1950). They predicted that for 90% approach to steady state, the time is approximately $0.096 \frac{R^2}{\kappa'}$ where κ' is the thermal diffusivity of the drop phase. The above comments pertain to the case when the physical properties remain constant.

The variation of physical properties with temperature implies that true steady state cannot be reached in these experiments. Most of the physical properties (such as the density, heat capacity, thermal conductivity, and the rate of change of interfacial tension with temperature) are nearly constant within the cell in the region where the velocity measurements were made. The viscosities of the silicone oil and the Fluorinert FC-75 change substantially with temperature, however. This causes the objects to generally accelerate through the cell until they are close to the hot wall, at which point the wall retards their motion. Therefore, one can only expect quasi-steady behavior at best, after completion of the initial transient period. If the quasi-steady assumption truly holds, as a first approximation, the instantaneous velocity of a bubble or drop should be proportional to the inverse of the local viscosity. This implies that the product of the instantaneous velocity of a bubble or drop and the local viscosity should be constant over the path of that bubble or drop. The actual behavior is not this simple, because the Marangoni number increases during the traverse of any given object. Numerical solutions predict a decrease in this product with increasing Marangoni number for bubbles. In the case of drops, the prediction is for a decrease at low to moderate Marangoni numbers, followed by an increase at large values of this quantity. When we analyzed the data, we found that the product of the velocity and the local viscosity behaved in a complex way, sometimes at odds with the trend predicted by the numerical solution. This leads to the conclusion that the variation of viscosity with temperature produces basic unsteadiness in the velocity that cannot be captured by a quasi-steady picture.

Data from both the IML-2 and LMS experiments on isolated bubbles are displayed in Figure 4.16.3. The abscissa is the Marangoni number based on the continuous phase properties, and the ordinate is the ratio of the velocity of the bubble, designated V_∞, divided by the value it should have if convective transport effects are negligible, designated V_{YGB}. This latter result is given in Equation (4.2.28). Data from the LMS

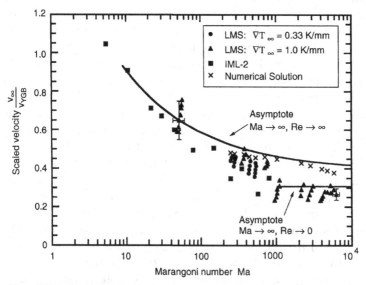

Figure 4.16.3 Scaled velocity of air bubbles migrating in two DC-200 silicone oils, measured in reduced gravity experiments conducted aboard the space shuttle; also included are predictions from a quasi-steady numerical solution from Ma (1998) and asymptotic results for $Re \to \infty$, $Ma \to \infty$, from Equation (4.13.46), and for $Re \to 0$, $Ma \to \infty$, from Equation (4.13.69). (Reproduced with permission from Hadland, P. H., Balasubramaniam, R., Wozniak, G., and Subramanian, R. S. Thermocapillary migration of bubbles and drops at moderate to large Marangoni number and moderate Reynolds number in reduced gravity. *Experiments in Fluids* 26, 240–248. Copyright 1998, Springer-Verlag GmbH & Co. KG.)

experiments are shown as the bubbles traverse a distance of approximately 10 mm in the middle of the cell, displaying the unsteady behavior of the velocity. Only a single velocity value was obtained for each bubble in the IML-2 experiments, and these are included in the figure. Also shown in the figure are quasi-steady predictions from a numerical solution of the governing equations from Ma (1998), and asymptotic results that have been given in Section 4.13 for large values of the Marangoni number. The Prandtl number of the silicone oil used in the IML-2 experiments varied from 371 to 567, and that of the silicone oil in the LMS experiments was in the range 59.4 to 92.9. The ratio of the thermal conductivity of air to that of the silicone oil, β, is approximately 0.21, and a similar ratio of viscosities, α, is much smaller, varying from approximately 0.003 to 0.004. In the asymptotic results shown, both of these parameters are assumed equal to zero. In the numerical solution, however, the appropriate values of α and β were used.

The data from the IML-2 and LMS experiments are consistent with each other where there is overlap, even though the Prandtl number of the continuous phase is significantly different. The reader will recall from Section 4.15 that the scaled migration velocity is only slightly sensitive to the Reynolds number, which explains this observation. The data are reasonably consistent with the numerical predictions, in spite of the transient nature of the migration process and the use of the quasi-steady assumption in obtaining the numerical solution. Also, for large values of the Marangoni number, the velocity is seen to approach the asymptotic prediction for $Ma \to \infty$ and $Re = 0$. It is remarkable that the asymptotic prediction for $Ma \to \infty$ and $Re \to \infty$ describes the behavior of the data well, even at moderate values of the Marangoni number. The reader will recall that in Section 4.13, we were able to obtain the scaled velocity to $o(Ma)$ in this case. Given the relatively small values of the Reynolds number in the experiments, if the asymptotic result for the scaled velocity could have been calculated to $o(Ma)$ in the Stokes flow case, one might conjecture that such a prediction would do at least as well as the result for $Re \to \infty$.

If true quasi-steady conditions prevailed during the migration, the cluster of data points for each bubble would follow the trend of the numerical solution, with the scaled velocity decreasing with increasing values of the Marangoni number. In fact, the trend is exactly opposite, with the scaled velocity increasing with increasing values of Ma for any given bubble, with the exception of the two largest bubbles in the large temperature gradient of 1 K/mm.

Now we turn our attention to experiments on Fluorinert drops. Figure 4.16.4 provides typical plots of the velocity of the drops against position in the cell. The figure illustrates two different behaviors. In the case of a small drop, if physical properties are constant, a quasi-steady condition can be reached early in the traverse. A relatively large drop is likely to be in transient motion throughout. From the figure, we see a knee in the plot for the larger drop. That is, the velocity first increases as the drop accelerates from rest, then reaches a plateau that persists for a short period, and then begins increasing once again. Several explanations are considered by Hadland et al. (1999), and the one most likely to be correct is the following. Within the thermal boundary layers, the temperature gradient achieves a pseudo-steady (as opposed to quasi-steady) distribution by the time the drop moves a distance of the order of its diameter. The average temperature of the interior of the drop is different from that of the continuous phase, however, and the temperature field is gradually changing as energy is conducted across streamlines within the drop. This process takes place over a time scale of approximately $0.096 \frac{R^2}{\kappa'}$ as mentioned earlier. So, the drop quickly achieves a pseudo-steady velocity, which explains the plateau. As the temperature field within the drop gradually

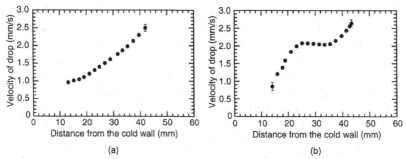

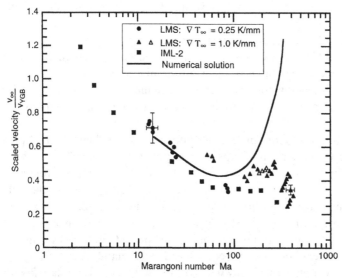

Figure 4.16.4 Typical plots of velocity against distance from the cold wall for a Fluorinert FC-75 drop migrating in 10 cs DC-200 silicone oil in reduced gravity (a) radius = 2.2 milli-meters, (b) radius = 5.35 millimeters; $\nabla T_\infty = 1$ K/mm in both cases. (Reproduced with permission from Hadland, P. H., Balasubramaniam, R., Wozniak, G., and Subramanian, R. S. Thermocapillary migration of bubbles and drops at moderate to large Marangoni number and moderate Reynolds number in reduced gravity. *Experiments in Fluids* **26**, 240–248. Copyright 1998, Springer-Verlag GmbH & Co. KG.)

undergoes adjustment, however, the distribution in the boundary layers also adjusts over that time scale, leading to the transient seen at a later time in the case of the larger drop. Of course, the continually decreasing viscosity along the path of the drop also must contribute to the increase in velocity seen after the plateau.

For relatively small drops, if the time spent during the traverse is significantly larger than $0.096 \frac{R^2}{\kappa'}$ at a location that is sufficiently away from the end walls, the motion may be expected to be quasi-steady. Data from such drops are displayed in Figure 4.16.5, along with data from IML-2 experiments. In the LMS experiments, data points for each

Figure 4.16.5 Scaled velocity of Fluorinert FC-75 drops migrating in two DC-200 silicone oils, measured in reduced gravity experiments conducted aboard the space shuttle. Also, predictions from a quasi-steady numerical solution from Ma (1998) are included for comparison. (Reproduced with permission from Hadland, P. H., Balasubramaniam, R., Wozniak, G., and Subramanian, R. S. Thermocapillary migration of bubbles and drops at moderate to large Marangoni number and moderate Reynolds number in reduced gravity. *Experiments in Fluids* **26**, 240–248. Copyright 1998, Springer-Verlag GmbH & Co. KG.)

drop cover a 10 mm traverse distance in the middle of the cell. The largest drop for which data are included in the figure has a radius of 2.85 mm. The Prandtl number in the continuous phase for the LMS data on drops ranges from 82.9 to 104, whereas the corresponding range in the IML-2 experiments is 327 to 556. Because the maximum Reynolds number for the LMS data displayed in Figure 4.16.5 is 4.67, one cannot expect the asymptotic prediction for large Re and large Ma from Section 4.14 to do well in this situation. Therefore, it is not included here. Such a prediction, which was smaller than the correct result by a factor of 4, was included in the corresponding figure in Hadland et al. (1999). Figure 4.16.5 includes predictions from a numerical solution by Ma (1998), who assumed quasi-steady conditions to prevail. We can infer from this figure that the LMS data on Fluorinert drops are consistent with data from IML-2. The behavior at Marangoni numbers up to about 90 appears to be in agreement with predictions from the numerical solution. Note how the scaled velocity of a given drop during its traverse decreases with increasing values of Ma for the four drops that fall in this range. Beyond this value of the Marangoni number, however, the numerical solution shows a sharp increase of the scaled velocity with further increase in the Marangoni number. The data for drops with values of Ma above approximately 100 show an opposite trend to those for smaller drops. In the case of each drop, the scaled velocity increases with increasing Marangoni number along the path of that drop, which is similar to the trend observed in the case of bubbles. Although superficially the increase is encouraging because it is the same trend as that displayed by the numerical solution, the data fall further away from the numerical solution as the value of the Marangoni number is increased. We suspect the reason for the increasing discrepancy at the larger values of Ma is the increasing relative importance of transient processes. Hadland et al. also provided transient migration data for drops larger than those included in Figure 4.16.5.

As mentioned earlier, shearing interferometry with a Wollaston prism was used in our reduced gravity experiments to provide some information on the variation of the refractive index and, therefore, the temperature in the cell. Fluorinert drops dissolve slightly in silicone oil. Because the refractive index depends on both temperature and composition, the interferometry information from the experiments on Fluorinert drops cannot be directly translated into temperature profiles. Even in the case of air bubbles, when the temperature variation was too large, the technique failed to provide useful data. In a gentle gradient of 0.33 K/mm, it was possible to infer the temperature distribution around the moving bubble assuming it to be axially symmetric. Figure 4.16.6 shows isotherms in a typical case, which are qualitatively consistent with those predicted from a numerical solution. The radius of the bubble is 4.1 mm, and its average velocity is approximately 2.7 mm/s. Note the crowding of the isotherms ahead of the bubble and the appearance of a long thermal wake behind the bubble.

A few words are appropriate regarding shape deformation. No deformation of shape has been reported by any of the authors in the experiments described earlier. In our IML-2 and LMS experiments, within the resolution limits of the data, there was no measurable deviation from a spherical shape in the case of even the largest Fluorinert drops. A slight trend toward deformation in the LMS experiments was noted by Hadland et al., however, in the case of air bubbles of radius ≥ 6 mm. These bubbles showed a gradual decrease in size along the direction of the temperature gradient while, at the same time, the size in the transverse direction increased; thus, the bubbles appeared to become oblate spheroids as they migrated.

We have discussed the present state of the available experimental data on the thermocapillary migration of isolated drops and bubbles in this section. It is evident upon

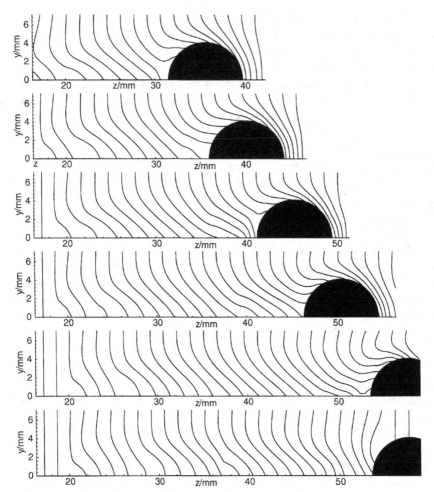

Figure 4.16.6 Isotherms from reduced gravity experiments conducted aboard the space shuttle on an air bubble of radius 4.1 mm migrating in a temperature gradient of 0.33 K/mm maintained in a 10 cs DC-200 silicone oil; distances in millimeters are from an arbitrary origin, and the applied temperature gradient points in the z-direction. The panels represent increasing values of time from the top to the bottom. In the penultimate panel, the bubble has just reached the hot wall. The isotherms are plotted at $0.5°C$ intervals, and in each panel, the first isotherm corresponds to $30°C$. (Reproduced with permission from Wozniak, G., Balasubramaniam, R., Hadland, P. H., and Subramanian, R. S. Temperature fields in a liquid due to the thermocapillary motion of bubbles and drops. *Proceedings of the International Conference on Microgravity Fluid Physics and Heat Transfer* eds. V. Dhir, J. Straub, and Y. Fujita (pp. 160–166), September 1999, Hawaii; Begell House, Inc.)

examination of these observations that the predictions of the linear theory have been substantiated by experiments. Also, for moderate values of the Marangoni number and small values of the Reynolds number, the data are consistent with theoretical predictions that assume quasi-steady migration, at least in the trend displayed. For gas bubbles, the data at relatively large values of the Marangoni number continue to exhibit agreement with such predictions. This is not the case for liquid drops. We suspect this is a consequence of the transient behavior of the motion, due to the large time scale required for thermal transients within the drop to decay. Also, the relatively complex behavior of the transient data on the scaled velocities of both bubbles and drops can be attributed to the influence of the variation of viscosity with temperature along the path. Predicting the unsteady motion of bubbles and drops, while accounting for the variation of physical properties with temperature, is an important problem that has not yet been solved. We postpone discussion of interactions with boundaries and other objects in some of these experimental studies to Chapter Seven.

4.17 Flow Structures

4.17.1 Introduction

Thermocapillary migration is unique not only in the fact that the entire driving force resides on the interface, but also in the types of flow fields induced in the surrounding fluid due to the motion of the drop or bubble. When a drop moves, other relatively small objects in the vicinity (such as drops, bubbles, or particles) are affected by the motion of the continuous phase, which is induced by the moving drop. Whether they approach the drop or recede from it will depend on the nature of this flow, as well as on other forces acting on these objects. In the context of interactions of a moving drop with neighboring boundaries, it is useful to be aware of the nature of the flow field generated by the motion of an isolated drop because the rate of decay of the velocity field with distance from the drop will determine the extent of the interaction with the boundary. In illustrating sample flow fields, we confine our attention to axisymmetric flows, which can be completely visualized by means of streamlines drawn in a meridian plane which contains the axis of symmetry.

Because it is the flow in the continuous phase that possesses interesting structure, it will prove sufficient in most cases to consider the motion of a gas bubble with α and β set equal to zero. When these ratios are nonzero, the flow in the interior of the drop is interesting in one case for which we provide a sample drawing. The flow structures discussed here have been taken principally from Merritt, Morton, and Subramanian (1993), with some earlier material from Shankar and Subramanian (1988) and Subramanian (1992). We begin with a consideration of flow structures encountered in purely thermocapillary migration. Then, we discuss the richer collection of flow structures that result when a bubble or drop moves under the combined influence of gravity and a vertical temperature gradient. Inertial effects are not considered, but we comment on the role of convective transport of energy in influencing the structure.

4.17.2 Motion Driven by Thermocapillarity

Consider the motion of an isolated gas bubble subject to all the assumptions used in Section 4.2. Figure 3.1.1 shows a sketch of the system. The z-direction points downward so that $\mathbf{g} = g\mathbf{k}$ and $\nabla T_\infty = |\nabla T_\infty|\mathbf{k}$. It is useful to display the structure of the basic flow driven by thermocapillarity and contrast it with that when motion occurs due to buoyancy. Because we shall consider motion under the combined influence of gravity and thermocapillarity later in this section, the scaled streamfunction in that problem in a frame of reference attached to the laboratory is given below.

$$\psi(r, s) = \frac{1}{2}\left[Gr - \frac{1}{2r} \right](1 - s^2).$$

(4.17.1)

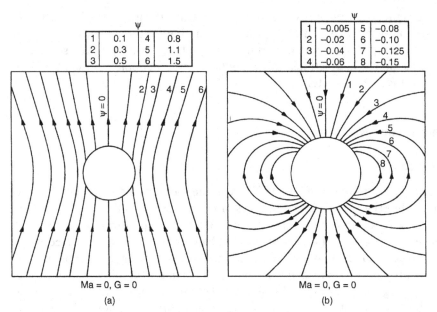

1	0.1	4	0.8
2	0.3	5	1.1
3	0.5	6	1.5

1	−0.005	5	−0.08
2	−0.02	6	−0.10
3	−0.04	7	−0.125
4	−0.06	8	−0.15

Ma = 0, G = 0   Ma = 0, G = 0

(a)   (b)

Figure 4.17.1 Streamlines in the laboratory reference frame: (a) gravity driven motion, (b) thermocapillary migration (Courtesy of Academic Press).

The scaled bubble velocity can be written as

$$\mathbf{v}_\infty = \left(\frac{1}{2} - G\right)\mathbf{k}. \tag{4.17.2}$$

Here, the quantity G is defined in Section 4.2.3, r is the scaled radial coordinate, and $s = \cos\theta$, where θ is the polar angle measured from the direction of the applied temperature gradient. In defining scaled variables, we use the same reference quantities that were employed in Section 4.2.

In the example chosen, the bubble moves upward in the negative z-direction because of buoyancy and downward because of thermocapillarity. The results for both the streamfunction and the velocity of the bubble are obtained by superposing results from the two separate problems because they are linear. The first term in the square bracket in Equation (4.17.1) for the streamfunction stands for the contribution from gravity and the second term to that from thermocapillarity. The flow caused by buoyant motion is a Stokeslet, and the flow caused by thermocapillary migration is equivalent to one caused by a combination of a source and sink of fluid called a *dipole*. As noted earlier, the velocity field for purely thermocapillary motion also is a potential flow. A Stokeslet arises when there is a nonzero hydrodynamic force exerted by the object on the fluid. Because the hydrodynamic force in purely thermocapillary migration is zero, the Stokeslet is absent in that solution. The velocity field due to a Stokeslet is proportional to $\frac{1}{r}$, whereas that from a potential dipole is proportional to $\frac{1}{r^3}$. The reader might recall from Section 3.1 that the flow that arises from the body force driven motion of a drop contains both a Stokeslet and a potential dipole contribution in the laboratory reference frame. For a gas bubble, however, the coefficient multiplying the dipole term is zero.

The flow in the continuous phase due to thermocapillary migration is contrasted with that caused by buoyant rise of the bubble in Figure 4.17.1. All the streamline drawings in this section are in a meridian plane, which contains the axis of symmetry. For clarity, streamlines are not shown at equal values of increment in the streamfunction, but rather at selected values of ψ which are reported in the figures. In calculating the values of ψ for gravity driven motion, we have used a different reference velocity from that given in Equation (4.1.1). This reference velocity is

$$v_1 = \frac{R^2 \Delta \rho g}{3\mu}. \tag{4.17.3}$$

The contrast between the two drawings is indeed striking. The fore and aft regions are qualitatively similar. In both cases, fluid moves into the region being vacated by the bubble, and the bubble pushes fluid away as it moves ahead. The motion in the equatorial region is quite different, however. For motion caused by the temperature gradient, the thermocapillary stress on the bubble surface drives this motion, which is aligned against the applied temperature gradient, and therefore, against the direction of bubble motion. In the case of body force driven motion, the fluid moves basically in the direction of bubble motion. For a liquid drop, wherein the exterior flow in body force driven motion is a superposition of a Stokeslet and a potential dipole, the qualitative features remain the same. The Stokeslet dominates the motion in the far field. The reader might wonder about the appearance of streamlines drawn for purely thermocapillary migration and purely gravity driven motion in a reference frame attached to the moving bubble. We already have displayed such streamlines for the gravity driven case in Figure 3.1.2, and the streamlines for thermocapillary migration are qualitatively similar in appearance. The differences become noticeable only when we examine the motion as seen by an observer in the laboratory.

Now, we consider thermocapillary migration when the Reynolds number still is negligibly small so that we can assume Stokes flow, but the Marangoni number Ma is not negligible. Streamlines drawn in a reference frame traveling with the bubble are similar to those for the case $Ma = 0$ and can be qualitatively sketched without solving the problem in detail. Streamlines in a laboratory reference frame displayed in Figure 4.17.2 for the case $Ma = 5$ reveal an unusual flow pattern, however. Note the occurrence of a reverse flow wake region, separated from the main flow around the bubble by a dividing streamline (separatrix) on which $\psi = 0$ and the associated saddle

	ψ		
1	−0.005	6	−0.08
2	−0.01	7	−0.10
3	−0.02	8	0.001
4	−0.04	9	0.0025
5	−0.06	10	0.004

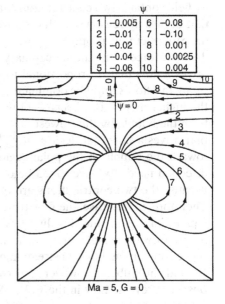

Figure 4.17.2 Streamlines in the laboratory reference frame when convective transport of energy is included (Courtesy of Academic Press).

Ma = 5, G = 0

point. This feature is common to numerous problems in thermocapillary migration, and it will be seen that it arises as a consequence of breaking the symmetry present in the case displayed in Figure 4.17.1(b). From the values of the streamfunction in the table accompanying the figure, it can be seen that the strength of the flow in the vicinity of the separatrix is comparable on both sides of it. It has been shown by Shankar and Subramanian (1988) that the saddle point moves closer to the bubble as the Marangoni number is increased. Balasubramaniam and Subramanian (1996) have provided a drawing that shows the asymptotic position of this point in the limit $Ma \to \infty$. The magnitude of the velocity in the reverse flow region will naturally depend on position. Generally, the velocity in this region is substantially smaller in magnitude than that in the vicinity of the bubble. The velocities in the vicinity of the separatrix would increase as the saddle point moves closer to the bubble. One should not expect significant changes beyond $Ma \approx 200$, however, because the saddle point is almost at its asymptotic position by the time the Marangoni number reaches that value.

We now provide an explanation for the occurrence of the reverse flow region, which is ubiquitous in thermocapillary migration. The driving force for the motion is the gradient of the interfacial tension, which is proportional to the gradient of the temperature field on the bubble surface. In general, the temperature field on the surface can be written as a weighted sum of Legendre Polynomials $P_n(s)$. Because Stokes's equation is linear, the actual motion can be considered a superposition of motion driven by each pure Legendre mode of the surface temperature. This can be seen from Equations (4.5.2) through (4.5.9), where each term in the infinite series arises from a scaled tangential stress discontinuity, which is the surface gradient of a single Legendre mode of the scaled surface temperature. In the pure conduction case, for an isolated bubble or drop, the surface temperature is seen from the solution given in Equations (4.2.16) and (4.2.17) to be proportional to the $P_1(s)$ mode. This drives the flow displayed in Figure 4.17.1(b). Introduction of convective energy transport excites higher Legendre modes, and the $P_n(s)$-mode for $n \geq 2$ drives a flow that is n-cellular. The flows induced by the Legendre modes of the surface temperature corresponding to $n = 2, 3$, and 4 are illustrated in Figure 4.17.3. The scaled surface temperature distribution driving the flow is $P_n(s)$.

In the general case for a surface temperature that is proportional to $P_n(s)$, the velocity field contains two contributions that decay as $\frac{1}{r^n}$ and $\frac{1}{r^{n+2}}$. Of these, the contribution that decays more slowly with distance is that proportional to $\frac{1}{r^n}$. For $n = 1$, however, the Stokeslet flow proportional to $\frac{1}{r}$ is absent because of the condition of zero hydrodynamic force in the thermocapillary migration problem. Therefore, the P_1-mode only drives a flow that is proportional to $\frac{1}{r^3}$. Because the Marangoni number is relatively small, this flow should be dominant everywhere, corrected by small contributions from the higher modes. But because of its rapid decay, it is overwhelmed at some distance from the bubble by the slowest decaying contribution, namely that from the P_2-mode that drives a flow that decays as $\frac{1}{r^2}$. Figure 4.17.3(a) displays this flow, which occurs toward the bubble in the polar regions and away from it at the equator. The actual direction needs to be reversed before superposition with the flow from the P_1-mode because the two contributions are of opposite sign for the situation in Figure 4.17.2. Therefore, in that case, the flow from the P_2-mode brings fluid in from far away in the equatorial region, and this fluid flows away from the bubble along the poles. Hence, in the forward region, the two flows reinforce each other, and there is no qualitative change in the far-field streamline structure from that in the pure conduction case. Behind the bubble, however, the two flows oppose each other, which leads to the observed reverse flow in the wake. Although the example shown here illustrates one

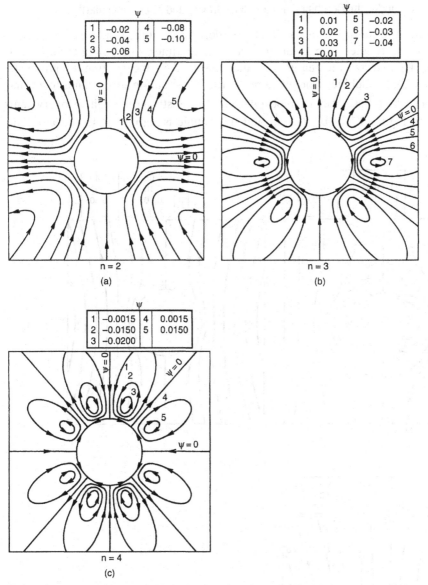

	ψ		
1	−0.02	4	−0.08
2	−0.04	5	−0.10
3	−0.06		

	ψ		
1	0.01	5	−0.02
2	0.02	6	−0.03
3	0.03	7	−0.04
4	−0.01		

n = 2
(a)

n = 3
(b)

	ψ		
1	−0.0015	4	0.0015
2	−0.0150	5	0.0150
3	−0.0200		

n = 4
(c)

Figure 4.17.3 Streamlines in the laboratory reference frame for flow driven by the scaled temperature field $P_n(\cos\theta)$ on the surface of the bubble: (a) $n = 2$, (b) $n = 3$, (c) $n = 4$. (Reproduced with permission from Subramanian, R. S. "The motion of bubbles and drops in reduced gravity," in *Transport Processes in Bubbles, Drops, and Particles*, edited by R. P. Chhabra and D. De Kee (pp. 1–42). New York: Hemisphere. Copyright 1992, Taylor and Francis, Inc.)

way in which higher order Legendre modes of the surface temperature field come into play, interaction with boundaries or other objects in the vicinity is sufficient to excite them even when convective transport effects are entirely negligible. Examples of this will be seen in Chapter Seven. Also, when a surfactant is adsorbed onto the interface, higher modes of the interfacial tension distribution can be excited, leading to the same flow structure, as illustrated by Chen and Stebe (1997). Similar flow structures also can be observed in the case of an isolated bubble for nonzero Reynolds number, as noted by Balasubramaniam and Lavery (1989), even though superposition of flows driven by pure modes cannot be used as an explanation because of the nonlinear nature of the problem.

4.17.3 Motion under the Combined Influence of Gravity and Thermocapillarity

Now, consider the motion of a gas bubble under the combined influence of gravity and a downward temperature gradient as in the experiments of Young et al. (1959). Figure 4.17.4 illustrates the features of the flow field in the continuous phase. In Figure 4.17.4(a), the bubble is moving downward. This situation corresponds to a value of $G < \frac{1}{2}$, in which case the thermocapillary effect is strong enough to overwhelm buoyancy. In Figure 4.17.4(b), $G = \frac{1}{2}$, and the bubble is stationary. In this case, the

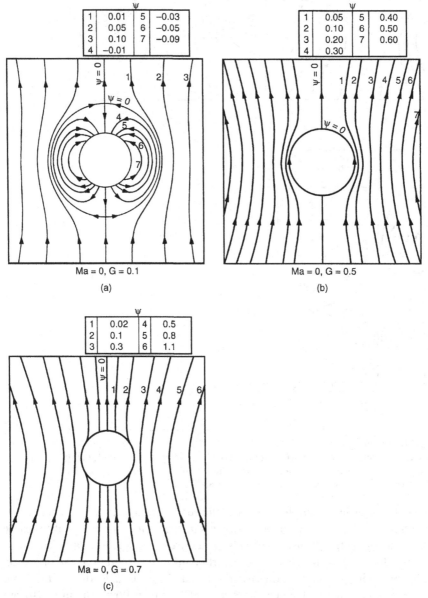

Figure 4.17.4 Streamlines in the laboratory reference frame when a bubble moves under the combined action of gravity and thermocapillarity: (a) bubble moving downward, (b) stationary bubble, (c) bubble moving upward (Courtesy of Academic Press).

thermocapillary effect precisely balances the gravitational effect. As pointed out in Sections 4.2 and 4.16, however, it is not possible to physically realize this situation for a gas bubble. We have included the drawing here for two reasons. First, it is possible to hold liquid drops stationary using such a balance, and the exterior flow is of the same form as that shown here. Second, it represents a logical intermediate situation. Figure 4.17.4(c) displays streamlines when $G > \frac{1}{2}$ in which case the bubble moves upward.

Note the appearance of a dividing streamline separating regions of flow exhibiting different features in the case when the bubble moves downward. This is a spherical dividing streamsurface in three dimensions, shown as a circle in meridian section. In the vicinity of the bubble, the flow structure is similar to that in Figure 4.17.1(b), but the thermocapillary contribution to the flow field in the surrounding fluid decays rapidly. The far field flow is dominated by the Stokeslet from the gravitational contribution because the velocity field from the Stokeslet decays more slowly. This flow is directed upward everywhere as illustrated in Figure 4.17.1(a) because it is driven by the hydrodynamic force exerted by the bubble on the fluid. Along the forward and rear stagnation streamlines, however, the flow in the vicinity of the bubble is directed downward. The result of this opposition is the appearance of two saddle points in the continuous phase, along with a dividing streamline, on which $\psi = 0$, which separates the two regions of flow. If the value of G is increased, the dividing streamline will move closer to the bubble surface. When the bubble is stationary, as shown in Figure 4.17.4(b), the dividing streamline collapses onto the bubble surface. In Figure 4.17.4(c), where buoyancy dominates over the thermocapillary effect, the bubble moves upward. Hence, the flow along the stagnation streamlines is in the same direction as that from the Stokeslet. The flow field is simple in this case.

From Equation (4.17.1), it can be seen that $\psi = 0$ when $\theta = 0$ or π. Also, $\psi = 0$ for two values of r given by

$$r = \pm \frac{1}{\sqrt{2G}}. \tag{4.17.4}$$

Both are real roots, but the negative root is physically meaningless. When $G < \frac{1}{2}$, the positive root lies outside the bubble, and hence a dividing streamline, $\psi = 0$, occurs in the form of a circle at the location $r_D = \frac{1}{\sqrt{2G}}$. When $G = \frac{1}{2}$, this dividing streamline will coincide with the surface of the bubble. When $G > \frac{1}{2}$, the bubble moves upward. The dividing streamline disappears because its predicted location is within the bubble, and the streamfunction in Equation (4.17.1) is only valid in the continuous phase.

Streamlines in a reference frame attached to the center of mass of the moving bubble are illustrated in Figure 4.17.5. In this reference frame, the bubble is stationary. When thermocapillarity dominates, which is the case for $G < \frac{1}{2}$, the fluid streams upward past the bubble, as shown in Figure 4.17.5(a). A dividing streamline appears when gravity dominates, as in Figure 4.17.5(b), which is drawn for the case $G = 0.7$. Recirculation driven by the thermocapillary stress is seen within the region bounded by the dividing streamline. This dividing streamline approaches the bubble surface as G increases, as displayed in Figure 4.17.5(c). The intermediate case $G = \frac{1}{2}$ already has been illustrated in Figure 4.17.4(b). Because the velocity of the bubble is zero, that drawing is applicable in both reference frames.

The location of the dividing streamline can be predicted in a straightforward manner by considering Equation (4.17.1) after modification to include a uniform stream to

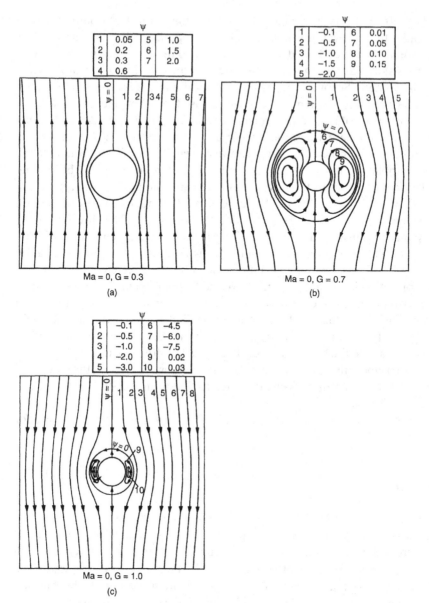

Figure 4.17.5 Streamlines in a reference frame attached to a bubble: (a) moving downward;
(b) moving upward, $G = 0.7$; (c) moving upward, $G = 1.0$ (Courtesy of Academic Press).

accommodate the change in reference frame. Again, the streamfunction is zero along
$\theta = 0$ and π. Also, it is zero when

$$G(r - r^2) - \frac{1}{2}\left(\frac{1}{r} - r^2\right) = 0. \tag{4.17.5}$$

This cubic equation has three roots:

$$r_1 = 1, \tag{4.17.6}$$

$$r_2 = \frac{1 + (8G - 3)^{\frac{1}{2}}}{2(2G - 1)}, \tag{4.17.7}$$

and

$$r_3 = \frac{1 - (8G - 3)^{\frac{1}{2}}}{2(2G - 1)}. \tag{4.17.8}$$

The first root, $r_1 = 1$, is expected. The third root r_3 is complex when $G < \frac{3}{8}$, and equal to -2 at $G = \frac{3}{8}$. For $G > \frac{3}{8}$, this root is always negative and has no physical significance. The second root r_2 is the one of interest. This root also is complex for $G < \frac{3}{8}$. For $\frac{3}{8} \leq G < \frac{1}{2}, r_2$ is negative. At $G = \frac{1}{2}$, this root approaches $\mp \infty$ depending on the direction of approach. For the range $\frac{1}{2} < G < \frac{3}{2}$, it remains positive and larger than unity. It is in this region that the dividing streamline shown in Figures 4.17.5(b) and 4.17.5(c), with a radius $r_D = r_2$, is observed. When $G = \frac{3}{2}, r_2 = 1$, leading to the collapse of the dividing streamline onto the bubble surface. For $G > \frac{3}{2}, r_2 < 1$ and therefore is of no physical significance. In this situation, the flow structure becomes simple.

The streamfunction given in Equation (4.17.1) is a linear combination of a Stokeslet and a potential dipole. Because this is the right form for the Stokes motion of a rigid sphere, it is logical to ask whether the streamfunction will mimic that result in a special case. This happens when $G = \frac{3}{2}$. In this situation, the bubble moves upward at a velocity that is twice that it would have if it were to move downward only under the action of the temperature gradient. The surface tension gradient along the surface is exactly of the form to make the surface appear rigid, and the situation is equivalent to that involving a stagnant surfactant cap that occupies the entire bubble surface.

One might wonder if new qualitative features are observed in the flow patterns within the bubble. In the general case of a drop, the streamfunction within the drop can be constructed from a superposition of results given in Sections 3.1 and 4.2. Results in the laboratory reference frame for the scaled streamfunctions in the continuous phase and within the drop, along with an expression for the scaled velocity of the drop, are given below.

$$\psi(r, s) = \left[\frac{v_\infty}{4} \left(\frac{\alpha}{1+\alpha} \frac{1}{r} - \frac{2+3\alpha}{1+\alpha} r \right) + \frac{I_2}{4(1+\alpha)} \left(\frac{1}{r} - r \right) \right] (1 - s^2), \quad (4.17.9)$$

$$\psi'(r, s) = \left[\frac{1}{4(1+\alpha)} (I_2 - v_\infty)(r^2 - r^4) - \frac{v_\infty}{2} r^2 \right] (1 - s^2), \tag{4.17.10}$$

$$\mathbf{v}_\infty = -\frac{1}{2 + 3\alpha} [I_2 + 2(1 + \alpha)G] \mathbf{k}. \tag{4.17.11}$$

Here,

$$I_2 = -\frac{2}{2 + \beta}. \tag{4.17.12}$$

In the discussion below, we assume that $G > 0$ for definiteness. Analogous inferences can be made from symmetry considerations when $G < 0$. The case of a positive G corresponds to a drop that is less dense than the continuous phase, and we continue to assume the temperature gradient to point downward. When the above equations are used to generate streamline drawings for the continuous phase, the results are qualitatively no different from those displayed for the continuous phase in the bubble limit. The main difference is that transition from downward to upward motion does not occur at $G = \frac{1}{2}$, but rather at $G = \frac{1}{(1+\alpha)(2+\beta)}$. Results analogous to those given earlier for the location of the dividing streamlines that occur in the continuous phase may be obtained in a straightforward way. We note a new feature in the flow patterns within

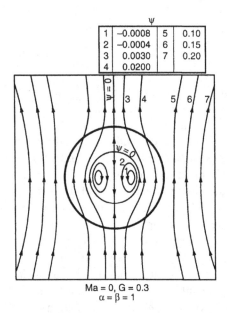

1	−0.0008	5	0.10
2	−0.0004	6	0.15
3	0.0030	7	0.20
4	0.0200		

$Ma = 0, G = 0.3$
$\alpha = \beta = 1$

Figure 4.17.6 Streamlines in the laboratory reference frame for a liquid drop moving upward; the outline of the drop is shown by the thick circle (Courtesy of Academic Press).

the drop phase, however. In a reference frame riding with the drop, one observes recirculation within the drop. We find this to be the case whether the drop moves downward or upward. In the laboratory reference frame, however, when gravity dominates so that the drop moves upward, a circular dividing streamline appears, on which $\psi' = 0$ when $\frac{1}{(1+\alpha)(2+\beta)} < G < \frac{5}{(3+2\alpha)(2+\beta)}$. It separates two distinct regions of flow within the drop phase. An example is displayed in Figure 4.17.6 in which this dividing streamline may be observed at $r = 0.488$, along with the expected saddle points. Its location can be established as

$$r_D = \sqrt{\frac{2G + 5v_\infty}{2G + 3v_\infty}}. \tag{4.17.13}$$

When the drop moves downward, v_∞ is positive and the location of the dividing streamline falls outside the drop where it has no physical significance because the result in Equation (4.17.13) is based on setting the streamfunction ψ', valid within the drop phase, to a value of zero. The flow structure shown in Figure 4.17.6 may be contrasted with that in Figure 4.17.7, which shows a typical set of streamlines in the laboratory reference frame, when the drop moves downward in the direction of the temperature gradient. In both drawings, the drop outline has been thickened for clarity.

In Figures 4.17.6 and 4.17.7, the general details of the flow in the continuous phase are qualitatively the same as those for the gas bubble; the only new feature when contrasted with the earlier drawings is the detail regarding the flows within the drop. The continuity of the velocity vector across the interface, and hence that of $\nabla\psi$, leads to the apparent smooth transition of the streamlines from one phase to the other. Of course, the tangential stress balance at the interface leads to a discontinuity in the second derivatives of the streamfunction, but this cannot be easily discerned from such drawings.

Finally, in Figure 4.17.8, we display a typical set of streamlines for motion under the combined influence of gravity and thermocapillarity when convective energy transport effects are accommodated. The results used to make this drawing were obtained by solving the energy equation using the method of finite differences, while employing an

ψ			
1	−0.055	5	0.005
2	−0.040	6	0.015
3	−0.020	7	0.025
4	−0.005		

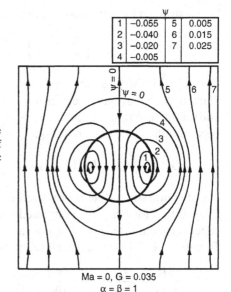

Figure 4.17.7 Streamlines in the laboratory reference frame for a liquid drop moving downward; the outline of the drop is shown by the thick circle (Courtesy of Academic Press).

Ma = 0, G = 0.035
$\alpha = \beta = 1$

analytical solution of the Stokes flow equations for the velocity field. Some details are given in Merritt and Subramanian (1992), and additional information can be found in Merritt (1988).

The drawing in Figure 4.17.8 should be contrasted with the corresponding one for purely thermocapillary migration given in Figure 4.17.2. Here, the velocity field is composed of a Stokeslet in addition, and this is the slowest decaying component. The Stokeslet thus dominates the far-field flow, producing a streamline structure that resembles Figure 4.17.4(a) rather than 4.17.2. Interestingly, the dividing streamline appears to be almost circular even though it no longer shares the same center as the bubble. The approximate radius of the dividing streamline can be predicted, at least for small values of the Marangoni number, by considering the result for the streamfunction, which can

ψ			
1	0.01	5	−0.01
2	0.05	6	−0.03
3	0.10	7	−0.05
4	0.15	8	−0.09

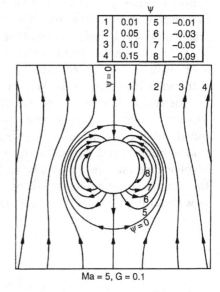

Figure 4.17.8 Streamlines in the laboratory reference frame for a bubble moving downward under the combined action of gravity and thermocapillarity showing the effect of convective transport of energy (Courtesy of Academic Press).

Ma = 5, G = 0.1

be constructed by superposition of results from Sections 3.1 and 4.5.

$$\psi(r,s) = \frac{1-s^2}{2}\left(\frac{I_2}{2r}+Gr\right)+\sum_{n=3}^{\infty}\frac{n(n-1)}{4}I_n\left[\frac{1}{r^{n-1}}-\frac{1}{r^{n-3}}\right]C_n(s). \qquad (4.17.14)$$

Here $C_n(s)$ is the Gegenbauer Polynomial encountered in Section 3.1. For small Ma, as a first approximation, we can truncate the infinite series in Equation (4.17.14) at $n=3$ and set the resulting expression for ψ equal to 0. The radial locations at which this would occur are the roots of a cubic equation:

$$Gr^3 + \frac{I_2}{2}r + \frac{3}{2}I_3 s(1-r^2) = 0. \qquad (4.17.15)$$

The dividing streamline is predicted to be a circle only if the term $\frac{3}{2}I_3 s$ is neglected. If the solution for r is large compared with unity, neglecting this term and retaining the other term multiplied by I_3 will be a good approximation. The approximate version of Equation (4.17.15) then can be rearranged in the following form:

$$\frac{9I_3^2}{16G^2} - \frac{I_2}{2G} = r^2 + \frac{9I_3^2}{16G^2} - 2rs\left(\frac{3I_3}{4G}\right). \qquad (4.17.16)$$

This is the equation of a circle whose center is shifted downward from the center of the bubble by a scaled distance of $\frac{3I_3}{4G}$. The radius of this circle is

$$r_D = \sqrt{\frac{9I_3^2}{16G^2} - \frac{I_2}{2G}}. \qquad (4.17.17)$$

For the values of the parameters used in Figure 4.17.8, $I_2 = -0.81$, and $I_3 = 0.073$. Using these values, Equation (4.17.17) predicts a scaled radius of 2.09, and the actual value is 1.96. Another approximation to the radius of the circular dividing streamline can be written by completely neglecting the contribution from the term multiplied by I_3, because I_3 is small. This yields a scaled radius of 2.01, which is closer to the actual value, but this approximation predicts no shift in the center of the circle. In the pure conduction case, the scaled radius of the dividing streamline is 2.24. Thus, the inclusion of convective energy transport in the model leads to a reduction in the extent of the region that displays the flow structure characteristic of thermocapillarity. The prediction for the scaled downward displacement of the center of the dividing streamline from the center of the bubble is not as good. It yields a value of 0.55 for this displacement, whereas the actual value is 0.39. Merritt et al. (1993) found the value of this displacement to increase with increasing values of the Marangoni number.

The results presented here should serve to convince the reader that common intuition, based on problems involving body force driven motion, is inadequate in making qualitative predictions regarding the flow patterns caused by the thermocapillary motion of bubbles and drops. The reverse flow region noted here is found in problems involving interactions that will be illustrated in Chapter Seven. Also, when gravity and thermocapillarity simultaneously influence the motion of a drop, we have seen that the flow patterns can exhibit rich structure.

INTERACTIONS OF BUBBLES AND DROPS

In Part Three, we consider situations in which a bubble or drop is affected by a neighboring boundary or by other bubbles or drops in the vicinity. In designing experiments to test theoretical predictions, it is normally possible to minimize such complicating interactions. To avoid boundary interactions, however, one must have some estimate of the influence of these boundaries on drop motion, so that the experimental cell can be chosen to be sufficiently large. Also, interactions are common in practical systems that rarely involve a single drop far removed from any boundaries or other drops. Finally, the approach of two drops is a necessary prerequisite to their ultimate coalescence. Therefore, one is often interested in predicting whether two drops will approach each other, or recede from each other, under a given set of conditions.

There are two principal complications in modeling an interaction problem, whether it involves drops and neighboring surfaces or a collection of drops. First, consider the motion of a drop in the vicinity of a rigid plane surface caused by the action of gravity. It is necessary to satisfy the boundary conditions, both on the surface of the moving drop and on that of the fixed plane. A steady problem cannot be posed in this situation, except when the drop settles near a vertical surface in the absence of inertia, by using either the laboratory reference frame or a reference frame moving with the drop. This inherent time dependence, which arises out of the unsteadiness of the configuration of the system, is not present when the same drop moves in the absence of external boundaries. The second difficulty is that of finding a suitable coordinate system in which the problem can be solved. It is not possible to describe the surface of the plane conveniently in a spherical polar coordinate system with its origin at the center of the drop, nor is it convenient to use rectangular cartesian coordinates to describe the surface of the drop. Fortunately, as will be seen in Section 5.1, there is a conformal transformation, which maps the region between the drop surface and the plane into a rectangular domain, in which the boundary value problem can be solved conveniently. We describe below how the problem of unsteadiness is circumvented.

In Section 4.1, the governing conservation equations were nondimensionalized in the thermocapillary migration problem by using typical length, velocity, and time scales. This process leads to Equations (4.1.2) through (4.1.7), which are applicable in a reference frame attached to the moving drop. When one reverts to a laboratory reference frame, the same equations are obtained, with the difference that the terms involving the acceleration of the drop are absent. These equations apply regardless of the nature of the boundaries of the system and can be used as a starting point for our discussion. The time scale in these equations was chosen as the viscous relaxation time in the continuous phase over the length scale of the drop. Now, imagine that we wish to observe the process on a time scale that is of the same order as that required for the drop to move a distance equal to L. Rescaling using a reference time $t'_R = \frac{L}{v_0}$, while continuing to use the radius of the drop R_0 as the length scale, we find that the Navier–Stokes equation remains unchanged except that the product $\frac{R_0}{L} Re$ now multiplies the unsteady term. Therefore, in the limit as $Re \to 0$, the unsteady term, along with the convective acceleration term, can be neglected. One might consider L to be representative of the distance between a drop and a surface or between two drops. When this distance is of the order of the radius of the drop used for scaling lengths or larger, $Re \ll 1$ is sufficient for neglecting the unsteady term. When a drop is relatively close to a surface, and the ratio $\frac{R_0}{L} \gg 1$, the condition on the Reynolds number is more stringent. In this case, the condition $Re \ll \frac{L}{R_0}$ must be satisfied for neglecting the unsteady term. The physical implication here is that when the Reynolds number is small, the time taken by the velocity and stress fields to achieve steady representations, for a particular configuration of the system, is small compared with the time taken for the configuration to change appreciably. Sometimes, one speaks of a *quasi-static* problem, implying that the system is virtually static in configuration during the period in which the field variables in the boundary value problem achieve "steady state." Mathematically, the problem in the limit $Re \to 0$ reduces to one of solving Stokes's equation with the associated boundary conditions.

In the corresponding thermocapillary migration problem of a drop interacting with a plane surface, the same argument extends to the energy equation. In this case, when we rescale time using t'_R as the reference time, the product $\frac{R_0}{L} Ma$ multiplies the unsteady term in that equation. Therefore, to neglect the unsteady term, the condition $Ma \ll \frac{L}{R_0}$ must be satisfied. Physically, when the Marangoni number is small, the temperature field for a given system configuration achieves a steady representation in a negligible amount of time compared with that required for the configuration to change appreciably. In such a problem, the temperature field satisfies Laplace's equation. Principally due to the limitations discussed above, analytical solutions of interaction problems are restricted to the quasi-static limit of $Re \to 0$ and, where the energy equation is involved, $Ma \to 0$.

Because rigid objects can rotate as well as translate in a fluid, an analysis of any interaction problem involving rigid bodies moving in a fluid must consider both the force and the torque on these objects. On the other hand, fluid drops normally do not rotate as rigid objects because the distance between any two elements of fluid within the drop need not be a fixed quantity as in rigid bodies. Therefore, when considering interaction effects on fluid drops, one deals only with the hydrodynamic force experienced by the drop when translating through another fluid.

In Chapter Five, we introduce general solutions of Stokes's and Laplace's equations in bispherical coordinates, which have been used widely in solving interaction problems, and provide a physical discussion of the method of reflections, which is another common technique employed for handling interactions. The techniques introduced in this chapter are applicable regardless of the motivating force for the motion. In Chapter Six, results are provided in the case of body force driven motion. Solutions of thermocapillary migration problems are presented and discussed in Chapter Seven.

CHAPTER FIVE

General Solutions

5.1 Solutions of Laplace's and Stokes's Equations in Bispherical Coordinates

5.1.1 Introduction

In this section, we first introduce bispherical coordinates suitable for solving problems involving two spheres or a sphere and a plane surface. We then provide solutions given by Jeffery (1912), Stimson and Jeffery (1926), and O'Neill (1964). We conclude with a few brief remarks about an alternative technique, known as the method of reflections, which provides useful approximate solutions.

5.1.2 Bispherical Coordinates

When solving boundary value problems, it is convenient to be able to specify each boundary by fixing the value of one of the coordinates. This is the reason for the use of cylindrical and spherical polar coordinates in handling problems involving cylindrical and spherical boundaries, respectively. When one deals with a sphere and a plane, or two spheres either external to each other or with one placed within the other, we cannot use the common coordinate systems. Instead, a special set of coordinates known as bipolar or bispherical coordinates must be used. The properties of the conformal transformation, that permits one to map the domains involved into rectangular ones, are well described by Moon and Spencer (1971) and Happel and Brenner (1965). Figure 5.1.1 provides a sketch showing the important features and a sampling of spheres shown in meridian section. The transformation is given by

$$z + i\omega = ic \cot \frac{1}{2}(\eta + i\xi), \tag{5.1.1}$$

where (ω, ϕ, z) represent cylindrical polar coordinates in Part Three. We shall use $(\mathbf{i}_\omega, \mathbf{i}_\phi, \mathbf{i}_z)$ to designate the associated unit vectors. The unit vectors, $(\mathbf{i}_\xi, \mathbf{i}_\eta)$, corresponding to the directions of the new coordinates, are shown in Figure 5.1.1.

By separately equating the real and imaginary parts from the two sides of Equation (5.1.1), it is possible to write (z, ω) in terms of the new coordinates (ξ, η):

$$z = \frac{c \sinh \xi}{\cosh \xi - \cos \eta}. \tag{5.1.2}$$

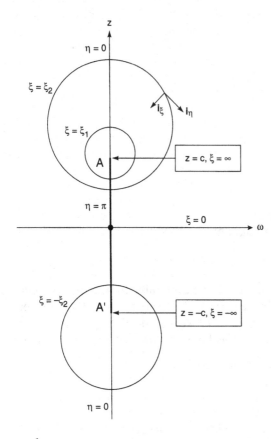

Figure 5.1.1 Sketch of bispherical coordinates, also showing the associated cylindrical polar coordinates.

and

$$\omega = \frac{c \sin \eta}{\cosh \xi - \cos \eta}. \tag{5.1.3}$$

The azimuthal coordinate continues to be the same in the new system. We can eliminate η between Equations (5.1.2) and (5.1.3) to yield the relation

$$(z - c \coth \xi)^2 + \omega^2 = \frac{c^2}{\sinh^2 \xi}, \tag{5.1.4}$$

which shows that a surface on which $\xi = \xi_1$, where ξ_1 is a positive constant, represents a sphere of radius $R_1 = \frac{c}{\sinh \xi_1}$, with its center on the z-axis ($\omega = 0$), at $z = c \coth \xi_1$. When ξ_1 is finite and positive, $\coth \xi_1 > 1$, and the center of the sphere $\xi = \xi_1$ lies above the point A, which is located at $z = c$ on the z-axis. Varying ξ_1 leads to the generation of a coaxial family of spheres above the plane $z = 0$, and a second sphere for $\xi = \xi_2 < \xi_1$ is shown in Figure 5.1.1. Because $\coth \xi \to 1$ as $\xi \to \infty$, this limit corresponds to a sphere of zero radius located at point A, which is known as a limiting point of the bispherical coordinate system. Negative values of ξ correspond to points below the plane $z = 0$, and a family of coaxial spheres is obtained for negative constant values of ξ, which are mirror images of those above the plane. The radius of such a sphere is now given by $\frac{c}{(-\sinh \xi)}$, so that the regions above and below the plane $z = 0$ can be covered by writing the radius as $\frac{c}{|\sinh \xi|}$. A second limiting point A' is located at $z = -c$, which corresponds to $\xi \to -\infty$. The plane $\xi = 0$ corresponds to $z = 0$ and represents a sphere of infinite radius. To summarize, as ξ begins from 0 and increases, we start from the plane at $z = 0$ and approach the limiting point for $\xi = +\infty$ through a series of nonconcentric, coaxial, spheres of decreasing radii. Similar results are obtained below the plane $\xi = 0$ in Figure 5.1.1 for negative values of ξ starting from zero and approaching $-\infty$.

In a like manner, we can eliminate ξ from Equations (5.1.2) and (5.1.3) to obtain

$$z^2 + (\omega - c \cot \eta)^2 = \frac{c^2}{\sin^2 \eta}. \tag{5.1.5}$$

By using this result, it is possible to demonstrate that curves on a meridian plane corresponding to constant values of η represent circular arcs centered on the plane $z = 0$, which terminate at the limiting points A and A'. When rotated through an angle 2π in the azimuthal direction, these produce spindle-shaped surfaces when $\eta < \frac{\pi}{2}$, apple-shaped surfaces when $\eta > \frac{\pi}{2}$, and a spherical surface of radius c when $\eta = \frac{\pi}{2}$. On the z-axis, $\eta = \pi$ for $|z| < c$, and $\eta = 0$ when $|z| > c$. Because we shall not have use for these surfaces on which η is constant, to retain clarity, we do not show the meridian sections of such surfaces in the figure. The only values of η that will be used in boundary conditions are 0 and π, which correspond to the z-axis.

A geometrical interpretation of the coordinates (ξ, η) can be offered. For any point B, two straight lines can be drawn joining the given point to the limiting points A and A'. Then, η is the angle between the lines AB and $A'B$, defined to lie between 0 and π. The coordinate $\xi = \log(\frac{A'B}{AB})$, where $A'B$ and AB represent the lengths of the two straight lines.

Given two spheres of unequal radii R_1 and R_2 with their centers a distance d apart, one can always define a z-axis passing through the centers and erect a bispherical coordinate system. If the spheres are external to each other, the plane $z = 0$ will be located between them. In this case, the parameters ξ_1 and ξ_2, corresponding to spheres 1 and 2 respectively, are related to the physical parameters as follows:

$$\cosh \xi_1 = \frac{\epsilon^2(\lambda - 1) + (1 + \lambda)}{2\epsilon\lambda}, \tag{5.1.6}$$

and

$$\cosh \xi_2 = \frac{\epsilon^2(\lambda - 1) - (1 + \lambda)}{2\epsilon}. \tag{5.1.7}$$

Here,

$$\lambda = \frac{R_1}{R_2}, \tag{5.1.8}$$

and

$$\epsilon = \frac{D}{\lambda - 1}, \tag{5.1.9}$$

and D is a scaled separation distance, given by

$$D = \frac{d}{R_2}. \tag{5.1.10}$$

Sphere 2 is assumed to be located below the plane $z = 0$, and the corresponding value of the bispherical coordinate that describes its surface, ξ_2, is negative. The value of the distance c can be obtained by using the result for one of the radii. For example, $c = R_1 \sinh \xi_1$. When the spheres are of equal radii, $R_1 = R_2 = R$, and the result for ϵ in Equation (5.1.9) becomes singular. Substitution of the definition of ϵ in Equations (5.1.6) and (5.1.7) yields the correct result upon taking the limit $\lambda \to 1$, however. In view of the symmetry of this situation, $\xi_2 = -\xi_1$, and $\cosh(\xi_1) = \frac{d}{2R}$. It follows that for a single sphere of radius R, located with its center at a distance h above the plane $z = 0$, if the value of ξ on the sphere surface is designated ξ_1, we can write

$$\cosh \xi_1 = \frac{h}{R},$$

(5.1.11)

and the distance c can be calculated from

$$c = R \sinh \xi_1.$$

(5.1.12)

Results for the case when one sphere is located nonconcentrically within another are very similar to those given in Equations (5.1.6) through (5.1.10) for spheres external to each other and are provided in Section 7.10, where the thermocapillary migration of a drop containing an eccentrically located droplet is analyzed.

5.1.3 Solution of Laplace's and Stokes's Equations in Bispherical Coordinates

In this section, some useful solutions are given. In the case of Laplace's equation, Jeffery (1912) provided a general solution. For Stokes's equation, Stimson and Jeffery (1926) obtained the axisymmetric solution. Subsequently, O'Neill (1964) provided the results needed in the case when the motion of a sphere occurs parallel to a plane boundary. The same solution applies when two spheres move perpendicular to the line joining their centers, henceforth abbreviated to the *line of centers*. Because Stokes's equation is linear, superposition of solutions can be used, in principle, to handle the general case of motion in an arbitrary direction with respect to the line of centers.

Laplace's Equation. If $T(\xi, \eta, \phi)$ satisfies Laplace's equation, $\nabla^2 T = 0$, the general solution, obtained by separation of variables, can be written as follows:

$$T = \sqrt{\cosh \xi - \zeta} \sum_{m=0}^{\infty} \sum_{n=m}^{\infty} \cos(m\phi + \chi_m) \left[A_{mn} \cosh\left(n + \frac{1}{2}\right)\xi \right.$$

$$\left. + B_{mn} \sinh\left(n + \frac{1}{2}\right)\xi \right] \left[C_{mn} P_n^m(\zeta) + D_{mn} Q_n^m(\zeta) \right].$$

(5.1.13)

Here, $\zeta = \cos \eta$. The sets of constants, χ_m, A_{mn}, B_{mn}, C_{mn}, and D_{mn} are arbitrary and are determined by using the boundary conditions appropriate to a given problem. The functions $P_n^m(\zeta)$ and $Q_n^m(\zeta)$ are the associated Legendre functions of the first and second kinds, respectively. Their properties can be found in MacRobert (1967).

In the axisymmetric case where the solution is symmetric about the z-axis, the dependence on the azimuthal coordinate ϕ vanishes, reducing the sum over m to only the term corresponding to $m = 0$. Also, because T needs to be finite at least on some segment of the axis of symmetry, the coefficients multiplying the Legendre functions of the second kind must vanish, because these functions are unbounded on the axis. This leads to the following simpler version of the solution, where the arbitrary constants A_n and B_n need to be determined from the boundary conditions.

$$T = \sqrt{\cosh\xi - \zeta} \sum_{n=0}^{\infty} \left[A_n \cosh\left(n + \frac{1}{2}\right)\xi + B_n \sinh\left(n + \frac{1}{2}\right)\xi \right] P_n(\zeta) \qquad (5.1.14)$$

It is worth commenting on the behavior of the solution in Equation (5.1.14) as the distance from the origin approaches infinity. This is the limit $(\cosh\xi - \zeta) \to 0$. Because $\cosh\xi \geq 1$, whereas $\zeta \leq 1$, this limit can be attained only when $\xi \to 0$ and $\zeta \to 1$ simultaneously. The solution given in Equation (5.1.14) approaches zero in this limit. In the thermocapillary migration problems that we consider, the undisturbed temperature distribution prevailing at infinity is linear in z and is a solution of Laplace's equation. Therefore, it must be added to the solution in Equation (5.1.14), when necessary.

Stokes's Equation. First, we give the solution for the scaled Stokes streamfunction in the axisymmetric case. The physical streamfunction is scaled by using the product $c^2 v_R$ where v_R is a suitable reference velocity. If $\psi(\xi, \eta)$ satisfies Stokes's equation, which is given in Chapter Three as Equation (3.1.6), the solution can be written as follows:

$$\psi = \frac{1}{(\cosh\xi - \zeta)^{\frac{3}{2}}} \sum_{n=-1}^{\infty} \left[A_n \cosh\left(n - \frac{1}{2}\right)\xi + B_n \sinh\left(n - \frac{1}{2}\right)\xi \right.$$

$$\left. + C_n \cosh\left(n + \frac{3}{2}\right)\xi + D_n \sinh\left(n + \frac{3}{2}\right)\xi \right] C_{n+1}(\zeta). \qquad (5.1.15)$$

As in Equations (5.1.13) and (5.1.14), $\zeta = \cos\eta$. The Gegenbauer Polynomials $C_{n+1}(\zeta)$ were introduced in Section 3.1. In writing the general solution, Stimson and Jeffery (1926) did not specify the range of values taken by the summation index in Equation (5.1.15), even though it is clear that the upper limit must be infinity. In their subsequent solution for the motion of two rigid spheres, they used a lower limit of $n = 1$, and this is satisfactory for most applications. The solutions for $n = -1$ and 0 introduce line sources along the axis of symmetry and should be discarded in problems that do not contain such a source, as noted by Oğuz and Sadhal (1987). These authors noted that these terms, included with suitable connections between their coefficients, can be used to accommodate a point source. In the problems considered in Chapters Six and Seven, we only need to use the range $n = 1$ to ∞ in the summation in Equation (5.1.15). As with the solution of Laplace's equation, one must consider the limit approached by the solution in Equation (5.1.15) as the distance from the origin approaches infinity. It can be shown that the ratio $\frac{\psi}{\omega^2} \to 0$ and that the ratio $\frac{\psi}{\omega}$ remains bounded in that limit. This is satisfactory in the problems we consider here because the fluid in the continuous phase is quiescent as the distance from the origin approaches infinity in these problems.

Stimson and Jeffery (1926) gave a result for the hydrodynamic force exerted by the fluid on a sphere, based on the above solution for the streamfunction. The symbol F_D is used to designate this hydrodynamic force nondimensionalized with $\mu c v_R$, where μ is the viscosity of the continuous phase. When the sphere lies above the plane $z = 0$, the value of ξ corresponding to the surface of the sphere is positive, and we can write F_D as follows:

$$F_D = 2\pi\sqrt{2} \sum_{n=1}^{\infty} (A_n + B_n + C_n + D_n). \qquad (5.1.16)$$

When the value of ξ corresponding to the sphere surface is negative, the following result should be used:

$$F_D = 2\pi\sqrt{2}\sum_{n=1}^{\infty}(A_n - B_n + C_n - D_n).$$

(5.1.17)

When two drops move perpendicular to their line of centers, the problem is not axisymmetric, and a Stokes streamfunction cannot be used. Instead, the solution of Stokes's equation needs to be obtained for the velocity components and pressure. This is accomplished by writing the velocity components in the cylindrical polar coordinate system of Figure 5.1.1. In this case, O'Neill (1964) showed that the dependence on the azimuthal angular coordinate, ϕ is especially simple:

$$p = f_1 \cos\phi,$$

(5.1.18)

$$v_\omega = \frac{1}{2}[\omega f_1 + f_2 + f_3]\cos\phi,$$

(5.1.19)

$$v_z = \frac{1}{2}[z f_1 + 2 f_4]\cos\phi,$$

(5.1.20)

and

$$v_\phi = \frac{1}{2}[f_2 - f_3]\sin\phi.$$

(5.1.21)

In the above, velocities are scaled by a reference velocity v_R, pressure is scaled by the viscous stress $\frac{\mu v_R}{c}$, and distances are scaled using the reference length c. The functions f_1 to f_4 depend only on ξ and η. They are given below.

$$f_1 = \sqrt{(1-\zeta^2)(\cosh\xi - \zeta)}\sum_{n=1}^{\infty}G_{n1}(\xi)P_n'(\zeta),$$

(5.1.22)

$$f_2 = (1-\zeta^2)\sqrt{\cosh\xi - \zeta}\sum_{n=2}^{\infty}G_{n2}(\xi)P_n''(\zeta),$$

(5.1.23)

$$f_3 = \sqrt{\cosh\xi - \zeta}\sum_{n=0}^{\infty}G_{n3}(\xi)P_n(\zeta),$$

(5.1.24)

$$f_4 = \sqrt{(1-\zeta^2)(\cosh\xi - \zeta)}\sum_{n=1}^{\infty}G_{n4}(\xi)P_n'(\zeta),$$

(5.1.25)

$$G_{nk} = A_{nk}\cosh\left(n+\frac{1}{2}\right)\xi + B_{nk}\sinh\left(n+\frac{1}{2}\right)\xi.$$

(5.1.26)

The set of arbitrary constants, A_{nk} and B_{nk}, are to be determined by using the boundary conditions. From a result given by O'Neill (1964), the hydrodynamic force exerted by the continuous phase on the sphere, when scaled in the same manner as in the axisymmetric case, can be written as follows:

$$F_D = -\pi\sqrt{2}\sum_{n=0}^{\infty}[n(n+1)B_{n1} + B_{n3}].$$

(5.1.27)

Whenever solutions in bispherical coordinates can be written, they should prove adequate for most cases. This is true even when the distance of separation between a given sphere and another, or that between a sphere and a plane surface (occasionally called the *gap*), is small relative to the radius of the sphere. More terms in the infinite series need to be included as this distance becomes small, and the required summations can be conveniently performed on a computer. If the computations become difficult, lubrication theory must be used, which is applicable when the gap becomes small. Also, lubrication theory provides useful asymptotic information regarding the situation as the gap size approaches zero. We shall provide additional comments on this topic when discussing an example problem at a later stage.

5.1.4 Solution by the Method of Reflections

The method of reflections is an approximate technique used to handle problems in Stokes flow involving interactions between objects, or between an object and a boundary, from a knowledge of the solution for the flow resulting from an arbitrary velocity field prescribed on the surface of the object. When such a solution is not known, the method still can be used with approximate solutions. This method leads to relatively simple results, and it can be used in cases involving more than two spheres, wherein bispherical coordinates are no longer useful. A detailed discussion of the mathematical approach used, as well as example problems involving the motion of two spheres, may be found in the book by Happel and Brenner (1965). Here, we present only the physical basis behind the method.

Consider the case of two rigid spheres A and B moving in an otherwise quiescent fluid under conditions of Stokes flow. We use $\mathbf{x}$ to designate the position vector defined from a suitable origin in the vicinity of the spheres. First, focus attention on sphere A and assume that the velocity field can be calculated if sphere B is not present. This solution satisfies the boundary condition $\mathbf{v}(\mathbf{x}_A) = \mathbf{U}_A$ where $\mathbf{x}_A$ stands for the surface of sphere A and $\mathbf{U}_A$ is its velocity. Also, the solution for the velocity approaches 0 as $|\mathbf{x}| \to \infty$. Let us use the symbol $\mathbf{v}_1(\mathbf{x})$ to designate this solution. Because sphere B is present, $\mathbf{v}_1$ will not necessarily satisfy the boundary conditions on the surface of sphere B, which require that $\mathbf{v}(\mathbf{x}_B) = \mathbf{U}_B$. Here $\mathbf{x}_B$ is used to designate the surface of sphere B. We correct the solution by adding a *reflection* from sphere B, namely a solution $\mathbf{v}_2(\mathbf{x})$ that satisfies the boundary condition $\mathbf{v}_2(\mathbf{x}_B) = \mathbf{U}_B - \mathbf{v}_1(\mathbf{x}_B)$. It is clear that the sum $\mathbf{v}_1 + \mathbf{v}_2$ should satisfy the correct boundary condition on the surface of sphere B. It will no longer satisfy the boundary condition on the surface of sphere A, however, being in error by $\mathbf{v}_2(\mathbf{x}_A)$. The remedy is to add a third solution $\mathbf{v}_3(\mathbf{x})$, which corrects this error, but it will of course lead to an error in satisfying the boundary condition on the surface of sphere B. Thus, we reflect the solution from the surface of each sphere to that of the other to correct the error, hence the name of the technique. It is not evident from this description that this process will actually produce a series that will be useful. In practice, because Stokes solutions contain terms in the velocity proportional to $\frac{1}{r}$, where r is distance from the center of the sphere, and higher powers of the same quantity, the solutions will contain terms of the type $\frac{R_1}{d}$ and $\frac{R_2}{d}$, where R_1 and R_2 are the radii of the two spheres, and d is the distance between their centers. Corrections at increasingly higher order contain terms of increasingly higher powers of these fractions, both of which are less than unity. Therefore, including the first few corrections can provide an adequate result for the velocity and pressure fields, as well as global quantities such as the drag experienced by each sphere.

The same ideas can be extended to the case of drops. The boundary conditions are a bit more involved because of the possibility of the motion of the fluid on the interface and the need to apply the stress balance at each interface. Also, to apply this method to thermocapillary migration problems, the solution of Laplace's equation must be constructed in a similar manner. Finally, the problem of a sphere and a stationary surface is a special case of that of two spheres. The method of reflections has been used in this problem as well.

CHAPTER SIX

Interactions When Motion Is Driven by a Body Force

6.1 Motion Normal to a Plane Surface

6.1.1 Introduction

In Stokes flow, the motion of a drop near a surface can be decomposed into motion normal to the surface and that parallel to the surface. If the solutions of each of these problems can be written separately, they can be superposed to obtain the solution in the more general situation. We shall consider only the case of motion normal to the surface that is caused by a body force because this problem can be analyzed without too much algebraic difficulty. The problem of motion parallel to a surface can be solved in principle using the general solution given by O'Neill (1964), who also specialized it for the case of a rigid sphere. In the case of drops, however, the algebra is too lengthy to be included here. Meyyappan provided some algebraic details for the motion of a gas bubble parallel to a plane surface in his doctoral thesis (1984). Tabulated results for this case can be found in an appendix to the article by Meyyappan and Subramanian (1987).

6.1.2 Theoretical Analysis

The geometry is depicted in Figure 6.1.1. A drop of fluid 1 of radius R is located in fluid 2 at a distance h from a plane surface at $z = 0$. The drop is moving in a direction normal to the plane surface with a velocity U^* in the z-direction due to the action of a body force such as gravity. As we shall see, the effect of the surface is always such as to increase the hydrodynamic resistance experienced by the moving drop. This is a consequence of velocity gradients introduced by the presence of the surface, which otherwise would be absent, and the resulting increase in dissipation.

Bart (1968) gave results for the general situation in which fluid 3 is present on the other side of the plane surface. Here, two limiting cases are considered. In the first, the plane surface is assumed to be rigid. In the second, the plane surface is assumed to be a gas-liquid interface, termed a *free surface* henceforth, where the tangential stress is assumed to be negligible. A fluid surface will deform because of the motion, but we neglect this deformation, assuming a suitably defined Capillary number to be small compared with unity. If fluid 3 is present on the other side of the surface $z = 0$, the above limiting cases correspond to $\alpha_2 \to 0$ and $\alpha_2 \to \infty$, respectively, where $\alpha_2 = \frac{\mu_2}{\mu_3}$ represents the ratio of the viscosities of fluids 2 and 3. We shall also give the result from Bart that is valid for any value of α_2. For convenience, we shall use the symbol α without

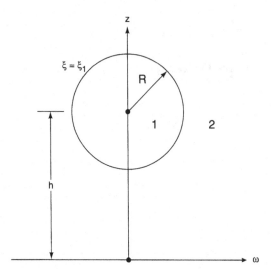

Figure 6.1.1   Sketch showing a drop near a plane surface.

a subscript for the viscosity ratio $\frac{\mu_1}{\mu_2}$. The motion is assumed to occur at sufficiently small Reynolds number that the entire inertia term can be neglected, so that Stokes's equation is applicable. The objective of the analysis is to calculate the hydrodynamic force on the drop and determine how it is influenced by the presence of the plane surface. As noted in Chapter Two, Stokes flows are reversible. Therefore, the direction of motion of the drop, whether away from the plane surface or toward it, does not influence the magnitude of the hydrodynamic force, merely altering its direction. The drop is assumed to be spherical. It is convenient to use bispherical coordinates (ξ, η) described in Section 5.1. Equations (5.1.11) and (5.1.12) relate the value of $\xi = \xi_1$ corresponding to the drop surface and the length c to R and h. The physical streamfunction is scaled using the product $c^2 v_R$, where v_R is a convenient reference velocity. It can be taken to be the velocity of the drop for estimating the magnitudes of relevant dimensionless groups such as the Reynolds and Capillary numbers.

We can use the solution of Stokes's equation in bispherical coordinates, given in Equation (5.1.15), within the drop phase and in the continuous phase as a starting point:

$$\psi_i = \frac{1}{(\cosh \xi - \zeta)^{\frac{3}{2}}} \sum_{n=1}^{\infty} X_{ni}(\xi) C_{n+1}(\zeta), \quad i = 1, 2. \tag{6.1.1}$$

Here, $\zeta = \cos \eta$, and the functions X_{n1} and X_{n2} are written as follows:

$$X_{n1}(\xi) = b_{n1} e^{-(n-\frac{1}{2})\xi} + d_{n1} e^{-(n+\frac{3}{2})\xi}, \tag{6.1.2}$$

$$X_{n2}(\xi) = A_n \cosh\left(n - \frac{1}{2}\right)\xi + B_n \sinh\left(n - \frac{1}{2}\right)\xi + C_n \cosh\left(n + \frac{3}{2}\right)\xi$$

$$+ D_n \sinh\left(n + \frac{3}{2}\right)\xi. \tag{6.1.3}$$

In writing the result for X_{n1}, use has been made of the fact that the streamfunction must remain bounded as $\xi \to \infty$.

The sets of arbitrary constants appearing in the solutions can be evaluated by the application of the boundary conditions relevant to this problem. At the plane surface, the kinematic condition $\mathbf{v} \cdot \mathbf{n} = 0$, where $\mathbf{n}$ is the unit normal to this surface, leads to

$$v_\xi(0, \eta) = 0. \tag{6.1.4}$$

In the case of a rigid plane surface, the tangential velocity also must vanish.

$$v_\eta(0, \eta) = 0. \tag{6.1.5}$$

For the free plane surface, instead we set the tangential stress to zero.

$$\tau_{\xi\eta}(0, \eta) = 0. \tag{6.1.6}$$

At the drop surface, we can use the continuity of tangential velocity, the kinematic condition, and the balance of tangential stress.

$$v_{\eta1}(\xi_1, \eta) = v_{\eta2}(\xi_1, \eta), \tag{6.1.7}$$

$$v_{\xi1}(\xi_1, \eta) = v_{\xi2}(\xi_1, \eta) = U\mathbf{i}_z \bullet \mathbf{i}_\xi, \tag{6.1.8}$$

$$\tau_{\xi\eta1}(\xi_1, \eta) = \tau_{\xi\eta2}(\xi_1, \eta). \tag{6.1.9}$$

Here, the velocity components are scaled using the reference v_R, and the stress is scaled using $\mu_2 \frac{v_R}{c}$. The scaled velocity of the drop is $U = \frac{U^*}{v_R}$. Application of the boundary conditions to the solutions given above is straightforward but tedious. The following results are useful in this process.

$$v_\xi = \frac{(\cosh\xi - \zeta)^2}{\sqrt{1-\zeta^2}} \frac{\partial\psi}{\partial\eta}, \tag{6.1.10}$$

$$v_\eta = -\frac{(\cosh\xi - \zeta)^2}{\sqrt{1-\zeta^2}} \frac{\partial\psi}{\partial\xi}, \tag{6.1.11}$$

$$\tau_{\xi\eta} = \frac{(\cosh\xi - \zeta)^2}{\sqrt{1-\zeta^2}} \left\{ (1-\zeta^2) \left[(\cosh\xi - \zeta)\frac{\partial^2\psi}{\partial\zeta^2} - 3\frac{\partial\psi}{\partial\zeta} \right] \right.$$

$$\left. - \left[(\cosh\xi - \zeta)\frac{\partial^2\psi}{\partial\xi^2} + 3\sinh\xi \frac{\partial\psi}{\partial\xi} \right] \right\}$$

$$= \frac{(\cosh\xi - \zeta)^{\frac{3}{2}}}{\sqrt{1-\zeta^2}} \left\{ (1-\zeta^2)\frac{\partial^2\Phi}{\partial\zeta^2} - \frac{\partial^2\Phi}{\partial\xi^2} + \frac{3\psi}{4\sqrt{\cosh\xi - \zeta}} \right.$$

$$\left. \times [(\cosh\xi - \zeta)^2 + 2\sinh^2\xi] \right\}, \tag{6.1.12}$$

where

$$\Phi = (\cosh\xi - \zeta)^{\frac{3}{2}}\psi. \tag{6.1.13}$$

Because

$$\mathbf{i}_z \bullet \mathbf{i}_\xi = -(\cosh\xi - \zeta)\frac{\partial\omega}{\partial\eta}, \tag{6.1.14}$$

the kinematic condition at the drop surface reduces to

$$\left[\frac{\partial}{\partial \eta}\left(\psi + \frac{U\omega^2}{2}\right)\right]_{\xi=\xi_1} = 0, \tag{6.1.15}$$

which can be integrated immediately to yield

$$\psi(\xi_1, \eta) = -\frac{U(1 - \zeta^2)}{2(\cosh \xi_1 - \zeta)^2}. \tag{6.1.16}$$

In applying this condition and the stress balance, use is made of the following results, obtained from an integral given by Jeffery (1912) and reported in Dean and O'Neill (1963):

$$(2n + 1)\int_{-1}^{+1}\frac{P_n(\zeta)}{(\cosh \xi - \zeta)^{\frac{1}{2}}}d\zeta = \sinh|\xi|\int_{-1}^{+1}\frac{P_n(\zeta)}{(\cosh \xi - \zeta)^{\frac{3}{2}}}d\zeta$$

$$= \frac{3\sinh^2 \xi}{2\coth|\xi| + 2n + 1}\int_{-1}^{+1}\frac{P_n(\zeta)}{(\cosh \xi - \zeta)^{\frac{5}{2}}}d\zeta$$

$$= 2\sqrt{2}e^{-(n+\frac{1}{2})|\xi|}. \tag{6.1.17}$$

The final objective is to determine the hydrodynamic force experienced by the drop. When this is scaled using $\mu_2 c v_R$, we can use the result given in Equation (5.1.16), which is reproduced below for convenience.

$$F_D = 2\pi\sqrt{2}\sum_{n=1}^{\infty}(A_n + B_n + C_n + D_n). \tag{6.1.18}$$

The kinematic condition at the plane surface, given in Equation (6.1.4), leads to the vanishing of the streamfunction at that surface. As a consequence, we find that

$$A_n + C_n = 0. \tag{6.1.19}$$

This simplifies the result for the scaled hydrodynamic force to

$$F_D = 2\pi\sqrt{2}\sum_{n=1}^{\infty}(B_n + D_n). \tag{6.1.20}$$

After some algebra, this force can be written as follows:

$$F_D = -2\sqrt{2}U\sum_{n=1}^{\infty}\frac{n(n+1)}{(2n-1)(2n+3)}\left(\frac{\Delta_{n1} + \alpha\Delta_{n2}}{\Delta_{n3} + \alpha\Delta_{n4}}\right). \tag{6.1.21}$$

The constants appearing in Equation (6.1.21) depend on whether the plane surface is rigid or free. Results for these constants for a rigid plane surface are given below.

$$\Delta_{n1} = 2e^{-(2n+1)\xi_1} + 2\cosh 2\xi_1 + (2n+1)\sinh 2\xi_1, \tag{6.1.22}$$

$$\Delta_{n2} = 2\sinh(2n+1)\xi_1 + (2n+1)\sinh 2\xi_1 - 4\sinh^2\left(n + \frac{1}{2}\right)\xi_1$$

$$+ (2n+1)^2\sinh^2 \xi_1, \tag{6.1.23}$$

$$\Delta_{n3} = 2\sinh(2n+1)\xi_1 - (2n+1)\sinh 2\xi_1, \tag{6.1.24}$$

$$\Delta_{n4} = 4\sinh^2\left(n + \frac{1}{2}\right)\xi_1 - (2n+1)^2\sinh^2 \xi_1. \tag{6.1.25}$$

In the case of a free plane surface, the constants are defined as follows:

$$\Delta_{n1} = -2e^{-(2n+1)\xi_1} + 2\cosh 2\xi_1 + (2n+1)\sinh 2\xi_1, \tag{6.1.26}$$

$$\Delta_{n2} = -2\sinh(2n+1)\xi_1 + (2n+1)\sinh 2\xi_1 + 4\cosh^2\left(n+\frac{1}{2}\right)\xi_1$$

$$+ (2n+1)^2 \sinh^2 \xi_1, \tag{6.1.27}$$

$$\Delta_{n3} = 2\left[\cosh(2n+1)\xi_1 - \cosh 2\xi_1\right], \tag{6.1.28}$$

and

$$\Delta_{n4} = 2\sinh(2n+1)\xi_1 - (2n+1)\sinh 2\xi_1. \tag{6.1.29}$$

It is useful to know how the hydrodynamic force in the presence of a plane surface relates to that on an isolated drop, under otherwise identical conditions. In the case of thermocapillary migration under similar circumstances, the hydrodynamic force is zero, and such a comparison cannot be made. Instead, one compares the quasi-steady velocity of the drop in the presence of the plane surface with that of an isolated drop. We use the same basis here in the body force driven motion problem for consistency. Therefore, we define an interaction parameter, Ω as follows:

$$\Omega = \frac{U^*}{U^*_{isolated}}. \tag{6.1.30}$$

Here, U^* is obtained by setting the net force on the drop in the presence of the plane surface to zero, and $U^*_{isolated}$ is obtained in a similar manner as shown in Equation (3.1.33). This definition of Ω then can be used in both the thermocapillary migration and the body-force-driven motion problems. The following expression can be written for the interaction parameter in the body-force-driven motion problem:

$$\Omega = \frac{2 + 3\alpha}{4(1+\alpha)\sinh\xi_1 \sum_{n=1}^{\infty} \frac{n(n+1)}{(2n-1)(2n+3)}\left(\frac{\Delta_{n1}+\alpha\Delta_{n2}}{\Delta_{n3}+\alpha\Delta_{n4}}\right)}. \tag{6.1.31}$$

The set of constants Δ_{ni} corresponding to the rigid or free plane surfaces must be substituted here as appropriate.

Sample results calculated from Equation (6.1.31) are displayed in Figure 6.1.2 in the form of the interaction parameter, Ω, plotted against a scaled separation distance, $H = \frac{h}{R}$, for both a rigid surface and a free surface. In each case, values of $\alpha = 0.1, 1$, and 10 are used for illustration. The curve for $\alpha = 0$ corresponding to the gas bubble limit is only a slight distance away from that for $\alpha = 0.1$; therefore, it is not included. Similarly the curve for $\alpha \to \infty$, which represents the case of a rigid sphere, is close to that for $\alpha = 10$ and is omitted. It is evident that a rigid surface offers more resistance to drop motion than a free surface. Also, for a given type of surface, increase in the viscosity of the drop, relative to that of the fluid surrounding it, increases the influence of the surface in retarding its motion. One would expect that $\Omega \to 1$ as $H \to \infty$. At a distance of 15 drop radii, a rigid plane surface still reduces the velocity of the drop from that of an isolated drop by about 6.2% when $\alpha = 1$. This far-reaching influence of the surface can be traced to the behavior of the velocity field in the continuous phase. The reader will recall from Section 3.1 that an isolated drop moving in a second fluid produces a disturbance velocity in that fluid that decays as $\frac{1}{r}$, where r is the radial distance measured from the center of the drop. This slow decay of the Stokeslet contribution to

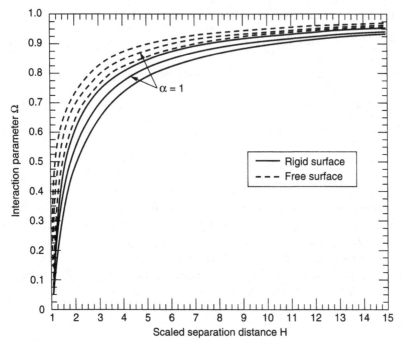

Figure 6.1.2 Interaction parameter for the body-force-driven motion of a drop normal to a plane surface plotted against scaled separation distance. The curves for $\alpha = 1$ are identified in the drawing, and in each case, the curve for $\alpha = 0.1$ lies above this curve and that for $\alpha = 10$ lies below.

the velocity field is a consequence of the hydrodynamic force experienced by the drop, which is required for balancing the hydrostatic force on it. Even at a distance of 15 drop radii, this disturbance flow is significant, and a plane surface at that distance still makes a contribution to the resistance experienced by the drop. When $\alpha = 1$, a rigid plane surface must be placed at a distance greater than 93.4 drop radii to obtain a reduction in velocity of less than 1%. The situation is slightly better in the opposite limit of a free plane surface, which needs to be placed only 50 drop radii away to achieve a similar approach to the velocity of an isolated drop.

The result obtained by Bart (1964) in the more general case can be recast in the present notation. We merely need to replace the ratio of the linear combinations of the constants Δ_{ni} appearing in Equation (6.1.31) by a ratio of a weighted sum of the numerator in the two limiting cases to a similarly weighted sum of the denominator in the same limiting cases. If we write

$$\Delta_{n5} = (\Delta_{n1} + \alpha \Delta_{n2})_{rigid\ surface} + \alpha_2 (\Delta_{n1} + \alpha \Delta_{n2})_{free\ surface} \tag{6.1.32}$$

and

$$\Delta_{n6} = (\Delta_{n3} + \alpha \Delta_{n4})_{rigid\ surface} + \alpha_2 (\Delta_{n3} + \alpha \Delta_{n4})_{free\ surface}, \tag{6.1.33}$$

then the interaction parameter in the general case is obtained as follows:

$$\Omega = \frac{2 + 3\alpha}{4\,(1 + \alpha)\sinh \xi_1 \sum_{n=1}^{\infty} \frac{n(n+1)}{(2n-1)(2n+3)} \left(\frac{\Delta_{n5}}{\Delta_{n6}} \right)}. \tag{6.1.34}$$

Bart (1964) provided tabulated results for the ratio of the hydrodynamic forces, which is the inverse of the interaction parameter defined above. One should be careful when using Bart's table because the ratio reported by Bart is that between the hydrodynamic force on a drop in the presence of a plane interface and that on a rigid sphere moving in the absence of boundaries. Bart also demonstrated that his general results approach various limiting cases as expected, including that of a rigid sphere moving normal to a plane surface, first solved independently by Brenner (1961) and Maude (1961).

When the gap between the drop and the plane surface is extremely small, asymptotic analysis can be used to extract the result for the force as illustrated in the case of a rigid sphere by Cox and Brenner (1967), who also provided an analysis of the problem for small values of a Reynolds number based on the gap width. The authors showed that for Stokes motion of a rigid sphere normal to a plane surface, the drag on the sphere increases in inverse proportion to the gap width. It is evident that the drag becomes extremely large as the gap width becomes small. This is because of the large hydrodynamic pressure that naturally arises in the gap to squeeze out the fluid. A strict proportionality of the hydrodynamic force to the inverse of the gap width implies that under the action of a constant body force such as gravity, the quasi-steady velocity of the sphere would be proportional to the gap width. As a consequence, one can write $\log(h - R) \sim \chi_1 - \chi_2 t^*$, where χ_1 and χ_2 are positive constants, and t^* is time. This means that it will take infinite time for the sphere and the plane surface to touch. In practice, attractive van der Waals forces come into play when the liquid film reaches a critical thickness and lead to rupture of the film. Cox and Brenner (1967) also found that the effect of fluid inertia is to increase the drag when motion is toward the surface and decrease it when the drop is moving away.

Available results for the force experienced by a spherical fluid drop moving normal to a plane surface, in the limit as the gap width approaches zero, are summarized conveniently by Loewenberg and Davis (1993a). We provide these results later in Section 7.1.5. When a drop moves toward a rigid surface, as the gap width approaches zero, the resistance is inversely proportional to the gap width, as is the case for a rigid sphere. The result is different for a drop approaching a free plane surface. Here, the resistance is inversely proportional to the square root of the gap width in the limit as that width approaches zero. If the viscosity of the drop is large compared with that of the continuous phase, the resistance first increases in inverse proportion to the gap width but ultimately reverts to the dependence just mentioned. This relieves the logarithmic singularity in the time needed for complete drainage of the film noted earlier. For a spherical fluid drop approaching a free plane surface, the drainage of the film should occur in a finite time. When a drop is close to a surface, however, deformation of shape becomes important, and results for a spherical drop are only of limited utility. Recent articles that consider deformation effects include those of Chi and Leal (1989), Yiantsios and Davis (1990, 1991), and Manga and Stone (1995a). All the authors employed the boundary integral method to handle the deformed interfaces. Chi and Leal (1989) analyzed the deformation and the associated film drainage problem, when the drop moves toward a fluid layer that has the same chemical composition as the drop. Yiantsios and Davis (1990) used a lubrication analysis in the gap and concluded that accommodating deformation leads to the prediction of an infinite time for film drainage. They also included van der Waals forces in their analysis, however, and showed that these are sufficient to cause rupture of the film at a finite thickness. Manga and Stone (1995a) analyzed the buoyant motion of a drop through a stably stratified fluid-fluid interface. The Bond numbers were permitted to be large, leading to substantial deformation of the shapes of the drop and the interface. They also provided photographic results from experiments on the rise of an air bubble through such an interface.

O'Neill's (1964) analysis for the motion of a rigid sphere parallel to a surface using a solution in bispherical coordinates is complemented by the lubrication theory treatment of O'Neill and Stewartson (1967) and Goldman, Cox, and Brenner (1967), who dealt with the problem of small gap widths. In this instance, the hydrodynamic resistance is proportional to the logarithm of the gap width.

6.1.3 Comparison with Experimental Results

Now, we make a few remarks regarding experimental work. In addition to providing a theoretical analysis, Bart (1964) also performed experiments on the movement of drops of glycerine and ethylene glycol, as well as air bubbles, in a UCON fluid. UCON fluids, manufactured and distributed by Union Carbide Corporation, are water-soluble copolymers of ethylene/propylene oxides, which are made in a range of molecular weights and, therefore, viscosities. In some cases Bart used an acrylic plate as a solid plane surface; in others, one of the above drop phase fluids was used to obtain a fluid-fluid interface. Also, some experiments were conducted using solid spheres. Bart calculated trajectories by integrating the result for the velocity as a function of separation distance, so that errors due to differentiation of data on position versus time could be avoided. He found reasonable agreement in the cases of a solid sphere or a glycerin drop approaching a free surface but reported some departure from the predictions in the other cases. In fact, he commented that Mackay, Suzuki, and Mason (1963) observed better agreement in the case of a solid sphere and a rigid surface than he did. The discrepancies are explained based on buoyant convection caused by heating induced by the illuminating flood lights, even though care was taken to minimize this effect. Bart also reported negligible deformation of the shape of the drops until the drops were touching the interface. Similarly, little deformation of a fluid interface was observed except when the gap was small. More recently, the correctness of the theoretical prediction for motion toward a rigid surface has been confirmed by Merritt and Subramanian (1989) and Barton and Subramanian (1991), who performed experiments on relatively small air bubbles in silicone oils and on ethylsalicylate drops in diethylene glycol, respectively. These authors studied the broader problem of motion under the simultaneous influence of gravity and thermocapillarity. In Figure 6.1.3, we display experimental results from Barton and Subramanian (1991) for the interaction parameter for drops of ethylsalicylate settling due to gravity in diethylene glycol at low Reynolds number toward a rigid plane surface. These experiments were performed at a temperature of 23°C, using a very slight upward temperature gradient of 0.03 K/mm to maintain stable conditions and minimize stray convection currents. At this temperature gradient, the thermocapillary effect makes a negligible contribution to the motion of the drops. The drops varied in radius from 200 to 330 μm. Also included in Figure 6.1.3 is the prediction from Equation (6.1.31) with the constants Δ_{ni} pertinent to the rigid plane surface case, obtained from Equations (6.1.22) to (6.1.25). It is evident that the experimental data are in good agreement with the prediction.

6.2   Interaction between Two Drops

6.2.1 Introduction

The interaction between two drops moving because of applied forces in a third fluid, with which they are both immiscible, can be analyzed by using bispherical coordinates

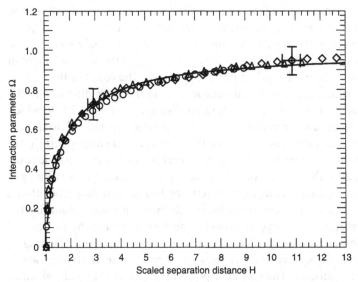

Figure 6.1.3 Data on the interaction parameter, for drops of ethylsalicylate settling toward a rigid horizontal surface in diethylene glycol under the action of gravity, plotted against scaled separation distance. The prediction from Equation (6.1.31) is shown as a solid curve (Courtesy of Academic Press).

in a manner similar to that employed for dealing with a drop and a plane surface. The same set of assumptions needs to be made regarding inertial effects being negligible and spherical shapes being maintained. The general problem of the motion of each drop in an arbitrary direction with respect to the line of centers can be decomposed into simpler problems involving motion along the line of centers and that normal to the line of centers. The former is called the axisymmetric problem, and the latter is commonly termed the *asymmetric* problem, even though a situation that is not axisymmetric in this context is, by necessity, asymmetric. The solution of Stokes's equation provided in Equation (5.1.15) can be used in the former case, and that in Equations (5.1.18) through (5.1.26) can be specialized in the problem of motion normal to the line of centers. Hydrodynamic forces can be obtained from the equations given in Section 5.1 for each case. Although the process is straightforward, the algebra can be lengthy when dealing with drops. Thus, we do not undertake analyses of such problems, instead providing a survey of available studies. First, we discuss the axisymmetric case and then the more general situation, concluding with remarks on how the results from pair interactions can be put to use in dealing with a dilute collection of drops and on available experimental work.

6.2.2 Motion along the Line of Centers

Stimson and Jeffery (1926) specialized the general solution given in Equation (5.1.15) to the case of two rigid spheres of possibly unequal size. In this situation, only one fluid is involved and, therefore, only one set of constants needs to be evaluated. The authors provided expressions for the constants, noting that when the two spheres are of equal size, the constants B_n and D_n vanish. They then gave numerical results for the correction to the Stokes force on an isolated rigid sphere when two identical spheres move in the same direction along their line of centers. The word *correction* is used in the literature on interacting objects with the implication that it is a multiplicative factor, which adjusts the drag on an isolated object to yield the drag when it interacts with one or more objects. The most important aspect of the results obtained by Stimson and Jeffery is that the hydrodynamic force on a given sphere is reduced by the motion of the second sphere in the same direction. This can be explained on physical grounds. The result of the motion of a given sphere in a certain direction is to induce motion in the same direction in the neighboring fluid. This is illustrated by the streamlines displayed in Figure 3.1.3c, which are close in appearance to the streamlines for the gravitational settling of a rigid sphere. As a result, even if there is no motivating force for the motion of a second sphere, it will be dragged along in the same direction by this flow. Therefore, the hydrodynamic resistance offered by the fluid to the motion of the second sphere, at any given velocity in the same direction, is reduced when compared with the resistance that would be offered if that sphere were to move at the same velocity in otherwise quiescent fluid. By the same reasoning, the first sphere is dragged along by the flow generated by the motion of the second sphere. Thus, the two spheres assist each other, each reducing the drag on the other. As a corollary, if the spheres were to move in opposite directions, they would tend to retard each other's motion. The second aspect of the numerical results given by Stimson and Jeffery is that even at a separation distance of 20 radii, there is an approximately 7% reduction in drag due to the presence of a second identical sphere. The reason for this far-reaching influence already has been discussed in the previous section. When the spheres are apart by about 0.26 radii, the reduction in drag is slightly over one-third.

Since the appearance of the paper by Stimson and Jeffery in 1926, a number of articles have been published on the axisymmetric case in which the constants in their general solution have been specialized for various situations. The case of rigid spheres can be considered to be a limiting situation of two drops when each is very viscous compared with the continuous phase. At the opposite extreme, when the viscosity of each drop is negligibly small compared with that of the continuous phase, we speak of the gas bubble limit. Results in the general case of arbitrary viscosity ratios were reported by Rushton and Davies (1973), and Haber, Hetsroni, and Solan (1973) in independent efforts. In both articles, results were given for the hydrodynamic force on each sphere. Because of the linearity of the problem, the force on each sphere can be written as a linear combination of the velocities of the two spheres, using coefficients of proportionality that depend on the radii of the two spheres, the separation distance between them, and the viscosities of the three fluids involved. From these results, a multiplicative correction factor, to be used with the result given in Section 3.1 for the hydrodynamic force on an isolated fluid drop, can be calculated. Being dimensionless, this factor is a function of the radius ratio, scaled separation distance, two viscosity ratios, and the ratio of the velocities of the two drops. It also depends on whether the drops are moving in the same direction or in opposite directions. The algebraic details are lengthy and therefore omitted here. The reader may refer to either article for such details. The results of Haber et al. contain minor typos pointed out by Zinchenko (1983). Keh and Chen (1990) mentioned that some typographical errors are present in the results of Rushton and Davies but do not give further details. Haber et al. provided drawings illustrating the role of each of the parameters in affecting the coefficients in the result for the hydrodynamic force for a few fixed values of the other parameters. Rushton and Davies, on the other hand, provided one comprehensive drawing, in which they plot the correction factor to be used with the hydrodynamic force on an isolated drop, which illustrates several limiting cases. They also listed expressions in various asymptotic cases, demonstrating that the hydrodynamic force approaches the correct limiting forms in those cases analyzed by other authors in the past. As an example, when one drop is very large compared with the other, it approximates a plane surface, in which case the results should reduce to those given by Bart (1964). In a later article, Rushton and Davies (1978) presented additional results for the correction factor, as a function of the parameters involved, and also showed several streamline drawings illustrating the flow features. Details of the flow in the region between the drops are illustrated in another article by Rushton and Davies (1974).

It is useful to consider the situation where the width of the gap between the two drops approaches zero. We can define $\epsilon_1 = \frac{h_0}{R_1}$, where h_0 is the gap between the surfaces of the drops along their line of centers and R_1 is the radius of drop 1. In the axisymmetric case, Zinchenko (1979, 1982) has examined the asymptotic behavior of the solution in bispherical coordinates as $\epsilon_1 \to 0$. He found that the hydrodynamic forces on spherical drops scale as $\frac{A}{\sqrt{\epsilon_1}} + B \log \epsilon_1$, where A and B are constants that depend on the parameters. This is in contrast to the case of two rigid spheres, wherein the hydrodynamic forces scale as $\frac{A^*}{\epsilon_1} + B^* \log \epsilon_1$ with different constants A^* and B^*, according to Jeffrey (1982), who provided higher order terms in this expansion for rigid spheres as well. Davis, Schonberg, and Rallison (1989) and Barnocky and Davis (1989) used lubrication theory, in conjunction with a boundary integral method, to obtain numerical results in the axisymmetric case involving spherical drops approaching each other. These authors demonstrated that when two drops are very viscous, the behavior of the resistance scales in the form suitable for rigid spheres when the gap is relatively large but switches to the scaling for fluid drops as the gap becomes thinner. This is the same shift in scaling discussed in Section 6.1 in the context of a drop and a plane surface. Also, as noted there, inclusion of drop deformation makes a difference in the asymptotic behavior in the film drainage problem. The case of two drops is analyzed by Yiantsios and Davis (1991), who concluded that deformed drop interfaces do not come into contact in a finite time and that an attractive force is necessary for achieving rupture of the intervening film of the continuous phase fluid.

6.2.3 Motion Normal to the Line of Centers

When each of two spheres moves along the line of centers, it is straightforward to see from symmetry that the hydrodynamic force on each sphere must act along the line of centers as well. Therefore, lateral motion of the spheres perpendicular to the line of centers is not possible because of this symmetry. Now, consider the motion of each of two spheres normal to their line of centers. In this case, the hydrodynamic force on each sphere will still act opposite to the direction of motion of each sphere, and therefore, perpendicular to the line of centers. Hence, the spheres will not move toward or away from each other along the line of centers. We can see that the plane containing the line of centers and the direction of motion of the spheres is a symmetry plane, which precludes the appearance of a hydrodynamic force that would tend to move the spheres in a direction normal to that plane. But why should the spheres show no tendency to move toward or away from each other? The reason is the reversibility of Stokes flows. Time does not appear explicitly in the Stokes equations, and the flows are perfectly reversible. This means that a Stokes motion can be traced backward. Now, imagine that two spheres settling downward, perpendicular to their line of centers, experience forces that cause the distance of separation between them to tend to increase with time. Run the experiment backward in time. The spheres will tend to come together. This is not possible, however, because there is nothing intrinsically different about the spheres moving upward and moving downward. If they will move apart from each other when moving in one direction perpendicular to their line of centers, they must also move apart from each other when moving in the opposite direction. Therefore, they will experience no tendency to move toward or away from each other. For the same reason, a sphere settling parallel to a plane wall will not move toward or away from it if it is in Stokes motion. In reality, there always will be some inertial effects that cause such lateral drift for a single sphere. Similarly, inertia will induce a hydrodynamic force along the line of centers when two spheres move in a direction normal to their line of centers.

For reasons similar to those advanced earlier, the motion of each sphere in a certain direction will cause a draft of fluid in that direction, which will help the other sphere move along the same direction. Therefore, the hydrodynamic resistance experienced by each sphere should be reduced by the motion of the other.

Moving on to available results, Dean and O'Neill (1963) obtained the solution for the rotation of a rigid sphere about an axis parallel to a plane surface, and O'Neill (1964) provided the solution for translation parallel to a plane surface. Numerical errors in the results of Dean and O'Neill were pointed out by Goldman, Cox, and Brenner (1967). Even though the problems that initially motivated Dean and O'Neill (1963) and O'Neill (1964) involved a sphere and a plane surface, the general solutions given in these articles also can be used for problems involving two spheres. Using these solutions, Wakiya (1967), Davis (1969), and O'Neill and Majumdar (1970a, 1970b) analyzed the translation and rotation of rigid spheres. Asymptotic analyses were reported by the latter authors for the situation when the gap width is small. Zinchenko (1981) worked out the details of the solution for the case of two drops translating at different velocities normal to their line of centers. He noted that the algebraic equations for the sets of constants appearing in the solution cannot be solved analytically. Rather, he reduced the system to a set of difference equations, which he solved numerically. As noted above, the hydrodynamic force acts in a direction opposite to that of the motion of each sphere, and Zinchenko wrote the hydrodynamic force in the form of a linear combination of the velocities of the two drops in a manner similar to that employed in the axisymmetric case. For numerical calculations, he assumed the viscosities of both drops to be the same. He provided a table for the coefficients in this relationship for several distances of separation, three values of the radius ratio of the drops, 0.1, 0.5, and 1, and viscosity ratios of 0, 10, 30, and 10^7. The last value of the viscosity ratio approximates the limiting situation of a rigid object.

Another approach for handling the case of arbitrary orientation of the line of centers with respect to the direction of motion of the objects is to use the method of reflections. Lorentz (1907) analyzed the motion of a rigid sphere normal to a plane surface using this method, and Smoluchowski (1911) used the approach to determine the interaction effects between two rigid spheres. Happel and Brenner (1965) discussed the problem of two rigid spheres in considerable detail in their book. The method of reflections was used by Hetsroni and Haber (1978) to analyze the case of two fluid spheres moving at velocities U_1 and U_2, which are arbitrarily oriented with respect to their line of centers. They obtained the first few terms in a series for the hydrodynamic forces F_1 and F_2 experienced by the drops in powers of $r_1 = \frac{R_1}{d}$ and $r_2 = \frac{R_2}{d}$, where R_1 and R_2 are the radii of the two drops, and d is the distance between their centers. Typographical errors in their results are noted by Zinchenko (1981). Later, Zinchenko (1983) developed a scheme to extend the method of reflections to a high order by numerical means. In view of its simplicity, we report the first few terms in the reflections result from Haber and Hetsroni, as corrected by Zinchenko. We have verified that the results given here are correct, by using information on mobilities given by Kim and Karrila (1991). For reasonable separation distances between the drops, these should prove adequate for the calculation of the interaction effects. It is possible to write the hydrodynamic forces experienced by the two drops as follows:

$$\mathbf{F}_1 = -2\pi\mu R_1 p_1 [\mathbf{A}_{11} \bullet \mathbf{U}_1 + \mathbf{A}_{12} \bullet \mathbf{U}_2], \tag{6.2.1}$$

and

$$\mathbf{F}_2 = -2\pi\mu R_2 p_2 [\mathbf{A}_{21} \bullet \mathbf{U}_1 + \mathbf{A}_{22} \bullet \mathbf{U}_2]. \tag{6.2.2}$$

The tensors $\mathbf{A}_{11}$ and $\mathbf{A}_{12}$ are defined as follows:

$$\mathbf{A}_{11} = X_{11}\mathbf{ee} + Y_{11}(\mathbf{I} - \mathbf{ee}), \tag{6.2.3}$$

and

$$\mathbf{A}_{12} = X_{12}\mathbf{ee} + Y_{12}(\mathbf{I} - \mathbf{ee}). \tag{6.2.4}$$

Here $\mathbf{e}$ is a unit vector in the direction from drop 1 to drop 2, and $\mathbf{I}$ is the identity tensor whose dot product with any vector is the same vector, and

$$X_{11} = 1 + \frac{p_1 p_2}{4}\left(r_1 r_2 + r_1 r_2^3\right) - \frac{p_2 q_1}{2}r_1^3 r_2 + \frac{p_1^2 p_2^2}{16}r_1^2 r_2^2, \tag{6.2.5}$$

$$X_{12} = -\frac{p_2}{2}r_2 + \frac{q_2}{2}r_2^3 + \frac{p_2 q_1}{2p_1}r_1^2 r_2 - \frac{p_1 p_2^2}{8}r_1 r_2^2, \tag{6.2.6}$$

$$Y_{11} = 1 + \frac{p_1 p_2}{16}r_1 r_2 + \frac{p_1 q_2}{8}r_1 r_2^3 + \frac{p_2 q_1}{8}r_1^3 r_2 + \frac{p_1^2 p_2^2}{256}r_1^2 r_2^2, \tag{6.2.7}$$

and

$$Y_{12} = -\left[\frac{p_2}{4}r_2 + \frac{q_2}{4}r_2^3 + \frac{p_2 q_1}{4p_1}r_1^2 r_2 + \frac{p_1 p_2^2}{64}r_1 r_2^2\right]. \tag{6.2.8}$$

The parameters p_i, q_i, and α_i are defined as follows:

$$p_i = \frac{2 + 3\alpha_i}{1 + \alpha_i}, \tag{6.2.9}$$

$$q_i = \frac{\alpha_i}{1 + \alpha_i}, \tag{6.2.10}$$

and

$$\alpha_i = \frac{\mu_i}{\mu}. \tag{6.2.11}$$

In the above definitions, i can take on the values 1 or 2. The symbol μ represents the viscosity of the continuous phase, and μ_i, $i = 1, 2$, stand for the viscosities of the two drop phases. From symmetry considerations, the tensor $\mathbf{A}_{21}$ can be constructed from $\mathbf{A}_{12}$ by interchanging the subscripts 1 and 2. By a similar interchange, the tensor $\mathbf{A}_{22}$ can be obtained from $\mathbf{A}_{11}$.

Fuentes, Kim, and Jeffrey (1988, 1989) treated the axisymmetric and asymmetric problems for two drops, by constructing expansions from solutions for the image velocity fields near a drop for a series of singularities, starting with a Stokeslet. A small drop interacting with a larger drop is approximated by a set of such singularities. The end result is a linear connection between the velocities of the drops and the forces acting on them in each case through matrices known as *mobility* matrices, the elements in which are given as nested expansions in r_1 and $\frac{r_2}{r_1}$. These mobility matrices may be inverted to obtain matrices of resistance coefficients, which would be similar to those reported above. Because of the nature of the approximation being made, the authors indicated that the results are useful when the gap between the surfaces of the drops is of the order of or larger than the diameter of the smaller of the two drops. For more information, the reader should consult Kim and Karrila (1991), who have provided a wealth of detail on interaction problems involving rigid objects, as well as fluid drops.

Finally, we mention the work of Manga and Stone (1993) who treated the axisymmetric problem for two deforming drops rising in line, in the Stokes flow limit, using the boundary integral method. In addition, they obtained approximate results for the nonaxisymmetric situation. In a subsequent article, Manga and Stone (1995b) provided results from fully three-dimensional calculations for collections of two to four drops.

6.2.4 Concluding Remarks

Interactions among drops in a large collection is usually handled by making the assumption of a dilute system and considering only pairwise interactions. Furthermore, the techniques used depend upon superposition, so that inertial effects must be neglected. One objective of this type of analysis is to predict collision rates. If collisions are assumed to lead to coalescence in some fraction of cases, usually assumed to be unity, the model then can be used to predict the time evolution of the size distribution in a cloud of spherical drops, as illustrated by Zhang, Wang, and Davis (1993), and in a cloud of deformable drops, as shown by Manga and Stone (1995b). The latter authors pointed out that deformation can substantially enhance coalescence rates by aligning drops that are initially offset horizontally. For a discussion of the role of gravitational settling and Brownian motion in affecting collision rates, the reader may consult Zhang and Davis (1991). Brownian motion will have a negligible influence on the system if the drops are significantly larger than a micrometer in radius. A second aspect in a collection of drops is the prediction of the influence of the other drops on the settling velocity a test drop. The objective is to obtain the correction to the velocity of an isolated drop as a function of the fractional volume occupied by the drops. Calculation of this influence is nontrivial as shown by Batchelor (1972), who made a careful evaluation for the case of identical rigid spheres. An important technical difficulty is that divergent integrals are encountered because of the slow decay of the disturbance velocity from a moving sphere. Batchelor showed how they can be evaluated by a technique known as renormalization. His analysis was extended to more general situations, including the case of identical fluid drops, by Reed and Anderson (1980). Brief discussions of the above two problems can be found in Section 7.6, where we consider a collection of drops.

Another aspect of a collection of drops is the behavior of the entire suspension. Examples of properties of interest are the rheology of the suspension, including its Newtonian viscosity as a function of the concentration of the suspended objects in the dilute limit and its thermal properties. Such topics are outside the scope of the present book.

There are only a small number of experimental studies on the interaction of drops moving because of gravity. Zhang, Davis, and Ruth (1993) have reported on experiments performed with castor oil drops settling in a silicone oil and on drops of a Union Carbide UCON fluid settling in heavy paraffin oil. Although a drop, released exactly above a smaller settling drop, will collide with the second drop, this is not the case when there is horizontal offset. The authors found that the larger drop in this case goes around the smaller drop. They compared predicted and observed trajectories, as well as other quantities extracted from the data, such as a suitably defined time spent together by the drops and mobility functions. The observed trajectories are consistent with those predicted for spherical drops except when the drops are close to each other, in which case the authors suggested that drop deformation plays a significant role. Overall, drop deformation otherwise is unimportant in the study of Zhang et al. The influence of deformation was studied by Manga and Stone (1993), who performed experiments on air bubbles rising in corn syrup. Even if a pair of vertically separated bubbles are initially offset horizontally, deformation leads to their acquiring a horizontal velocity such that vertical alignment is promoted. Therefore, a large trailing bubble collides and coalesces with a smaller leading bubble. Another mechanism that leads to coalescence involves a large bubble rising past a small bubble and deforming such that the smaller bubble is entrained into the wake of the large bubble. Additional experimental observations from this system are reported in Manga and Stone (1995b). The articles mentioned here also provide a brief review of the literature on the subject.

Wei and Subramanian (1994) performed experiments on pairs of air bubbles moving in a viscous silicone oil under the combined action of gravity and a downward temperature gradient. In some of these experiments, no temperature gradient was applied, and the bubbles were permitted to rise due to buoyancy in asymmetric configurations. Small bubbles in the range 20 to 140 μm in radius were used. Because the silicone oil was slightly undersaturated with air at room temperature, the experiments were performed at an elevated temperature of 33°C to minimize dissolution of the bubbles. The authors compared the observed velocities with predictions from the method of reflections, as well as those from the method of images. The velocities were generally consistent with the predictions, even though the gap separating the surfaces of the bubbles in some instances was less than 15% of the sum of the radii. But the authors noted that the predictions suggested a greater sensitivity of the velocities to the separation distance than was observed.

CHAPTER SEVEN

Interactions When Motion Is Driven by Thermocapillarity

7.1 Motion Normal to a Plane Surface

7.1.1 Introduction

In this chapter, we consider problems in thermocapillary migration that involve interactions with a neighboring surface or object. The situation to be analyzed in the present section is that of a drop moving away from a plane surface which can be rigid or free and is analogous to that treated in Section 6.1 for body-force-driven motion. Here, we assume zero body force. Instead, referring to Figure 6.1.1, a uniform temperature gradient ∇T_{∞} is imposed in the z-direction in the undisturbed continuous phase fluid and causes the motion of the drop in that direction. The drop, of radius R, is located a distance h from the plane surface. All the assumptions made in Section 6.1 apply here. In addition, we assume that the physical properties are independent of temperature, with the exception of the interfacial tension. As in Chapter Four, the rate of change of interfacial tension with temperature, σ_T, is assumed to be a negative constant. Because the Reynolds number is assumed negligible, for fixed Prandtl number, the Marangoni number also is negligible. Therefore, the problem can be considered quasi-steady. The objective is to determine the scaled migration velocity U of the drop. Lengths are scaled using the reference length c in bispherical coordinates, defined in Section 5.1. Velocity is scaled using $v_0 = -\dfrac{\sigma_T|\nabla T_{\infty}|c}{\mu}$, where μ is the viscosity of the continuous phase. The present problem was analyzed by Barton and Subramanian (1990), and the development given here is based on that article. An independently obtained solution in bispherical coordinates was also given by Chen and Keh (1990). We also report on results from a lubrication analysis for the case when the drop is close to the surface, given by Loewenberg and Davis (1993a).

7.1.2 Analysis

We designate variables within the drop phase by the subscript 1, and those in the continuous phase by the subscript 2. We can write the solution for the scaled streamfunction field in this axisymmetric problem as in Section 6.1. Equations (6.1.1) through (6.1.8) are applicable here. The only change is in the tangential stress balance, which should be modified to account for the discontinuity in tangential stress at the surface of the drop. Therefore, the following equation replaces Equation (6.1.9):

$$\tau_{\xi\eta 2}(\xi_1, \eta) - \tau_{\xi\eta 1}(\xi_1, \eta) = -(\cosh \xi_1 - \zeta)\frac{\partial T_2}{\partial \eta}(\xi_1, \eta). \tag{7.1.1}$$

Here, $\zeta = \cos \eta$, and $T(\xi, \eta)$ represents temperature scaled by subtracting its value T_0 at $z = 0$ and then dividing by the product of the applied temperature gradient in the z-direction and the length scale c. Because the Marangoni number is negligible, the temperature fields satisfy Laplace's equation. The boundary conditions on the temperature fields include continuity of temperature and heat flux across the drop surface. At the plane surface, the scaled temperature is zero. In addition, the temperature must remain bounded within the drop.

It is straightforward to specialize the general solution of Laplace's equation, given in Equation (5.1.14), in the drop phase and in the continuous phase, using the above conditions. The resulting temperature field can be used in the tangential stress balance, so that the arbitrary constants in the solution for the streamfunction fields in Equations (6.1.1) to (6.1.3) can be evaluated. Meyyappan, Wilcox, and Subramanian (1981) analyzed the present problem in the gas bubble limit, $\alpha = \beta = 0$, and used numerical quadrature to evaluate a set of integrals in the solution. Sadhal (1983) showed how this numerical integration can be avoided. For this, the scaled heat flux in each phase $\mathbf{q}_i$ is written as

$$\mathbf{q}_i \equiv \nabla T_i = \nabla \times \left(\frac{\Phi_i}{\omega} \mathbf{i}_\phi \right), \quad i = 1, 2, \tag{7.1.2}$$

where $\mathbf{i}_\phi$ is the unit vector in the ϕ-direction. The components of $\mathbf{q}_i$ are related to derivatives of the functions $\Phi_i(\xi, \eta)$ as follows:

$$q_{i\xi} = (\cosh \xi - \zeta) \frac{\partial T_i}{\partial \xi} = \frac{(\cosh \xi - \zeta)^2}{\sqrt{1 - \zeta^2}} \frac{\partial \Phi_i}{\partial \eta} \tag{7.1.3}$$

and

$$q_{i\eta} = -(\cosh \xi - \zeta) \frac{\partial T_i}{\partial \eta} = -\frac{(\cosh \xi - \zeta)^2}{\sqrt{1 - \zeta^2}} \frac{\partial \Phi_i}{\partial \xi}. \tag{7.1.4}$$

The divergence of the curl of any vector field vanishes identically. Therefore, $\nabla \bullet \mathbf{q}_i = \nabla^2 T_i = 0$, and Laplace's equation is satisfied. To obtain an equation for the functions $\Phi_i(\xi, \eta)$, we note that the curl of the gradient of any scalar field vanishes identically. Therefore, $\nabla \times \mathbf{q}_i = \nabla \times \nabla T_i = \mathbf{0}$. Substituting for $\mathbf{q}_i$ in this result, from Equation (7.1.2), leads to the following equation for Φ_i:

$$E^2 \Phi_i = 0, \quad i = 1, 2. \tag{7.1.5}$$

The operator E^2 was introduced in Chapter Two. The general solution of Equation (7.1.5) in bispherical coordinates was given by Jeffery (1912). A solution that is bounded everywhere can be written as follows:

$$\Phi = \frac{1}{\sqrt{\cosh \xi - \zeta}} \sum_{n=1}^{\infty} \left[A_n e^{(n+\frac{1}{2})\xi} + B_n e^{-(n+\frac{1}{2})\xi} \right] C_{n+1}(\zeta). \tag{7.1.6}$$

The Gegenbauer Polynomials $C_{n+1}(\zeta)$ were introduced in Chapter Three. It now remains to write the boundary conditions on Φ_i. In the half-plane $z \geq 0$, at large distances from the drop, the temperature gradient must approach its uniform value in the undisturbed continuous phase fluid. Because the drop is at a finite distance from the origin, the limit as the distance from the drop approaches infinity is the same as that when the distance from the origin approaches infinity, which is the limit $(\cosh \xi - \zeta) \to 0$. As noted in Section 5.1, this limit can be attained only when $\xi \to 0$ and $\zeta \to 1$ simultaneously. Therefore,

$$\lim_{\xi \to 0, \zeta \to 1} \Phi_2 = \Phi_{2\infty} = \frac{\omega^2}{2} = \frac{1 - \zeta^2}{2(\cosh \xi - \zeta)^2}. \tag{7.1.7}$$

Here, ω represents distance from the z-axis as shown in Figure 5.1.1. Note that Φ_2 grows with this distance from the symmetry axis at large distances, whereas the solution in Equation (7.1.6) approaches zero at the point at infinity. Therefore, we must add $\Phi_{2\infty}$, which itself is a solution of $E^2\Phi = 0$, to the solution given in Equation (7.1.6) when using it for the continuous phase.

At the surface $z = 0$ ($\xi = 0$), the tangential component of the scaled heat flux, $q_{2\eta}$, vanishes. Therefore,

$$\frac{\partial \Phi_2}{\partial \xi}(0, \zeta) = 0. \tag{7.1.8}$$

At the surface of the drop, the temperature and heat flux must be continuous. Continuity of temperature implies continuity of the derivative of the temperature in the η-direction as well. This permits us to write the following conditions:

$$\frac{\partial \Phi_1}{\partial \xi}(\xi_1, \zeta) = \frac{\partial \Phi_2}{\partial \xi}(\xi_1, \zeta) \tag{7.1.9}$$

and

$$\beta \frac{\partial \Phi_1}{\partial \eta}(\xi_1, \zeta) = \frac{\partial \Phi_2}{\partial \eta}(\xi_1, \zeta). \tag{7.1.10}$$

Application of the boundary conditions to the general solution leads to the following final results for the functions Φ_i:

$$\Phi_1(\xi, \eta) = \frac{1}{\sqrt{\cosh \xi - \zeta}} \sum_{n=1}^{\infty} A_n e^{-(n+\frac{1}{2})(\xi - \xi_1)} C_{n+1}(\zeta) \tag{7.1.11}$$

and

$$\Phi_2(\xi, \eta) = \Phi_{2\infty} + \frac{1}{\sqrt{\cosh \xi - \zeta}} \sum_{n=1}^{\infty} \left[B_n \cosh\left(n + \frac{1}{2}\right)\xi \right] C_{n+1}(\zeta). \tag{7.1.12}$$

An additive arbitrary constant that appears in each solution is omitted because only derivatives of Φ_i are used.

Using Equation (7.1.10), the sets of constants A_n can be written in terms of the constants B_n as follows:

$$\beta A_n = B_n \cosh\left(n + \frac{1}{2}\right)\xi_1 + E_n, \tag{7.1.13}$$

where E_n are given by

$$E_n = \sqrt{2}\, n(n+1) e^{-(n+\frac{1}{2})\xi_1}. \tag{7.1.14}$$

The only constants yet to be determined are B_n. To find them, we use Equation (7.1.9), along with Equation (7.1.13). This process yields the following set of equations:

$$-(n+1)F_{n-1}B_{n-1} + \left[(1-\beta)\sinh \xi_1 \cosh\left(n+\frac{1}{2}\right)\xi_1 + (2n+1)F_n \cosh \xi_1\right] B_n$$

$$- nF_{n+1}B_{n+1}$$

$$= (n+1)E_{n-1} - [(2n+1)\cosh \xi_1 + (1-4\beta)\sinh \xi_1]E_n + nE_{n+1}. \qquad (7.1.15)$$

Here, the constants F_n are defined as follows:

$$F_n = \cosh\left(n+\frac{1}{2}\right)\xi_1 + \beta \sinh\left(n+\frac{1}{2}\right)\xi_1. \qquad (7.1.16)$$

Equation (7.1.15) applies for $n = 1, 2, 3, \ldots$ with the convention that $B_0 = 0$. For numerical evaluation, the set of equations must be closed at some large value N by setting the next constant $B_{N+1} = 0$. Because the series in Equation (7.1.12) is a Fourier series in a complete basis set, it is convergent. It is known that $|C_{n+1}|_{max}$ approaches zero as $n \to \infty$ approximately as $n^{-\frac{3}{2}}$. Because the hyperbolic function increases exponentially with n, it follows that the constants B_n must decrease exponentially with n as $n \to \infty$, for sufficiently large values of n.

Using the results obtained for Φ_i above, the solution of Stokes's equation given in Equation (5.1.15) can be specialized to the boundary conditions applicable here, along with the condition that the hydrodynamic force on the drop must be zero. This permits us to obtain results for the migration velocity of the drop. For a rigid plane surface, U can be written as

$$U = \frac{\sum_{n=1}^{\infty} \frac{I_n G_n}{(2n+1)(\Delta_{n3}+\alpha \Delta_{n4})}}{\sqrt{2} \sum_{n=1}^{\infty} \frac{n(n+1)}{(2n-1)(2n+3)}\left(\frac{\Delta_{n1}+\alpha \Delta_{n2}}{\Delta_{n3}+\alpha \Delta_{n4}}\right)}, \qquad (7.1.17)$$

where the constants Δ_{ni}, $i = 1$ to 4, are defined in Equations (6.1.22) through (6.1.25), and the constants G_n are defined below.

$$G_n = \cosh\left(n-\frac{1}{2}\right)\xi_1 - \cosh\left(n+\frac{3}{2}\right)\xi_1. \qquad (7.1.18)$$

The constants I_n are obtained by solving the following set of equations, with the convention $I_0 = 0$:

$$2\left[\frac{n+1}{2n-1}I_{n-1} - \cosh \xi_1 I_n + \frac{n}{2n+3}I_{n+1}\right]$$

$$= 4E_n \sinh \xi_1 + (n+1)B_{n-1}\sinh\left(n-\frac{1}{2}\right)\xi_1$$

$$+ B_n\left[\sinh \xi_1 \cosh\left(n+\frac{1}{2}\right)\xi_1 - (2n+1)\cosh \xi_1 \sinh\left(n+\frac{1}{2}\right)\xi_1\right]$$

$$+ nB_{n+1}\sinh\left(n+\frac{3}{2}\right)\xi_1. \qquad (7.1.19)$$

As in the case of the set of equations for B_n, this set for I_n also must be closed at some large N by choosing $I_{N+1} = 0$. The constants I_n are related to the scaled temperature distribution at the drop surface through the following expansion:

$$\sqrt{\frac{1-\zeta^2}{\cosh\xi_1-\zeta}}\frac{\partial T_2}{\partial\eta}(\xi_1,\eta)=\sqrt{\cosh\xi_1-\zeta}\frac{\partial\Phi_2}{\partial\xi}(\xi_1,\eta)=\sum_{n=1}^{\infty}I_nC_{n+1}(\zeta). \qquad (7.1.20)$$

It is the set of constants I_n that Meyyappan et al. (1981) evaluated by numerical quadrature. Because the Fourier series in Equation (7.1.20) must converge, we can expect $|I_nC_{n+1}|$ to approach 0 as $n\to\infty$. Because the magnitude of the Gegenbauer Polynomials approaches 0 as $n\to\infty$, however, there is no guarantee that $I_n\to 0$ in this limit. The fact that the constant I_n decreases in magnitude as n increases was established by Barton and Subramanian (1990) numerically. In the gas bubble limit $\beta\to 0$, it is possible to write the following expression for I_n:

$$I_n=-\frac{n(n+1)(2n+1)}{\sqrt{2}\cosh\left(n+\frac{1}{2}\right)\xi_1}. \qquad (7.1.21)$$

The above result illustrates that for the gas bubble case, $|I_n|$ decays exponentially with increasing n for sufficiently large values of n. Barton and Subramanian found that indeed $|I_n|$ decayed with increasing n for nonzero values of β and did so more rapidly than in the gas bubble limit.

Results for the scaled migration velocity U in the case of a free plane surface can be calculated from Equation (7.1.17) by using the expressions for Δ_{ni}, $i=1$ to 4, given in Equations (6.1.26) through (6.1.29) for that case, with the following substitution for G_n:

$$G_n=\sinh\left(n-\frac{1}{2}\right)\xi_1-\sinh\left(n+\frac{3}{2}\right)\xi_1. \qquad (7.1.22)$$

7.1.3 Some Results from the Analysis

An interaction parameter Ω is defined as in Equation (6.1.30) in the body force driven motion problem. It is the ratio of the velocity of the drop in the presence of the plane surface to that when isolated. The former is obtained from Equation (7.1.17), and the latter obtained from Equation (4.2.26), can be written as $\frac{2}{(2+3\alpha)(2+\beta)\sinh\xi_1}$, when scaled with the reference velocity used in this section. The list of parameters on which Ω depends includes the scaled distance of the drop from the plane surface, defined as $H=\frac{h}{R}$, and the property ratios α and β. Values of Ω as a function of H in the gas bubble limit, $\alpha=\beta=0$, were tabulated and discussed in Meyyappan et al. (1981). Results provided by Barton and Subramanian (1990) for representative sets of values of α and β, including the gas bubble limit, are reproduced here in Tables 7.1.1 and 7.1.2.

Figure 7.1.1 illustrates the behavior of Ω plotted against the scaled separation distance H, for values of the thermal conductivity ratio β spanning a good range when the viscosity ratio $\alpha=1$. The case of a rigid plane surface is shown in Figure 7.1.1(a); that of the free surface is shown in Figure 7.1.1(b). As observed in the body-force-driven motion problem, a rigid plane surface offers more resistance to the motion of a nearby drop than a free plane surface. Two important features unique to the thermocapillary migration problem can be discerned from Figure 7.1.1. First, note that the interaction parameter is practically at its asymptotic value of unity in all cases when the drop is located at a distance of approximately 3 radii from the surface. This is in contrast to the cases illustrated in Figure 6.1.2 for the body force driven motion problem

Table 7.1.1 Interaction Parameter, Ω, at Selected Values of H, α, and β, for a Drop Moving Normal to a Rigid Plane Surface

		β				
H	α	0	0.1	1	10	100
1.00125	0	0.00704	0.00726	0.00909	0.02066	0.04339
1.03141		0.14905	0.15289	0.18207	0.28611	0.34942
1.12763		0.42578	0.43308	0.48303	0.60659	0.65629
1.29468		0.65955	0.66621	0.70889	0.79707	0.82706
1.54308		0.81317	0.81764	0.84527	0.89754	0.91401
2.35241		0.95076	0.95211	0.96028	0.97481	0.97915
3.76220		0.98819	0.98853	0.99053	0.99406	0.99511
1.00125	0.1	0.00768	0.00792	0.00996	0.02291	0.04849
1.03141		0.15570	0.15978	0.19078	0.30141	0.36879
1.12763		0.43303	0.44055	0.49195	0.61899	0.67007
1.29468		0.66376	0.67053	0.71382	0.80315	0.83352
1.54308		0.81511	0.81961	0.84744	0.90005	0.91661
2.35241		0.95105	0.95240	0.96058	0.97514	0.97948
3.76220		0.98822	0.98856	0.99057	0.99410	0.99514
1.00125	1	0.01144	0.01185	0.01531	0.03817	0.08482
1.03141		0.18219	0.18745	0.22743	0.37118	0.45917
1.12763		0.45651	0.46497	0.52270	0.66480	0.72178
1.29468		0.67654	0.68373	0.72962	0.82381	0.85572
1.54308		0.82095	0.82560	0.85429	0.90831	0.92527
2.35241		0.95196	0.95333	0.96158	0.97623	0.98060
3.76220		0.98833	0.98867	0.99068	0.99422	0.99526
1.00125	10	0.01630	0.01707	0.02379	0.07489	0.19201
1.03141		0.19170	0.19824	0.24843	0.43237	0.54639
1.12763		0.46254	0.47196	0.53605	0.69298	0.75572
1.29468		0.68037	0.68799	0.73645	0.83528	0.86862
1.54308		0.82315	0.82795	0.85749	0.91286	0.93019
2.35241		0.95243	0.95381	0.96213	0.97688	0.98127
3.76220		0.98840	0.98873	0.99075	0.99429	0.99534
1.00125	100	0.01454	0.01534	0.02257	0.08428	0.24041
1.03141		0.18774	0.19452	0.24668	0.43946	0.55958
1.12763		0.46095	0.47057	0.53598	0.69595	0.75985
1.29468		0.68010	0.68781	0.73680	0.83655	0.87018
1.54308		0.82324	0.82807	0.85778	0.91341	0.93082
2.35241		0.95249	0.95387	0.96220	0.97697	0.98137
3.76220		0.98841	0.98874	0.99076	0.99431	0.99535

where, even at 10 drop radii, the surface exerts a significant influence. The difference can be attributed to the nature of the decay of the disturbance velocity field due to the motion of the drop. As mentioned in Section 3.1, in the absence of the plane surface, the disturbance flow due to a drop moving because of a body force decays as $\frac{1}{r}$, where r is distance measured from the center of the drop. In contrast, when the hydrodynamic force on the drop is zero, this slow decay is absent. We noted in Section 4.2 that, when convective transport effects are negligible, an isolated drop moving due to the

Table 7.1.2 Interaction Parameter, Ω, at Selected Values of H, α, and β, for a Drop Moving Normal to a Free Plane Surface

H	α	β				
		0	0.1	1	10	100
1.00125	0	0.38521	0.39611	0.48654	1.04875	2.13857
1.03141		0.60447	0.61869	0.72655	1.10994	1.34249
1.12763		0.76312	0.77522	0.85826	1.06486	1.14823
1.29468		0.86425	0.87248	0.92543	1.03567	1.07335
1.54308		0.92595	0.93083	0.96119	1.01898	1.03726
2.35241		0.98039	0.98177	0.99009	1.00493	1.00937
3.76220		0.99528	0.99562	0.99764	1.00118	1.00223
1.00125	0.1	0.29397	0.30249	0.37333	0.81678	1.68181
1.03141		0.57806	0.59192	0.69714	1.07158	1.29890
1.12763		0.75461	0.76674	0.84992	1.05666	1.14004
1.29468		0.86157	0.86984	0.92303	1.03367	1.07146
1.54308		0.92516	0.93007	0.96053	1.01849	1.03681
2.35241		0.98034	0.98172	0.99005	1.00490	1.00935
3.76220		0.99528	0.99562	0.99763	1.00118	1.00223
1.00125	1	0.13056	0.13492	0.17166	0.41316	0.90461
1.03141		0.47501	0.48771	0.58430	0.93083	1.14250
1.12763		0.71675	0.72909	0.81354	1.02244	1.10645
1.29468		0.84941	0.85791	0.91239	1.02507	1.06343
1.54308		0.92163	0.92664	0.95762	1.01635	1.03487
2.35241		0.98010	0.98148	0.98986	1.00478	1.00924
3.76220		0.99527	0.99560	0.99762	1.00117	1.00222
1.00125	10	0.05095	0.05324	0.07333	0.22705	0.58267
1.03141		0.37009	0.38183	0.47184	0.80083	1.00434
1.12763		0.67490	0.68762	0.77435	0.98775	1.07330
1.29468		0.83639	0.84516	0.90121	1.01644	1.05552
1.54308		0.91797	0.92308	0.95464	1.01423	1.03298
2.35241		0.97986	0.98125	0.98967	1.00466	1.00913
3.76220		0.99525	0.99559	0.99761	1.00117	1.00222
1.00125	100	0.03214	0.03382	0.04894	0.17761	0.50307
1.03141		0.34505	0.35656	0.44508	0.77085	0.97323
1.12763		0.66551	0.67832	0.76565	0.98025	1.06621
1.29468		0.83362	0.84246	0.89886	1.01466	1.05390
1.54308		0.91722	0.92235	0.95403	1.01381	1.03260
2.35241		0.97981	0.98120	0.98964	1.00463	1.00911
3.76220		0.99525	0.99559	0.99761	1.00117	1.00221

action of a temperature gradient produces a disturbance flow that decays as $\frac{1}{r^3}$. This is the principal reason why the plane surface has to be close to the drop to influence its motion due to thermocapillarity. Of course, we should also investigate the conditions when the plane surface will disturb the temperature gradient field significantly because this affects the driving force for motion in the present problem. In contrast, for a drop moving because of gravity, the driving force is the hydrostatic force on it, which is not

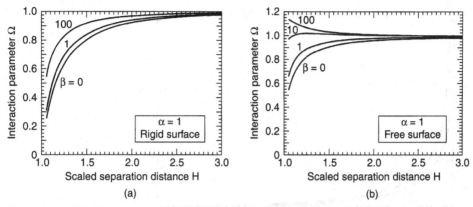

Figure 7.1.1 Interaction parameter for thermocapillary migration of a drop normal to a plane surface plotted against scaled separation distance for a single viscosity ratio and different values of the thermal conductivity ratio: (a) rigid surface, (b) free surface (Courtesy of Academic Press).

affected by the presence of the plane surface. For an isolated drop placed in a uniform temperature gradient field, in the pure conduction limit, the disturbance to the temperature field caused by the presence of the drop decays as $\frac{1}{r^2}$, and its spatial gradient decays as $\frac{1}{r^3}$. Therefore, it is clear that the driving force for thermocapillary migration also is significantly affected only when the plane surface is close.

The above discussion of the temperature gradient brings us to the second remarkable feature displayed in Figure 7.1.1(b). When the thermal conductivity of the drop is larger than that of the continuous phase fluid, we see, both from this drawing and the results reported in Table 7.1.2, that it is possible for a drop to move more rapidly near a surface than when it is isolated. The reason, of course, is that the driving force for motion is enhanced by the presence of the surface, which distorts the isotherms. Such enhancement is independent of whether the surface is rigid or free. The added resistance to motion from the rigid plane surface appears to overwhelm the increase in driving force in that case, however, leading to values of Ω less than unity even when $\beta = 100$. The ratio, λ, of the temperature difference between the poles of the drop (farthest from and closest to the plane surface) to that which would exist in the absence of the plane surface is displayed against the scaled separation distance in Figure 7.1.2. This clearly illustrates the enhancement of the temperature difference when the thermal conductivity of the drop exceeds that of the continuous phase.

Chen (1999a) recently has obtained an approximate solution of this problem by using the method of reflections. The result is a series for the interaction parameter Ω in inverse powers of H, the leading term being unity. The first correction occurs at $O(H^{-3})$. The coefficient of this correction only depends on the thermal conductivity ratio β. The viscosity ratio α influences the interaction parameter only at $O(H^{-5})$ in the case of a rigid surface and at $O(H^{-6})$ in the case of a free plane surface. Based on a comparison of the solution for the interaction parameter from the method of reflections with that from the solution in bispherical coordinates, Chen established the adequacy of the former as long as the separation distance is not too small. In the case of the rigid plane surface, the first correction in the series for Ω at $O(H^{-3})$ is always negative, whereas in the case of the free surface, it becomes positive when the thermal conductivity ratio β exceeds 4. Of course, the precise value of β at which the interaction parameter is greater than unity in this case will depend on the viscosity ratio as well as the value of H.

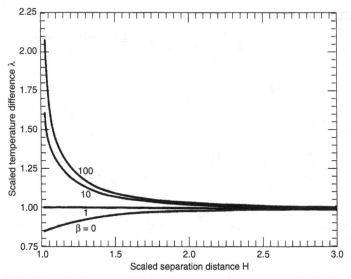

Figure 7.1.2 Temperature difference over the drop surface scaled by the corresponding difference for an isolated drop plotted against scaled separation distance for a set of values of the thermal conductivity ratio (Courtesy of Academic Press).

In Figure 7.1.3, a comparison is made between body force driven motion and thermocapillary migration, using representative streamline drawings for the case $\alpha = \beta = 1$ for a rigid plane surface. The drawings show the drop moving away from the plane surface. Because Stokes flows are reversible, however, the same structure would be seen when the drop moves toward the surface, with all the arrows reversed. The reference

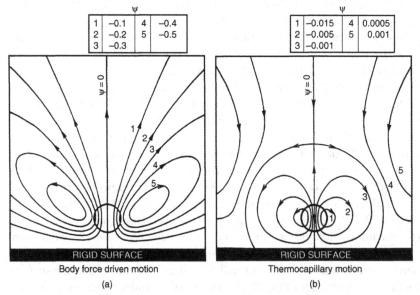

Figure 7.1.3 Streamlines in the laboratory reference frame for a drop moving away from a rigid plane surface for $H = 2.5$ and $\alpha = 1$: (a) body-force-driven motion, (b) thermocapillary motion, $\beta = 1$.

for scaling the streamfunction in Figure 7.1.3(a) is $U^* R^2$ where U^* is the magnitude of the steady rise velocity of an isolated drop due to gravity, given in Equation (3.1.33). In Figure 7.1.3(b) the streamfunction is scaled using $-\frac{\sigma_T |\nabla T_\infty| R^3}{\mu}$. For clarity of display, the streamlines do not correspond to equal increments of the streamfunction in Figure 7.1.3(b). The most noteworthy feature in the case of thermocapillary motion is the appearance of a region of reverse flow where fluid is approaching the plane surface while the drop is moving away from it. Such a flow structure is absent in the gravitational rise problem. The explanation for the complex flow structure in the thermocapillary case is similar to that given in Section 4.17. For an isolated drop under these conditions, the surface temperature field is proportional to $P_1(\cos\theta)$, where θ is the polar angle in a spherical coordinate system with its origin at the center of the drop. When this is disturbed by the presence of a neighboring surface in this case, the higher Legendre modes of the surface temperature field are excited. Because the flow driven by the P_2-mode decays away from the drop most slowly, it determines the nature of the pattern of flow far from the drop, which is opposite in direction in this case from that driven by the P_1-mode at such locations. The resulting opposition of flows produces a dividing streamline (separatrix) and the reverse flow pattern seen in the drawing. Although not shown here, the separatrix appears farther from the drop in the case of the free plane surface, indicating a weaker relative contribution from the P_2-mode, when compared with the case of a rigid plane surface. The above explanation is somewhat simplified. In the presence of the plane surface, each pure mode of the surface temperature will actually drive a mixture of flow modes, unlike in the case of an isolated drop. Nonetheless, the explanation given here is adequate for our purposes. The locations at which the separatrix intersects the axis of symmetry and the plane surface are tabulated by Chen and Keh (1990) for a range of values of the parameters.

As mentioned earlier, the results of this section apply equally well when the drop moves toward the surface. This will not hold, however, when convective transport of either momentum or energy becomes sufficiently important. No results are available for such problems at this time.

7.1.4 Some Related Problems

It is important to comment on the role of drop deformation in influencing the results because the analysis given here ignores such deformation. The extent of deformation is controlled by the magnitude of the Capillary number, which was defined in Section 4.1. The problem of a drop moving toward a rigid plane surface, subject to the same assumptions used here but permitting the drop to deform, was solved by Ascoli and Leal (1990). These authors used the boundary integral formulation, mentioned briefly in Section 3.2, to numerically obtain the shape of the drop and its velocity. The gas bubble limit was approximated by using $\alpha = \beta = 0.01$, and other sets of values of these parameters also were considered. Illustrations of the gradual deformation of a bubble as it approaches a surface at two different values of the Capillary number are given. The surface of an initially spherical bubble or drop facing the rigid plane is found to flatten in shape as it approaches the plane. Also, the back of the drop farthest from the plane is blunted in shape, even though this does not occur to the same extent as in the forward segment. The authors indicated that when the drop is close to the plane, the velocity of the flattened surface facing the plane is smaller than that of regions of the surface on the side and back. As a consequence, the center of mass moves toward the plane more rapidly than that of an equivalent spherical drop. This happens only when the drop is close to the surface. A drop that is sufficiently separated from the plane surface moves at a slightly smaller velocity than that of an equivalent spherical drop. For the range of parameters considered by the authors, the transition appears to occur when the center of mass of the deformed drop is approximately 1.3 equivalent spherical radii from the plane. Ascoli and Leal also provided the maximum temperature difference over the surface of the drop as a function of the distance from the plane surface for representative values of the parameters.

Guelcher et al. (1998) studied an interesting situation that occurs in electrolysis. Oxygen bubbles that evolve at an electrode during the electrolysis of a potassium hydroxide solution were found to be attracted to each other. The authors noted that a temperature gradient normal to the plane surface arises naturally in this system and used a solution in bispherical coordinates for a bubble that is held stationary a small distance away from a planar surface for making predictions. The resulting thermocapillary flow brings neighboring bubbles toward the stationary bubble. Furthermore, the stationary bubble alters the local temperature gradient field in such a way as to add a thermocapillary migration velocity in the right direction. Another application that involves interactions is explored by Leshansky, Golovin, and Nir (1997), who considered the motion of a rigid spherical particle that is at a different temperature from the surrounding fluid when the particle is present near a fluid interface. The temperature difference between the particle and the fluid produces temperature variations in the fluid. The resulting thermocapillary stress at the fluid interface drives a flow in the fluid which drags the particle along with it. As a consequence, the particle will move toward the surface if it is warmer than the fluid, provided σ_T is negative.

7.1.5 The Case When the Drop Is Very Close to the Plane Surface

It becomes progressively more difficult to calculate results from the solution in bispherical coordinates as the scaled distance between the drop and the plane surface is reduced. Therefore, Loewenberg and Davis (1993a) analyzed this problem in the limit as the gap between the surface of the drop and the plane surface approaches zero. The authors considered a nonconducting spherical drop approaching a plane surface, so that $\beta = 0$ in their analysis. They permitted a second fluid to be present on the other side of the plane surface. When the scaled width of the gap, $\epsilon = H - 1$, satisfies $\epsilon \ll 1$, the solutions of Laplace's and Stokes's equations for a drop in contact with the plane surface yield adequate first approximations of the fields everywhere, except in the gap. These are obtained using tangent-sphere coordinates, which are described in Moon and Spencer (1971). Overall, the drop must move at zero hydrodynamic force. The authors expressed this as the sum of equal and opposite contributions, termed the thermocapillary contact force and the viscous force. The former is the force that would act on a stationary drop in the same configuration, and the latter is the hydrodynamic force on a drop moving at the same velocity in an isothermal fluid. This decomposition is possible because of the linearity of the governing equations. The hydrodynamic force is equal to the lubrication resistance in the gap, to within a correction of $O(\epsilon)$. The term *lubrication resistance* refers to the hydrodynamic force calculated by accommodating the flow in the gap through the method of matched asymptotic expansions. An example of this type of analysis can be found in Cooley and O'Neill (1969). Loewenberg and Davis showed that the temperature gradient on the surface of the drop in the gap region is negligible compared with the applied temperature gradient. In fact, the authors found that the ratio of the two gradients is, at most, of $O(\exp\{-\frac{1}{\sqrt{\epsilon}}\})$. This permitted them to use the results for the lubrication resistance from the isothermal case in the thermocapillary migration problem. The results for the lubrication resistance depend on the values of two parameters m and m_I. These are termed *mobilities* by Davis et al. (1989), and are defined as follows:

$$m = \frac{1}{\alpha\sqrt{\epsilon}} \tag{7.1.23}$$

and

$$m_I = \frac{1}{\alpha_I\sqrt{\epsilon}}. \tag{7.1.24}$$

Here, m stands for the mobility of the interface between the drop and the continuous phase, and m_I represents a similar quantity at the planar interface. The symbol α_I is used to designate the ratio of the viscosity of the second fluid on the other side of the interface to that of the fluid in which the drop is present. The situation where the plane surface is rigid corresponds to $m_I = 0$.

We use the symbol F_L to designate the lubrication resistance, scaled by μRU^*, where U^* is the velocity of the drop. Loewenberg and Davis attributed the asymptotic results they report for this force in the isothermal case to Zinchenko (1983), Davis et al. (1989), and Barnocky and Davis (1989). We reproduce these below.

$$
\begin{aligned}
F_L &= \frac{3\pi}{2\epsilon}\left[1 + 3f\left(\frac{2}{3}m\right)\right], \quad m_I \ll 1, \\[6pt]
&= \frac{6\pi}{\epsilon}f(m), \quad m_I = m \ll -\frac{1}{\epsilon\log\epsilon}, \\[6pt]
&= 6\pi\left[\frac{\pi^2}{16}\alpha\sqrt{\frac{2}{\epsilon}} - \frac{3-\alpha^2}{9}\log\epsilon + C_1(\alpha)\right], \quad m_I = m \gg 1, \\[6pt]
&= \frac{3\pi}{2\epsilon}f\left(\frac{1}{2}m\right), \quad m \ll -\frac{1}{\epsilon\log\epsilon} \ll m_I, \\[6pt]
&= 6\pi\left[\frac{\pi^2}{32}\alpha\sqrt{\frac{2}{\epsilon}} - \frac{3-\alpha^2}{9}\log\epsilon + C_2(\alpha)\right], \quad m \gg 1,\ m_I \gg -\frac{1}{\epsilon\log\epsilon}. \tag{7.1.25}
\end{aligned}
$$

The function f in the above results is given by Davis et al. (1989) and, with the constants slightly modified, appears in the following form in Loewenberg and Davis (1993a):

$$f(x) = \frac{1 + 0.402x}{1 + 1.711x + 0.461x^2}. \tag{7.1.26}$$

The constants C_1 and C_2 are of $O(1)$ and are obtained from an integral expression reported in Zinchenko (1983). In the gas bubble limit, $\alpha = 0$, $C_1(0) = C_2(0) \approx 0.616$.

The thermocapillary contact force, which balances the lubrication resistance, is evaluated from the solution for a stationary drop, obtained by using tangent-sphere coordinates. By using this approach, Loewenberg and Davis arrived at a result for the contact force, scaled with the thermocapillary force on an isolated stationary drop, which is

$-\frac{2\pi R^2 \sigma_T |\nabla T_\infty|}{1+\alpha}$. We designate this scaled contact force as F_C. The result for F_C is given below.

$$F_C = (1+\alpha) \int_0^\infty \frac{1 + \alpha_I \tanh x}{(\alpha + \alpha_I)(\cosh x \sinh x - x) + (1 + \alpha\alpha_I)\sinh^2 x - \alpha\alpha_I x^2} x^2 \, dx.$$

$$(7.1.27)$$

The integral in Equation (7.1.27) has to be evaluated numerically in the general case. The authors obtained several asymptotic results, however, that are valid in limiting situations involving the viscosity ratios α and α_I. These are given below.

$$F_C = \frac{\pi^2}{6} + O(\alpha), \quad \alpha, \alpha_I \ll 1,$$

$$= 2.135619\ldots + O(\alpha), \quad \alpha \ll 1, \quad \alpha_I \gg 1,$$

$$= \frac{3}{2}\log\alpha + \log 2 - \frac{1}{2} + O\left(\frac{1}{\alpha}\right), \quad \alpha \gg 1, \quad \alpha_I \ll 1,$$

$$= 3\log\alpha - 2.619\ldots + O\left(\frac{1}{\alpha}\right), \quad \alpha = \alpha_I \gg 1,$$

$$= 3\log\alpha - 2\log 2 + O\left(\frac{1}{\alpha}\right), \quad \alpha_I \gg \alpha \gg 1. \qquad (7.1.28)$$

Loewenberg and Davis provided plots of the scaled contact force, F_C, as a function of the viscosity ratio, α, for $\alpha_I = 0$, corresponding to a free plane surface, $\alpha_I \to \infty$, representing a rigid plane surface and an intermediate situation wherein $\alpha_I = \alpha$. In all cases, the contact force is larger than the thermocapillary force on an isolated drop. It is relatively insensitive to changes in α when $\alpha \le O(1)$ but increases proportionally to $\log\alpha$ for $\alpha \gg 1$, as can be seen form the asymptotic results given above.

By equating the contact force with the lubrication resistance, Loewenberg and Davis obtained results for the scaled velocity of the drop, which we have defined as the interaction parameter, Ω, against the scaled gap width, ϵ, for a wide range of values of the viscosity ratio α, in the three cases $\alpha_I = 0, \alpha$, and ∞. The most important observation made by the authors is that the thermocapillary migration process is more efficient than motion due to a body force in draining the fluid in the gap. For body-force-driven motion, the driving force is the same regardless of the relative distance between the drop and the plane surface. In the thermocapillary migration case, however, the thermocapillary contact force is larger than the thermocapillary force on an isolated stationary drop, at least in the case of the nonconducting drop analyzed by the authors. As a result, the interaction parameter Ω, or equivalently, the scaled migration velocity of the drop, is larger in the thermocapillary migration case than that for body-force-driven motion, for all values of ϵ. The physical explanation for the behavior of the interaction parameter is based on the traction exerted by the interfacial fluid on the neighboring continuous phase fluid around the drop surface. The resulting flow in the interfacial region helps to drain the gap more effectively than in the corresponding case of motion driven by a body force. The flow patterns can be seen from Figure 7.1.3 upon reversing the direction of the arrows because the drop is moving toward the surface.

We note that if the interface of the drop and that of a fluid-fluid plane surface are not allowed to deform, drainage of the liquid in the gap will occur in a finite amount of time. This is because the mobilities of both the drop surface and the plane interface, defined in Equations (7.1.23) and (7.1.24), respectively, will eventually become large as the gap width approaches zero, even if they are small initially. The consequence is that the lubrication resistance will ultimately scale as $\frac{1}{\sqrt{\epsilon}}$, as can be seen from Equation (7.1.25). This implies drainage in a finite interval of time and is similar to the behavior in the case of body force driven motion, discussed in Section 6.1.2. Drainage due to thermocapillarity will be more efficient for the reason mentioned above. When the plane surface is rigid, the lubrication resistance scales as $\frac{1}{\epsilon}$, and complete drainage cannot be accomplished in a finite amount of time without invoking nonhydrodynamic forces. As noted in Section 6.1.2, when the interfaces are allowed to deform, the hydrodynamic drainage time will become unbounded, but nonhydrodynamic forces can cause the rupture of the liquid film in the gap in a finite amount of time.

Loewenberg and Davis also noted an interesting feature about the dependence of the thermocapillary migration velocity of the drop on the viscosity ratio α. In the case of a free plane surface and in the intermediate situation, wherein a fluid of the same viscosity as that of the drop is present on the other side of the interface, the scaled velocity decreases with increasing α at sufficiently small values of the scaled gap width, ϵ. At large values of α, one can see a reversal in this behavior at some intermediate gap width. The results in the case of a rigid plane surface show a different trend. Here, the tendency is for the scaled velocity to increase with increasing values of α, at sufficiently small gap widths. This trend is seen for small to large α but is reversed when α increases from 100 to 1000. Also, a reversal of the behavior is seen as the gap width increases, in the same range of α values. Some of these features in the behavior of the interaction parameter can be discerned from the results given here in Tables 7.1.1 and 7.1.2, even when $\beta \neq 0$.

7.1.6 Comparison of Theoretical Predictions with Experimental Data

Experiments on the interaction between a horizontal rigid plane surface and air bubbles moving in three Dow-Corning silicone oils, of nominal viscosity 200, 500, and 1000 centistokes, were performed by Merritt (1988) in the same apparatus in which he studied the motion of isolated bubbles under the combined action of gravity and a downward temperature gradient. His work on isolated bubbles is mentioned in Section 4.16. The results on bubbles interacting with a plane surface are reported in Merritt and Subramanian (1989). Air bubbles of radius ranging from 30 to 150 μm were injected into the liquid about 10 radii away from the lower surface of the cell after establishing a downward temperature gradient ranging from 5.7 to 10.4 K/mm in the liquid. The Reynolds and Marangoni numbers were very small. Therefore, convective transport effects can be considered negligible in these experiments. When introduced into the liquid, the bubbles moved downward toward the bottom surface. Only the latter part of the trajectory was used for obtaining the reported results to avoid interaction effects with the tip of the injector, as well as to allow time for the disturbances from the injection process to dissipate. The bubbles grew as they moved toward the bottom of the cell because of the supersaturated nature of the liquid in the vicinity of the bubbles. In other experiments, bubbles were injected near the top horizontal surface of the cell, and they came to rest at the surface while shrinking in size. When they became sufficiently small, they moved downward. After this point, data were also gathered from these downward trajectories. The authors considered the motion quasi-steady and used local physical properties and the instantaneous size at a given location in interpreting the data. They extracted results for the thermocapillary contribution to the migration velocity from the data, assuming the buoyancy driven contribution to be in accord with the predictions given in Section 6.1. This allowed them to plot this result in the form of the interaction parameter against scaled distance and verify that indeed the thermocapillary contribution was in excellent agreement with the predictions reported in the present section. In particular, they showed that the thermocapillary contribution was reduced only when the bubble was within approximately three radii from the surface. The bubble velocity under the combined influence of thermocapillarity and buoyancy was found to be in excellent agreement with the prediction obtained from the solution in bispherical coordinates. Merritt and Subramanian also showed an example of a bubble present near the upper surface that moved in a direction opposite to that which would be anticipated for an isolated bubble. This bubble moved downward with a velocity of 0.5 μm/s, whereas it would have risen at a velocity of 35 μm/s if isolated. They correctly attributed this to the stronger role played by the surface in retarding the upward motion due to buoyancy, in contrast to that in retarding the thermocapillary drift, which would be downward. We comment further on this point in the context of Barton's experiments on liquid drops.

Barton (1990), whose work on drops of ethyl salicylate in diethylene glycol was mentioned in Sections 4.16 and 6.1, also made observations on interaction effects with the upper horizontal rigid plane boundary in his cell. The drops were slightly more dense than the continuous phase, and sank under isothermal conditions. In the presence of an upward temperature gradient, the drops rose instead toward the upper surface. Convective transport effects were negligible in these experiments. By adjusting the conditions, Barton was able to study virtually pure thermocapillary migration of the drops toward the upper horizontal surface in some experiments. For this purpose, drops ranging from 40 to 160 μm in radius were used in an upward temperature gradient of 6.5 K/mm. Under these conditions, isolated drops were predicted to move at a velocity of 100 to 350 μm/s. The data on nine drops, reported in Barton and Subramanian (1990), are displayed in Figure 7.1.4, along with predictions from the solution given here in bispherical coordinates. During the period of observation, the drops changed slightly in size. The largest drop decreased by 1.2% in radius, whereas a drop of intermediate size shrank by 4.1%, and the smallest drop decreased 11.8% in radius. At each location of a drop, the velocity of an equivalent isolated drop was obtained by using the instantaneous size and the local viscosity and density of each phase. Even though the gravitational contribution to the motion was minimal, it was included. As a result, the theoretical prediction for each drop falls on a separate curve. To avoid cluttering, only the predictions for the largest and the smallest drops are shown in the figure. Predictions for the remaining drops fall between these two curves.

The data in Figure 7.1.4 for the larger drops are consistent with the prediction to within the measurement uncertainty. A trend is evident, however, especially in the case of the smaller drops. Measured velocities are lower than those predicted, or the velocity of an equivalent isolated drop is overestimated. The authors suggested that slight contamination by surface active chemicals could have caused the observed trend. The smaller drops dissolved much more than the larger drops, so that any trace contaminant present on their surface would become more concentrated as the drops moved toward the plane surface. In any case, the figure shows that the drops behave as though they are isolated, except when are within approximately three radii of the plane surface. Also, the theory predicts the trend in the data correctly.

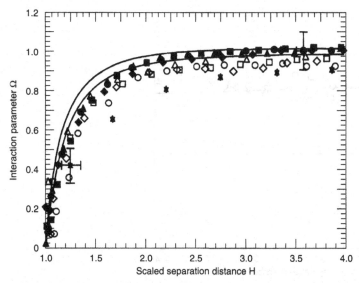

Figure 7.1.4 Data on the interaction parameter, for drops of ethylsalicylate moving vertically upward in diethylene glycol toward a horizontal surface under the influence of a temperature gradient $\nabla T_\infty = 6.5$ K/mm, plotted against scaled separation distance. Data on nine drops, each represented by a different symbol, are displayed. Filled triangles represent data from the largest drop of radius 163 μm, and filled stars correspond to data from the smallest drop of radius 43 μm. The upper curve represents the theoretical prediction for the largest drop, and the lower curve represents that for the smallest drop (Courtesy of Academic Press).

Barton and Subramanian also presented an interesting observation regarding drops for which the influence of gravity is comparable to that of thermocapillarity. For this experiment, they used a more gentle temperature gradient of 1.3 K/mm. Data on three drops, each of average radius 168 μm, are presented in Figure 7.1.5. The data are seen to be consistent with the theoretical prediction, which is included in the drawing.

The most remarkable feature of Figure 7.1.5 is the fact that the interaction parameter is above unity during most of the traverse. This implies that the drops are migrating at velocities greater than that of an equivalent isolated drop. Intuition suggests that the plane surface should retard the drop, and we saw that this is indeed the case in the experimental results in Figure 7.1.4, as well as those shown in Section 6.1 for motion driven purely by gravity. Yet when the two driving forces act in conjunction, it appears possible that a drop can move more rapidly in the presence of a neighboring surface than when isolated. The explanation is simple. In this linear problem, the rise velocity of these drops can be considered as the vector sum of the downward settling velocity caused by gravity and the upward velocity caused by thermocapillarity. Each contribution is reduced by the plane surface as the drop approaches it. We already know that the effect of the plane surface extends much deeper into the liquid in the case of motion due to gravity. Therefore, as the drop moves to within a few radii of the plane surface, the downward settling velocity is reduced substantially, whereas the upward velocity due to thermocapillarity would be virtually unaffected until the drop comes close to the surface. Therefore, at such intermediate locations, the net velocity upward should be larger than that of an isolated drop. Of course, when the drop is near the surface, both

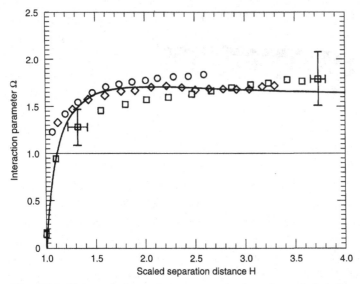

Figure 7.1.5 Data on the interaction parameter for three drops of ethylsalicy-late moving vertically upward in diethylene glycol toward a horizontal surface under the influence of a temperature gradient $\nabla T_\infty = 1.3\,\text{K/mm}$, plotted against scaled separation distance. All three drops have the same radius of $168\,\mu\text{m}$, and the curve represents the theoretical prediction (Courtesy of Academic Press).

contributions are strongly retarded, and the interaction parameter is seen to lie below unity in that case. As noted earlier, Merritt and Subramanian (1989) had observed the same phenomenon in their work with air bubbles in silicone oils in a different context, where they showed that the effect is sufficiently strong to reverse the direction of motion of a bubble.

7.2 Motion of a Bubble in an Arbitrary Direction Near a Rigid Plane Surface

7.2.1 Introduction

In this section, we consider the effect of a rigid plane surface on the motion of a nearby bubble, which occurs because of a temperature gradient oriented in an arbitrary direction in the absence of any body force. In the bubble limit, the ratio of the viscosity of a drop to that of the continuous phase, α, and a similar ratio of thermal conductivities, β, are both set equal to zero. This permits us to obtain results by considering the transport problem only in the continuous phase, which provides a substantial simplification in algebraic detail. The method employed can be extended in a straightforward manner to the case when either of these ratios is nonzero, as well as to plane fluid interfaces. In the case of a plane interface with another fluid, because the temperature is no longer constant along that interface, thermocapillary flow will arise in the fluid body from the resulting interfacial tension gradient. This does not happen in the case of motion perpendicular to such an interface, considered in Section 7.1, because the plane fluid interface in that section is isothermal. Our development here follows that of Meyyappan and Subramanian (1987). All the physical assumptions employed in Section 7.1 continue to hold, with the addition of those noted above regarding the values of α and β.

7.2.2 Analysis

We can continue to use Figure 6.1.1 for reference purposes. We assume the bubble has a radius R and that it is located at a distance h from the plane surface. Let $\mathbf{w}$ be a unit vector in some arbitrary direction. The applied temperature gradient is given by $\nabla T_\infty = |\nabla T_\infty| \mathbf{w}$. For consistency, the projected undisturbed temperature gradient along the plane surface is used to define the boundary condition on the temperature distribution along that surface. The bubble will not, in general, move in the direction $\mathbf{w}$, the two exceptions being when $\mathbf{w}$ is oriented either perpendicular or parallel to the plane surface. Lengths are scaled with the characteristic distance, c, in bispherical coordinates. Velocities are scaled using $-\frac{\sigma_T |\nabla T_\infty| c}{\mu}$, where σ_T represents the rate of change of the surface tension of the liquid with temperature, and μ is its viscosity. Temperature is scaled by first subtracting a reference temperature T_0 existing at the origin of coordinates and then dividing by the product $c |\nabla T_\infty|$. The scaled temperature gradient along the plane surface will satisfy $\nabla T \bullet \mathbf{t} = \mathbf{w} \bullet \mathbf{t}$. Here, $\mathbf{t}$ is an arbitrary vector lying on the plane surface. The governing equations are Stokes's and Laplace's equations for the velocity and temperature fields, respectively, in addition to the continuity equation. Because the problem is linear, we can write the scaled velocity of the bubble, $\mathbf{U}$, as follows:

$$\mathbf{U} = \mathbf{u} \bullet \mathbf{w}. \tag{7.2.1}$$

Here, the second-order tensor $\mathbf{u}$ is given by

$$\mathbf{u} = u_\| (\mathbf{I} - \mathbf{kk}) + u_\perp \mathbf{kk}, \tag{7.2.2}$$

where $\mathbf{I}$ is the identity tensor and $\mathbf{k}$ is a unit vector in the z-direction. The symbol $u_\|$ represents the magnitude of the scaled velocity when the bubble moves because of a unit scaled temperature gradient parallel to the plane surface, and similarly $u_\perp$ stands for the magnitude of the scaled velocity when motion occurs because of a unit scaled temperature gradient in the z-direction. Both $u_\|$ and $u_\perp$ are functions of the scaled separation distance $H = \frac{h}{R}$. Results for $u_\perp$ already have been obtained in Section 7.1. Only the problem of motion parallel to the plane surface needs to be solved. In this situation, the relative geometric configuration does not change with time, so that the quasi-steady assumption need not be invoked with respect to the configuration. Nonetheless, because the physical properties change with temperature, we still must employ the quasi-steady state assumption to justify treating the properties as being constant.

Let x be measured in some direction along the plane surface. Consider the motion of a gas bubble due to a temperature gradient that is aligned in the x-direction, so that $\nabla T_\infty = |\nabla T_\infty| \mathbf{i}$. Define the azimuthal angle ϕ, such that it is measured from the positive x-direction. Then, the scaled undisturbed temperature field is given by $\omega \cos \phi$, where ω is the distance from the z-axis, scaled by c. The solution of Stokes's equation in the continuous phase can be written in the general form given in Equations (5.1.18) through (5.1.26). Similarly, the solution for the scaled temperature field can be written as $\omega \cos \phi + T_1$, where the disturbance temperature field T_1 is written in the general form of Equation (5.1.13). The undisturbed temperature field, which is a solution of Laplace's equation, is added to the solution given in Equation (5.1.13) because the latter approaches zero as the distance from the origin approaches infinity. It only remains to choose the arbitrary constants appearing in these solutions, so as to satisfy the boundary conditions on the fields and the condition of zero hydro-dynamic force on the drop. The boundary conditions are similar to those given in the previous section. They include the requirement that the disturbance temperature at the plane surface is zero, and the condition of zero heat flux normal to the surface of the bubble.

$$T_1(0, \eta) = 0,$$ (7.2.3)

$$\frac{\partial T_1}{\partial \xi}(\xi_1, \eta) = \frac{\sinh \xi_1 \sin \eta \cos \phi}{(\cosh \xi_1 - \zeta)^2}.$$ (7.2.4)

Here, $\zeta = \cos \eta$, and $\xi = \xi_1$ represents the bubble surface. In view of these boundary conditions, only the term involving $m = 1$ in Equation (5.1.13) needs to be retained, and boundedness conditions can be used to eliminate the constants D_{mn} altogether. In the end, the disturbance temperature field reduces to the simple form given below.

$$T_1 = \sqrt{\cosh \xi - \zeta} \sin \eta \cos \phi \sum_{n=1}^{\infty} \left[B_n \sinh\left(n + \frac{1}{2}\right) \xi \right] P_n'(\zeta).$$ (7.2.5)

The constants B_n satisfy the following set of equations, which is closed by setting $B_{N+1} = 0$ for a sufficiently large value of N, in a manner similar to that described in Section 7.1.

$$[(n + 1)B_n - (n + 2)B_{n+1}] \cosh\left(n + \frac{3}{2}\right) \xi_1$$

$$+ [nB_n - (n - 1)B_{n-1}] \cosh\left(n - \frac{1}{2}\right) \xi_1 = 4\sqrt{2} \sinh \xi_1 e^{(n+\frac{1}{2})\xi_1}.$$ (7.2.6)

Meyyappan and Subramanian (1987) established numerically that the constant B_n approaches 0 with increasing n.

The velocity field must satisfy the no slip boundary condition at the rigid surface.

$$v_\omega(0, \eta) = 0,$$ (7.2.7)

$$v_\phi(0, \eta) = 0.$$ (7.2.8)

The kinematic condition at the rigid surface yields

$$v_z(0, \eta) = 0.$$ (7.2.9)

Far from the bubble, the fluid must be stationary. This is naturally satisfied by the form of the solution. At the bubble surface, the kinematic condition is written as follows:

$$v_\xi(\xi_1, \eta) = -u_\parallel \frac{\sinh \xi_1 \sin \eta \cos \phi}{\cosh \xi_1 - \zeta}.$$ (7.2.10)

The tangential stress balance leads to two conditions:

$$\tau_{\xi\eta}(\xi_1, \eta) = -(\cosh \xi_1 - \zeta) \sin \eta \frac{\partial T}{\partial \zeta}(\xi_1, \eta)$$ (7.2.11)

and

$$\tau_{\xi\phi}(\xi_1, \eta) = \frac{\cosh \xi_1 - \zeta}{\sin \eta} \frac{\partial T}{\partial \phi}(\xi_1, \eta).$$ (7.2.12)

The stress components $\tau_{\xi\eta}$ and $\tau_{\xi\phi}$ are related to the velocity components as follows:

$$\tau_{\xi\eta} = \frac{\partial}{\partial\xi}[(\cosh\xi - \zeta)v_\eta] + \frac{\partial}{\partial\eta}[(\cosh\xi - \zeta)v_\xi] \tag{7.2.13}$$

and

$$\tau_{\xi\phi} = \frac{\partial}{\partial\xi}[(\cosh\xi - \zeta)v_\phi] + \frac{\cosh\xi - \zeta}{\sin\eta}\frac{\partial v_\xi}{\partial\phi}. \tag{7.2.14}$$

The above boundary conditions, along with the condition that the hydrodynamic force on the bubble given by Equation (5.1.27) is zero, are used to specialize the constants in the velocity field and determine $u_\parallel$ as a function of H. The work is laborious but straightforward and details may be found in the doctoral thesis of Meyyappan (1984).

7.2.3 Results

We define an interaction parameter, Ω, using Equation (6.1.30). It is the ratio of the velocity of a given bubble in the presence of the rigid surface to that in its absence. From the solution, we can calculate Ω as a function of the scaled separation distance H. The results are reported in Table 7.2.1 and plotted in Figure 7.2.1. The results for small separations, included in Figure 7.2.1, are displayed more clearly in Figure 7.2.2, by using a magnified scale for the abscissa. The weak influence exerted by the plane surface in this case, when compared with motion normal to a plane surface, is evident from the figures and the results in the table. The reduction in velocity is only 1.83% when the

Table 7.2.1 Interaction Parameter, Ω, at Selected Values of H for a Gas Bubble Moving Parallel to a Rigid Plane Surface

H	Ω
1.04534	0.74314
1.08107	0.84781
1.12763	0.89967
1.18547	0.92698
1.25517	0.94236
1.33743	0.95217
1.43309	0.95953
1.50000	0.96354
1.54308	0.96581
2.00000	0.98172
2.35241	0.98808
3.00000	0.99392
3.76220	0.99681
5.00000	0.99860
6.13229	0.99922
10	0.99982
15	0.99995
25	0.99999

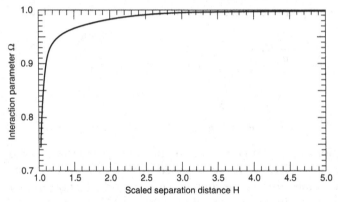

Figure 7.2.1 Interaction parameter for thermocapillary motion of a bubble parallel to a rigid plane surface plotted against scaled separation distance (Courtesy of Academic Press).

bubble is just two radii away from the rigid surface. At slightly greater distances, the surface exerts virtually no influence on the motion of the bubble.

Figure 7.2.3 pictorially displays the velocity of the bubble, normalized using the magnitude of the velocity in the absence of the plane surface. The results correspond to motion caused by a temperature gradient in the direction **w**, oriented at an angle θ with respect to the normal to the plane surface pointing into the fluid. We have shown the case where the bubble is located "below" the plane surface, but the results are identical for the mirror image situation. Results are shown for three different scaled separation distances for various orientation angles of the undisturbed temperature gradient. The figure provides graphical illustration of the contrast between the manner in which the plane surface retards motion parallel to it and the manner in which it retards motion normal to it. The orientation angle of the vector velocity of the bubble, defined in the same way as θ, is greater than or at least equal to θ.

Meyyappan and Subramanian (1987) also reported results for motion due to a body force such as gravity. Two significant differences are that the tangential stress is set equal to zero at the bubble surface in that problem, and the hydrodynamic force on the bubble is not zero. The authors provided a table of values of a multiplicative factor to be used with the result for an isolated bubble given in Section 3.1, which is the inverse of the interaction parameter Ω defined in Equation (6.1.30). From their results, it is evident

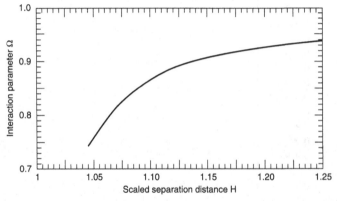

Figure 7.2.2 A magnified view of Figure 7.2.1 for small values of the scaled separation distance (Courtesy of Academic Press).

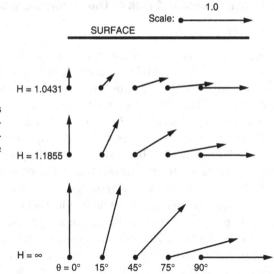

Figure 7.2.3 Pictorial illustration of the effects of a rigid plane surface on thermocapillary migration. The undisturbed temperature gradient makes an angle θ with the normal to the plane surface (Courtesy of Academic Press).

that the motion of a bubble parallel to a rigid plane surface due to a body force is retarded more significantly, at any given distance from that surface, than thermocapillary motion parallel to that surface. This behavior is similar to that noted in Section 7.1 for motion normal to the surface, and the explanation is the same as that offered there.

Recently, Chen (1999b) has provided an approximate solution of a related problem obtained by the method of reflections. The author analyzed the case of a drop moving parallel to an adiabatic plane surface in response to an undisturbed temperature gradient that is parallel to the plane surface. The thermal boundary condition at the plane surface is different from the one used here. Chen permitted the plane surface to be either rigid or free and found that a gas bubble can move slightly more rapidly than when it is isolated, when under the influence of the free surface. The interaction parameter in this case is given by $\Omega = 1 + \frac{1}{128H^8}$, to the order of the approximation made. A rigid plane surface retards the bubble, yielding the approximation $\Omega = 1 - \frac{1}{16H^3} - \frac{13}{256H^6} - \frac{1}{128H^8}$. Results in the general case when α and β are nonzero can be found in the article by Chen (1999b).

A related problem was considered by Chen, Dagan, and Maldarelli (1991). These authors analyzed the situation where a spherical drop migrates along the centerline of an insulated cylindrical tube filled with a different fluid, because of the action of a uniform axial temperature gradient imposed in that fluid. The authors mentioned some potential applications and noted that the problem provides a good model for studying the competition between the thermal and hydrodynamic interactions between the drop and the wall of the tube. The analysis is restricted to negligible values of the Reynolds and Marangoni numbers. When the thermal conductivity of the drop is small relative to that of the continuous phase, the steady heat flux along the cylinder must be transmitted around the drop, mostly bypassing it. This sharpens the temperature gradient along the surface of the drop, when compared with the case where the same drop is present in an unbounded continuous phase. This effect, which should lead to an increase in the migration velocity, is countered by the hydrodynamic interaction with the wall of the tube. Chen et al. (1991) reported that the net consequence is to always reduce the migration velocity when compared with that in an unbounded medium. They provided detailed results from their calculations for various values of the parameters and discussed the hydrodynamic and thermal interactions in depth.

7.3 Axisymmetric Motion of Two Interacting Drops

7.3.1 Introduction

Now, we move on to problems involving two spheres. The simplest case is that of two drops moving along their line of centers, as depicted in Figure 7.3.1. The radii of the drops are R_1 and R_2, and the smallest separation distance between their surfaces at a given instant is labeled h. The distance between the centers of the drops is $d = h + R_1 + R_2$. A temperature gradient ∇T_∞ is applied in the z-direction, and drops 1 and 2 move with instantaneous velocities $U_1^* \mathbf{i}_z$ and $U_2^* \mathbf{i}_z$ respectively. All the assumptions made in Section 7.1 continue to hold. Because the Reynolds and Marangoni numbers are both negligible, the problem is quasi-steady. The objective is to determine the change in the velocity of a drop due to the presence of the other. We pose the problem for two drops of unequal viscosity and thermal conductivity and designate the ratio of the viscosity of drop phase i to that of the continuous phase as α_i, and a similar ratio of thermal conductivities as β_i. The algebra is cumbersome in this problem. Therefore, we restrict the presentation of results to the limiting case of a pair of gas bubbles, for which one can set $\alpha_1 = \alpha_2 = 0$, $\beta_1 = \beta_2 = 0$. In view of this, we use a single symbol, σ_T, to represent the rate of change of the interfacial tension with temperature at both interfaces. It is straightforward to modify the problem statement to accommodate the case wherein the values of σ_T are different at the two interfaces.

The analysis and results presented in this section are based on the work of Meyyappan, Wilcox, and Subramanian (1983) and Wei and Subramanian (1993). Additional details can be found in the doctoral theses of Meyyappan (1984) and Wei (1994). We also provide some useful results from alternative approaches from Anderson (1985), who used the method of reflections, and from Wang, Mauri, and Acrivos (1994), who

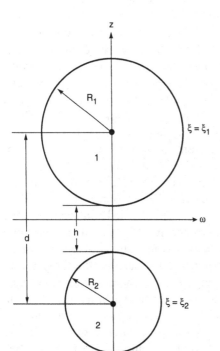

Figure 7.3.1   Schematic showing a pair of drops

employed twin multipole expansions. In addition, we briefly summarize the results from a lubrication analysis given by Loewenberg and Davis (1993b).

7.3.2 Analysis in Bispherical Coordinates

Lengths are scaled using the characteristic distance, c, in bispherical coordinates. Velocities are scaled using $v_0 = -\frac{\sigma_T|\nabla T_\infty|c}{\mu}$, as in Section 7.1. Here, μ is the viscosity of the continuous phase. Because the fields are axisymmetric, a Stokes streamfunction, scaled using $c^2 v_0$, can be used. It is labeled $\psi(\xi, \eta)$ in the continuous phase, and $\psi_i(\xi, \eta)$ within drop i. The symbol $T(\xi, \eta)$ represents temperature in the continuous phase, scaled by subtracting its value T_0 at $z = 0$ and then dividing by the product $c|\nabla T_\infty|$. Scaled temperatures, $T_i(\xi, \eta)$, are defined within drop phase i in a similar manner. The governing equations are Stokes's and Laplace's equations, and we can begin with the general solutions given in Equations (5.1.14) and (5.1.15). Because the fluid is quiescent far from the drops, the behavior of the solution for the streamfunction in the continuous phase is satisfactory, and we can use Equation (5.1.15) directly. The undisturbed scaled temperature far from the drops is given by $T = z$. This is a solution of Laplace's equation and needs to be added to that given in Equation (5.1.14) for the temperature field in the continuous phase because that solution vanishes as the distance from the origin approaches ∞.

$$\psi = \frac{1}{(\cosh\xi - \zeta)^{\frac{3}{2}}} \sum_{n=1}^{\infty} \left[a_n \cosh\left(n - \frac{1}{2}\right)\xi + b_n \sinh\left(n - \frac{1}{2}\right)\xi \right.$$
$$\left. + c_n \cosh\left(n + \frac{3}{2}\right)\xi + d_n \sinh\left(n + \frac{3}{2}\right)\xi \right] C_{n+1}(\zeta), \tag{7.3.1}$$

$$T = z + \sqrt{\cosh\xi - \zeta} \sum_{n=0}^{\infty} \left[A_n \cosh\left(n + \frac{1}{2}\right)\xi + B_n \sinh\left(n + \frac{1}{2}\right)\xi \right] P_n(\zeta). \tag{7.3.2}$$

In the above, $\zeta = \cos\eta$, as usual.

Within drop phase i, specialized versions of the above solutions are written, which take advantage of the fact that the fields must be bounded everywhere inside a drop.

$$\psi_i = \frac{1}{(\cosh\xi - \zeta)^{\frac{3}{2}}} \sum_{n=1}^{\infty} \left[p_{ni} e^{-(n-\frac{1}{2})|\xi|} + q_{ni} e^{-(n+\frac{3}{2})|\xi|} \right] C_{n+1}(\zeta), \tag{7.3.3}$$

$$T_i = z + \sqrt{\cosh\xi - \zeta} \sum_{n=0}^{\infty} C_{ni} e^{-(n+\frac{1}{2})|\xi|} P_n(\zeta). \tag{7.3.4}$$

The following boundary conditions can be used to specialize the arbitrary constants in the above solution. First, the kinematic condition holds at the surface of each drop.

$$\mathbf{v}(\xi_i, \eta) \bullet \mathbf{i}_\xi = \mathbf{v}_i(\xi_i, \eta) \bullet \mathbf{i}_\xi = U_i \mathbf{i}_z \bullet \mathbf{i}_\xi, \quad i = 1, 2. \tag{7.3.5}$$

Here, U_i represents the magnitude of the scaled velocity of drop i. Continuity of tangential velocity at the surface of each drop yields

$$\mathbf{v}(\xi_i, \eta) \bullet \mathbf{i}_\eta = \mathbf{v}_i(\xi_i, \eta) \bullet \mathbf{i}_\eta \quad i = 1, 2. \tag{7.3.6}$$

The tangential stress balance at the surface of each drop is written as follows:

$$\tau_{\xi\eta}(\xi_1, \eta) - \tau_{\xi\eta 1}(\xi_1, \eta) = -(\cosh\xi_1 - \zeta)\frac{\partial T}{\partial \eta}(\xi_1, \eta) \tag{7.3.7}$$

and

$$\tau_{\xi\eta}(\xi_2, \eta) - \tau_{\xi\eta 2}(\xi_2, \eta) = (\cosh \xi_2 - \zeta)\frac{\partial T}{\partial \eta}(\xi_2, \eta). \tag{7.3.8}$$

The stress components appearing in the above equations are scaled using $(-\sigma_T |\nabla T_\infty|)$. The connection between $\tau_{\xi\eta}$ and derivatives of ψ is given in Equation (6.1.12). For evaluating the constants in the temperature field, we use the condition that at the surface of each drop, the temperature and the heat flux normal to the surface must be continuous.

$$T(\xi_i, \eta) = T_i(\xi_i, \eta), \quad i = 1, 2, \tag{7.3.9}$$

$$\frac{\partial T}{\partial \xi}(\xi_i, \eta) = \beta_i \frac{\partial T_i}{\partial \xi}(\xi_i, \eta), \quad i = 1, 2. \tag{7.3.10}$$

Finally, the hydrodynamic force on each drop is set equal to zero. Results for the force, in terms of the constants in the solution for the streamfunction in Equation (7.3.1), are given in Equations (5.1.16) and (5.1.17). Evaluation of the constants requires numerical solution of sets of tridiagonal equations. The equations to be solved are reported by Keh and Chen (1990). Because they are lengthy, we do not reproduce them here. Even in the special case of two gas bubbles ($\alpha_1 = \alpha_2 = 0$ and $\beta_1 = \beta_2 = 0$), some numerical evaluation is required, as noted by Meyyappan, Wilcox, and Subramanian (1983). We give the results only in that special case. For gas bubbles, the momentum and energy transport problems in the continuous phase are decoupled from those for the gas phase within the bubbles. Therefore, only the constants a_n, b_n, c_n, and d_n in the streamfunction field in Equation (7.3.1) can be evaluated.

$$a_n = h_1\left[I_n(\xi_1)\sinh\left(n - \frac{1}{2}\right)\xi_2 - I_n(\xi_2)\sinh\left(n - \frac{1}{2}\right)\xi_1\right]$$
$$+ h_3\left[U_1 e^{-(n-\frac{1}{2})\xi_1}\sinh\left(n - \frac{1}{2}\right)\xi_2 - U_2 e^{(n-\frac{1}{2})\xi_2}\sinh\left(n - \frac{1}{2}\right)\xi_1\right], \tag{7.3.11}$$

$$b_n = h_1\left[I_n(\xi_2)\cosh\left(n - \frac{1}{2}\right)\xi_1 - I_n(\xi_1)\cosh\left(n - \frac{1}{2}\right)\xi_2\right]$$
$$+ h_3\left[U_2 e^{(n-\frac{1}{2})\xi_2}\cosh\left(n - \frac{1}{2}\right)\xi_1 - U_1 e^{-(n-\frac{1}{2})\xi_1}\cosh\left(n - \frac{1}{2}\right)\xi_2\right], \tag{7.3.12}$$

$$c_n = h_2\left[I_n(\xi_2)\sinh\left(n + \frac{3}{2}\right)\xi_1 - I_n(\xi_1)\sinh\left(n + \frac{3}{2}\right)\xi_2\right]$$
$$+ h_4\left[U_2 e^{(n+\frac{3}{2})\xi_2}\sinh\left(n + \frac{3}{2}\right)\xi_1 - U_1 e^{-(n+\frac{3}{2})\xi_1}\sinh\left(n + \frac{3}{2}\right)\xi_2\right], \tag{7.3.13}$$

$$d_n = h_2\left[I_n(\xi_1)\cosh\left(n + \frac{3}{2}\right)\xi_2 - I_n(\xi_2)\cosh\left(n + \frac{3}{2}\right)\xi_1\right]$$
$$+ h_4\left[U_1 e^{-(n+\frac{3}{2})\xi_1}\cosh\left(n + \frac{3}{2}\right)\xi_2 - U_2 e^{(n+\frac{3}{2})\xi_2}\cosh\left(n + \frac{3}{2}\right)\xi_1\right]. \tag{7.3.14}$$

Here,

$$I_n(\xi_i) = \mp\frac{n(n+1)(2n+1)}{2}\int_{-1}^{+1}\frac{\frac{\partial T}{\partial \xi}(\xi_i, \zeta)C_{n+1}(\zeta)\,d\zeta}{\sqrt{\cosh \xi_i - \zeta}}, \tag{7.3.15}$$

where the minus sign applies at $\xi = \xi_1$, and the plus sign should be used at $\xi = \xi_2$. The scaled velocities U_1 and U_2 are obtained from the solution of the pair of linear equations given below.

$$U_1 \sum_{n=1}^{\infty} m_1 f_1 + U_2 \sum_{n=1}^{\infty} m_1 f_2 = \sum_{n=1}^{\infty} m_2 [g_1 I_n(\xi_1) + g_2 I_n(\xi_2)], \qquad (7.3.16)$$

$$U_1 \sum_{n=1}^{\infty} m_1 f_2 + U_2 \sum_{n=1}^{\infty} m_1 f_3 = \sum_{n=1}^{\infty} m_2 [g_3 I_n(\xi_1) + g_4 I_n(\xi_2)]. \qquad (7.3.17)$$

The various constants appearing in the above solution are defined below.

$$f_1 = (2n - 1) e^{-\gamma_2} \sinh \gamma_3 - (2n + 3) e^{-\gamma_1} \sinh \gamma_4, \qquad (7.3.18)$$

$$f_2 = (2n + 3) e^{-\gamma_3} \sinh \gamma_4 - (2n - 1) e^{-\gamma_4} \sinh \gamma_3, \qquad (7.3.19)$$

$$f_3 = (2n - 1) e^{\gamma_2} \sinh \gamma_3 - (2n + 3) e^{\gamma_1} \sinh \gamma_4, \qquad (7.3.20)$$

$$g_1 = e^{-(n+\frac{3}{2})\xi_2} \sinh \gamma_3 - e^{-(n-\frac{1}{2})\xi_2} \sinh \gamma_4, \qquad (7.3.21)$$

$$g_2 = e^{-(n-\frac{1}{2})\xi_1} \sinh \gamma_4 - e^{-(n+\frac{3}{2})\xi_1} \sinh \gamma_3, \qquad (7.3.22)$$

$$g_3 = e^{(n-\frac{1}{2})\xi_2} \sinh \gamma_4 - e^{(n+\frac{3}{2})\xi_2} \sinh \gamma_3, \qquad (7.3.23)$$

$$g_4 = e^{(n+\frac{3}{2})\xi_1} \sinh \gamma_3 - e^{(n-\frac{1}{2})\xi_1} \sinh \gamma_4, \qquad (7.3.24)$$

$$h_1 = \frac{1}{2(2n + 1) \sinh \gamma_3}, \qquad (7.3.25)$$

$$h_2 = \frac{1}{2(2n + 1) \sinh \gamma_4}, \qquad (7.3.26)$$

$$h_3 = \frac{n(n + 1)}{\sqrt{2}(2n - 1) \sinh \gamma_3}, \qquad (7.3.27)$$

$$h_4 = \frac{n(n + 1)}{\sqrt{2}(2n + 3) \sinh \gamma_4}, \qquad (7.3.28)$$

$$m_1 = \frac{n(n + 1)}{\sqrt{2}(2n - 1)(2n + 3) \sinh \gamma_3 \sinh \gamma_4}, \qquad (7.3.29)$$

$$m_2 = -\frac{1}{2(2n + 1) \sinh \gamma_3 \sinh \gamma_4}, \qquad (7.3.30)$$

$$\gamma_1 = \left(n - \frac{1}{2}\right)(\xi_1 + \xi_2), \qquad (7.3.31)$$

$$\gamma_2 = \left(n + \frac{3}{2}\right)(\xi_1 + \xi_2), \qquad (7.3.32)$$

$$\gamma_3 = \left(n - \frac{1}{2}\right)(\xi_1 - \xi_2), \qquad (7.3.33)$$

$$\gamma_4 = \left(n + \frac{3}{2}\right)(\xi_1 - \xi_2). \qquad (7.3.34)$$

Finally, the constants A_n and B_n appearing in the temperature field in Equation (7.3.2) can be obtained by solving the following system of equations, with closure achieved by setting $A_{N+1} = B_{N+1} = 0$ for a sufficiently large value of N.

$$(n+1)(A_n - A_{n+1})\sinh\left(n + \frac{3}{2}\right)\xi_i + n(A_n - A_{n-1})\sinh\left(n - \frac{1}{2}\right)\xi_i$$

$$+ (n+1)(B_n - B_{n+1})\cosh\left(n + \frac{3}{2}\right)\xi_i + n(B_n - B_{n-1})\cosh\left(n - \frac{1}{2}\right)\xi_i$$

$$= \frac{2\sqrt{2}}{\sinh\xi_i}\left[\left\{(n+1)e^{-(n+\frac{3}{2})\xi_i} + ne^{-(n-\frac{1}{2})\xi_i}\right\}\cosh\xi_i - (2n+1)e^{-(n+\frac{1}{2})\xi_i}\right],$$

$$i = 1, 2, \quad n = 0, 1, 2, \dots. \quad (7.3.35)$$

The convention that $A_{-1} = B_{-1} = 0$ is used here. By using the approach outlined in Section 7.1 following Sadhal (1983), it should be possible to calculate the constants $I_n(\xi_i)$ without resort to numerical integration of the scaled surface temperature gradient. Meyyappan et al. were not aware of this possibility, however, and performed their calculations using such integration. They provide results for typical values of the governing parameters, and we shall present and discuss some results in this problem in Section 7.3.4. Results in the analogous case of two drops are presented by Keh and Chen (1990).

The motion of two drops due to a temperature gradient aligned normal to the line of centers is amenable to analysis using bispherical coordinates. Each drop will move in the direction of the applied temperature gradient because of symmetry and the reversibility of Stokes flows. The arguments are analogous to those outlined in Section 6.2.3. The solution procedure is similar to that of Zinchenko (1981), who analyzed the analogous problem for body force driven motion, but the algebra is lengthy. Ultimately, numerical work is necessary to evaluate the coefficients appearing in the solution. Therefore, no results obtained by using a solution in bispherical coordinates are available for this problem. The alternative is to obtain an approximate solution by some other means. Such solutions also are useful in the axisymmetric case. Therefore, we discuss these next.

7.3.3 Alternative Approaches

In presenting results in the axisymmetric case, one can use scalar interaction parameters that are defined as the ratio of the velocity of a given drop divided by its value when isolated because the direction of motion is unaffected by the presence of a second drop in this configuration. When the line of centers is oriented arbitrarily with respect to the applied temperature gradient, however, both the direction and the magnitude of the velocity of each drop will be affected by the presence of the other. In this case, one can conveniently use the following formulation, which employs mobility coefficients. The results are expressed in terms of the velocity of each drop when isolated, which can be obtained from Equation (4.2.28).

$$\mathbf{U}_1^* = \mathbf{M}_{11} \bullet \mathbf{U}_{1,isolated}^* + \mathbf{M}_{12} \bullet \mathbf{U}_{2,isolated}^*, \quad (7.3.36)$$

$$\mathbf{U}_2^* = \mathbf{M}_{21} \bullet \mathbf{U}_{1,isolated}^* + \mathbf{M}_{22} \bullet \mathbf{U}_{2,isolated}^*. \quad (7.3.37)$$

Here, physical velocities are identified with an asterisk, and the tensors $\mathbf{M}_{ij}$ are called *mobility* tensors. We can write $\mathbf{M}_{11}$ and $\mathbf{M}_{12}$ as follows:

$$\mathbf{M}_{11} = T_{11}\mathbf{ee} + W_{11}(\mathbf{I} - \mathbf{ee}) \quad (7.3.38)$$

and

$$\mathbf{M}_{12} = T_{12}\mathbf{ee} + W_{12}(\mathbf{I} - \mathbf{ee}). \tag{7.3.39}$$

As in Section 6.2, $\mathbf{e}$ is a unit vector in the direction from drop 1 to drop 2, and $\mathbf{I}$ is the identity tensor. The coefficients T_{11} and T_{12} correspond to the axisymmetric case, and W_{11} and W_{12} are obtained from the asymmetric case wherein the temperature gradient is perpendicular to the line of centers. The tensors $\mathbf{M}_{21}$ and $\mathbf{M}_{22}$ can be obtained by exchanging the symbols corresponding to drops 1 and 2 in the results for $\mathbf{M}_{12}$ and $\mathbf{M}_{11}$, respectively.

The reader will recall that the method of reflections, mentioned in Section 5.1, provides simple results that should be adequate as long as the drops are not very close to each other. An advantage of this method is that it is convenient to use when the problem is not axisymmetric. Anderson (1985) has obtained the solution for two drops of the same fluid moving in a second fluid that is subjected to a uniform temperature gradient, using the method of reflections. The orientation of the line of centers with respect to the temperature gradient is arbitrary. The results from this solution for the coefficients in Equations (7.3.38) and (7.3.39) are given below.

$$T_{11} = 1 - 2p\frac{\lambda^3}{D^3} + \left[4p^2 - \frac{3(2+5\alpha)}{2(1+\alpha)}\right]\frac{\lambda^3}{D^6}, \tag{7.3.40}$$

$$T_{12} = \frac{\lambda^3}{D^3} - \left[2p - \frac{9(2+3\alpha)(1-\beta)}{2(1+\alpha)(3+2\beta)}\right]\frac{\lambda^3}{D^6}, \tag{7.3.41}$$

$$W_{11} = 1 + p\frac{\lambda^3}{D^3}\left(1 + \frac{p}{D^3}\right), \tag{7.3.42}$$

$$W_{12} = -\frac{\lambda^3}{2D^3}\left(1 + \frac{p}{D^3}\right). \tag{7.3.43}$$

Here, the radius ratio λ, the scaled distance between centers D, and the symbol p, are defined as follows:

$$\lambda = \frac{R_2}{R_1}, \tag{7.3.44}$$

$$D = \frac{d}{R_1}, \tag{7.3.45}$$

and

$$p = \frac{1-\beta}{2+\beta}. \tag{7.3.46}$$

The correction to the above results appears at $O(\frac{1}{D^8})$. Because the drops are of the same fluid, the subscript on the physical property ratios has been omitted. It is seen from Equations (7.3.36) and (7.3.37) that when the temperature gradient is aligned perpendicular to the line of centers, implying that $\mathbf{U}^*_{1,isolated} \bullet \mathbf{e} = \mathbf{U}^*_{2,isolated} \bullet \mathbf{e} = 0$, the drops move in the direction of the applied temperature gradient, as mentioned earlier.

Another approach, which provides similar expansions, is exemplified by the work of Satrape (1992) and Wang et al. (1994). These authors solved the problem for two gas bubbles, wherein $\alpha = \beta = 0$, when the line of centers is oriented arbitrarily with respect to the applied temperature gradient, by using twin multipole expansions. Such expansions are written in two cases, namely the axisymmetric situation considered in this section and the case where the line of centers is perpendicular to the applied temperature gradient. The general case, where the line of centers is oriented arbitrarily with respect to the applied temperature gradient, is handled by superposition of these two cases as shown in Equations (7.3.36) and (7.3.37). The method of twin multipole expansions involves writing the solutions of Laplace's and Stokes's equations as a superposition of two series solutions in each case. Each of these is a general solution appropriate for the domain outside one bubble in spherical polar coordinates, with the origin at the center of that bubble. The arbitrary constants are then evaluated by applying the boundary conditions. Satrape truncated the series at a suitable point and solved for the constants numerically. Instead, Wang et al. expanded the constants in power series in $\frac{\lambda}{D}$ and $\frac{1}{D}$. The final results from Wang et al. can be written as follows:

$$T_{11} = 1 - \frac{\lambda^3}{D^3} - 2\frac{\lambda^3}{D^6} - 3\frac{\lambda^5}{D^8} - 4\frac{\lambda^6}{D^9} - 4\frac{\lambda^7}{D^{10}} - \frac{9\lambda^6 + 6\lambda^8}{D^{11}} - \frac{8\lambda^6 + 5\lambda^9}{D^{12}},$$

(7.3.47)

$$T_{12} = 1 - T_{11},$$

(7.3.48)

$$T_{21} = 1 - T_{22},$$

(7.3.49)

$$T_{22} = 1 - \frac{1}{D^3} - 2\frac{\lambda^3}{D^6} - 3\frac{\lambda^3}{D^8} - 4\frac{\lambda^3}{D^9} - 4\frac{\lambda^3}{D^{10}} - \frac{6\lambda^3 + 9\lambda^5}{D^{11}} - \frac{5\lambda^3 + 8\lambda^6}{D^{12}},$$

(7.3.50)

$$W_{11} = 1 + \frac{\lambda^3}{2D^3} + \frac{\lambda^3}{4D^6} + \frac{\lambda^5}{4D^8} + \frac{\lambda^6}{8D^9} + \frac{\lambda^7}{4D^{10}} + \frac{4\lambda^6 + \lambda^8}{8D^{11}} + \frac{\lambda^6 + 4\lambda^9}{16D^{12}},$$

(7.3.51)

$$W_{12} = 1 - W_{11},$$

(7.3.52)

$$W_{21} = 1 - W_{22},$$

(7.3.53)

and

$$W_{22} = 1 + \frac{1}{2D^3} + \frac{\lambda^3}{4D^6} + \frac{\lambda^3}{4D^8} + \frac{\lambda^3}{8D^9} + \frac{\lambda^3}{4D^{10}} + \frac{\lambda^3 + 4\lambda^5}{8D^{11}} + \frac{4\lambda^3 + \lambda^6}{16D^{12}}.$$

(7.3.54)

The authors reported T_{11}, T_{12}, W_{11}, and W_{12}. For convenience, we have provided the complete set of functions needed to construct the results for either bubble. Wang et al. indicated that the corrections in Equations (7.3.47) and (7.3.51) appear at $O(\frac{1}{D^{13}})$. Also, they provided a table of representative values of T_{11} and W_{11}, obtained by including terms up to $O(\frac{1}{D^{20}})$. Note that when we set $\alpha = \beta = 0$ in Equations (7.3.40) through (7.3.43), the results reduce to the terms up to and including $O(\frac{1}{D^6})$ in the expansions given above.

Now, we briefly discuss the situation when the drops are very close to each other. Loewenberg and Davis (1993b) analyzed this problem in the axisymmetric situation, when the thermal conductivity of the drops is negligible, compared with that of the continuous phase. They used ideas similar to those discussed already in Section 7.1 in the context of a drop approaching a plane surface. They assumed that the drops are spherical in shape, and that the values of the Reynolds and Marangoni numbers in each phase are negligible. The viscosities of the drops are equal but can be different from that of the continuous phase. If the drops are very close to each other, macroscopically they appear to move together at some velocity as a pair. Each drop moves at a slightly different velocity from this pair velocity, the difference being determined by the requirement of zero hydrodynamic force on each drop, and the magnitude of the lubrication resistance to the flow of fluid in the gap between the two drops. The authors wrote the zero hydrodynamic force on a given drop as the algebraic sum of a thermocapillary force, which is the force on that drop if it is held fixed in the same temperature gradient, a hydrodynamic force, which is that experienced by the same drop when the drop translates at the pair velocity through an isothermal continuous phase, and a contact force between the two drops. It is implied here that the forces mentioned are those acting on each drop in the presence of the other drop in the given geometric configuration in the hypothetical problems. The contact force exerted on one drop is equal and opposite to that on the other drop. The lubrication resistance is described by expressions similar to those in Equation (7.1.25). The temperature field around two nonconducting spheres in point-contact is obtained by solving Laplace's equation using tangent-sphere coordinates. Together with a solution of Stokes's equation for the flow generated by the drops, when held fixed, this permited the authors to obtain a result for the thermocapillary force on each drop. The hydrodynamic force exerted by the fluid on each of the two drops in near-contact is obtained from available results. Using all of this information, the velocity of the pair, and the relative velocity of one drop with respect to the other, are calculated for a range of values of the radius ratio λ and the viscosity ratio α. Asymptotic formulae are given for these quantities in the case when $\lambda \ll 1$. Loewenberg and Davis made the following observations.

The pair migration velocity is smaller than that of isolated drops, but the reduction is modest, being no more than 12% of the velocity of the larger member of the pair, when isolated. In the case of gas bubbles, for which $\alpha = 0$, when the bubbles are of equal size, the pair moves at the same velocity as that of an isolated bubble, a result that holds for any separation distance as will be seen in Section 7.3.4. When $\alpha \neq 0$, for equal-sized drops, the velocity of the pair is slightly smaller than that of an isolated drop. The pair velocity shows little sensitivity to the viscosity ratio over the entire range of values of the radius ratio, and the extent of variation is within 3%. The relative velocity decreases with increasing α, for any given separation distance, and with decreasing separation distance, for fixed α. When the velocity of each drop is compared to the pair velocity, the larger drop always moves more rapidly than the pair, whereas the smaller drop lags the pair. The differences between the individual drop velocities and the pair velocity are most pronounced in the gas bubble limit when $\alpha = 0$ and decrease in magnitude with increasing α. The velocity of the larger of the two drops is considerably closer to that of the pair at any given α. For a fixed value of α, the difference in velocity between the members of the pair decreases as the gap width is reduced.

7.3.4 Results in the Axisymmetric Case

We now return to discuss some results in the gas bubble limit, in the axisymmetric case. For the purpose of presentation of results, we define a scaled separation distance between the surfaces of the two bubbles as

$$H = \frac{h}{R_1}. \tag{7.3.55}$$

An interaction parameter Ω_i for bubble i is defined as

$$\Omega_i(H, \lambda) = \frac{U_i^*}{U_{i,isolated}^*}. \tag{7.3.56}$$

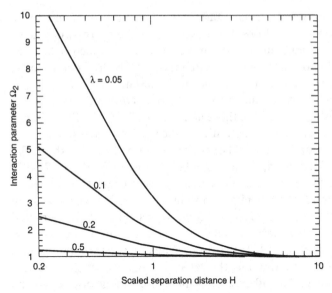

Figure 7.3.2 Interaction parameter for the small bubble in the pair plotted against the scaled separation distance between the surfaces for a set of values of the radius ratio.

The interaction parameters in the axisymmetric case are related in a simple way to the mobility functions T_{11} and T_{12}. For example, we can write

$$\Omega_1 = T_{11} + \lambda T_{12} \tag{7.3.57}$$

and

$$\Omega_2 = \frac{1}{\lambda} T_{21} + T_{22}, \tag{7.3.58}$$

because the velocity of a bubble, when isolated, is proportional to its radius. The interaction parameters provide a physically convenient way to display the results because they show the impact of the presence of the second bubble on the first in a direct manner. In the following, the larger of the two bubbles is chosen as bubble 1. Therefore, the parameter λ varies from 0 to 1. In Figure 7.3.2 the interaction parameter Ω_2, for the smaller bubble in the pair, is plotted as a function of the scaled separation distance between the surfaces H, for a few illustrative values of the radius ratio λ. Meyyappan et al. (1983) provided a table of values of Ω_1 and Ω_2 calculated from the solution in bispherical coordinates for various values of λ and H. We have checked the predictions obtained using Equations (7.3.57) and (7.3.58), in conjunction with the results from twin multipole expansions, against this table, and found that the two sets of results for the smaller bubble in the pair agree to within 1% for $H \geq 0.15$ and $0.05 \leq \lambda \leq 0.9091$. In fact, for $H \geq 0.5$, the results are within 0.05% of each other. Therefore, in preparing Figure 7.3.2 we have used the results from twin multipole expansions that are easy to evaluate.

The figure shows that the smaller of the two bubbles always moves more rapidly than when isolated. As λ decreases, implying a larger contrast in size, the increase in Ω_2 becomes quite substantial. The principal contribution to the motion of the small bubble in this case comes from the flow induced by the large bubble, which carries the small bubble along. In contrast to the results shown in Figure 7.3.2, the influence of the small bubble on the larger one is minor. The greatest reduction in the velocity of the large bubble is less than approximately 3% over a wide range of size ratios and separation distances, as illustrated in Meyyappan et al. (1983). Therefore, we have not included a drawing here for Ω_1 as a function of λ and H. The interactions observed in the case of drops with nonzero α and β are qualitatively similar, as noted by Keh and Chen (1990). These authors also commented that the ratio of thermal conductivities, β, plays a stronger role in influencing the interaction, when compared with that played by α, the ratio of viscosities. This is evident from the reflections solution as well.

As a corollary from the results presented above, note that bubbles of equal size move at the velocity of an isolated bubble regardless of the separation distance. This can be seen from the results in Equations (7.3.48) and (7.3.49) upon setting $\lambda = 1$ in Equations (7.3.57) and (7.3.58). The fact that the solution of Equations (7.3.16) and (7.3.17) yields the same result was demonstrated by Feuillebois (1989). This remarkable result is the consequence of a cancellation of two interaction effects. The result of the thermal interaction is to reduce the effective temperature gradient over the surface of each bubble, and that of the hydrodynamic interaction is to reduce the resistance experienced by each bubble. As shown by Acrivos, Jeffrey, and Saville (1990), this exact cancellation holds, regardless of the orientation of the line of centers of the bubbles with respect to the applied temperature gradient. We can understand this from the solution given at the end of Section 4.4. For the purpose of this discussion, physical quantities are used. We mentioned in that section that an isolated bubble of radius R can be held fixed in a fluid with a temperature gradient $\nabla T_\infty(\mathbf{x})$ by using a Stokes flow $\mathbf{u}_\infty(\mathbf{x}) = \Lambda \nabla T_\infty(\mathbf{x})$ where the negative constant, Λ, is given by

$$\Lambda = \frac{\sigma_T R}{2\mu}. \tag{7.3.59}$$

The solution for the velocity field in the presence of the bubble, $\mathbf{u}(\mathbf{x})$, is then related to the temperature gradient field $\nabla T(\mathbf{x})$ in that situation in the same way:

$$\mathbf{u}(\mathbf{x}) = \Lambda \nabla T(\mathbf{x}). \tag{7.3.60}$$

Note that this disturbance flow produced by the bubble also is a potential flow, which satisfies Stokes's equation with a uniform pressure field, just like the undisturbed flow. The temperature field satisfies Laplace's equation, just as the undisturbed temperature field does. Therefore, if we introduce a second bubble of the same radius somewhere else in the fluid, it also will be stationary and will produce a new velocity field that is again related to the temperature field by Equation (7.3.60). Moreover, the effect of the presence of the first bubble on this field will be to produce a reflection which will follow the same rule. It is then straightforward to see that all the reflections from the two bubbles will obey this rule, and both bubbles will be stationary. This argument continues to hold if we introduce any number of additional bubbles of the same size, as long as the fluid extends to infinity, so that all of them will be stationary. If we now consider a uniform temperature gradient field ∇T_∞ in the undisturbed state, a uniform velocity $\mathbf{U} = \Lambda \nabla T_\infty$ will keep all the bubbles fixed, or in a quiescent fluid, each bubble will move with the velocity $-\Lambda \nabla T_\infty$, which is precisely the result for an isolated bubble from Equation (4.2.28). The above argument is given by Wang et al. (1994).

Now, we present the flow structure in the problem of a pair of bubbles executing axisymmetric motion. Figure 7.3.3 contrasts streamlines in a meridian section in the laboratory reference frame for thermocapillary migration with streamlines for

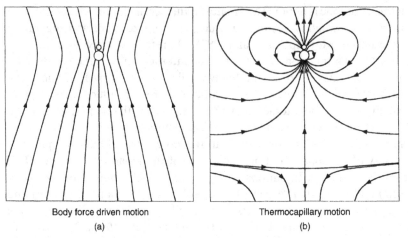

Body force driven motion   Thermocapillary motion

(a)   (b)

Figure 7.3.3 Streamlines in the laboratory reference frame for a pair of bubbles; $\lambda = 0.5$, $H = 0.25$. (Reprinted with permission from Wei, H., and Subramanian, R. S. Thermocapillary migration of a small chain of bubbles. *Phys. Fluids A* 5, No. 7, 1583–1595. Copyright 1993, American Institute of Physics.)

body-force-driven motion in a representative case. This figure is taken from Wei and Subramanian (1993), who obtained the solutions in these two problems using a boundary collocation technique that is described in Section 7.5. The authors calculated the velocity components, and not the streamfunction. The streamlines in Figure 7.3.3 have been drawn such that the velocity vector is tangent to these curves everywhere and are not shown for equal increments of the streamfunction. We suggest that the sketch for the case of body force driven flow could have been drawn from intuition without making detailed calculations. It would not be as simple to draw the sketch for the thermocapillary case, without the benefit of an analysis. This type of flow structure, showing a separated reverse flow region and a saddle point, already has been noted in Section 4.17 in the case of isolated bubbles, and the explanation for its appearance is similar. It arises because the presence of each bubble excites all the Legendre modes of the temperature field on the surface of the other bubble. For more details, the reader should consult Section 4.17. The flow in the separated wake is weak compared with the flow in the vicinity of the bubbles. Wei and Subramanian (1993) also evaluated the location of the saddle point as the relative radius of the smaller bubble in the pair and the scaled distance of separation between the surfaces of the two bubbles are varied. The saddle point moves closer to the bubble pair when the separation distance between the bubbles is reduced. The authors provided a graph, which shows that for a given separation distance, the location is relatively insensitive to the radius ratio over a good range of values, especially when H is relatively small. A minimum in the magnitude of the scaled distance from the center of the large bubble in the pair to the saddle point is noted in all cases as the radius ratio λ is varied. When $\lambda = 0$, there is no reverse flow region, and the saddle point begins at an infinite distance behind the pair. As λ is gradually increased, it moves closer to the pair. As λ approaches unity, the two radii approach equality, and the saddle point again moves away to the point at infinity behind the pair. When the two bubbles are equal in size, the reverse flow region disappears due to an exact cancellation of the contributions from each bubble. Because Stokes flows are reversible, the entire flow pattern would be reversed if the large bubble leads the pair.

Therefore, the saddle point and the separated flow region would be in front of the pair in that case.

7.3.5 Related Problems

In concluding this section, we mention some related problems. Using the general solutions in bispherical coordinates given in Chapter Five, Golovin (1995) analyzed the motion of a rigid sphere and a gas bubble that arises because of a temperature difference between the rigid sphere and the continuous phase. The resulting temperature gradient in the fluid causes thermocapillary migration of the bubble and a drift of the sphere in the resulting flow. The sphere and the bubble approach each other if the sphere is warmer than the continuous phase. Golovin et al. (1995) considered the quasi-steady axisymmetric motion of a pair of drops due to interfacial tension gradients arising from surfactant concentration gradients using solutions in bispherical coordinates. The novelty of the problem lies in the fact that the surfactant concentration in the continuous phase is uniform in the undisturbed state. The surfactant is permitted to transfer from the drops to the continuous phase. An isolated drop, when placed in the continuous phase, would not move. When two drops are present, however, the concentration of surfactant is larger in the gap between the drops than elsewhere. The resulting interfacial tension gradients cause motion that brings the drops together. In the opposite case, when surfactant transfers into the drops, the drops move away from each other. If surfactant transfers out of one drop and into the other, the drops move in the same direction.

7.4 Motion of a Pair of Bubbles under the Combined Action of Gravity and Thermocapillarity

7.4.1 Introduction

In this section, we consider the situation when a pair of bubbles moves under the combined influence of the gravitational force and a vertical temperature gradient and predict some interesting features in this problem. As before, it is assumed that the Reynolds and Marangoni numbers are negligible and that the bubbles are spherical. Due to the linearity of the problem, the velocity of each bubble can be calculated by superposition of the results for thermocapillary migration and for buoyancy driven motion. We shall use an approximate solution from the method of reflections. Therefore, although most of the discussion will pertain to the axisymmetric case, some relevant aspects of the asymmetric case, wherein the line of centers is horizontal at a given instant, also will be considered. A detailed examination of thermocapillary migration in the asymmetric case is undertaken in the next section, with the help of a numerical solution. Most of the qualitative aspects of the present problem can be illustrated by a consideration of the motion of gas bubbles, treated by Wei and Subramanian (1995), which minimizes the number of parameters involved. Therefore, we only analyze the case of bubbles and note that the extension to drops is straightforward. At the end of this section, we provide brief comments regarding relevant experimental results.

When an upward temperature gradient is used with gas bubbles, thermocapillarity and buoyancy assist each other, and no noteworthy new phenomena can be expected. Hence, we only treat the case when a downward temperature gradient is applied, so that

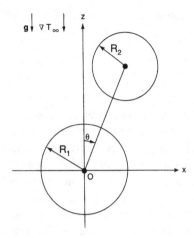

Figure 7.4.1   Sketch showing the orientation of a pair of bubbles.

the effect of thermocapillarity on the bubbles is to cause them to move downward and that of buoyancy is to push them upward. In this situation, we pointed out in Section 4.2 that an isolated bubble can achieve zero velocity when the dynamic Bond number G, defined in Equation (4.2.30), takes on the value $\frac{1}{(1+\alpha)(2+\beta)}$, given in Equation (4.2.32). This reduces to $\frac{1}{2}$ in the gas bubble limit $\alpha = \beta = 0$. We also noted that this motionless state can become unstable to small perturbations of the position of the bubble upward or downward under certain conditions. Here, we explore the possibility that two bubbles aligned vertically can be motionless, or that they may move at the same velocity even though their radii may be different. The downward temperature gradient is assumed to cause a sufficiently gentle adverse density gradient that the onset of cellular convection is avoided. It is possible to achieve such conditions experimentally as mentioned in Section 4.16.

A sketch of the system is provided in Figure 7.4.1. The drawing shows a general orientation of the pair with respect to the gravity vector. The axisymmetric case corresponds to $\theta = 0$, and the asymmetric case of two bubbles in a horizontal line corresponds to $\theta = \frac{\pi}{2}$. For definiteness, bubble 1 is located below bubble 2. Results for the reverse orientation, where bubble 2 is below bubble 1, can be obtained by reversing subscripts where needed or by considering values of $\theta > \frac{\pi}{2}$.

The radii of the bubbles are R_1 and R_2, and the distance between their centers is d. When the bubbles are unequal in size, bubble 1 is chosen to be the larger of the two. The radius ratio λ, and the scaled separation distance D, are defined as in Section 7.3.

$$\lambda = \frac{R_2}{R_1}, \tag{7.4.1}$$

$$D = \frac{d}{R_1}. \tag{7.4.2}$$

Therefore, values of λ are confined to the range 0 to 1. A dynamic Bond number can be defined for each bubble as follows:

$$G_i = \frac{g R_i (\rho' - \rho)}{3\sigma_T |\nabla T_\infty|}, \quad i = 1, 2. \tag{7.4.3}$$

Here, g is the magnitude of the acceleration due to gravity, ρ' and ρ correspond to the densities of the gas and the liquid respectively, σ_T is the rate of change of the interfacial tension with temperature, and $\nabla T_\infty = -|\nabla T_\infty| \mathbf{i}_z$ is the applied temperature gradient. When σ_T is negative, G_1 and G_2 will be positive because $(\rho' - \rho)$ is negative. The ratio $\frac{G_2}{G_1} = \frac{R_2}{R_1} = \lambda$. It will be evident from the discussion that the phenomena of interest mostly occur when the distance between the centers of the two bubbles is relatively large. Therefore, all the results presented in this section are based on the solution from the method of reflections, which is adequate for this purpose. The thermocapillary contribution to the velocity of each bubble can be obtained by using Equations (7.3.40) through (7.3.46) in conjunction with Equations (7.3.36) through (7.3.39). The gravitational contribution can be calculated from Equations (6.2.1) through (6.2.11), after equating the hydrodynamic force on each bubble to the hydrostatic force. It is possible to obtain more accurate answers in the thermocapillary case by using the solution from twin multipole expansions. However, the reflections result for thermocapillary migration is good to $O(\frac{1}{D^6})$, whereas that from the gravitational motion problem is only good to $O(\frac{1}{D^4})$. Therefore, there is no gain to be achieved by using the more accurate solution in the thermocapillary migration problem.

In the next two subsections, we present results based on these velocities for bubbles of equal size and for unequal pairs. These results are obtained for two orientations, the axisymmetric case wherein the bubbles are aligned vertically and the asymmetric case of two bubbles in a horizontal line. The motion of each bubble will be in the vertical direction for both orientations, from symmetry considerations and the reversibility of Stokes flows, as discussed in Section 6.2.3. Therefore, the physical velocity of bubble i is designated as $\mathbf{U}_i^* = U_i^* \mathbf{i}_z$. The scaled difference between the magnitudes of the two velocities ΔU is defined as follows:

$$\Delta U = \frac{U_1^* - U_2^*}{U_1^{*T(0)}}, \tag{7.4.4}$$

where $U_1^{*T(0)}$ is the purely thermocapillary migration velocity of bubble 1 when isolated. In the case of vertical alignment, a positive value of ΔU corresponds to bubbles approaching each other.

It is important to note that when the line joining the centers of the bubbles is oriented in an arbitrary direction with respect to the gravity vector, the bubbles will not necessarily move in a vertical direction. Also, the two velocities need not be in the same direction.

7.4.2 Predictions for Bubbles of Equal Size

First, consider the case of two identical bubbles. They will move with identical velocities $U\mathbf{i}_z$, provided the physical properties of the liquid remain constant. When each is isolated, this velocity will be zero when the dynamic Bond number corresponding to it takes on the value $\frac{1}{2}$ as noted in the introduction. This means that for a given pair of liquid and gas densities, and given value of σ_T, for any specified value of the radius of the bubble, there is a certain applied temperature gradient that will lead to a prediction of zero velocity. In the presence of a second bubble, the value of this temperature gradient will be altered, leading to a value of G different from $\frac{1}{2}$. In Figure 7.4.2, we display the value of the dynamic Bond number G at which the velocity is predicted to be zero, as a function of the scaled separation distance D between the two bubbles. The figure includes a curve for the axisymmetric configuration wherein the bubbles are aligned vertically and a second curve for the asymmetric configuration in which they are on a horizontal line. We note that the interaction is relatively weaker in the asymmetric case. In both

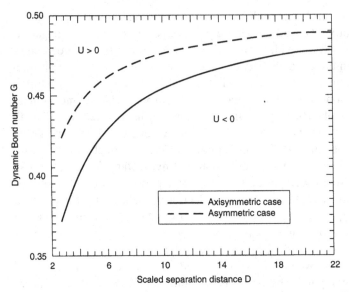

Figure 7.4.2 The critical value of the dynamic Bond number as a function of the scaled separation distance, for a pair of identical motionless bubbles for the axisymmetric and asymmetric cases. In each case, the bubbles move upward for conditions corresponding to points above the curve and downward for points below the curve (Courtesy of Academic Press).

cases, the value of G approaches $\frac{1}{2}$ asymptotically as the separation distance becomes large, albeit very slowly. The dynamic Bond number G is less than $\frac{1}{2}$ when the bubbles interact. Therefore, the temperature gradient required to achieve zero velocity for pair of identical bubbles is larger than that needed to achieve the same condition for an isolated bubble of the same size. This can be explained as follows: In the purely thermocapillary case, the interaction yields no effect on the velocity of each bubble because the hydrodynamic and thermal interaction effects lead to a precise cancellation. In the gravitational case, the presence of a second bubble causes a reduction in the hydrodynamic resistance experienced by each bubble. The reduction is greater when the bubbles are closer together. As a consequence, if two bubbles were to move only under the influence of buoyancy, each would move more rapidly than when isolated. To compensate for this increased upward velocity, we must apply a greater downward temperature gradient.

The stationary configurations predicted in the axisymmetric case are unstable to small vertical perturbations. Recall that bubble 1 is below bubble 2. Assume that for a given value of G, they are at the separation distance corresponding to zero velocities. If bubble 1 moves slightly downward, or if bubble 2 moves slightly upward, the separation distance will increase. This will reduce the buoyant contribution to the velocity of each bubble. Because the thermocapillary contribution is unaffected, the perturbation will lead to a downward motion of both bubbles in the pair at a small nonzero velocity, which will maintain this new separation distance. The variation of physical properties with distance will enhance this effect. As the two bubbles move downward, they will move into warmer liquid, which should reduce the value of G. The reason is that the reduction in the density of a liquid due to an increase in temperature is typically much larger than that of a gas. Also, the bubbles will be moving in less viscous liquid. Therefore continued downward motion at an increasing velocity can be expected. A small size decrease, which will contribute to a corresponding reduction in the value of G, can be anticipated from the increase in hydrostatic pressure, as discussed in Section 4.2.3. An increase in the size of both the bubbles caused by the increase in temperature and by the entry of dissolved gas into them will ultimately overwhelm this motion, however. Such growth in size will lead to an increase in the values of G and reverse the direction of bubble 1, and eventually that of bubble 2 as well.

Similar arguments can be advanced in the opposite case when the separation distance is slightly reduced, causing the bubbles to move upward with equal nonzero velocities. We should add that there is yet another effect of the variation in density with height due to the temperature difference. Because of this variation, one cannot apply the results of Figure 7.4.2 in the axisymmetric case to bubbles of strictly equal size. Because the density of the liquid is smaller around the lower bubble in a pair aligned vertically, it will experience a smaller buoyant force than the upper bubble. This means that it should be slightly larger than the upper bubble in order for G_1 to be equal to G_2, and this must be taken into account in calculating the velocity of each bubble in the pair.

When the bubbles are in a horizontal line, a change in the separation distance between them will induce vertical velocities that can be upward or downward, depending on the direction of the perturbation in the separation distance. They still can remain aligned horizontally, however. It is more difficult to interpret the consequence of a vertical perturbation in the position of one bubble in the pair because this will change the orientation and affect the interaction between the two. For the same reason, we have not discussed lateral perturbations in the axisymmetric orientation. Finally, we note that even at separation distances that are 20 radii, the presence of each bubble affects the other. This is evidence of the far-reaching influence of the Stokeslets from each bubble.

7.4.3 Predictions for Bubbles of Unequal Size

Now we consider two bubbles of unequal size and first focus on the axisymmetric case. As before, we require all physical properties to be spatially uniform and the bubble radii to be independent of time. With these restrictions, Figure 7.4.3 illustrates the scaled relative velocity between the two bubbles as a function of the scaled separation distance, for a representative value of $G_1 = 0.5$ at various values of the radius ratio λ. The parameter G_2 is not independent because it is given by $G_2 = \lambda G_1$, so that ΔU depends on three parameters, λ, D, and G_1. The most important result revealed by the figure is that when a small bubble is located above a large bubble, the scaled relative velocity is negative for small separations and positive for large separations. Therefore, at large separations the bubbles will approach each other regardless of the direction in which each moves, and at small separations they will move apart from each other. This means that for a given pair, there is a critical separation distance D_c at which the relative velocity is zero, and the bubbles will subsequently maintain this separation. Also, note that D_c decreases with increasing values of λ, even though the relative velocity at large separations first increases with increasing λ and then begins to decrease with further increase in λ. In the limit $\lambda \to 1$, the bubbles are of equal size. In this situation, for $G = G_1 = G_2 = \frac{1}{2}$, both bubbles will move upward for all finite values of the separation distance, maintaining whatever initial separation they have. So, every value of D can be regarded as a solution for D_c. In the limit $D \to \infty$, both bubbles will be stationary.

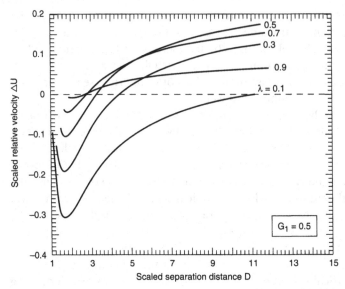

Figure 7.4.3 Scaled relative velocity between a pair of unequal bubbles plotted against the scaled separation distance for $G_1 = 0.5$ (Courtesy of Academic Press).

The existence of a critical separation distance, at which each bubble in an unequal pair moves at the same velocity, is worthy of further consideration. First, it is important to recognize that for any given radius ratio λ, zero relative velocity between the two bubbles can be realized only in a certain range of G_1 values. This fact is illustrated in Figure 7.4.4(a), in which two solid curves separate the $G_1 - \lambda$ plane into three regions. Above the upper solid curve, $\Delta U > 0$ and similarly below the lower solid curve, $\Delta U < 0$. Therefore, it is possible to have $\Delta U = 0$ only in the region between the two curves. Within this region, the dashed curve corresponds to the case when both bubbles remain stationary. Above this dashed curve but below the upper solid curve, it is possible to achieve $U_1 = U_2$, with both bubbles rising. Similarly, below the dashed curve but above the lower solid curve, one can achieve $\Delta U = 0$, with both bubbles moving downward.

The surface $\Delta U = 0$, corresponding to a critical separation D_c, is illustrated in Figure 7.4.4(b). Sections of this critical surface, at fixed values of the radius ratio λ, are displayed in Figure 7.4.4(c). Also, we show the case of stationary bubbles, $U_1 = U_2 = 0$, in this figure using a dashed curve. From the shape of this curve, we see that at relatively small values of λ, the value of G_1 required to keep the bubbles motionless is insensitive to the value of the radius ratio, remaining close to 0.5. As expected from intuition, a relatively small bubble does not influence the value of G required to hold a large bubble stationary. The small bubble will itself become stationary at a distance from the large bubble that decreases as the radius ratio increases. Note that around $\lambda \approx 0.7$, the scaled critical separation distance D_c reaches a minimum value and begins to increase as λ is increased further. The figure also shows that it is possible to achieve a given critical separation distance for two different values of G_1 under certain conditions, but each G_1 is associated with a different radius ratio.

We now discuss the question of stability of the critical separation between two unequal bubbles in a pair. For simplicity in discussing the stability issue, we treat the density and viscosity as being uniform in the liquid, in spite of the temperature variation.

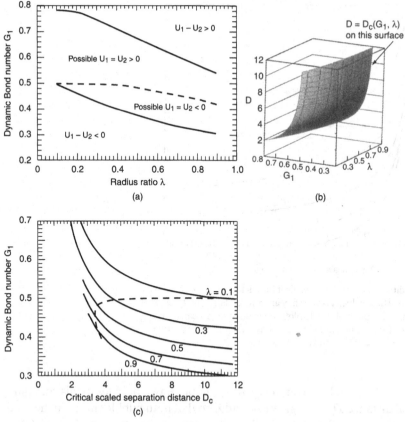

Figure 7.4.4 Conditions under which a pair of unequal-sized bubbles can move at the same velocity in the axisymmetric case: (a) dynamic Bond number for bubble 1 plotted against the radius ratio. In the region between the upper and lower curves, it is possible for $U_1 = U_2$ for a critical separation distance D_c. (b) The critical surface in the three-dimensional space of the parameters λ, G_1, and D. (c) Dynamic Bond number for bubble 1 plotted against the critical separation distance for selected values of the radius ratio. The dashed curves in (a) and (c) correspond to the conditions for $U_1 = U_2 = 0$ (Courtesy of Academic Press).

Regardless of the direction of motion of each bubble, the behavior is determined by the slope of the plot of ΔU against D in the vicinity of $D = D_c$. When the separation is slightly increased, $D > D_c$, and a positive ΔU is necessary to return to the equilibrium state. Similarly, when the separation is slightly decreased, $D < D_c$, and ΔU must become negative for stability. This type of behavior occurs when the slope of the plot is positive, which is the case when the small bubble is located above a large bubble, regardless of the direction of motion. When bubble 1 is the smaller of the two bubbles, the plot of ΔU against D will be reversed such that ΔU will be negative at large separations, crossing zero at an intermediate separation, and becoming positive at small separations. The slope is opposite in sign to the earlier case, and the critical separation will be unstable.

When buoyancy causes an unequal pair to rise, the case when the orientation angle $\theta = \frac{\pi}{2}$, which implies that the line of centers is horizontal, will occur only instantaneously. The orientation angle will change continuously with time. The same is true for the purely thermocapillary motion of a pair of bubbles in this situation. When the two

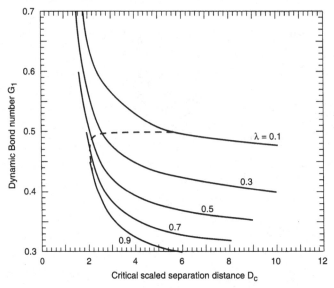

Figure 7.4.5 Conditions under which a pair of bubbles of unequal size can move at the same vertical velocity when they are aligned horizontally. The dynamic Bond number for bubble 1 is plotted against the critical scaled separation distance for a set of values of the radius ratio; the dashed curve corresponds to $U_1 = U_2 = 0$ (Courtesy of Academic Press).

effects oppose each other, however, it is possible to place two bubbles in this configuration at a critical distance D_c, for a given G_1 and λ, so that both bubbles move in the vertical direction at the same velocity. Figure 7.4.5 illustrates this result. As in Figure 7.4.4(c), the dashed curve corresponds to the case when the pair remains motionless. The value of D_c for this asymmetric case differs from that for the axisymmetric configuration. The critical separation is not stable, however. Any disturbance that alters the separation distance will lead to relative motion between the bubbles in the vertical direction. This will alter the orientation angle, which will continue to change in the same direction because there is no restoring effect. It follows that in the general case, when the orientation angle θ between the bubbles is not equal to 0, $\frac{\pi}{2}$ or π, any given orientation will change with time. This is true even if the separation distance at a given instant is such that it yields identical instantaneous vertical velocities for the bubbles.

We have neglected several physical effects in the discussion of the stability issue. The temperature dependence of the densities of the liquid and the gas, as well as that of the viscosity of the liquid, effects of changes in hydrostatic pressure, and the growth and dissolution of the bubbles will all influence the stability of the critical separation. Therefore, the results should not be interpreted literally, but only as providing information about trends.

7.4.4 Flow Structures

We now present some flow structures which arise in the motion of a pair of bubbles due to the combined influence of buoyancy and a downward vertical temperature gradient. The results displayed here were obtained by Wei and Subramanian (1995). These authors used the technique of boundary collocation, described in Section 7.5, for solving the

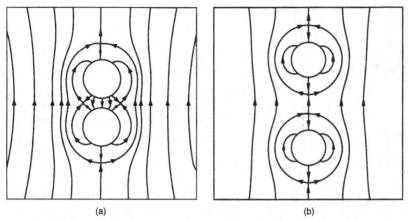

(a)   (b)

Figure 7.4.6 Streamlines in the laboratory reference frame for the axisymmetric migration of two identical bubbles for $G = 0.1$: (a) $D = 2.5$, (b) $D = 5$ (Courtesy of Academic Press).

governing equations. The streamlines in the axisymmetric case are in meridian section, and those for bubbles oriented in a horizontal line are in the x–z plane. The streamlines are not plotted for equal increments in the values of the streamfunction, and only a sufficient number are displayed for bringing forth the qualitative features. First, considering the axisymmetric case, Figure 7.4.6 shows illustrative streamlines when both bubbles in an identical pair move downward at equal velocities. When the bubbles are relatively far from each other, as in Figure 7.4.6(b), the structures in the vicinity of each bubble resemble those from Figure 4.17.4(a) for an isolated bubble. When the bubbles are close, as in Figure 7.4.6(a), we see a logical merging of the two structures, with the appearance of a ring of saddle points between the two bubbles. A single peanut-shaped dividing streamline separates the region dominated by the thermocapillary influence from the far field flow structure, which is driven by the pair of Stokeslets from the two bubbles. As we increase the value of G, this dividing streamline will approach the bubbles and eventually disappear when the bubbles begin to move upward. The flow structure for upward motion is illustrated in Figure 7.4.7 for $G = 0.7$. On the left is the flow as seen

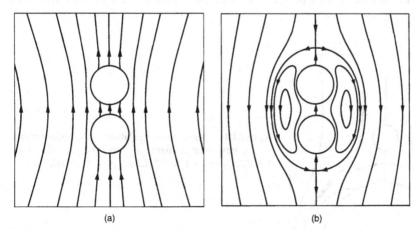

(a)   (b)

Figure 7.4.7 Streamlines for the axisymmetric migration of two identical bubbles for $G = 0.7$, $D = 2.5$: (a) in the laboratory reference frame and (b) in a reference frame attached to the bubbles (Courtesy of Academic Press).

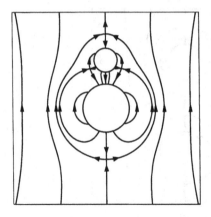

Figure 7.4.8 Streamlines in the laboratory reference frame for the axisymmetric migration of two unequal bubbles for $\lambda = 0.5$, $G_1 = 0.1$, $D = 2$ (Courtesy of Academic Press).

by an observer in the laboratory, whereas the drawing on the right shows the flow as seen by an observer riding with the pair. These streamline drawings are reminiscent of those seen in the case of an isolated bubble in Figures 4.17.4(c) and 4.17.5(b) and should be contrasted with them.

Figure 7.4.8 shows representative streamlines in the laboratory reference frame when the bubbles are unequal in size. The dividing streamline and reverse flow region, noted in Section 7.3 for purely thermocapillary migration, are absent in the present problem. This is because the far-field flow is dominated by the Stokeslets from the two bubbles. It can be seen that the near-field structure is similar to that noted for the motion of identical bubbles. For sufficiently large values of the separation distance, the single dividing streamline will break into a pair of nearly circular dividing streamlines, each enveloping a bubble, and the drawing will be analogous to Figure 7.4.6(b).

As anticipated, a large bubble will cause a substantial change in the behavior of a neighboring small bubble. Figure 7.4.9 displays streamlines that illustrate this feature. Here, $G_2 = 0.06$. In the absence of the large bubble, the small bubble would move downward, its motion being dominated by thermocapillarity. Instead, the flow in the vicinity of the small bubble is governed by the structure imposed by the large bubble. Even though the large bubble with $G_1 = 0.3$ is moving downward, it causes an upward draft of fluid in its vicinity due to the Stokeslet contribution. This sweeps the small bubble upward.

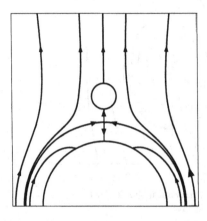

Figure 7.4.9 Streamlines in the laboratory reference frame for the axisymmetric migration of two unequal bubbles for $\lambda = 0.2$, $G_1 = 0.3$, $D = 1.7$ (Courtesy of Academic Press).

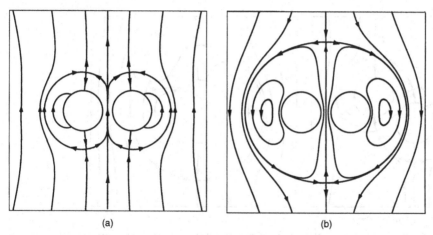

(a) (b)

Figure 7.4.10 Streamlines in the x–z plane for the migration of a pair of identical bubbles, with $\theta = \frac{\pi}{2}$, $D = 2.5$: (a) in the laboratory reference frame, $G = 0.1$ and (b) in a reference frame attached to the bubbles, $G = 0.7$ (Courtesy of Academic Press).

When the bubbles are separated horizontally, the fields are no longer axisymmetric, and the flow is three-dimensional. Therefore, for illustrating flow patterns, we have chosen the x–z plane. On this plane, v_ϕ is zero, and therefore the velocity vectors lie on it. We display the streamlines in the x–z plane in Figures 7.4.10 and 7.4.11 for some representative situations. In the case shown in Figure 7.4.10(a), the streamlines are in the laboratory reference frame for a pair of identical bubbles migrating downward. The figure shows how the streamline structure observed in Figure 4.17.4(a), in the case of an isolated bubble, is altered when two bubbles migrate side by side. When G is increased, the streamline that divides the far-field flow dominated by the Stokeslets from that near the migrating pair eventually disappears. The case when the bubbles move upward for $G = 0.7$ is shown in Figure 7.4.10(b), in a reference frame attached to the moving pair. The flow structure is reminiscent of that observed in the case of an isolated bubble in Figure 4.17.5(b), with the expected

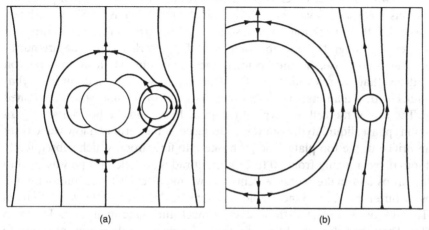

(a) (b)

Figure 7.4.11 Streamlines in the x–z plane in the laboratory reference frame for the migration of two bubbles of unequal size with $\theta = \frac{\pi}{2}$: (a) $\lambda = 0.5$, $G_1 = 0.1$, $D = 2$ and (b) $\lambda = 0.2$, $G_1 = 0.3$, $D = 1.7$ (Courtesy of Academic Press).

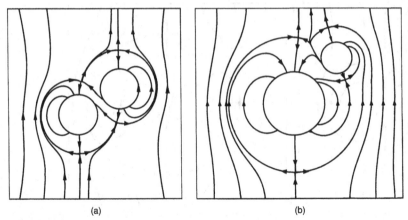

(a)   (b)

Figure 7.4.12 Streamlines in the x–z plane in the laboratory reference frame for the migration of a pair of bubbles with $\theta = \frac{\pi}{3}$, $G_1 = 0.1$:   (a) identical bubbles, $\lambda = 1$, $D = 2.5$, (b) bubbles of unequal size, $\lambda = 0.5$, $D = 2$ (Courtesy of Academic Press).

distortions from the interaction. In Figure 7.4.11, streamlines in the laboratory reference frame are shown for unequal bubbles, illustrating the changes that occur as the size ratio is altered. The anticipated distortion of the symmetric structure in Figure 7.4.10(a) is noted in the structure displayed in Figure 7.4.11(a). Figure 7.4.11(b) shows the flow around the small bubble being dominated by the flow structure induced by the large bubble.

For completeness, we have included Figure 7.4.12, which shows streamline drawings in the x–z plane for a pair of bubbles oriented arbitrarily with respect to the gravity vector. It is clear that the structures displayed in the axisymmetric case have been distorted and rotated. Saddle points are still observed, but the dividing streamlines are no longer symmetrical and do not envelop the pair.

7.4.5 Experimental Results

There are only two articles in which experimental results from the types of problems discussed in this section are available. Maris, Seidel, and Williams (1987) were interested in measuring the nucleation rates of solid hydrogen from supercooled drops of liquid hydrogen. Although the principal objective of the work was to measure nucleation rates of the solid phase, the authors mentioned some remarkable observations on interacting drops that are relevant here. The experiments were carried out in a cylindrical chamber with a diameter of 50 mm, and a height of 65 mm, which contained liquid ^4He. The top of the cell was typically kept at 15 K, and the bottom at 7 K. In this continuous phase, liquid hydrogen was introduced in the form of a low velocity jet through an orifice in the top plate. The jet broke into fine drops, which subsequently coalesced into drops ranging from 20 to 250 μm in radius. A microscope was used to observe the drops and make measurements. As we might expect, the authors noted that drops of different radii found stable positions at different vertical locations in the cell. The authors interpreted these data in much the same manner as Delitzsch et al. (1984). They then described the qualitative features of interaction phenomena observed in some of the experiments. When two drops were of comparable size, they were aligned in a horizontal line. When the drops were within a few diameters of each other, they were found to accelerate toward each other and collide. After collision, instead of forming a single drop, a small satellite drop was usually ejected on the side of the coalesced drop where the smaller of the two drops was originally located. In other experiments, a large drop located vertically below a smaller drop was observed to pull the small drop toward it. The small drop in this case reached a stable final position above the large drop. The authors mentioned vertical columns of as many as four drops in line, with the largest drop at the bottom and the smallest drop at the top. These observations are similar to the predictions for bubbles made here. The Reynolds and Marangoni numbers were not negligible in these experiments, however. Therefore, predicting the behavior observed by these authors would require the inclusion of convective transport effects. Stable motionless configurations such as these are worthy of consideration in producing a controlled distribution of inclusions in materials processing applications.

In Section 6.2, we mentioned experimental results reported by Wei and Subramanian (1994) on a pair of bubbles in the context of isothermal motion. These authors also permitted pairs of air bubbles to migrate in a Dow-Corning DC-200 series silicone oil, of nominal viscosity 1000 centistokes, under the action of a downward temperature gradient. The apparatus used was similar to that of Merritt (1988). The authors performed experiments on bubbles ranging in radius from 15 to 100 μm, in downward temperature gradients of 7.3, 9.1, and 10.9 K/mm. The largest Reynolds number encountered in the experiments was less than 10^{-6}, and the largest Péclet number was less than 10^{-3}. Therefore, the results of a quasi-steady analysis can be used with confidence in interpreting the observations. The authors displayed a plot of observed velocities versus those predicted from the reflections solution, which shows good agreement both for the large and the small bubble in the pair. Also, results from individual runs are shown, which demonstrate similar agreement. A remarkable observation made by the authors is that a small bubble, located below and displaced to one side of a large bubble, was found to move upward against the applied temperature gradient when a bubble of that size would move downward if isolated. As noted earlier, the influence of a large bubble is to produce a significant flow in its vicinity, which can drastically alter the motion of a neighboring small bubble. The streamlines in the vicinity of a large bubble moving downward in the experiments would be similar to those displayed in Figure 7.4.11(b). Therefore, outside the separatrix, there is an upward draft everywhere arising from the Stokeslet. In the experiments, the orientation angle was in the range $127 \leq \theta \leq 134°$. The separation distance was such that the small bubble was caught in the strong upward draft from the large bubble. This explanation is only approximate because there also would have been a distortion of the local temperature field around the large bubble. The small bubble would have experienced this field, instead of the undisturbed linear temperature distribution in the liquid. This effect is minor, however, when compared with that from the flow induced by the large bubble. Wei and Subramanian did not observe stable configurations such as those predicted here. The likely reason is that the injection of the bubbles occurred from the sides of the cell, and the bubbles were very small. This made it difficult to align the bubbles one below the other in a precise manner in these experiments.

7.5 Interactions between Two Bubbles Oriented Arbitrarily with Respect to the Temperature Gradient

7.5.1 Introduction

We mentioned in Section 6.2 that the Stokes motion of a pair of drops normal to the line of centers due to the action of a body force was analyzed by Zinchenko (1981), who used the solution given by O'Neill (1964) in bispherical coordinates. It is possible to perform a similar analysis in the case of thermocapillary migration, but the algebraic details are lengthy. Therefore, Wei and Subramanian (1993) chose instead to solve this problem for a pair of gas bubbles using boundary collocation. Because the method is general and can be used for handling any number of bubbles, these authors also calculated results for a chain of three bubbles with their centers on a straight line. The analogous problem for drops was considered by Keh and Chen (1992), who also used boundary collocation to treat the axisymmetric case involving two and three drops, followed by a generalization in Keh and Chen (1993) to the nonaxisymmetric situation. In this section, we present the analytical development and some results from Wei and Subramanian (1993) for the case of a pair of gas bubbles. The results for drops involve more free parameters, but all the qualitative features can be discerned from the problem involving bubbles. We only consider the case when the temperature gradient is normal to the line of centers of the bubbles. Other arbitrary orientations can be handled by superposition as implied in Equations (7.3.36) through (7.3.39).

The method of boundary collocation is useful in linear problems in geometries in which an analytical solution is not feasible. The idea behind the method is described briefly here. Lamb (1932) provided general three-dimensional solutions for the velocity and pressure fields in Stokes flow problems. This solution is given in Happel and Brenner (1965) in a form that is convenient to use. Using this solution in spherical polar coordinates for each sphere in a finite collection of rigid spheres, Ganatos, Pfeffer, and Weinbaum (1978) employed superposition to construct the solution for the motion of the fluid. The individual solutions are in the form of infinite series, and need to be truncated at a suitable number of terms. The boundary conditions at the surface of each sphere are not satisfied exactly. Instead, they are satisfied at a specific number of points called *collocation points*. The total number of these points is such that a proper linear system of equations can be written. As a consequence, a unique solution is found for the unknown constants in the solution. Later, the technique was significantly generalized by Hassonjee, Ganatos, and Pfeffer (1988), who removed some of the difficulties associated with the earlier calculations of Ganatos et al. One important idea in their work was to satisfy the boundary conditions on a discrete set of rings on the surface of each sphere instead of satisfying them at a discrete set of points. The authors noted that their method is superior to that of reflections and can handle spheres that are very close to each other.

Wei and Subramanian (1993) applied the ideas put forth by Hassonjee et al. to the case of gas bubbles undergoing thermocapillary migration. They provided results in the case of two bubbles, both for the axisymmetric problem and for problems in which the line of centers is normal to the direction of the applied temperature gradient. Meyyappan et al. (1983) had solved the axisymmetric problem earlier using bispherical coordinates, as discussed in Section 7.3. Wei and Subramanian solved it by using boundary collocation, both to provide validation of the numerical procedure and to

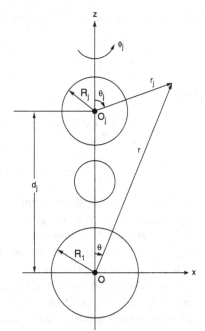

Figure 7.5.1   Sketch of a chain of bubbles.

obtain results showing the appearance of a reverse flow region and a saddle point in the streamline structure associated with this reverse flow. We have discussed this feature in Section 7.3. The authors then went on to provide results for three bubbles in the axisymmetric case and when the line of centers is normal to the applied temperature gradient. Because we shall be reporting results for a chain of three bubbles in the next section, the analysis is presented here in a form suitable for the general case of J bubbles.

A sketch of the system is provided in Figure 7.5.1, which shows three representative bubbles in line. Any number of such bubbles can be handled by the collocation method, and a general bubble is labeled bubble j. The distance between the center of the jth bubble and that of the bubble labeled 1 is designated d_j. The bubbles can be of different radii R_j as shown. We set the ratio of the viscosity of the gas in the bubbles to that of the liquid, α, and a similar ratio of thermal conductivities, β, equal to zero. A temperature gradient ∇T_∞, oriented arbitrarily with respect to the z-axis, is imposed in the liquid. We neglect convective transport of momentum and energy and assume deformation from the spherical shape to be negligible. The former assumption requires that the Reynolds and Marangoni numbers associated with the motion of each bubble be negligibly small, and the latter implies small values of the Capillary numbers, defined in the usual manner.

7.5.2   Analysis

Scaled variables are used in the analysis. Lengths are scaled by the radius R_1 of bubble 1. Velocities are scaled by using a reference velocity $v_0 = -\frac{\sigma_T |\nabla T_\infty| R_1}{\mu}$ where σ_T is the rate of change of interfacial tension with temperature, and μ is the viscosity of the liquid. Pressure and viscous stresses are scaled using $-\sigma_T |\nabla T_\infty|$. The governing equations in the continuous phase are Stokes's equation for the velocity field, and Laplace's equation for the temperature field. General solutions for the pressure and velocity fields, obtained by Lamb (1932), are given in Happel and Brenner (1965). In the continuous phase, they are written as follows:

$$p = \sum_{j=1}^{J} \sum_{n=1}^{\infty} P_{-(n+1),j} \tag{7.5.1}$$

and

$$v = \sum_{j=1}^{J} \sum_{n=1}^{\infty} \left[\nabla \times (\mathbf{r}_j \chi_{-(n+1),j}) + \nabla \Phi_{-(n+1),j} \right.$$
$$\left. - \frac{(n-2)}{2n(2n-1)} r_j^2 \nabla P_{-(n+1),j} + \frac{(n+1)}{n(2n-1)} \mathbf{r}_j P_{-(n+1),j} \right]. \tag{7.5.2}$$

If drops are to be considered, one would write results for the fields within each drop using superposition of solutions given in Happel and Brenner (1965) that are nonsingular within that drop. In Equations (7.5.1) and (7.5.2), the functions $\chi_{-(n+1),j}$, $\Phi_{-(n+1),j}$, and $P_{-(n+1),j}$ are solid spherical harmonics of order $-(n+1)$ in a spherical polar coordinate system (r_j, θ_j, ϕ_j) centered in the jth bubble, and $\mathbf{r}_j$ is the position vector measured from this origin. The length of the vector $\mathbf{r}_j$ is r_j. In writing Equation (7.5.2), we already have satisfied the condition that the velocity approach zero at a large distance from the spheres. A solid spherical harmonic function such as $\chi_{-(n+1),j}$ is written as the sum

$$\chi_{-(n+1),j}(r_j, \theta_j, \phi_j) = \sum_{m=0}^{n} \frac{1}{r_j^{n+1}} P_n^m(s_j)[A_{njm} \cos m\phi_j + B_{njm} \sin m\phi_j], \tag{7.5.3}$$

where $P_n^m(s_j)$ is the associated Legendre function, and $s_j = \cos \theta_j$. A similar equation is written for the solid harmonic $\Phi_{-(n+1),j}$, replacing the sets of constants A_{njm} and B_{njm} by C_{njm} and D_{njm}, respectively, and for $P_{-(n+1),j}$, replacing A_{njm} and B_{njm} by E_{njm} and F_{njm}, respectively. For spheres moving only in the x-z plane, symmetry requires that $A_{njm} = D_{njm} = F_{njm} = 0$. Next, the coordinates (r_j, θ_j, ϕ_j) are transformed to a single global spherical coordinate system which, without losing generality, is chosen as one centered in sphere 1. The relations between the spherical coordinates with the origin at the center of the jth bubble and the global spherical coordinates (r, θ, ϕ) are given below.

$$r^2 = r_j^2 + D_j^2 + 2r_j D_j \cos \theta_j, \tag{7.5.4}$$

$$\cos \theta = \frac{r^2 + D_j^2 - r_j^2}{2r D_j}, \tag{7.5.5}$$

$$\phi = \phi_j. \tag{7.5.6}$$

In the above, D_j is the scaled z-coordinate of the center of the jth bubble in the global coordinates. Because we have chosen the global coordinates to be centered in bubble 1, this is consistent with the definition of d_j as a separation distance in Figure 7.5.1.

The components of the velocity field can be expressed using the global spherical coordinate system. The algebra is lengthy, and therefore, the results are not reproduced here. The required relations can be found in Appendix A of Wei's doctoral thesis (1994). After this is done, the coefficients B_{njm}, C_{njm}, and E_{njm} can be computed by satisfying the boundary conditions at N_v collocation rings on each bubble surface and replacing the infinite series $\sum_{n=1}^{\infty}$ by truncated series $\sum_{n=1}^{N_v}$. These boundary conditions are similar to those written in Section 7.3 suitably specialized to the gas bubble limit. The boundary condition of a quiescent state far from the bubbles already has been satisfied by the form of the solution that has been chosen. At the surface of each bubble, we can use the kinematic condition and the tangential stress balance wherein the tangential stress in the gas phase is set equal to zero. With drops, one would use suitably generalized versions of the tangential stress balances and also the continuity of tangential velocity across each interface. Also, the hydrodynamic force on each bubble is set equal to zero.

The temperature field in the continuous phase is scaled by subtracting a reference temperature T_0, and then dividing by $R_1 |\nabla T_\infty|$. The solution of Laplace's equation for the scaled temperature field is written as follows:

$$T = T_\infty + \sum_{j=1}^{J} \sum_{n=1}^{\infty} T_{-(n+1),j}. \tag{7.5.7}$$

Here, $T_{-(n+1),j}$ is another solid spherical harmonic function, which can be written in the same form as $\chi_{-(n+1),j}$, replacing the sets of constants A_{njm} and B_{njm} by a_{njm} and b_{njm}, respectively. The function T_∞ represents the undisturbed scaled temperature field in the liquid far from the bubbles. Therefore, the chosen form of the solution already satisfies the condition that the temperature field approach the undisturbed field at a large distance from the bubbles. At the surface of each bubble, the normal heat flux is set equal to zero. For drops, solutions of Laplace's equation that are nonsingular within each drop would be used in addition, and one would require continuity of the temperature and heat flux at the surface of each drop.

The infinite series in n is replaced by truncated series $\sum_{n=1}^{N_T}$, and the boundary conditions on the surfaces of the bubbles are satisfied on N_T collocation rings. This permits the determination of the unknown coefficients a_{njm} and b_{njm} by solving a linear system of equations. Once these coefficients are known, the temperature field anywhere in the liquid, including that at the surface of the bubbles, can be determined. Therefore, the gradient of this field along each bubble surface can be evaluated and used in the tangential stress balance, which is applied on collocation rings on the surface of each bubble. When the computations are carried out either for the axisymmetric case, or the asymmetric case when the temperature gradient is perpendicular to the line of centers, further simplification of the equations is possible. This is discussed next.

Axisymmetric Migration. In this case, the velocity and temperature fields are independent of ϕ, and therefore, we need to use only the $m = 0$ mode in the solutions for these fields. For the temperature field in the axisymmetric case, $T_\infty = z = r \cos \theta$; therefore,

$$T - z = T - r \cos \theta = \sum_{j=1}^{J} \sum_{n=1}^{N_T} a_{nj0} \frac{P_n(s_j)}{r_j^{n+1}}. \tag{7.5.8}$$

The $J N_T$ unknown coefficients a_{nj0} can be obtained by satisfying the boundary conditions

$$\frac{\partial T}{\partial r_j} = 0 \quad \text{at} \quad r_j = \lambda_j, \quad j = 1, 2, 3, \ldots, J \tag{7.5.9}$$

at N_T rings on each bubble surface. Here, λ_j is the scaled radius of the jth bubble. In view of the involved connection between the global coordinate system and the local coordinates centered in each bubble, we have chosen the expedient of specifying the boundaries in this section in the manner shown above.

In the velocity field, $\chi_{-(n+1),j} = 0$, and we need to solve only for the coefficients C_{nj0} and E_{nj0}. For J bubbles, the number of unknown coefficients in the velocity field is $2JN_v$. Adding the set, $\{U_j\}$, of the J scaled velocities of the bubbles leads to a total of $J(2N_v + 1)$ unknowns.

For the velocity field given in Equation (7.5.2), the hydrodynamic force on the jth bubble is

$$\mathbf{F}_j = -4\pi \nabla (r_j^2 p_{-2,j}),$$ (7.5.10)

and therefore, setting it to zero leads to

$$E_{1j0} = 0, \quad j = 1, 2, 3, \ldots, J.$$ (7.5.11)

This reduces the total number of unknowns to $2JN_v$. The condition of impenetrability at N_v rings on each bubble surface yields JN_v equations:

$$\cos(\theta_j - \theta)v_r + \sin(\theta_j - \theta)v_\theta = U_j \cos\theta_j \quad \text{at} \quad r_j = \lambda_j, \quad j = 1, 2, 3, \ldots, J.$$ (7.5.12)

The tangential stress balance in the θ_j direction at N_v rings on each bubble surface yields another JN_v equations. In global coordinates, this condition is written as

$$\frac{1}{2}(\tau_{\theta\theta} - \tau_{rr})\sin[2(\theta_j - \theta)] + \tau_{r\theta}\cos[2(\theta_j - \theta)] = \nabla T \bullet \mathbf{i}_{\theta_j} \quad \text{at} \quad r_j = \lambda_j,$$
$$j = 1, 2, 3, \ldots, J, \quad (7.5.13)$$

where $\tau_{\theta\theta}$, τ_{rr}, and $\tau_{r\theta}$ are appropriate components of the scaled stress tensor.

We have precisely the same number of equations as unknowns. The linear system of equations can be solved to obtain the unknowns, which include the scaled velocities of the bubbles.

Migration Normal to the Line of Centers. In this case, $T_\infty = x$, and the temperature gradient points in the x-direction. Due to symmetry about the plane $y = 0$, the bubbles cannot move in the y-direction. As discussed in Section 6.2.3, the reversibility of Stokes flows precludes the bubbles from approaching each other, or moving apart from each other, in the z-direction. Therefore, the bubbles will move in the x-direction. From symmetry considerations, it can be seen that only the $m = 1$ mode needs to be considered in the solutions. We can write

$$T - x = T - r\sin\theta\cos\phi = \sum_{j=1}^{J}\sum_{n=1}^{N_T} a_{nj1}\frac{P_n^1(s_j)}{r_j^{n+1}}\cos\phi_j.$$ (7.5.14)

The JN_T unknown coefficients can be obtained by satisfying the zero heat flux condition at the surface of each of the J bubbles, with N_T collocation rings on each bubble surface.

The number of unknown coefficients, B_{nj1}, C_{nj1}, and E_{nj1}, in the truncated series $\sum_{n=1}^{N_v}$ for the velocity field components, is $3JN_v$. Adding the set, $\{U_j\}$, of the unknown scaled velocities of the bubbles, leads to $J(3N_v + 1)$ unknowns. The condition of zero net force on each bubble gives

$$E_{1j1} = 0, \quad j = 1, 2, 3, \ldots, J,$$ (7.5.15)

which reduces the total number of unknowns to $3JN_v$. The condition of impenetrability at N_v rings on each bubble surface gives JN_v equations:

$$\cos(\theta_j - \theta)v_r + \sin(\theta_j - \theta)v_\theta = U_j \sin\theta_j \cos\phi_j \quad \text{at} \quad r_j = \lambda_j,$$

$$j = 1, 2, 3, \ldots, J. \quad (7.5.16)$$

The tangential stress balance in the θ_j direction provides another JN_v equations, similar to the axisymmetric case, as shown in Equation (7.5.13). The tangential stress balance in the ϕ_j direction, at N_v rings on each bubble surface, yields an additional JN_v equations. In the global spherical polar coordinates, this condition is written as follows:

$$\tau_{r\phi}\cos(\theta_j - \theta) + \tau_{\theta\phi}\sin(\theta_j - \theta) = \nabla T \bullet \mathbf{i}_{\phi_j} \quad \text{at} \quad r_j = \lambda_j, \quad j = 1, 2, 3, \ldots, J.$$

$$(7.5.17)$$

Here, $\tau_{r\phi}$ and $\tau_{\theta\phi}$ are appropriate components of the scaled stress tensor. Solution of the linear system of equations yields the scaled velocities of the bubbles, as well as the unknown constants in the velocity field.

Calculation of numerical results involves choosing the location and number of the collocation rings suitably. This is discussed in detail by Hassonjee et al. (1988) for the case of rigid spheres and by Wei and Subramanian (1993) in the present problem. Overall, between 20 and 30 collocation rings on the surface of each bubble appear to be adequate. Placing rings at $\theta_j = 0, \frac{\pi}{2}$, or π leads to singular matrices in the axisymmetric case, and similarly, the degenerate rings at $\theta_j = 0$ and π produce singular matrices in the asymmetric case. Therefore, care must be used in the choice of the number and location of the collocation rings to obtain an accurate solution.

7.5.3 Results

As mentioned in the introduction, results will be given in the present section for a pair of bubbles moving under the action of a temperature gradient applied normal to their line of centers and, in the next section, for a chain of three bubbles. We adopt the convention used in Section 7.3 and define a scaled separation distance between the surfaces of the two bubbles, H, and an interaction parameter, Ω_i, for bubble i, using Equations (7.3.55) and (7.3.56). In the asymmetric orientation under consideration, the interaction parameters are related to the mobility functions W_{ij}:

$$\Omega_1 = W_{11} + \lambda W_{12} \quad (7.5.18)$$

and

$$\Omega_2 = \frac{1}{\lambda}W_{21} + W_{22}. \quad (7.5.19)$$

Here, $\lambda = \frac{R_2}{R_1}$ is the radius ratio. As before, bubble 1 is chosen to be the larger of the two bubbles. The effect of the interaction is far more pronounced on the small bubble. This is shown in Figure 7.5.2, where the interaction parameter Ω_2 is plotted against the scaled separation distance between the surfaces, for a set of values of the radius ratio. The most important result is that in contrast to the axisymmetric case, the small bubble is retarded by the presence of the large bubble. The retardation can be so severe as to make it come to a standstill or move backward against the applied temperature gradient, as implied by negative values of Ω_2 in the figure. The

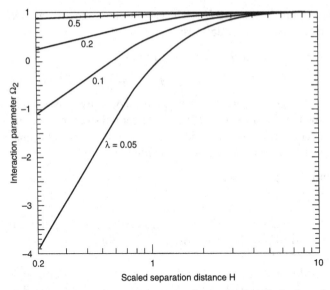

Figure 7.5.2 Interaction parameter for the smaller bubble in the pair plotted against the scaled separation distance between the surfaces of the two bubbles for a set of values of the radius ratio.

reason for this behavior is that the thermocapillary flow generated by the large bubble along its equatorial plane, in the absence of the small bubble, is directed against the temperature gradient, and the small bubble is caught in this draft. The large bubble moves more rapidly than when isolated, again in contrast to the predictions in the axisymmetric problem. The effect is even smaller than that noted in the axisymmetric case, however. For all practical purposes, the large bubble may be assumed to move as though it is isolated. Therefore, we do not display the interaction parameter Ω_1 graphically. As pointed out in Section 7.3, bubbles of equal size move at the velocity they each would have when isolated, regardless of the separation distance, or their relative orientation with respect to the temperature gradient. The results from the numerical solution, for both Ω_1 and Ω_2, are practically indistinguishable from those predicted from the twin multipole expansion technique. The latter are given by Equations (7.3.51) through (7.3.54) used in conjunction with Equations (7.3.36) through (7.3.39), and the results plotted in Figure 7.5.2 were calculated from these equations.

The flow in the asymmetric case is fully three-dimensional. As mentioned in Section 7.4, however, v_ϕ is zero on the x–z plane, and therefore the velocity vectors lie on it. Hence, streamlines can be used effectively to illustrate features of the flow. Figure 7.5.3 contrasts flow features in the far-field for body force driven motion against those for thermocapillary migration; magnified views of the flow fields in the vicinity of the bubbles, in the region bounded by the small rectangle in the main view, are shown in the insets. The flow in the gap for body force driven motion appears straightforward. However, in the thermocapillary case, the flows generated by the thermocapillary stresses support each other but oppose the flow coming toward the gap and leaving it. This opposition produces saddle points and separatrices. If the bubbles were of equal size, the location of these saddle points would be symmetric in the gap. Note that a dividing streamline, which loops around the small bubble, encloses a region consisting

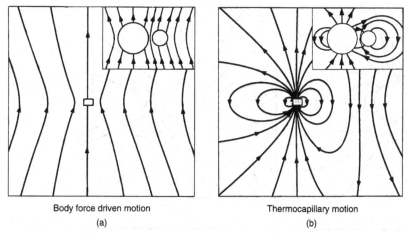

Body force driven motion   Thermocapillary motion
        (a)   (b)

Figure 7.5.3 Streamlines in the laboratory reference frame in the x–z plane for a pair of bubbles; $\lambda = 0.5$, $H = 0.25$. (Reprinted with permission from Wei, H., and Subramanian, R. S. Thermocapillary migration of a small chain of bubbles. *Phys. Fluids A* 5, No. 7, 1583–1595. Copyright 1993, American Institute of Physics.)

of streamlines that begin on the surface of the small bubble, loop around to the right, and end on its surface. To the left of the region bounded by this dividing streamline, there is another region that encloses streamlines that begin and end on the surface of the large bubble. To the right of the bubble pair, there is another dividing streamline, and the flow on the right side of that streamline is different in structure. On the left are streamlines that begin and end on the surface of the larger bubble. On the right, we see fluid being drawn in from infinity in the upper half plane and thrown out in the lower half plane. In contrast, such behavior is not observed to the left of the large bubble, no matter how far one looks.

The explanation for the far-field flow structure is simple and is modeled after that given in Section 4.17 for axisymmetric problems. The temperature field on the surface of a given bubble can be decomposed into pure surface harmonics in a coordinate system with its origin at the center of that bubble. The functions $P_n^1(\cos\theta)\cos\phi$ are purely imaginary, and this makes the coefficients multiplying them also purely imaginary, so that the solutions for the field variables are real valued. For the purpose of this discussion, we ignore this detail and use $P_n^1(\cos\theta)\cos\phi$ to imply the corresponding real-valued function. For an isolated bubble, in a coordinate system with the origin at its center, the only mode that is excited is the $P_1^1(\cos\theta)\cos\phi$-mode. The corresponding thermocapillary stress drives a potential dipole flow, which decays as $\frac{1}{r^3}$ as $r \to \infty$, where r is the distance measured from the center of the bubble. In the presence of a second bubble, all the higher modes with the appropriate symmetry are excited. The mode $P_n^1(\cos\theta)\cos\phi$ produces a flow that decays as $\frac{1}{r^n}$ as $r \to \infty$, which is similar to the axisymmetric case. As a consequence, the velocity field resulting from the $P_2^1(\cos\theta)\cos\phi$-mode decays least rapidly with distance (as $\frac{1}{r^2}$), and dominates the far-field flow structure. In Figure 7.5.4, the flow patterns in the x–z plane generated by the first four pure $P_n^1(\cos\theta)\cos\phi$-modes are displayed.

A delicate balance of the contributions from the pure modes determines the flow structure. In the following, we refer to the flows driven in the single bubble case by the $P_1^1(\cos\theta)\cos\phi$ and the $P_2^1(\cos\theta)\cos\phi$-modes as the first and second modes, respectively.

Figure 7.5.4 Streamlines in the laboratory reference frame in the x–z plane for motion driven by a scaled surface temperature field equal to $P_n^1(\cos\theta)\cos\phi$: (a) $n = 1$, (b) $n = 2$, (c) $n = 3$, (d) $n = 4$. (Reprinted with permission from Wei, H., and Subramanian, R. S. Thermocapillary migration of a small chain of bubbles. *Phys. Fluids A* 5, No. 7, 1583–1595. Copyright 1993, American Institute of Physics.)

In the vicinity of the bubbles, the flow is dominated by the first mode, especially from the larger bubble. As one moves away from the bubbles, however the contributions from the second modes can be expected to control the flow structure. The sign of this contribution, relative to that of the first mode, is positive for the bubble on the left and negative for that on the right. Therefore, the far-field flow structure will depend on the relative strength of the contributions from the second modes. The coefficients of the second modes are equal, but opposite in sign, for two identical bubbles. Thus one sees a cancellation in a manner similar to that noted in the axisymmetric case. In the case of unequal bubbles, the large bubble dominates in establishing the flow structure. Therefore, when the small bubble is placed on the right side of the large bubble, as shown in Figure 7.5.3(b), the flows from the first and second modes reinforce each other in the left half plane, $x < 0$. On the right half plane, an opposition occurs between these two flows, as is evident from Figures 7.5.4(a) and 7.5.4(b). This leads to a dividing streamline and a structure dominated by the residual effects from the second mode in this half plane. The position of this dividing streamline and its shape depend on the relative strengths of the two modes. As in the axisymmetric case, the net effects from the second mode are weak when the radius ratio is close to unity and when it is close to zero. Therefore, the greatest effects are observed in an intermediate situation. Wei and Subramanian (1993) provided a drawing illustrating the behavior of the distance of the dividing streamline from the center of the large bubble. The qualitative features of this behavior are similar to those discussed in the axisymmetric case in Section 7.3.

7.6 Interactions among Three or More Drops

7.6.1 Introduction

In the last two sections, we considered the motion of a pair of drops and provided results in the gas bubble limit. In this section, we discuss available results for a collection of three or more drops. As in the earlier sections, convective transport of momentum and energy are neglected, and the deformation of the drops from the spherical shape also is assumed to be negligible. When the suspension is dilute, which implies that the distance between the drops is relatively large compared with their size, a first approximation to the behavior of a test drop can be obtained by summing over the interactions with every other drop as though only that pair is present. Such a pairwise-additive approximation is constructed in this section, and it is shown that it provides an adequate description in the case of a chain of three bubbles, a case that was solved by Wei and Subramanian (1993) using boundary collocation. Finally, we briefly discuss the literature dealing with suspensions of drops undergoing thermocapillary migration and provide some results.

7.6.2 Pairwise-Additive Approximation

When more than two drops are present, a pairwise-additive approximation can be constructed that proves useful. It involves summing the relevant two-drop interactions, one pair at a time. For convenience, we use physical velocities and lengths here. All the drops are assumed to be of the same material, and the radius of drop i is R_i. A radius ratio λ_{ij} is defined as follows:

$$\lambda_{ij} = \frac{R_j}{R_i}. \tag{7.6.1}$$

If $\mathbf{U}_i$ stands for the velocity of drop i, it can be represented as follows:

$$\mathbf{U}_i = \mathbf{U}_i^{(0)} + \sum_{j \neq i} \Delta \mathbf{U}_i^{(j)}, \tag{7.6.2}$$

where the summation is over all the drops excluding drop i and

$$\Delta \mathbf{U}_i^{(j)} = \mathbf{U}_i^{(j)} - \mathbf{U}_i^{(0)}. \tag{7.6.3}$$

The symbols $\mathbf{U}_i^{(j)}$ and $\mathbf{U}_i^{(0)}$ represent the velocity of drop i when under the influence of drop j alone and when isolated, respectively. For the purpose of constructing an approximation, it is best to use simple analytical results. Therefore, we use the solution obtained from the method of reflections, given in Equations (7.3.40) through (7.3.43), in conjunction with the results in Equations (7.3.36) through (7.3.39). For gas bubbles, it would be possible to also construct a higher order approximation using the results in Equations (7.3.47) through (7.3.54), if desired. Because $\mathbf{U}_i^{(0)}$ and $\mathbf{U}_j^{(0)}$ are both in the direction of ∇T_∞, we can write $U_j^{(0)} = \frac{R_j}{R_i} U_i^{(0)}$ from Equation (4.2.28). Then, the reflections result for two arbitrarily oriented drops can be rearranged as follows:

$$\mathbf{U}_i^{(j)} = [(T_{11} + \lambda_{ij} T_{12})\, \mathbf{ee} + (W_{11} + \lambda_{ij} W_{12})\, (\mathbf{I} - \mathbf{ee})] \bullet \mathbf{U}_i^{(0)}. \tag{7.6.4}$$

Here, $\mathbf{I}$ is the identity tensor, $\mathbf{e}$ is the unit vector pointing in the direction from drop i to drop j, and the mobility coefficients are given below.

$$T_{11} = 1 - 2p\frac{R_j^3}{d_{ij}^3} + \left[4p^2 - \frac{3(2+5\alpha)}{2(1+\alpha)}\right]\frac{R_i^3 R_j^3}{d_{ij}^6}, \tag{7.6.5}$$

$$T_{12} = \frac{R_j^3}{d_{ij}^3} - \left[2p - \frac{9(2+3\alpha)(1-\beta)}{2(1+\alpha)(3+2\beta)}\right]\frac{R_i^3 R_j^3}{d_{ij}^6}, \tag{7.6.6}$$

$$W_{11} = 1 + p\frac{R_j^3}{d_{ij}^3}\left(1 + p\frac{R_i^3}{d_{ij}^3}\right), \tag{7.6.7}$$

$$W_{12} = -\frac{1}{2}\frac{R_j^3}{d_{ij}^3}\left(1 + p\frac{R_i^3}{d_{ij}^3}\right). \tag{7.6.8}$$

The symbol d_{ij} represents the distance of separation between drops i and j, α represents the ratio of the viscosity of the drop phase to that of the continuous phase, and β stands for a similar ratio of thermal conductivities. As in Section 7.3, p is given by

$$p = \frac{1-\beta}{2+\beta}. \tag{7.6.9}$$

7.6.3 Results for a Chain of Three Bubbles

The graphical results presented here are based on the boundary collocation solution obtained by Wei and Subramanian (1993) in the gas bubble limit $\alpha = \beta = 0$. The technique already has been described in Section 7.5, and a sketch of the system is provided there in Figure 7.5.1. The results, available only for three gas bubbles with their centers on a straight line, are organized as follows. First, we present results from the axisymmetric calculation for the scaled velocities of three bubbles in line. This will be followed by similar results in the asymmetric orientation where the line of centers is perpendicular to the applied temperature gradient. After this, flow structures from these problems will be presented and discussed. In the gas bubble limit, we can write the following simple results from the pairwise-additive approximation. For the axisymmetric case,

$$\Delta U_i^{(j)} = -\left[(1 - \lambda_{ij})\frac{R_j^3}{d_{ij}^3} + 2(1 - \lambda_{ij})\frac{R_i^3 R_j^3}{d_{ij}^6}\right]U_i^{(0)}; \tag{7.6.10}$$

in the situation where the line of centers is perpendicular to the applied temperature gradient,

$$\Delta U_i^{(j)} = \left[\frac{1}{2}(1 - \lambda_{ij})\frac{R_j^3}{d_{ij}^3} + \frac{1}{4}(1 - \lambda_{ij})\frac{R_i^3 R_j^3}{d_{ij}^6}\right]U_i^{(0)}. \tag{7.6.11}$$

In the case of three bubbles, the number of parameters doubles. We define a scalar interaction parameter $\Omega_i = \frac{U_i}{U^{(0)}}$ for bubble i in the usual manner. There are two independent radius ratios and two scaled separation distances. One way to restrict the

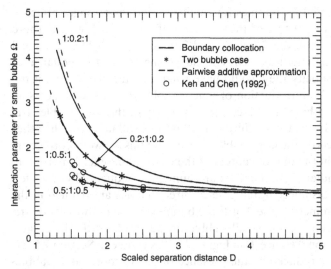

Figure 7.6.1 Axisymmetric motion of three bubbles with two being of equal size. The interaction parameter for the small bubble is plotted against scaled separation distance; the ratios next to the curves represent the relative scaled radii of the bubbles in the chain. (Reprinted with permission from Wei, H., and Subramanian, R. S. Thermocapillary migration of a small chain of bubbles. *Phys. Fluids A* 5, No. 7, 1583–1595. Copyright 1993, American Institute of Physics.)

number of parameters is to place one bubble between two other bubbles that are of equal size, but different in size from that bubble. In this case, there are only two radii, and only a single radius ratio parameter λ, which is defined as the ratio of these two radii such that it is less than or equal to 1. The bubbles are numbered 1, 2, and 3, starting from the bubble in the rear. There still are two distance parameters, D_1 and D_2, which are defined as the separation distances between bubbles 1 and 2 and bubbles 2 and 3, respectively, scaled by the radius of the largest bubble in the set. For illustration, we have chosen $D_1 = D_2 = D$. As in the two-bubble case, the larger bubbles are hardly affected by the interaction, whereas the smaller bubbles in the set experience significant change. Results for the interaction parameter for a small bubble, simply designated as Ω without a subscript, are displayed in Figure 7.6.1, which also includes results from Keh and Chen (1992) where available, results when only a pair of bubbles with the same radius ratio is present and predictions from the pairwise-additive approximation. For each of the two radius ratios used, two cases are considered in the three bubble problem. In one, the small bubble is in the middle and is flanked by two larger bubbles of equal size. In the other, the situation is reversed, with the large bubble in the middle, flanked by two small bubbles of equal size. The results from the pairwise-additive approximation are simple to use and are adequate for describing the scaled velocity of the small bubble when $\lambda = 0.2$, except when the bubbles are fairly close together. For $\lambda = 0.5$, the approximation is indistinguishable on the graph from the collocation result. The approximation does even better for the large bubble in the set as demonstrated by Wei (1994).

Just as in the case of a pair, the small bubble always moves more rapidly than it would when isolated, in all the instances shown. The scaled velocity falls off toward unity when the scaled separation distance, D, is approximately 5. Obvious effects, such as the greater influence of the larger bubbles for smaller λ, and the stronger effect of the larger bubbles when there are two of them instead of one, are observed. There is virtually no difference between the 2-bubble and 3-bubble cases when two small bubbles flank a large one in the middle, which is not surprising. Recall that equal-sized bubbles move at the same velocity they would have when isolated. Therefore, if the large bubble is not present, the introduction of a new bubble of the same size as an existing (small) bubble should have no effect on the velocity of the first bubble. The situation is more complicated because the third (small) bubble will influence the other small bubble through its effects on the large bubble as well, but this is a weak correction.

Consider the case wherein the line of centers of the chain of bubbles is normal to the applied temperature gradient. The bubbles are numbered sequentially from left to right, and the separation distances in the case of three bubbles are defined in the same manner as in the axisymmetric case. For three bubbles in line in the asymmetric case, the interaction parameter for the small bubble in the set, Ω, is displayed as a function of the scaled separation distance D in Figure 7.6.2. As noted in Section 7.5, the velocity of the small bubble is reduced by the presence of one or more large bubbles in this orientation. Comments already made in the context of the axisymmetric case apply regarding the connection to the two-bubble case and the performance of the pairwise-additive approximation.

Flow structures in the case of three or more bubbles in a chain, both in the axisymmetric and in the asymmetric case, can be divided into two categories based on the presence or absence of geometrical symmetry. In the geometrically symmetric case, a plane can be found normal to the line of centers such that the configuration has mirror image symmetry about this plane. For three bubbles, this occurs when the two outer bubbles are of the same size and the third bubble is located between them. When the

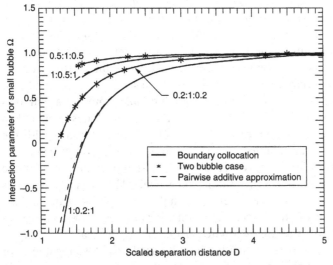

Figure 7.6.2 Motion of three bubbles perpendicular to their line of centers, with two bubbles being of equal size. The interaction parameter for the small bubble is plotted against scaled separation distance. The ratios next to the curves represent the relative scaled radii of the bubbles in the chain. (Reprinted with permission from Wei, H., and Subramanian, R. S. Thermocapillary migration of a small chain of bubbles. *Phys. Fluids A* 5, No. 7, 1583–1595. Copyright 1993, American Institute of Physics.)

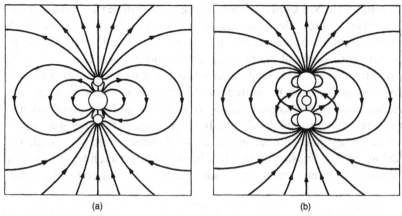

Figure 7.6.3 Streamlines in the laboratory reference frame in a meridian plane for the axisymmetric motion of three bubbles. The two outer bubbles have the same radius, and the bubble in the middle is twice that size in (a) and half that size in (b); $D_1 = D_2 = D = 2$. (Reprinted with permission from Wei, H., and Subramanian, R. S. Thermocapillary migration of a small chain of bubbles. *Phys. Fluids A* 5, No. 7, 1583–1595. Copyright 1993, American Institute of Physics.)

configuration lacks geometrical symmetry, the flow structures will resemble those displayed in Figure 7.3.3(b) in the axisymmetric case and those in Figure 7.5.3(b) in the asymmetric case. In geometrically symmetrical configurations, the dividing streamline that separates the flow into two regions is absent. Illustrations of the near-field flow structure in such cases are provided in Figure 7.6.3 in the axisymmetric case and in Figure 7.6.4 in the asymmetric case. In the asymmetric case, the saddle points and separatrices in the gap, mentioned in the context of Figure 7.5.3(b), are more clearly seen in Figure 7.6.4. Qualitatively, no new features can be expected when any number of additional bubbles are added in line.

Nas (1995) has solved the governing momentum and energy equations for a collection of drops, accommodating convective transport of momentum and energy and shape deformation of drops. This is the only work known in which a significant collection of drops has been considered while taking convective transport effects into account.

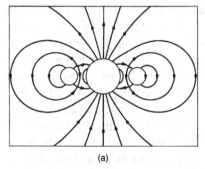 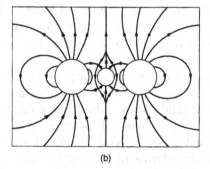

Figure 7.6.4 Streamlines in the laboratory reference frame in the x–z plane for the motion of three bubbles perpendicular to their line of centers. The two outer bubbles have the same radius, and the bubble in the middle is twice that size in (a) and half that size in (b); $D_1 = D_2 = D = 2$. (Reprinted with permission from Wei, H., and Subramanian, R. S. Thermocapillary migration of a small chain of bubbles. *Phys. Fluids A* 5, No. 7, 1583–1595. Copyright 1993, American Institute of Physics.)

To accomplish the solution, Nas used a front tracking finite difference method due to Unverdi and Tryggvason (1992), which was mentioned briefly in Sections 3.2 and 4.15. Because of limited computational resources, the drops were placed in a box with the top and bottom boundaries several bubble radii apart, and periodic conditions were imposed at lateral boundaries. Also, most of the calculations were performed on two-dimensional problems, with results presented for the three-dimensional problem only in a limited number of cases. Nas considered a pair of drops, and also a collection of nine drops, in the three-dimensional simulation. In the latter case, two runs were made, one with equal-sized drops and another in which two different sizes were used. The principal conclusion reached by Nas was that the drops displayed a tendency to line up in layers normal to the temperature gradient, whether moving as a pair or in a large collection.

7.6.4 Behavior of a Suspension of Drops Moving under the Influence of Gravity or Thermocapillarity

Two types of problems have been considered in the context of a collection of drops. In the first, the drops have a certain distribution of sizes, and the object is to determine how this distribution evolves with time as the drops move and encounter each other. It is usually assumed that if the trajectories of two drops intersect, they will coalesce. Therefore, a given size distribution should gradually coarsen with time. To determine the evolution of the size distribution, one needs to know the collision rate function, which is the rate at which drops of a given radius, R_1, collide with drops of another radius, R_2, per unit volume of the suspension. In a sufficiently dilute dispersion, only the interactions between pairs of drops needs to be considered because the chances of three drops being sufficiently close are negligibly small. Smoluchowski (1917) obtained a result for the collision rate due to Brownian motion, in the absence of hydrodynamic interactions or interparticle forces between the particles. In the case of deterministic motion of drops such as that arising from gravitational settling, a simple result can be written for the collision rate if the drops do not interact with each other. In practice, they do interact; commonly one drop will go around the other. Consider two drops, labeled 1 and 2, initially far apart so that each settles in the direction of the gravity vector. If drop 1 begins its relative motion with its center sufficiently close to a vertical line drawn through the center of drop 2, then it will arrive sufficiently close to drop 2 to collide with it. A critical trajectory for the relative motion of drop 1 with respect to drop 2 must be determined such that drop 1 will just graze the surface of drop 2 during their encounter. A rotation of this critical trajectory through an angle 2π about the vertical axis yields a surface which, along with the surface of drop 2, encloses an annular volume that contains the centers of all drops of the same size as drop 1 that can possibly collide with drop 2. In the absence of any hydrodynamic interactions, the critical trajectory is simply a vertical line positioned at a distance equal to the sum of the radii of the two drops from the vertical line through the center of drop 2. When hydrodynamic interactions are included, the critical trajectory is obtained by assuming that the motion at each instant is quasi-steady and integrating the results for the instantaneous velocity calculated for a series of suitable relative orientations of the two drops. This places a limitation that the Reynolds number be negligible. In the case of thermocapillary motion, the Marangoni number also must be negligible. These considerations limit the use of such analyses to small drops and gentle driving forces.

In the case of spherical drops, Zhang and Davis (1991) provided a good summary of available results on hydrodynamic interactions between drops for motion driven by a body force and a detailed calculation of collision rates due to both Brownian motion and gravitational settling. The results for collision rates are reported in the form of collision efficiencies. These are defined using an ideal standard for collision rates in each of the two cases considered. For Brownian motion, the result of Smoluchowski (1917), which was obtained ignoring hydrodynamic interactions and interparticle forces, is used as the standard. For deterministic motion, the standard is the collision rate determined by ignoring hydrodynamic interactions. The authors noted that nonzero collision rates are possible for undeformed drops even in the absence of attractive forces, whereas rigid spheres cannot come into contact unless an attractive force is present between them. This is because the rate at which the hydrodynamic resistance increases, as the distance between the particles approaches zero, is different in the two cases. Of course, drops will deform when sufficiently close, and such deformation must be considered in a precise calculation of collision rates. Manga and Stone (1995b) have studied this issue in detail in the case of bubbles. Their experimental results show that coalescence is more likely if the bubbles are of comparable size. From a model of the size evolution of a cloud of bubbles, they concluded that deformation has two important consequences. The first is that coalescence rates in a suspension of deformable bubbles can be an order of magnitude higher than the rates obtained in the case of nondeformable bubbles. The second is that the dilute system approximation, which permits one to consider only pair interactions, is superior in the case of highly deformable drops, and can be used at larger concentrations. This is because the rate of coalescence is mostly dependent on deformation in the limit of large Bond number, and the magnitude of the deformation decays as the inverse square of the scaled distance from a drop, whereas the Stokeslet from a moving sphere decays as the inverse of the scaled distance, so that the probability that a third drop will influence the dynamics is larger in the case of nondeformable drops. Manga and Stone also noted that spatial inhomogeneities in bubble concentration can evolve from the alignment and coalescence of deforming bubbles.

Returning to spherical drops, collision rates for the thermocapillary motion of such drops were calculated by Zhang and Davis (1992). In this work, the authors obtained collision rates using mobility functions from the solution by the method of reflections when the line of centers is perpendicular to the temperature gradient and the solution in bispherical coordinates in the axisymmetric case. Because the latter is computationally expensive when the drops are close, the results are fitted to an approximate expansion in powers of the square root of the separation distance between two drops when that distance is small. An interesting prediction made by Zhang and Davis is that when the conductivity of the drops is sufficiently larger than that of the continuous phase, the drops will not collide. In that situation, for motion along the line of centers, as the drops approach each other, the relative velocity vanishes at a critical separation distance and changes sign for smaller separation distances. This implies that if the drops are separated by a distance smaller than the critical separation distance, they would move away from each other. The critical separation distance increases with increase in either the viscosity ratio α or the thermal conductivity ratio β. If the large drop approaches the small drop from behind, this critical separation is stable. When it is in front of the small drop, the situation is unstable, meaning that a slight displacement will lead to an increase of that displacement in the same direction. The origin of this effect is the same as that leading to the behavior displayed in Figure 7.1.2, which shows the influence of a neighboring isothermal plane surface on the temperature difference over the surface of a drop. For two drops, the effect can be explained by consideration of the temperature difference between the poles of each drop. The authors provided sample results for a radius ratio $\lambda = \frac{1}{2}$ and $\alpha = 1$, showing that ΔT across the small drop is increased to a much greater extent than that across the large drop for $\beta > 1$, when the separation distance is very small. Therefore, although the velocity of each drop is increased, it is the larger increase in the velocity of the small drop, when the drops are very close, that leads to the prediction that the drops do not collide. This effect leads to a situation wherein the collision efficiency vanishes when either α or β is large compared with unity; therefore, for a given viscosity ratio, when the thermal conductivity ratio exceeds a critical value, there would be no collisions, and the suspension would be regarded as stable. This remarkable result does not have a counterpart in collisions induced by gravitational settling. In gravity-induced settling, collision efficiencies decrease as the viscosity ratio is increased or as the size ratio λ is decreased, but do not become zero, only approaching that value in asymptotic limiting situations. Also, collision efficiencies for thermocapillary migration are an order of magnitude larger than those for gravitational settling because of the weaker nature of the hydrodynamic interactions between drops in the case of thermocapillary motion. The exception, of course, occurs in the case of highly conducting drops for which the collision efficiency can become zero. Zhang and Davis (1992) also included van der Waals attraction forces in their calculations and showed that this naturally leads to an increase in the collision efficiency.

Zhang, Wang, and Davis (1993) used the results for collision rates, mentioned above, in a model for the temporal evolution of the size distribution in a collection of drops. Initially, the size distribution is assumed to be Gaussian. The drops are subjected to the simultaneous influence of gravity and thermocapillarity acting either parallel or antiparallel to each other. In the former case, the two influences support each other, whereas they oppose each other in the latter case. The authors also provided results on the evolution of the average drop size in the purely thermocapillary case. Zhang et al. made two interesting predictions. When gravity and thermocapillarity oppose each other and the relative velocity due to thermocapillarity at large separations is larger than that due to gravity, the drops approach each other. Even as they interact, however, the net relative velocity increases as the separation distance is reduced. This continues until the separation distance becomes small when the relative velocity begins to decrease. The physical explanation of this interesting prediction is precisely the same as that offered earlier in Section 7.1, in the context of Figure 7.1.5, for a drop moving normal to a plane surface under the combined action of gravity and thermocapillarity. It is the difference in the relative rates of decay of the disturbance velocity fields due to the two driving forces that causes this counter-intuitive interaction behavior. The second prediction made by the authors is that when gravity and thermocapillarity oppose each other, under the right conditions, a broad distribution of small drops can ultimately coalesce into a relatively narrow distribution of drops of a certain larger size, even though the size distribution can broaden initially. This happens because collision rates become zero under certain conditions. The authors suggested that this information has practical utility in that drops of a certain desired uniform size can be formed by beginning with smaller drops of a nonuniform size distribution and applying a judicious temperature gradient that opposes gravity-driven settling. Satrape (1992) independently considered the size evolution problem for a cloud in the limiting case of gas bubbles. He used his solution from twin multipole expansions, mentioned earlier, in calculating the mobility functions. He found that using the reflections solution does not lead to a large difference in the collision efficiency. Results are provided for the size evolution of a collection initially distributed in a Gaussian manner about a mean size. Also, Satrape provided a graph of the temporal evolution of the volume-averaged migration velocity of the bubbles.

A different problem is that of determining the velocity of a drop in a cloud of identical drops. Because the configuration of the drops with respect to each other can vary, one usually considers a statistically homogeneous suspension in which drops take all available (nonoverlapping) positions with equal probability and seeks the mean velocity over all possible realizations of the configuration. This is done only in the linear limit of negligible convective transport because of the need to superpose solutions. Although others had considered the problem earlier, Batchelor (1972) was the first to treat this problem carefully for a statistically homogeneous suspension of rigid spheres settling under gravity in a fluid, and he put forth several important ideas. He showed how statistical methods can be used to calculate average properties of suspensions and how these properties can be written in terms of convergent integrals. The difficulty associated with nonconvergent integrals that arise in these problems was known earlier. The reason can be traced to the slow decay of the disturbance flow caused by a sphere with distance from its center. Remedies had been proposed in prior work, but Batchelor's new approach, later termed *renormalization*, was adapted quickly to a variety of other problems. An important concept here is the difference between a cloud of particles surrounded by an infinite expanse of clear fluid and a suspension in a container with dimensions that far exceed those of the individual particle, the latter being the case tackled by Batchelor. When the container is large, its shape should play a negligible role in establishing the mean velocity of a particle, but the return flow caused by the presence of boundaries must be taken into account. The distinction between the two problems is not so much the fact that rigid boundaries are present, but that the statistical properties of the suspension are not uniform when a cloud is surrounded by clear fluid. The approach outlined by Batchelor leads to a correction to the mean settling velocity of $O(c)$, where c is the volume fraction of the particles in the suspension, assumed to be small compared with unity. This is in contrast to a correction proportional to $c^{\frac{1}{3}}$ that arises from models that assume a spatially regular distribution of the particles. The most important contribution in the correction to the velocity of a test particle is that from interaction with a single neighbor that has a probability of $O(c)$ of being present at a distance of the order of the particle radius R from the test particle. Simultaneous interactions among three particles would lead to corrections at $O(c^2)$ because the probability of two particles being present at such distances is proportional to c^2. Grouping interactions in this way leads naturally to a virial expansion for the mean property of interest. Batchelor showed that gravitational settling is hindered by effects arising both from return flow due to the displacement of the particles themselves and from return flow associated with the motion of the shell of fluid near each particle that is inaccessible to other particles because the radial extent of the shell is smaller than the radius of a particle. The latter makes a large contribution. Also, there are contributions from the overall flow in regions outside of the shell mentioned above, including that arising from interactions between particles. The net consequence is a reduction in the mean settling velocity.

Using the tools put forth by Batchelor, Anderson (1985) obtained the following approximate result for the mean thermocapillary migration velocity of a drop in a suspension of identical drops, as a function of the volume fraction c of the drops:

$$\frac{v^*}{v_\infty^*} = 1 - c\left[\frac{3}{2+\beta} + \frac{1}{8}\left(\frac{3\beta(1-\beta)}{(2+\beta)^2} + \frac{6\alpha + 10\beta + 19\alpha\beta}{2(1+\alpha)(3+2\beta)}\right)\right] + O(c^2). \quad (7.6.12)$$

In Equation (7.6.12), the left side is the ratio of the physical velocity of a drop v^* in the suspension, to the velocity the drop would have when isolated; the latter, represented by the symbol v_∞^*, is given by Equation (4.2.28). In writing this result, the interactions between any two drops are described by the reflections solution given in Equations (7.3.40) through (7.3.43), and the probability of finding a second drop outside of a spherical shell of radius $2R$ centered in a drop of radius R is assumed spatially uniform.

Acrivos et al. (1990) considered this problem in more detail for identical spherical bubbles. They first showed how any number of bubbles of equal size, subjected to a temperature gradient in an infinite expanse of clear liquid, will move at the same velocity they would have when isolated. For reasons outlined earlier, however, it does not follow that in a suspension that fills a large container, each bubble would move at the velocity it would have when isolated. Acrivos et al. discussed the case of a statistically homogeneous cloud of bubbles in a large container with the attendant return flow and considered the renormalization issue. They noted that an indeterminacy arises at the two-body interaction level. That is, one is unable to make a specific choice from an infinite set of different mathematically acceptable results. The authors therefore proceeded to the next level of three-body interactions to try to resolve the issue and showed that the three-body interaction problem for the hydrodynamic correction can be related to the two-body interaction problem for the effective thermal conductivity of the suspension. The indeterminacy is not resolved by going to the three-body interaction problem, however, as long as the bubbles have zero thermal conductivity. Only an argument based on conservation of mass permits one to make a specific choice. Finally, the authors obtained a remarkable result valid for any particle concentration. They showed that the mean velocity of a bubble, normalized by the velocity of an isolated bubble, is equal to the inner product between the effective thermal conductivity tensor, normalized by the conductivity of the liquid, and the direction of the applied temperature gradient. By using a result given by Jeffrey (1973) for the effective thermal conductivity of an isotropic well-mixed suspension, the authors obtained the following result:

$$\frac{v^*}{v_\infty^*} = 1 - \frac{3}{2}c + 0.59c^2 + o(c^2). \quad (7.6.13)$$

Note that the mean velocity in Equation (7.6.12) given to $O(c)$ is in agreement with that given in Equation (7.6.13) when one sets $\alpha = \beta = 0$, but Equation (7.6.13) is good to the next higher order in c in the special case mentioned. Acrivos et al. also presented an important analogy between the above problem and that of the motion of a collection of identical charged nonconducting rigid spheres under the action of a uniform applied electric field when the thickness of the electrical double layer is small compared with the size of the particle. Here, nonconducting refers to the conduction of electricity and not that of heat. Because of a similarity in the structure of the mathematical problem to that considered above, the authors were able to show that each sphere in a collection of N spheres would move at the same electrophoretic velocity it would have if isolated and that in the presence of recirculation, the above result in Equation (7.6.13) also applies to this case.

Wang et al. (1994) extended the work of Acrivos et al. to a bidisperse collection of bubbles, a term that implies a suspension of bubbles of two sizes. The authors showed that the mean velocity of a bubble of radius R_1 is given by

$$\frac{v^*}{v_\infty^*} = 1 - \frac{3}{2}c_1 - S(\lambda)c_2. \tag{7.6.14}$$

where the volume fraction of bubbles of radius R_1 is given by c_i for $i = 1, 2$ and $S(\lambda)$ is a function of the radius ratio $\lambda = \frac{R_2}{R_1}$, and a graph is provided in the article displaying the behavior of this function. The authors also provided asymptotic results for small and large values of λ. In the limit $\lambda \to 0$ they write

$$S(\lambda) \sim 1 - 0.57\lambda + O(\lambda^2); \tag{7.6.15}$$

in the opposite limit $\lambda \to \infty$, the following result is provided:

$$S(\lambda) \sim \lambda + \frac{1}{2} - \frac{0.84}{\lambda} + O\left(\frac{\log \lambda}{\lambda^2}\right). \tag{7.6.16}$$

Corrections to the result given in Equation (7.6.14) will appear at a quadratic order in the volume fractions.

7.7 Axisymmetric Motion of a Droplet within a Stationary Drop – Concentric Case

7.7.1 Introduction

In this and the next three sections, we present results for problems involving a drop containing a droplet. Such an object is called a *compound drop*. Compound drops are encountered in a variety of practical applications as mentioned in a review by Johnson and Sadhal (1985). Two examples are direct contact heat exchange between two fluids and liquid membranes used in separation processes. In the present section, we are concerned with a compound drop that is held fixed without contact by use of a suitable technique. One way for holding and manipulating a drop floating in a gas is to use acoustic forces as discussed by Wang, Saffren, and Elleman (1974). In their device, loudspeakers placed on three of the interior walls of a rectangular chamber generate standing sound waves. As a result, the time-average pressure is a minimum in the middle of the chamber, and objects introduced into it are pushed gently by the sound pressure field to the middle and held there. The attraction of this way of holding a drop is that a solid can be allowed to float without contact with a container and melted by heating it, for example, using a suitable laser. The molten material can be cooled subsequently and solidified while it floats. This technique offers the advantage of avoiding two problems introduced by contact with a container, contamination and heterogeneous nucleation of crystals. For example, Neilson and Weinberg (1977) suggested that certain substances that are difficult to cool into a glassy (amorphous) state because they crystallize readily at container walls would be good candidates for cooling without a container. The sound intensity required for levitating drops of any reasonable size against the gravitational pull of Earth is large, and Neilson and Weinberg put forth the idea that the processing be done in a reduced gravity environment. Even in a low gravity environment, as we saw in Chapter One, objects will experience residual accelerations. Therefore, it is necessary to use a gentle acoustic field to hold a drop fixed and prevent it from coming into contact with the chamber walls.

We mentioned in Chapter One that when a glass is made, gas bubbles are easily formed because of the gas trapped among the grains of the raw material and also from chemical reactions that produce gaseous products. These bubbles must be eliminated both for aesthetic reasons and for ensuring good mechanical properties of the final product. Even on Earth, bubbles are difficult to remove from the molten glass because of its large viscosity. In reduced gravity conditions, the problem would be worse. As discussed by Annamalai et al. (1982b), bubbles in a molten drop of glass in a space laboratory can be removed by selectively heating a spot on the surface of the liquid. The authors reasoned that the resulting thermocapillary flow within the drop will move the bubbles toward the heated spot and help eliminate them. This provided the motivation for the analyses reported in Shankar, Cole, and Subramanian (1981), who considered the concentric case, and Shankar and Subramanian (1983), who analyzed the eccentric case, wherein the droplet is a gas bubble.

The model problem can be stated as follows. A drop is held fixed by the application of a suitable noncontact force. A droplet, which can be a gas bubble, is present within this drop. In Section 9.8, we consider the case of a general temperature field applied on the surface of a drop when there is no droplet present within it. Here, in view of the type of application envisioned, a temperature field is applied on the drop surface that is symmetric about an axis passing through the center of the drop and that of the droplet, but otherwise arbitrary. This implies axisymmetric velocity and streamfunction fields, in addition to an axisymmetric temperature field within the drop and the droplet. The assumption of axial symmetry keeps the algebra tractable. As in the earlier sections of this chapter, we assume the Reynolds and Marangoni numbers to be negligible. Deformation of the drop, as well as that of the droplet, from the spherical shape is ignored, assuming that the Capillary number is sufficiently small in both cases. First, we analyze the concentric case, which occurs at the instant when the droplet passes through the center of the drop. This can be accomplished using spherical polar coordinates. In the next section, the more general problem, wherein the droplet is located eccentrically within the drop, is posed and solved. The simple results from the concentric case can be useful as a first approximation when the eccentricity of the drop is relatively small.

7.7.2 Analysis

Consider the geometry depicted in Figure 7.7.1. The radius of the outer drop is R_2, and that of the droplet present within it is R_1. The radius ratio κ is defined as follows, so that it assumes values in the range 0 to 1:

$$\kappa = \frac{R_1}{R_2}. \tag{7.7.1}$$

Lengths are scaled with the radius of the drop, R_2, and velocities are scaled using the thermocapillary reference velocity $v_0 = \frac{-\sigma_{T,drop}\Delta T_{ref}}{\mu}$. Here, $\sigma_{T,drop}$ stands for the rate of change of the surface tension at the surface of the drop with temperature, and ΔT_{ref} represents a characteristic temperature difference across the drop. The quantity μ represents the dynamic viscosity of the drop phase. Pressure and viscous stresses are scaled using $\frac{-\sigma_{T,drop}\Delta T_{ref}}{R_2}$, and temperature is scaled by subtracting a suitable reference

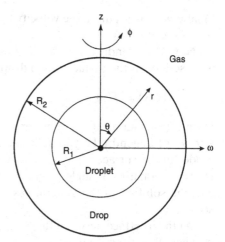

Figure 7.7.1 Sketch of a drop containing a droplet in the concentric case.

value T_0 and then dividing by ΔT_{ref}. In view of the axial symmetry, we can define a Stokes streamfunction, which is scaled with $R_2^2 v_0$. The symbol α represents the ratio of the viscosity of the droplet to that of the drop, and β is a similar ratio of thermal conductivities. The symbol $\delta = \frac{\sigma_{T,droplet}}{\sigma_{T,drop}}$, where $\sigma_{T,droplet}$ represents the rate of change of the interfacial tension at the interface between the droplet phase and the drop phase fluids with temperature. We use the symbol r to designate the scaled radial coordinate. As in earlier sections, we define $s = \cos\theta$ and express the dependencies of the fields in terms of r and s. The principal objective here is to obtain a result for the scaled velocity of the droplet, $V\mathbf{i}_z$, as a function of the parameters.

The scaled streamfunctions ψ and ψ' in the drop and droplet phases, respectively, will satisfy Stokes's equation, and the scaled temperature fields T and T' will satisfy Laplace's equation. We can use the general solutions for the streamfunctions in spherical polar coordinates, given in Equations (3.1.9) and (3.1.10). Similarly, the form of the solution in Equation (4.2.15) is used for the temperature field in the drop and that within the droplet. Therefore, all that remains is the evaluation of the arbitrary constants appearing in these solutions by using boundary conditions applicable to this problem. These are given below.

At the surface of the drop, the radial component of the velocity must vanish, and the usual tangential stress balance applies.

$$v_r(1, s) = 0, \tag{7.7.2}$$

$$\tau_{r\theta}(1, s) = -\frac{\partial T}{\partial \theta}(1, s). \tag{7.7.3}$$

At the interface between the drop and the droplet, $r = \kappa$, the kinematic condition, the continuity of tangential velocity, and the tangential stress balance are written as follows:

$$v_r(\kappa, s) = v'_r(\kappa, s) = V\cos\theta, \tag{7.7.4}$$

$$v_\theta(\kappa, s) = v'_\theta(\kappa, s), \tag{7.7.5}$$

and

$$\tau_{r\theta}(\kappa, s) - \tau'_{r\theta}(\kappa, s) = \frac{\delta}{\kappa}\frac{\partial T}{\partial \theta}(\kappa, s). \tag{7.7.6}$$

Finally, we must require the velocity to be bounded everywhere within the droplet and, therefore, at $r = 0$.

Next, the boundary conditions on the temperature field are written. The temperature is prescribed at the surface of the drop.

$$T(1, s) = f(s). \tag{7.7.7}$$

Here, $f(s)$ is an arbitrary function of s. When presenting results, we expand $f(s)$ in a series of Legendre Polynomials and investigate the flow arising from pure Legendre modes of the surface temperature. A pure mode means that the surface temperature is proportional to a single Legendre Polynomial, $P_n(s)$. Because the problem is linear, the solution in the general case is simply the superposition of such pure mode solutions.

At the interface between the drop and the droplet, we require that the temperature and heat flux be continuous.

$$T(\kappa, s) = T'(\kappa, s), \tag{7.7.8}$$

$$\frac{\partial T}{\partial r}(\kappa, s) = \beta \frac{\partial T'}{\partial r}(\kappa, s). \tag{7.7.9}$$

As with the velocity field, the temperature field must be bounded everywhere within the droplet and, therefore, at its center.

The solution, which satisfies the boundary conditions and the condition of zero hydrodynamic force on the droplet, can be obtained in a straightforward manner. The temperature fields in the drop and droplet phases can be written as follows:

$$T = \sum_{n=0}^{\infty} \left[E_n r^n + \frac{F_n}{r^{n+1}} \right] P_n(s) \tag{7.7.10}$$

and

$$T' = \sum_{n=0}^{\infty} E'_n r^n P_n(s). \tag{7.7.11}$$

Here,

$$E_n = [1 + n(1 + \beta)]H_n, \tag{7.7.12}$$

$$F_n = n(1 - \beta)\kappa^{2n+1} H_n, \tag{7.7.13}$$

$$E'_n = (2n + 1)H_n, \tag{7.7.14}$$

and

$$H_n = \frac{G_n}{1 + n(1 + \beta) + n(1 - \beta)\kappa^{2n+1}}. \tag{7.7.15}$$

The constants G_n are coefficients in the expansion of the temperature field imposed on the surface of the drop in a series of Legendre Polynomials.

$$T(1, s) = f(s) = \sum_{n=0}^{\infty} G_n P_n(s). \tag{7.7.16}$$

Therefore, we can relate G_n to $f(s)$ using the orthogonality property of Legendre Polynomials:

$$G_n = \frac{2n+1}{2} \int_{-1}^{+1} f(s) P_n(s) \, ds. \tag{7.7.17}$$

The solutions for the streamfunction fields can be written as follows:

$$\psi(r, s) = \sum_{n=2}^{\infty} (A_n r^n + B_n r^{-n+1} + C_n r^{n+2} + D_n r^{-n+3}) C_n(s) \tag{7.7.18}$$

and

$$\psi'(r, s) = \sum_{n=2}^{\infty} (A'_n r^n + C'_n r^{n+2}) C_n(s). \tag{7.7.19}$$

Here, $C_n(s)$ represent the Gegenbauer Polynomials, introduced in Section 3.1. The constants appearing in the solution are

$$A_n = -(B_n + C_n + D_n), \tag{7.7.20}$$

$$B_n = -\frac{1}{(2n-1) X_n} \left\{ Y_n + \frac{3\alpha V}{2\kappa^{n-1}} \delta_{n2} - \frac{n(n-1)}{2} G_{n-1} \right.$$
$$\left. \times \left[(\alpha - 1) \kappa^{2n-2} + \frac{(2n-3)\alpha}{2\kappa} + \kappa \left(1 - \frac{(2n-1)\alpha}{2} \right) \right] \right\}, \tag{7.7.21}$$

$$C_n = \frac{1}{(2n-1) X_n} \left\{ Y_n + \frac{3\alpha V}{2\kappa^{n-1}} \delta_{n2} - \frac{n(n-1)}{2} G_{n-1} \right.$$
$$\left. \times \left[\frac{1+\alpha}{\kappa^{2n}} + \frac{(2n-3)\alpha}{2\kappa} - \frac{1}{\kappa^3} \left(1 + \frac{(2n-1)\alpha}{2} \right) \right] \right\}, \tag{7.7.22}$$

$$D_n = \frac{\kappa^{n-1}}{\kappa^{2n-3} - 1} \left[V\delta_{n2} + \frac{B_n}{\kappa^{n+1}} + C_n \kappa^n - \kappa^{n-2} \frac{n(n-1)}{2(2n-1)} G_{n-1} \right], \tag{7.7.23}$$

$$A'_n = -\frac{1}{2} \left[n A_n - \frac{n-1}{\kappa^{2n-1}} B_n + (n+2)\kappa^2 C_n - \frac{n-3}{\kappa^{2n-3}} D_n + 4V\delta_{n2} \right], \tag{7.7.24}$$

and

$$C'_n = -\frac{1}{\kappa^2} [A'_n + V\delta_{n2}] \tag{7.7.25}$$

where δ_{n2} is the Kronecker delta, taking on a value of unity when $n = 2$ and 0 otherwise. The constants X_n and Y_n are given below:

$$X_n = \kappa^{2n-2}(\alpha - 1) + \kappa \left(1 - \frac{(2n-1)\alpha}{2} \right) - \frac{1+\alpha}{\kappa^{2n}} + \frac{1}{\kappa^3} \left(1 + \frac{(2n-1)\alpha}{2} \right) \tag{7.7.26}$$

and

$$Y_n = \delta \left(\kappa^{n-2} - \frac{1}{\kappa^{n-1}} \right) \frac{n(n-1)}{2} \left(E_{n-1} \kappa^{n-1} + \frac{F_{n-1}}{\kappa^n} \right). \tag{7.7.27}$$

The magnitude of the scaled velocity of the droplet is given by

$$V = G_1 \left[\frac{1}{3} + \frac{2\kappa}{[2(1-\kappa^5) + \alpha(3+2\kappa^5)]} \left(\frac{\delta(1-\kappa^5)}{[2+\kappa^3 + \beta(1-\kappa^3)]} - \frac{5}{6}\alpha\kappa \right) \right].$$

(7.7.28)

Note that Equation (7.7.28) shows that a droplet of negligible size relative to the drop ($\kappa \to 0$) behaves simply as a tracer, following the motion of the fluid along the axis at the drop center. This yields a scaled velocity of $\frac{G_1}{3}$ in this limit. A small drop would move with nearly this velocity. Corrections to this result can be obtained from Section 4.4, where we analyze the motion of a drop in a large body of fluid which is in Stokes flow and in which the temperature distribution satisfies Laplace's equation. We can set the hydrodynamic force on the drop, given in Equation (4.4.10), equal to zero, yielding the quasi-steady migration velocity of the droplet. The result should apply to a droplet that is sufficiently small compared with the drop. The correctness of this argument is demonstrated by an expansion of the result in Equation (7.7.28) for small values of κ that yields

$$V = G_1 \left[\frac{1}{3} + \frac{2\delta}{(2+3\alpha)(2+\beta)}\kappa - \frac{5\alpha}{3(2+3\alpha)}\kappa^2 \right.$$
$$\left. - \frac{2\delta(1-\beta)}{(2+3\alpha)(2+\beta)^2}\kappa^4 + O(\kappa^6) \right].$$

(7.7.29)

We can consider the premultiplier G_1 to be unity for the purpose of a brief discussion of this expansion. For simplicity, we also assume that $\sigma_{T,drop}$ and $\sigma_{T,droplet}$ are both negative, so that δ is positive. The features noted here are not materially altered if the sign of δ is reversed. The basic circulatory motion in the drop, which would occur in the absence of the droplet, produces a positive scaled droplet velocity of $\frac{1}{3}$. Physically, the circulation causes fluid to move toward the warm pole, and the droplet drifts with this motion. A positive correction, from the linear term in κ, applies for sufficiently small κ. This is the contribution of the thermocapillary stress at the surface of the droplet to its motion toward the warm pole, when κ is small. This linear term in κ results from the $(\nabla T_\infty)_0$ term in Equation (4.4.10). The quadratic term in κ, which is independent of δ, arises from the $(\nabla^2 \mathbf{u}_\infty)_0$ term which appears in Equation (4.4.10). The higher order terms correct for the finite size of the drop because Equation (4.4.10) applies to a continuous phase of infinite extent.

7.7.3 Results

It is seen from Equation (7.7.28) that only the P_1-mode of the temperature distribution on the drop surface makes a contribution to the motion of the droplet when it is concentric with the drop. The droplet moves toward the warm pole as one would expect. The other modes lead to interfacial tension gradients on both interfaces, which possess symmetry about the equator that precludes the possibility of causing motion of the droplet. When the drop is subjected to one of these higher order pure modes of the surface temperature field and the droplet is precisely at the center of the drop, it will not move. Of course, a slight displacement of the droplet from the center, along the axis of symmetry, would destroy the symmetry mentioned. In this case, it is possible for the droplet to either move back toward the center or move away from it, depending on

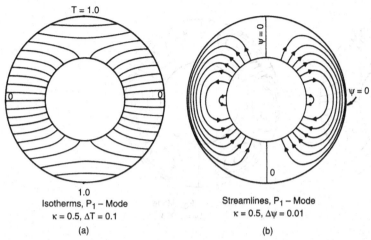

Figure 7.7.2 Isotherms (a) and streamlines (b) for a $P_1(\cos\theta)$-mode temperature distribution on the drop surface. (Reprinted from *Int. J. Multiphase Flow* 7, No. 6, Shankar, N., Cole, R., and Subramanian, R. S., Thermocapillary migration of a fluid droplet inside a drop in a space laboratory, pp. 581–594, Copyright 1981, with permission from Elsevier Science.)

the situation. Therefore, the motionless state of the droplet at the center of the drop, in the case of higher modes, may not always be stable. In the next section, we predict the velocity of a droplet when it is displaced from the center of the drop. From the direction of this velocity in a given case, it is possible to make an inference regarding the stability, or lack thereof, of the motionless concentric state.

Results for isotherms and streamlines for the first few pure Legendre modes of the surface temperature are displayed in Figures 7.7.2, 7.7.3, and 7.7.4. These results pertain to the gas bubble limit, $\alpha = \beta = 0$. The drawings are arranged in pairs, with the figure on the left showing the structure of isotherms and that on the right showing the

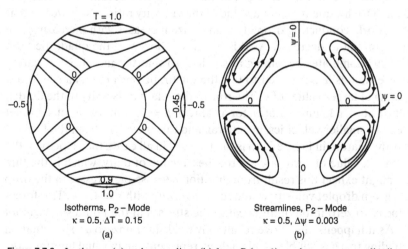

Figure 7.7.3 Isotherms (a) and streamlines (b) for a $P_2(\cos\theta)$-mode temperature distribution on the drop surface. (Reprinted from *Int. J. Multiphase Flow* 7, No. 6, Shankar, N., Cole, R., and Subramanian, R. S., Thermocapillary migration of a fluid droplet inside a drop in a space laboratory, pp. 581–594, Copyright 1981, with permission from Elsevier Science.)

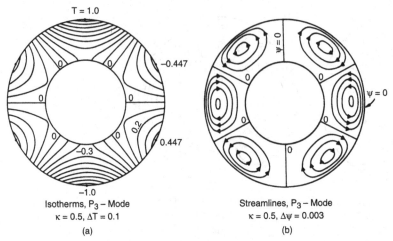

Figure 7.7.4 Isotherms (a) and streamlines (b) for a $P_3(\cos\theta)$-mode temperature distribution on the drop surface. (Reprinted from *Int. J. Multiphase Flow* 7, No. 6, Shankar, N., Cole, R., and Subramanian, R. S., Thermocapillary migration of a fluid droplet inside a drop in a space laboratory, pp. 581–594, Copyright 1981, with permission from Elsevier Science.)

corresponding streamline pattern. In preparing these drawings, the relevant coefficient, G_n, was set to unity for a specific value of $n = 1, 2,$ or 3, and all the other coefficients were set equal to zero. By this choice, the reference temperature difference ΔT_{ref} is selected as well. For example, when $n = 1$, ΔT_{ref} is the difference in temperature between one of the poles and the equator of the drop. Similar connections can be made in the case of the higher Legendre modes.

In Figure 7.7.5, some representative drawings are used to display the behavior of the scaled velocity of the droplet as a function of the radius ratio, κ, for a selected set of values of the other parameters. In making the calculations, the ratio $\delta = \frac{\sigma_{T,droplet}}{\sigma_{T,drop}}$ has been set equal to unity for convenience. Also, we have set $G_1 = 1$ because the velocity is simply proportional to this quantity. For a value of the viscosity ratio $\alpha = 1$, we see that the velocity of a poorly conducting droplet of small size increases with increasing size at first, before beginning to decrease. As β is increased, however, the quadratic term in Equation (7.7.29) dominates beginning at small κ, so that the initial rise is hardly noticeable. We see that as α is reduced to 0.1, the quadratic term exerts a significant influence beginning at larger values of κ. The role of the relative viscosity of the fluid in the droplet is illustrated in Figure 7.7.5(c), for a value of $\beta = 1$. A more viscous droplet phase should lead to smaller velocities, in general, and this is evident from the figure.

One noteworthy feature in Figure 7.7.5 is the fact that in all cases, the velocity of the droplet approaches a value of zero as κ approaches a value of unity, which is the thin shell limit. One might expect this result from intuition because the stresses in the drop phase fluid for a given droplet velocity would scale inversely with $\epsilon = 1 - \kappa$. This forces the droplet velocity to zero in the limit because the stresses in the liquid shell cannot be unbounded. As it happens, the above result only holds for nonzero α. The situation $\alpha = 0$, corresponding to a gas bubble within a drop, represents an unusual limiting case. If we set $\alpha = 0$ in Equation (7.7.28), we get the following result:

$$V = G_1\left[\frac{1}{3} + \frac{\delta\kappa}{[2 + \kappa^3 + \beta(1 - \kappa^3)]}\right]. \tag{7.7.30}$$

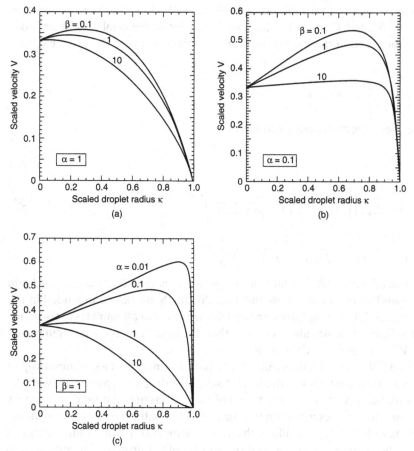

Figure 7.7.5 Scaled velocity of the droplet plotted against its scaled radius for $\delta = 1$ and for (a) $\alpha = 1$, (b) $\alpha = 0.1$, and (c) $\beta = 1$.

As $\kappa \to 1$, this yields a limiting value of $V = \frac{1+\delta}{3}$, which is not zero because the most suitable value for δ is unity in this situation. This is not physically reasonable. Therefore we must examine the situation further. If we expand the droplet velocity in Equation (7.7.28) for small values of $\epsilon = 1 - \kappa$, we see the source of this behavior. The first two terms in this expansion are shown below.

$$V = G_1 \left[\frac{2(1+\delta)}{3} \frac{\epsilon}{\alpha} - \frac{1}{3}[2\delta(2+\alpha\beta) + 4 - 3\alpha^2]\frac{\epsilon^2}{\alpha^2} + O\left(\frac{\epsilon^3}{\alpha^3}\right) \right]. \tag{7.7.31}$$

The terms are grouped to show the nature of the limit. As noted, for any nonzero value of α, Equation (7.7.31) shows that the velocity of the droplet approaches zero, as the thickness of the fluid between the two boundaries is decreased. It is not possible to make this inference when $\alpha = 0$, however, because this expansion breaks down in that limit. We find that the solution for the velocity field in the shell for $\alpha = 0$ becomes singular, as the thickness of the shell approaches zero. This can be seen by examining the behavior of the various constants in that limit, after setting $\alpha = 0$. Of prime interest here are the constants A_2, B_2, and C_2 because the scaled velocity of the bubble is given from the kinematic condition on the bubble surface, in conjunction with the force balance, $D_2 = 0$, as

$$V = -\left[A_2 + \frac{B_2}{\kappa^3} + C_2 \kappa^2 \right]. \tag{7.7.32}$$

The results for these constants are given below.

$$A_2 = -\frac{G_1}{3}, \tag{7.7.33}$$

$$B_2 = -\frac{G_1}{3(1 - \kappa^5)} \left[\kappa^5 + \frac{3\delta\kappa^4}{2 + \kappa^3 + \beta(1 - \kappa^3)} \right], \tag{7.7.34}$$

$$C_2 = \frac{G_1}{3(1 - \kappa^5)} \left[1 + \frac{3\delta\kappa^4}{2 + \kappa^3 + \beta(1 - \kappa^3)} \right]. \tag{7.7.35}$$

We see from the above equations that the result for V is bounded because the sum $\frac{B_2}{\kappa^3} + C_2\kappa^2$ is bounded, even though B_2 and C_2 individually become unbounded in the limit $\kappa \to 1$. For $n > 2$, it is straightforward to demonstrate that all four sets of constants A_n, B_n, C_n, and D_n become singular as $\kappa \to 1$; therefore, this limit is itself singular, and the result for V in this limit for the case $\alpha = 0$ is incorrect.

Mok and Kim (1987) provided a similar analysis for the motion of a spherical droplet within a rigid spherical container at the instant when the droplet is concentric with the container. The motion occurs because of the combined effects of gravity and an arbitrary temperature distribution prescribed on the surface of the outer rigid sphere. As in the present problem, only the $P_1(s)$-mode of the surface temperature distribution makes a contribution to the thermocapillary migration velocity of the droplet. The authors also obtained the temperature gradient needed to hold the droplet motionless at the center and provided illustrative numerical results for air bubbles moving in a silicone oil held in a spherical glass shell.

7.8 Axisymmetric Motion of a Droplet within a Stationary Drop – Eccentric Case

7.8.1 Introduction

In this section, we analyze the problem which was posed in Section 7.7 in the case where the droplet is located eccentrically within the drop, as shown in Figure 7.8.1. All the assumptions made in Section 7.7 continue to apply; only the geometry is altered. The arbitrary temperature field imposed on the drop surface is symmetric about the z-axis in the figure. As before, we consider the expansion of this field in a series of Legendre Polynomials, $P_n(s)$, where $s = \cos\theta$, and θ is the polar angle measured from the z-axis in a spherical coordinate system with its origin at the center of the drop. Because the problem is linear, we present results for each of the first few pure modes of the surface temperature field in the same manner as in Section 7.7. The problem is posed in general terms for the case of a droplet. As usual, α represents the ratio of the dynamic viscosity of the droplet to that of the drop, and β represents a similar ratio of thermal conductivities. To keep the algebra tractable, however, we provide

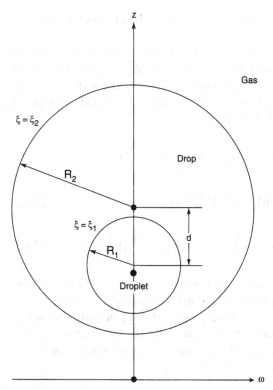

Figure 7.8.1 Sketch of a drop containing an eccentrically located droplet.

results only in the gas bubble limit $\alpha = \beta = 0$. The results given here are based on work reported in Shankar and Subramanian (1983). Additional details can be found in Shankar (1984).

7.8.2 Analysis

We shall continue to use the symbols specific to the problem of a droplet within a drop, which were introduced in Section 7.7. In bispherical coordinates, the surface of the droplet corresponds to $\xi = \xi_1$, and that of the drop to $\xi = \xi_2$. The distance between the centers of the droplet and the drop is d, and a scaled distance is defined as

$$D = \frac{d}{R_2}. \tag{7.8.1}$$

Here, R_2 is the radius of the drop, and the symbol R_1 is used to represent the radius of the droplet. As in the last section, we use $\kappa = \frac{R_1}{R_2}$ to designate the scaled radius of the droplet, and define $\delta = \frac{\sigma_{T,droplet}}{\sigma_{T,drop}}$. For convenience in working with the problem in bispherical coordinates, lengths are scaled with c, a characteristic length that was defined in Section 5.1. We continue to scale velocities by using the reference velocity $v_0 = \frac{-\sigma_{T,drop}\Delta T_{ref}}{\mu}$. Pressure and viscous stresses are scaled using $\frac{-\sigma_{T,drop}\Delta T_{ref}}{c}$, and temperature is scaled by subtracting a suitable reference value T_0 and then dividing by ΔT_{ref}. The Stokes streamfunction is now scaled with $c^2 v_0$. The scaled velocity of the droplet is designated Vi_z.

The scaled streamfunctions ψ and ψ' in the drop and droplet phases, respectively, will satisfy Stokes's equation, and the scaled temperature fields T and T' will satisfy Laplace's equation. For the drop phase, we can use the axisymmetric solution for the streamfunction field in bispherical coordinates given in Equation (5.1.15) and that for the temperature field given in Equation (5.1.14).

$$\psi = \frac{1}{(\cosh \xi - \zeta)^{\frac{3}{2}}} \sum_{n=1}^{\infty} \left[a_n \cosh\left(n - \frac{1}{2}\right)\xi + b_n \sinh\left(n - \frac{1}{2}\right)\xi \right.$$
$$\left. + c_n \cosh\left(n + \frac{3}{2}\right)\xi + d_n \sinh\left(n + \frac{3}{2}\right)\xi \right] C_{n+1}(\zeta), \tag{7.8.2}$$

$$T = \sqrt{\cosh \xi - \zeta} \sum_{n=0}^{\infty} \left[A_n \cosh\left(n + \frac{1}{2}\right)\xi + B_n \sinh\left(n + \frac{1}{2}\right)\xi \right] P_n(\zeta). \tag{7.8.3}$$

In the above, as usual, $\zeta = \cos \eta$. In the same way in which the solutions for the interior of a drop containing the point $\xi = \pm\infty$ were written in Equations (7.3.3) and (7.3.4) to accommodate the fact that the fields must remain bounded, we can write specialized versions here for the interior of the droplet. Because only the point at $\xi = +\infty$ is included in the domain, there is no need for the absolute sign used in those equations which were designed to handle the point at $\xi = -\infty$ as well.

$$\psi' = \frac{1}{(\cosh \xi - \zeta)^{\frac{3}{2}}} \sum_{n=1}^{\infty} \left[p_n e^{-(n-\frac{1}{2})\xi} + q_n e^{-(n+\frac{3}{2})\xi} \right] C_{n+1}(\zeta), \tag{7.8.4}$$

$$T' = \sqrt{\cosh \xi - \zeta} \sum_{n=0}^{\infty} \left[A'_n e^{-(n+\frac{1}{2})\xi} \right] P_n(\zeta). \tag{7.8.5}$$

Next, the boundary conditions on the velocity field are given. The kinematic condition and the tangential stress balance apply at the drop surface.

$$\mathbf{v}(\xi_2, \eta) \bullet \mathbf{i}_\xi = 0, \tag{7.8.6}$$

$$\tau_{\xi\eta}(\xi_2, \eta) = (\cosh \xi_2 - \zeta)\frac{\partial T}{\partial \eta}(\xi_2, \eta). \tag{7.8.7}$$

At the interface between the drop and the droplet, we can use the kinematic condition, continuity of tangential velocity, and the tangential stress balance.

$$\mathbf{v}(\xi_1, \eta) \bullet \mathbf{i}_\xi = \mathbf{v}'(\xi_1, \eta) \bullet \mathbf{i}_\xi = Vi_z \bullet \mathbf{i}_\xi, \tag{7.8.8}$$

$$\mathbf{v}(\xi_1, \eta) \bullet \mathbf{i}_\eta = \mathbf{v}'(\xi_1, \eta) \bullet \mathbf{i}_\eta, \tag{7.8.9}$$

$$\tau_{\xi\eta}(\xi_1, \eta) - \tau'_{\xi\eta}(\xi_1, \eta) = -\delta(\cosh \xi_1 - \zeta)\frac{\partial T}{\partial \eta}(\xi_1, \eta). \tag{7.8.10}$$

Next, the boundary conditions on the temperature field are written. The temperature is prescribed at the surface of the drop.

$$T(\xi_2, \eta) = f(s). \tag{7.8.11}$$

At the interface between the drop and the droplet, the temperature and heat flux must be continuous.

$$T(\xi_1, \eta) = T'(\xi_1, \eta),$$

(7.8.12)

$$\frac{\partial T}{\partial \xi}(\xi_1, \eta) = \beta \frac{\partial T'}{\partial \xi}(\xi_1, \eta).$$

(7.8.13)

The arbitrary constants a_n, b_n, c_n, d_n, p_n, q_n, A_n, B_n, and A'_n and the scaled velocity of the droplet V can be evaluated by straightforward use of the above boundary conditions and the requirement that the hydrodynamic force on the droplet should vanish for quasi-steady motion. We give results here only in the gas bubble limit $\alpha = \beta = 0$, to maintain simplicity in the algebraic details. In this case, because the transport problem within the droplet phase is not solved, only the constants a_n, b_n, c_n, and d_n in the stream-function field in the drop phase can be determined. They are the same as those given in Equations (7.3.11) to (7.3.14) in the problem of two gas bubbles external to each other. The magnitude of the scaled velocity can be written as

$$V = \frac{\sum_{n=1}^{\infty} m_2 \left[g_1 I_n (\xi_1) + g_2 I_n (\xi_2) \right]}{\sum_{n=1}^{\infty} m_1 f_1},$$

(7.8.14)

where the constants appearing on the right side are functions of the summation index n and the parameters ξ_1 and ξ_2, and expressions for them are given in Section 7.3. This result for the velocity of the bubble is the same as that obtained from Equation (7.3.16), upon setting $U_2 = 0$ and $U_1 = V$. This equation represents the force balance on the gas bubble in the present problem, which corresponds to bubble 1 in Section 7.3.

It remains to determine the solution for the temperature field in the drop phase, subject to the boundary conditions here, so that it can be used in evaluating the integrals in Equation (7.3.15). The constants A_n and B_n can be obtained by the application of the boundary conditions. We expand the temperature field prescribed on the drop surface in spherical polar coordinates, in the same form as the solution of Laplace's equation in bispherical coordinates, as follows:

$$f(s) = \sqrt{\cosh \xi_2 - \zeta} \sum_{n=0}^{\infty} G_n P_n(\zeta).$$

(7.8.15)

The constants G_n can be obtained by using the orthogonality property of the Legendre Polynomials:

$$G_n = \frac{2n + 1}{2} \int_{-1}^{+1} \frac{f(s) P_n(\zeta)}{\sqrt{\cosh \xi_2 - \zeta}} d\zeta.$$

(7.8.16)

It is useful to give the relationship between s and ζ in the above result:

$$s = \frac{\zeta \cosh \xi_2 - 1}{\cosh \xi_2 - \zeta}.$$

(7.8.17)

When $f(s) = P_1(s)$, the constants G_n can be written as

$$G_n = \sqrt{2} \coth \xi_2 \left[n e^{-(n-\frac{1}{2})\xi_2} + (n + 1) e^{-(n+\frac{3}{2})\xi_2} - \frac{2n + 1}{\cosh \xi_2} e^{-(n+\frac{1}{2})\xi_2} \right].$$

(7.8.18)

For the case $f(s) = P_2(s)$, the following result holds:

$$
\begin{aligned}
G_n = \frac{1}{\sqrt{2}} \frac{e^{-(n+\frac{1}{2})\xi_2}}{\sinh^2 \xi_2} \Bigg[& \left\{ (2n+1) + \frac{4n^3 + 6n^2 - 1}{(2n-1)(2n+3)} \cosh^2 \xi_2 \right\} \\
& \times \left\{ 2 \coth \xi_2 + 2n + 1 \right\} - \sinh^2 \xi_2 \Bigg] - \sqrt{2} \frac{\cosh \xi_2}{\sinh^2 \xi_2} \\
& \times \left[n e^{-(n-\frac{1}{2})\xi_2} (2 \coth \xi_2 + 2n - 1) + (n+1) e^{-(n+\frac{3}{2})\xi_2} (2 \coth \xi_2 + 2n + 3) \right] \\
& + \frac{\coth^2 \xi_2}{\sqrt{2}} \left[\frac{n(n-1)}{2n-1} e^{-(n-\frac{3}{2})\xi_2} (2 \coth \xi_2 + 2n - 3) \right. \\
& \left. + \frac{(n+1)(n+2)}{(2n+3)} e^{-(n+\frac{5}{2})\xi_2} (2 \coth \xi_2 + 2n + 5) \right].
\end{aligned}
\tag{7.8.19}
$$

For higher Legendre modes of the surface temperature, it is not practical to obtain G_n analytically, and numerical integration can be used as needed.

The results for the constants A_n and B_n are obtained by solving the following equations:

$$
A_n \cosh\left(n + \frac{1}{2}\right)\xi_2 + B_n \sinh\left(n + \frac{1}{2}\right)\xi_2 = G_n
\tag{7.8.20}
$$

and

$$
(n+1)(A_n - A_{n+1}) \sinh\left(n + \frac{3}{2}\right)\xi_1 + n(A_n - A_{n-1}) \sinh\left(n - \frac{1}{2}\right)\xi_1
$$

$$
+ (n+1)(B_n - B_{n+1}) \cosh\left(n + \frac{3}{2}\right)\xi_1 + n(B_n - B_{n-1}) \cosh\left(n - \frac{1}{2}\right)\xi_1 = 0,
$$

$$
n = 0, 1, 2, \ldots. \tag{7.8.21}
$$

In Equation (7.8.21), $A_{-1} = B_{-1} = 0$, and closure is achieved by setting $A_{N+1} = B_{N+1} = 0$ for a sufficiently large N.

In Section 4.4, we provided a result for the hydrodynamic force on a spherical drop subjected to an arbitrary velocity field in the continuous phase that satisfies Stokes's equation, and an arbitrary temperature field that satisfies Laplace's equation. The conditions were that the continuous phase be of unbounded extent and that convective transport effects be negligible. We can obtain an approximation for the velocity of the gas bubble in the present problem by setting the hydrodynamic force in Equation (4.4.10) equal to zero:

$$
V_{approx} = (-1)^{n+1} n D^{n-1} \left[\frac{(n+1)(1 - D^2)}{2(2n+1)} + \frac{\kappa}{2} \right].
\tag{7.8.22}
$$

Because the viscosity of the gas bubble is zero, the above result is simply a superposition of two contributions. The first term is the velocity in the drop phase fluid at the location of the bubble, but in its absence, and the second term is the thermocapillary migration velocity of the bubble in an otherwise quiescent drop phase fluid due to the temperature gradient evaluated at the location of the bubble. Because Equation (4.4.10) holds for an unbounded fluid, we can expect that Equation (7.8.22) may be of some utility in the limit $\kappa \to 0$. It will be seen that this result indeed proves useful for small values of κ.

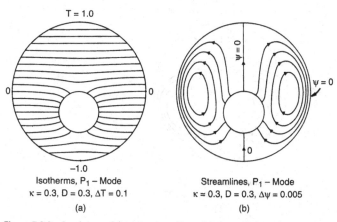

Isotherms, P_1 – Mode
$\kappa = 0.3$, $D = 0.3$, $\Delta T = 0.1$
(a)

Streamlines, P_1 – Mode
$\kappa = 0.3$, $D = 0.3$, $\Delta\psi = 0.005$
(b)

Figure 7.8.2 Isotherms (a) and streamlines (b) for a $P_1(\cos\theta)$-mode temperature distribution on the drop surface (Courtesy of Academic Press).

7.8.3 Results for the First Few Legendre Modes of the Surface Temperature

Now, we provide illustrative results, calculated from the solution, for the case when the temperature field on the drop surface is set equal to $P_n(s)$, for $n = 1, 2$, and 3. In Figures 7.8.2 through 7.8.6 isotherms and streamlines in a meridian plane are displayed. In each case, the isotherms are shown on the left, and the corresponding streamlines are displayed on the right. These drawings should be contrasted with similar drawings included in the previous section. Note the way in which the symmetry about the equator present in the concentric case is broken, as well as the change in the flow structure with increasing κ, shown in the case when $T(\xi_2, \eta) = P_2(s)$. The flow changes from one that is bicellular to one containing a single cell, for κ lying somewhere between 0.5 and 0.8. When $T(\xi_2, \eta) = P_3(s)$, the streamlines displayed in Figure 7.8.6(b) should be contrasted with those given in Figure 7.8.7, to see how the flow structure evolves with increasing κ in this case.

Results for the scaled velocity of the bubble, calculated from Equation (7.8.14), for the first three pure Legendre modes of the temperature field on the surface of the drop,

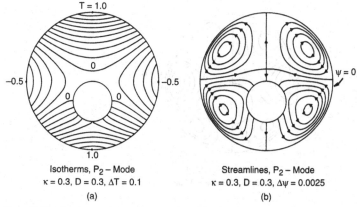

Isotherms, P_2 – Mode
$\kappa = 0.3$, $D = 0.3$, $\Delta T = 0.1$
(a)

Streamlines, P_2 – Mode
$\kappa = 0.3$, $D = 0.3$, $\Delta\psi = 0.0025$
(b)

Figure 7.8.3 Isotherms (a) and streamlines (b) for a $P_2(\cos\theta)$-mode temperature distribution on the drop surface (Courtesy of Academic Press).

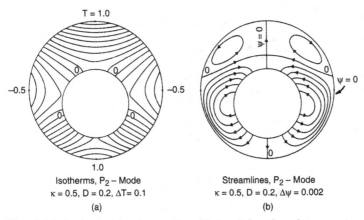

Figure 7.8.4 Isotherms (a) and streamlines (b) for a $P_2(\cos\theta)$-mode temperature distribution on the drop surface (Courtesy of Academic Press).

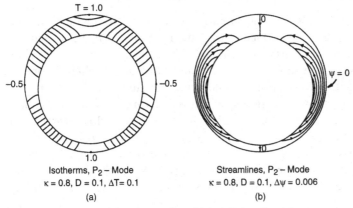

Figure 7.8.5 Isotherms (a) and streamlines (b) for a $P_2(\cos\theta)$-mode temperature distribution on the drop surface (Courtesy of Academic Press).

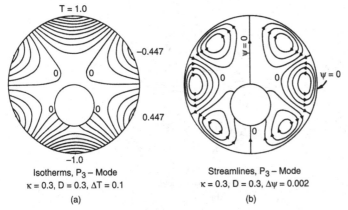

Figure 7.8.6 Isotherms (a) and streamlines (b) for a $P_3(\cos\theta)$-mode temperature distribution on the drop surface (Courtesy of Academic Press).

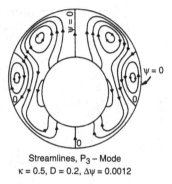

Figure 7.8.7 Streamlines for a $P_3(\cos\theta)$-mode temperature distribution on the drop surface (Courtesy of Academic Press).

Streamlines, P_3 – Mode
$\kappa = 0.5$, $D = 0.2$, $\Delta\psi = 0.0012$

are displayed in Figures 7.8.8, 7.8.9, and 7.8.10. Also shown is the approximation for a relatively small bubble, which is presented in Equation (7.8.22). In the case wherein $T(\xi_2, \eta) = P_1(s)$, we have included the result from the previous section for the concentric case when $D = 0$. The reason why this result does not approach the limiting value of 0 as $\kappa \to 1$ has been discussed in Section 7.7. In the eccentric case, the comparable situation is the limit $\kappa \to (1 - D)$. In that limit, the surfaces of the bubble and the drop would touch at a point, even though there would be drop phase fluid elsewhere in the space between the two. Because the drop surface is stationary, for kinematic consistency at the point of contact, the bubble should have a velocity of zero. The numerical evaluation of results becomes difficult when this limit is approached. Therefore, the curves in Figures 7.8.8, 7.8.9, and 7.8.10 terminate before the velocity becomes zero. In most cases, however, the approach to this asymptote can be observed. Also, note that for a P_1-mode surface temperature field, for $\kappa \le 0.5$, the results for the concentric case are close to those from the solution in bispherical coordinates for $D = 0.2$. Therefore, the results from the concentric case can be used as a first approximation, for small deviations from that configuration.

The reader will recall that, for the concentric configuration, the velocity of the bubble is zero for all the modes of the surface temperature field except $P_1(s)$. This is not the case in the eccentric situation, where we no longer have symmetry about the equator

Figure 7.8.8 Scaled migration velocity of the bubble plotted against the scaled radius of the bubble for a $P_1(\cos\theta)$-mode temperature distribution on the drop surface (Courtesy of Academic Press).

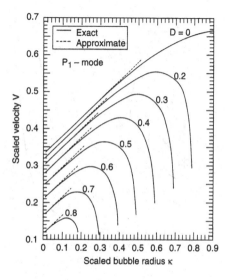

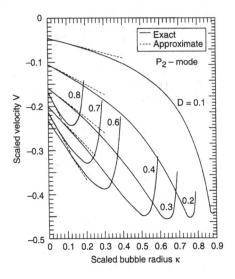

Figure 7.8.9 Scaled migration velocity of the bubble plotted against the scaled radius of the bubble for a $P_2(\cos\theta)$-mode temperature distribution on the drop surface (Courtesy of Academic Press).

of the drop. The scaled velocity of the bubble, V, is negative for a P_2-mode surface temperature field because the bubble moves in the negative z-direction in this case. The drawings show that when $\kappa = 0$, the velocity of the bubble, calculated from the solution in bispherical coordinates, approaches the correct value from the approximation given in Equation (7.8.22). It increases in magnitude with increasing κ for small κ, as predicted in Equation (7.8.22). The approximation is observed to be reasonable for $\kappa \leq 0.1$. As κ is increased further, for a given scaled separation distance D, the magnitude of the velocity reaches a maximum and then begins to decrease. The behavior of the velocity of the bubble when D is changed, while keeping κ fixed, is more complex. For a P_1-mode surface temperature field, the behavior is monotonic. As the separation distance is increased at a given radius ratio, the magnitude of V decreases. On the other hand, for the P_2- and P_3-modes, the magnitude first increases from its value of zero for a concentric configuration as the separation distance is increased. At larger values of D, there is a reversal. The magnitude starts decreasing with further increase in D. A

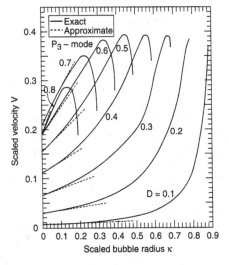

Figure 7.8.10 Scaled migration velocity of the bubble plotted against the scaled radius of the bubble for a $P_3(\cos\theta)$-mode temperature distribution on the drop surface (Courtesy of Academic Press).

qualitative explanation for this behavior follows. The explanation is based on the way in which the background flow and the temperature gradient in the drop, in the absence of the droplet, change with position along the axis of symmetry.

When $T(\xi_2, \eta) = P_1(s)$, the behavior of the background flow in the drop, induced by the temperature gradient at its surface, is as follows. The velocity along the axis of symmetry is a maximum at the center of the drop and decreases in magnitude as one moves away from the center. Except for the disturbance introduced by the bubble, the temperature gradient remains the same everywhere within the drop. For a relatively small bubble, we can consider the effects of the flow and the temperature gradient to be additive. So, we can see why the velocity of a bubble decreases as the distance from the center increases. The situation is qualitatively different for the higher modes. In those cases, the velocity along the axis of symmetry is zero at the center and also at the poles. For a surface temperature distribution $T(\xi_2, \eta) = P_n(s)$, this velocity reaches a maximum in magnitude at $r_{max} = \sqrt{\frac{n-1}{n+1}}$, where r is the radial coordinate from the center of the drop, scaled by the radius of the drop. The temperature gradient in the drop phase, in the absence of the bubble, is proportional to r^{n-1} along the symmetry axis. For a relatively small bubble, the appearance of a maximum in the magnitude of the background velocity, along the symmetry axis in the drop, leads to a maximum in the magnitude of the velocity of the bubble as D is increased. This maximum in the magnitude of the velocity in the drop occurs approximately between $D = 0.5$ and 0.6 for the P_2-mode and between 0.7 and 0.8 for the P_3-mode. The extremum in V occurs at $D > r_{max}$ because of the monotonic increase in the temperature gradient with r. Of course, these arguments only hold for small bubbles, which disturb the background flow and temperature gradient fields only slightly.

When the temperature on the drop surface is proportional to the P_n-mode, where $n \geq 2$, the direction of motion of the bubble can be used to make inferences about the stability of the motionless state of the bubble in the concentric configuration. The discussion is restricted to small displacements along the axis. Because the temperature is a maximum at the poles in the case of the P_2-mode, the bubble will move toward the nearest warm pole when displaced from the center, as seen in Figures 7.8.3, 7.8.4, and 7.8.5. On the other hand, if we were to impose the same scaled temperature field on the drop surface but with the sign reversed, so that the poles are cool and the equator is warm, the bubble would move toward the center when displaced from it. This would mean that the motionless concentric state is stable to displacements along the axis, for that temperature distribution on the drop surface. In the case of the P_3-mode, the bubble moves toward the center in Figure 7.8.6(b), but if it is placed above the equator in the figure, it would move away from the center. A sign reversal of the scaled temperature would lead to motion along the axis in the opposite direction. In both situations, for the P_3-mode, displacements in one direction along the axis bring the bubble back to the center, whereas those in the opposite direction take it away from the center. The conclusions drawn in the case of the P_2-mode are applicable to even-order higher modes. Similarly, the statements made about the P_3-mode are valid for odd-order higher modes.

Using the values for the quasi-steady velocities predicted here, it is possible to construct the trajectory of a bubble as it moves from its initial location toward a warm pole on the drop surface. This would provide an estimate of the approximate time

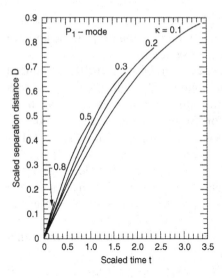

Figure 7.8.11 Scaled separation distance between centers plotted against scaled time for a $P_1(\cos\theta)$-mode temperature distribution on the drop surface (Courtesy of Academic Press).

required to remove a gas bubble by the application of spot-heating to a drop. The calculation for a pure mode is made by simply finding the area under the curve of the plot of $\frac{1}{V}$ against D, from some initial value D_1 to any given D, which yields a dimensionless time t required to reach that value of D. Here, the reference time scale used is $\frac{R_2}{v_0}$. In performing the integrations, we can begin right at the center of the drop when $T(\xi_2, \eta) = P_1(s)$. For the higher modes, because the velocity is zero in the concentric configuration, the integrations are begun at a very small nonzero value of D. In Figures 7.8.11, 7.8.12, and 7.8.13, we display trajectories for the first three Legendre modes of the temperature field on the drop surface.

In closing, we make the following observations. Results for the higher Legendre modes are not expected to reveal any qualitative differences. Also, we have placed the droplet within the drop in such a way that the upper half-plane is used in Figure 7.8.1 for the application of the solution in bispherical coordinates. By symmetry, the results will be similar if the droplet is placed such that its center is located above that of the drop in the sketch. The droplet will move along the axis toward the nearest warm pole

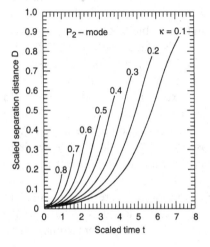

Figure 7.8.12 Scaled separation distance between centers plotted against scaled time for a $P_2(\cos\theta)$-mode temperature distribution on the drop surface (Courtesy of Academic Press).

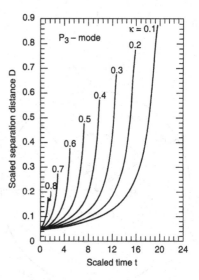

Figure 7.8.13 Scaled separation distance between centers plotted against scaled time for a $P_3(\cos\theta)$-mode temperature distribution on the drop surface (Courtesy of Academic Press).

in each case, with a velocity of the same magnitude as in the mirror image situation. Finally, when the interfacial tension gradients at the two interfaces are opposite in sign, it is possible for the droplet to reach a stationary location, where its velocity is zero. We discuss this situation in more detail in the case of a moving compound drop in Section 7.10.

7.9 Axisymmetric Motion of a Compound Drop – Concentric Case

7.9.1 Introduction

In this and the following section, we analyze the case of a compound drop that moves in a continuous phase in which a uniform temperature gradient has been established. The geometric configuration is assumed to be such that the fields are axisymmetric about the direction of the temperature gradient. The principal difference between this problem and the one considered in Sections 7.7 and 7.8 is that the entire object is free to move. Also, instead of prescribing the temperature distribution on the surface of the drop, it is determined as part of the solution, and only the temperature field in the undisturbed continuous phase is specified. In the present section, we develop results at the instant when the droplet is located concentrically within the drop. A sketch of the compound drop in the concentric configuration is provided in Figure 7.9.1. The uniform temperature gradient ∇T_∞ in the continuous phase fluid points in the positive z-direction. The droplet is designated phase 1, the drop is phase 2, and the exterior fluid forms phase 3. Both the drop and the droplet are assumed to retain their spherical shape. This is a good assumption when suitably defined Capillary numbers associated with these two interfaces are negligibly small. We also neglect the influence of convective transport of momentum and energy and consider the problem quasi-steady. The results presented in this section are drawn from Morton, Subramanian, and Balasubramaniam (1990).

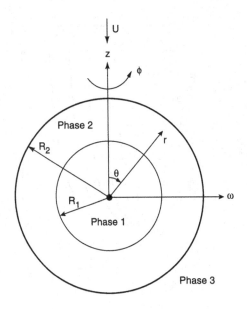

Figure 7.9.1   Sketch of a concentric compound drop.

7.9.2   Analysis

The symbol R_1 is used to represent the radius of the droplet, and R_2, the radius of the drop. As usual, $\kappa = \frac{R_1}{R_2}$ designates the scaled radius of the droplet, and $\delta = \frac{\sigma_{T,droplet}}{\sigma_{T,drop}}$, where $\sigma_{T,droplet}$ stands for the rate of change of the interfacial tension at the interface between the droplet phase and the drop phase fluids with temperature and $\sigma_{T,drop}$ represents a similar quantity at the drop-continuous phase interface. We shall need to use two viscosity ratios and two thermal conductivity ratios. The quantity $\alpha_1 = \frac{\mu_1}{\mu_2}$ is the ratio of the viscosity of the droplet to that of the drop, and $\alpha_2 = \frac{\mu_2}{\mu_3}$ represents the ratio of the viscosity of the drop phase to that of the continuous phase. In a similar way, the symbols $\beta_1 = \frac{k_1}{k_2}$ and $\beta_2 = \frac{k_2}{k_3}$ stand for corresponding thermal conductivity ratios. Lengths are scaled with the radius of the drop, R_2, and velocities are scaled using the thermocapillary reference velocity $v_0 = \frac{-\sigma_{T,drop}|\nabla T_\infty| R_2}{\mu_3}$. Pressure and viscous stresses are scaled using $-\sigma_{T,drop}|\nabla T_\infty|$, and temperature is scaled by subtracting a suitable reference value T_0 and then dividing by $|\nabla T_\infty| R_2$. In view of the axial symmetry, we can define a Stokes streamfunction in each phase scaled with $R_2^2 v_0$. As before, r stands for the scaled radial coordinate, and $s = \cos\theta$. The quasi-steady scaled velocity of the drop is $U\mathbf{i}_z$. In analyzing this problem, a reference frame is used that is attached to the drop. Therefore, in this reference frame, the fluid far from the drop approaches it with a uniform velocity $-U\mathbf{i}_z$. The scaled velocity of the droplet relative to the drop is designated $V\mathbf{i}_z$.

The scaled streamfunctions ψ_i will satisfy Stokes's equation, and the scaled temperature fields T_i will satisfy Laplace's equation. Here, the subscript i takes on the values 1, 2, and 3, corresponding to the identification number of the phase. We can use the general solutions for the streamfunctions in spherical polar coordinates given in Equations (3.1.9) and (3.1.10). Similarly, the form of the solution in Equation (4.2.15) is used for the temperature field in the drop and that within the droplet. The boundary conditions can be stated as follows.

Far from the compound drop, the velocity field must approach the uniform stream $-U\mathbf{i}_z$.

$$v_{3_\theta}(r \to \infty, s) \to U \sin \theta, \tag{7.9.1}$$

$$v_{3_r}(r \to \infty, s) \to -U \cos \theta. \tag{7.9.2}$$

At each interface, the kinematic condition, the continuity of tangential velocity, and the tangential stress balance are written as follows.

$$v_{3_r}(1, s) = v_{2_r}(1, s) = 0, \tag{7.9.3}$$

$$v_{3_\theta}(1, s) = v_{2_\theta}(1, s), \tag{7.9.4}$$

$$\tau_{r\theta3}(1, s) - \tau_{r\theta2}(1, s) = \frac{\partial T_2}{\partial \theta}(1, s), \tag{7.9.5}$$

$$v_{2_r}(\kappa, s) = v_{1_r}(\kappa, s) = V \cos \theta, \tag{7.9.6}$$

$$v_{2_\theta}(\kappa, s) = v_{1_\theta}(\kappa, s), \tag{7.9.7}$$

and

$$\tau_{r\theta2}(\kappa, s) - \tau_{r\theta1}(\kappa, s) = \frac{\delta}{\kappa} \frac{\partial T_2}{\partial \theta}(\kappa, s). \tag{7.9.8}$$

Also, we must require the velocity to be bounded at $r = 0$.

The boundary conditions on the temperature field are written as follows. Far from the compound drop, the temperature gradient must approach ∇T_∞ in the undisturbed continuous phase fluid. If we choose the reference temperature as that at $z = 0$, this yields

$$T_3(r \to \infty, s) \to z. \tag{7.9.9}$$

At each interface, we must require continuity of temperature and heat flux.

$$T_3(1, s) = T_2(1, s), \tag{7.9.10}$$

$$\frac{\partial T_3}{\partial r}(1, s) = \beta_2 \frac{\partial T_2}{\partial r}(1, s), \tag{7.9.11}$$

$$T_2(\kappa, s) = T_1(\kappa, s), \tag{7.9.12}$$

$$\frac{\partial T_2}{\partial r}(\kappa, s) = \beta_1 \frac{\partial T_1}{\partial r}(\kappa, s). \tag{7.9.13}$$

As with the velocity field, we require that the temperature field be bounded at the center of the droplet.

The solution of the governing equations, along with the boundary conditions and the condition of zero hydrodynamic force on the drop and on the droplet, is reported below. The temperature fields can be written as follows:

$$T_1 = A_t r \cos \theta, \tag{7.9.14}$$

$$T_2 = \left(C_t r + \frac{D_t}{r^2} \right) \cos \theta, \tag{7.9.15}$$

and

$$T_3 = \left(r + \frac{B_t}{r^2} \right) \cos \theta. \tag{7.9.16}$$

The constants appearing in the above solution are given below.

$$A_t = \frac{9}{P},$$ (7.9.17)

$$B_t = C_t + D_t - 1,$$ (7.9.18)

$$C_t = \frac{3(2 + \beta_1)}{P},$$ (7.9.19)

$$D_t = \frac{3\kappa^3(1 - \beta_1)}{P}.$$ (7.9.20)

Here,

$$P = (2 + \beta_1)(2 + \beta_2) + 2\kappa^3(1 - \beta_1)(1 - \beta_2).$$ (7.9.21)

The results for the streamfunction fields are given below, along with the various constants appearing in the solution.

$$\psi_1 = (A_1 r^2 + C_1 r^4)\frac{\sin^2 \theta}{2},$$ (7.9.22)

$$\psi_2 = \left(A_2 r^2 + \frac{B_2}{r} + C_2 r^4 \right)\frac{\sin^2 \theta}{2},$$ (7.9.23)

$$\psi_3 = U\left(r^2 - \frac{1}{r} \right)\frac{\sin^2 \theta}{2},$$ (7.9.24)

$$A_1 = -(V + \kappa^2 C_1),$$ (7.9.25)

$$C_1 = \frac{1}{10\alpha_2 \kappa^5}[3U(2 + 3\alpha_2) - 6U\kappa^5(1 - \alpha_2) - 2(1 + B_t)(1 - \kappa^5)],$$ (7.9.26)

$$A_2 = \frac{1}{3\alpha_2}[3U - (1 + B_t)],$$ (7.9.27)

$$B_2 = -\frac{1}{15\alpha_2}[3U(2 + 3\alpha_2) - 2(1 + B_t)],$$ (7.9.28)

$$C_2 = -\frac{1}{5\alpha_2}[3U(1 - \alpha_2) - (1 + B_t)].$$ (7.9.29)

The magnitudes of the scaled velocity of the droplet, and that of the drop, are written as follows:

$$V = \frac{1}{15\alpha_2\kappa^3}[(2 - 5\kappa^3 + 3\kappa^5)(3U - \{1 + B_t\}) + 9\alpha_2 U(1 - \kappa^5)]$$ (7.9.30)

and

$$U = \frac{10\delta\kappa^4 A_t + 2(1 + B_t)[(2 + 3\alpha_1) + 3\kappa^5(1 - \alpha_1)]}{3[(2 + 3\alpha_1)(2 + 3\alpha_2) + 6\kappa^5(1 - \alpha_1)(1 - \alpha_2)]}.$$ (7.9.31)

Because of the symmetry of the configuration, only the $P_1(s)$-mode of the temperature field on the drop surface is excited. The streamfunction field in the continuous phase is precisely the same as that obtained in the case when a drop containing no droplet migrates at a velocity $U\mathbf{i}_z$, under the action of the applied temperature gradient. Also, it can be verified that the streamfunction field in the drop phase reduces to the result given in Equation (4.2.22), when the scaled radius of the droplet $\kappa = 0$. In this case, the expression for the velocity of the drop given in Equation (7.9.31) reduces to the result in Equation (4.2.26).

We do not encounter a problem with singular behavior here in the limit $\kappa \to 1$ for the case when the droplet is a gas bubble, as we did in a related problem in Section 7.7. As $\kappa \to 1$, Equations (7.9.30) and (7.9.31) yield the following results at leading order:

$$V \sim \frac{6(1+\delta)}{(2+3\alpha_1\alpha_2)(2+\beta_1\beta_2)}(1-\kappa) + O((1-\kappa)^2) \qquad (7.9.32)$$

and

$$U \sim \frac{2(1+\delta)}{(2+3\alpha_1\alpha_2)(2+\beta_1\beta_2)} + O(1-\kappa). \qquad (7.9.33)$$

The asymptotic results show that the velocity of the droplet relative to the drop approaches zero as the thickness of the shell between the two surfaces approaches zero and that the velocity of the drop approaches a constant value. It can be shown that in the same limit, the shear stress $\tau_{r\theta 2}$ is finite and virtually constant across the thickness of the shell at any specified value of the angular coordinate. In the limit when the droplet is a gas bubble, α_1 and β_1 can be set equal to zero, and no special treatment is necessary.

The results for both U and V are easy to calculate from the expressions given here, but for convenience in making a quick estimate, we include an approximation for the scaled velocity of the droplet relative to the drop, obtained by using the technique described in Section 7.8.

$$V_{approx} = \frac{3}{2+\beta_2}\left[\frac{1}{2+3\alpha_2} + \frac{2\kappa\delta}{\alpha_2(2+3\alpha_1)(2+\beta_1)} - \frac{5\alpha_1\kappa^2}{(2+3\alpha_1)(2+3\alpha_2)}\right].$$

$$(7.9.34)$$

Just as in the problem considered in Section 7.8, this approximation can be expected to be useful when κ is relatively small compared to unity.

7.9.3 Results

There are six parameters in this problem. Therefore, it would be unwieldy to display graphical results for all possible situations. In any case, the principal results given in Equations (7.9.30) and (7.9.31) can be calculated easily when needed. In the next section, we present a few graphical results in the case when the droplet is a gas bubble located eccentrically within the drop. Therefore, to serve as a reference, similar results are displayed here for the case when the gas bubble is concentrically located. Besides setting $\alpha_1 = \beta_1 = 0$, we also set $\delta = 1$. These assumptions leave us with three parameters, which are the radius ratio κ and the property ratios α_2 and β_2. Representative results for the scaled velocity of the bubble are displayed in Figure 7.9.2. We have included the approximate result calculated from Equation (7.9.34) for comparison.

The following observations are made based on Figure 7.9.2. As the radius of the bubble approaches that of the drop, the velocity approaches the correct limiting value of zero. When the radius of the bubble approaches zero, the velocity approaches a different constant value in each case. This is simply the circulation velocity in the drop because a bubble of negligible size merely serves as a tracer. Next, we examine the role of the property ratios. First consider the role of α_2. As α_2 is increased, we see that

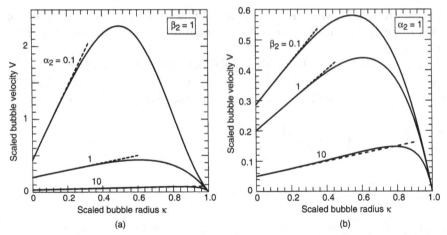

Figure 7.9.2 Scaled velocity of the bubble plotted against its scaled radius for $\delta = 1$ and compared with the approximation from Equation (7.9.34), shown as the dashed curves: (a) $\beta_2 = 1$, (b) $\alpha_2 = 1$.

the scaled velocity of the bubble decreases as expected because the drop phase fluid becomes more viscous compared with the continuous phase fluid. Next consider the role of the thermal conductivity ratio. Decreasing β_2 leads to an increase in the velocity of the bubble for any given value of κ. This is a consequence of sharper temperature gradients and increased circulation velocity in the drop. As observed from Equations (4.2.26) and (4.2.27), which apply to a drop containing no bubble, decreasing β_2 leads to an increase in the migration velocity of the drop because of the sharper temperature gradient at the surface. The resulting increase in the circulation velocity makes a contribution to the behavior observed in Figure 7.9.2(b). In addition, the sharper gradients have a direct impact on the thermocapillary migration velocity of the bubble. The velocity of a relatively large bubble is insensitive to the value of β_2. This is understandable because in this case, the temperature gradient at the surface of the drop is not influenced as much by β_2 as by the fact that the nonconducting bubble surface is close to the drop surface. Finally, we note that the approximation given in Equation (7.9.34) does remarkably well for relatively large values of α_2 and β_2. It becomes less useful as these property ratios decrease in value.

Now, we turn to the behavior of the scaled velocity of the drop. This is displayed for the same set of parameters in Figure 7.9.3. The scaled velocity of the drop is insensitive to the presence of a bubble within the drop unless the bubble is relatively large. When the bubble is relatively small, the dependence of the scaled velocity U on α_2 and β_2 is straightforward and follows directly from the result in Equations (4.2.26) and (4.2.27). It is interesting that the presence of a bubble always leads to an increase in the velocity of the drop. This is a final consequence of three effects. To understand the first effect, consider the situation when the interfacial tension gradient at the drop surface is zero. In this case, the motion of the bubble within the drop will cause the fluid at the surface of the drop to drag the neighboring continuous phase fluid. By reaction, the drop will propel itself in the opposite direction, which is toward warmer fluid. Secondly, the presence of the bubble within the drop sharpens the temperature gradient on the drop surface. The third effect involves the viscous stresses that resist motion in the shell between the two interfaces. For a given velocity of the drop, the stress is greater when a bubble is present because of sharper velocity gradients. As we see, the net effect of the presence of a gas

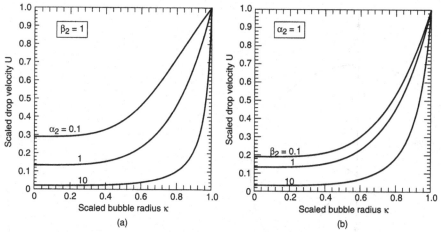

Figure 7.9.3 Scaled velocity of the drop plotted against its scaled radius for $\delta = 1$: (a) $\beta_2 = 1$, (b) $\alpha_2 = 1$.

bubble within a drop is to increase its velocity relative to the value in the absence of the bubble. Finally, note that in all cases when $\kappa \to 1$, $U \to 1$. This limit can be obtained from Equation (7.9.33), which yields $U \to \frac{1+\delta}{2}$ when we set $\alpha_1 = \beta_1 = 0$ and take the limit $\kappa \to 1$. This leads to a value of $U = 1$ when we set $\delta = 1$. It is useful to write a result for the physical velocity of the drop in the limit $\kappa \to 1$ using Equation (7.9.33):

$$U^*_{drop} \to \frac{-|\nabla T_\infty| R_2 (\sigma_{T,drop} + \sigma_{T,droplet})}{2\mu_3}. \tag{7.9.35}$$

Equation (7.9.35) shows that the thermocapillary migration velocity of a thin-walled shell of the drop-phase fluid is independent of either the viscosity or the thermal conductivity of the drop phase fluid. Also, the variation of interfacial tension at each interface contributes in an additive fashion to the total velocity.

Borhan, Haj-Hariri, and Nadim (1992) have analyzed this problem including the influence of a surfactant in the insoluble limit. The surfactant was permitted to adsorb on both interfaces. A perturbation expansion in the Elasticity number was used. Results are given for the retardation of the velocities of the drop and the droplet, as well as for small corrections to the spherical shape. Borhan et al. concluded that a droplet that is a gas bubble always deforms into a prolate spheroid whereas the drop is prolate for small values of κ and oblate otherwise.

7.10 Axisymmetric Motion of a Compound Drop – Eccentric Case

7.10.1 Introduction

In this section, we analyze the same problem that was considered in the previous section, namely, the thermocapillary migration of a compound drop. The only change is in the geometric configuration. The droplet is now allowed to be located anywhere within the drop instead of being concentric with it. A sketch of the system is provided in Figure 7.10.1. We define an eccentricity, ϵ, as follows:

$$\epsilon = \frac{d}{R_2 - R_1}. \tag{7.10.1}$$

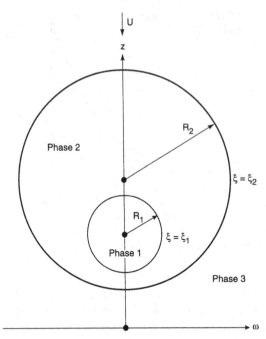

Figure 7.10.1 Sketch of an eccentric compound drop.

Here, d represents the distance between the centers of the drop and the droplet. As before, R_1 is the radius of the droplet, and R_2 is the radius of the drop. We restrict consideration to the axisymmetric case, which requires that the temperature gradient in the undisturbed continuous phase, ∇T_∞, be parallel to the line of centers joining the drop and the droplet. In Figure 7.10.1, the applied temperature gradient points in the z-direction. The physical assumptions made in Section 7.9 continue to apply, and the same system of designating variables in each phase is used, with a unique subscript ranging from 1 to 3. The results presented in this section also are drawn from Morton, Subramanian, and Balasubramaniam (1990), with more details given in the doctoral thesis of Morton (1990).

7.10.2 Analysis

No doubt as anticipated by the reader, the analysis of the eccentric case is carried out using bispherical coordinates. The parameters κ, δ, α_1, α_2, β_1, and β_2 are defined in the same way as in Section 7.9. The surface of the droplet corresponds to $\xi = \xi_1$, and that of the drop is represented by $\xi = \xi_2$ in the bispherical coordinate system. The geometrical parameters κ and ϵ can be used to calculate ξ_1 and ξ_2 by using the following relationships:

$$\cosh \xi_1 = \frac{1 + \kappa - \epsilon^2(1 - \kappa)}{2\epsilon\kappa} \tag{7.10.2}$$

and

$$\cosh \xi_2 = \frac{1 + \kappa + \epsilon^2(1 - \kappa)}{2\epsilon}. \tag{7.10.3}$$

Lengths are scaled with the characteristic length in bispherical coordinates, c, and velocities are scaled using the thermocapillary reference velocity $v_0 = \frac{-\sigma_{T,drop}|\nabla T_\infty|c}{\mu_3}$. Pressure and viscous stresses are scaled using $-\sigma_{T,drop}|\nabla T_\infty|$, and temperature is scaled by subtracting the reference value T_0 at $z = 0$ in the continuous phase, and then dividing by $|\nabla T_\infty|c$. We define a Stokes streamfunction in each phase that is scaled with $c^2 v_0$. The quasi-steady scaled velocity of the drop is $U\mathbf{i}_z$. In a reference frame attached to the moving drop, the fluid far from the drop approaches it with a uniform velocity $-U\mathbf{i}_z$. This is the reference frame in which the governing equations are written. As before, the scaled velocity of the droplet relative to the drop is designated $V\mathbf{i}_z$.

The axisymmetric solution of Stokes's equation for the scaled streamfunction fields ψ_i, and that of Laplace's equation for the scaled temperature fields T_i, can be used; these are given in Section 5.1. The subscript i is 1 for the droplet, 2 for the drop, and 3 for the continuous phase. In the case of the temperature field, to avoid numerical integration needed to calculate certain sets of constants in the solution, Morton et al. adopted Sadhal's (1983) approach and calculated the fields, $\Phi_i(\xi, \eta)$, defined in Section 7.1. The physical boundary conditions on the velocity fields in the three phases are precisely the same as those written in Section 7.9 in spherical polar coordinates, and one need only rewrite them in a form suitable for use in bispherical coordinates. The boundary conditions on the fields $\Phi_i(\xi, \eta)$ can be obtained from those on the temperature fields in the same manner as in Section 7.1. Using these boundary conditions and the conditions of zero hydrodynamic force on the drop and on the droplet the constants appearing in the general solution can be specialized. The algebraic details of the solution are too lengthy to include here. For these details, we refer the reader to the Appendix of the article by Morton et al. (1990), in which Equation (A41) needs to be corrected by including a plus sign before D_n^p, or to Morton's doctoral thesis. Here, we merely present and discuss some representative results from this solution.

Even when the droplet is a gas bubble, in which case we can set $\alpha_1 = \beta_1 = 0$, the results for U and V reported by Morton et al. require significant numerical computation. Therefore, we give an approximation for the velocity of the droplet obtained by the technique described in Section 7.8. This approximation should be useful when the radius of the droplet is relatively small compared with that of the drop.

$$\frac{V_{approx}}{V_{YGB}} = \frac{3}{2 + \beta_2}\left[1 + \frac{\alpha_2(1 - \epsilon^2\{1 - \kappa\}^2)(2 + 3\alpha_1)(2 + \beta_1)}{2\kappa\delta(2 + 3\alpha_2)} - \frac{5\alpha_1\alpha_2\kappa(2 + \beta_1)}{2\delta(2 + 3\alpha_2)}\right].$$

(7.10.4)

7.10.3 Results

In addition to the six parameters noted in the concentric configuration, we have an additional parameter, namely the eccentricity, ϵ. In presenting results from the solution in bispherical coordinates, we shall only consider the gas bubble limit by setting $\alpha_1 = \beta_1 = 0$. In addition, we set $\delta = 1$. These assumptions leave us with two geometric parameters, κ and ϵ, and two property ratios α_2 and β_2. Morton (1990) displayed graphical results for the scaled velocity of the drop, as well as that of the bubble, as a function of the radius ratio κ when $\epsilon = 0.8$, for $\beta_2 = 1$, varying α_2 from 0.1 to 10 and for $\alpha_2 = 1$, varying β_2 from 0.1 to 10. These results demonstrate that the qualitative influence of the parameters α_2 and β_2 is the same as in the concentric case. Thus, we can concentrate here on the effect of eccentricity for a set of values of κ in the representative case $\alpha_2 = \beta_2 = 1$. This is done in Figures 7.10.2 and 7.10.3 for the bubble and the drop, respectively. The length scale, c, used in the reference velocity, is convenient for analysis

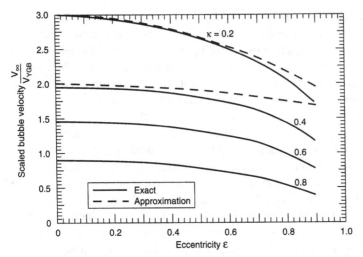

Figure 7.10.2 Scaled velocity of the bubble plotted against eccentricity for $\alpha_2 = \beta_2 = 1$ and selected values of κ and compared with the approximation for small eccentricities given in Equation (7.10.4). (Reprinted with permission from Morton, D. S., Subramanian, R. S., and Balasubramaniam, R. The migration of a compound drop due to thermocapillarity. *Phys. Fluids A* 2, No. 12, 2119–2133. Copyright 1990, American Institute of Physics.)

but is not physically as meaningful in presenting results. Therefore, Morton et al. used natural reference velocities for the bubble and the drop. For the bubble, this reference, V_{YGB}, is the velocity it would have, if present in a large body of the drop phase fluid, when subjected to the same temperature gradient that exists in the undisturbed continuous phase fluid. For the drop, the reference, U_{YGB}, is the velocity the drop would have in the absence of a bubble within it. These quantities can be obtained from Equation (4.2.28)

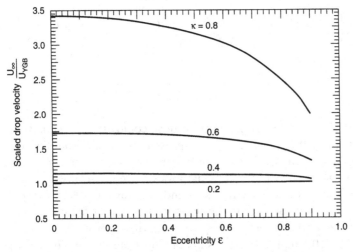

Figure 7.10.3 Scaled velocity of the drop plotted against eccentricity for $\alpha_2 = \beta_2 = 1$ and selected values of κ. (Reprinted with permission from Morton, D. S., Subramanian, R. S., and Balasubramaniam, R. The migration of a compound drop due to thermocapillarity. *Phys. Fluids A* 2, No. 12, 2119–2133. Copyright 1990, American Institute of Physics.)

by using the relevant symbols in each case. The case $\epsilon = 0$ is a singular limit for the bispherical coordinate system. Therefore, at $\epsilon = 0$, results from Section 7.9 are used. Also included in Figure 7.10.2 is the approximation given in Equation (7.10.4).

From Figure 7.10.2, we see that the scaled velocity of the bubble decreases with increasing eccentricity. This is consistent with the observations made in Section 7.8, where an analogous problem for a fixed compound drop was considered. The situation, where the prescribed temperature field on the drop surface is proportional to the first Legendre Polynomial $P_1(\cos\theta)$ in that problem, corresponds most closely to that considered here. The most useful feature in Figure 7.10.2 is the relative insensitivity of the velocity of the bubble to the eccentricity, when the eccentricity is small. This is fortunate, because the solution from bispherical coordinates requires some effort to calculate. Instead, the simple results of Section 7.9 can be used for small to moderate values of the eccentricity. From Figure 7.10.2, it is seen that up to an eccentricity $\epsilon \approx 0.4$, the results from the concentric case can be used with little error for large values of κ. Relatively small bubbles, in contrast, move at significantly lower velocities, as the eccentricity increases above approximately 0.2. This is due to the considerable influence of the circulation in the drop on the velocity of a small bubble. Fortunately, for small bubbles, the simple approximation given in Equation (7.10.4) can be used up to a substantial value of ϵ before incurring significant error. For example, it is adequate up to an eccentricity of 0.5, when $\kappa = 0.2$.

From Figure 7.10.3, we observe that a drop containing a bubble moves at a velocity larger than that of a drop of the same size not containing a bubble, just as in the concentric case; however, a small bubble exerts virtually no influence on the motion of the drop at all values of the eccentricity investigated. It requires a bubble of substantial size to increase the velocity of the drop significantly. It is noteworthy that the velocity of the drop is even more insensitive to eccentricity than that of the bubble. Therefore, the relatively simple results for the velocity of the drop from the concentric case have a wide range of applicability, even when the bubble is eccentrically placed within the drop.

In Figure 7.10.4, typical streamlines are shown in the laboratory reference frame, for a compound drop executing thermocapillary motion. Note the appearance of a

	ψ		
1	−0.085	6	0
2	−0.067	7	0.003
3	−0.047	8	0.006
4	−0.028	9	0.009
5	−0.009		

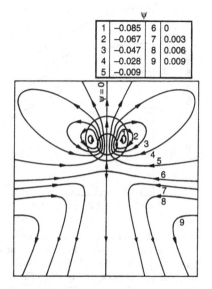

Figure 7.10.4 Streamlines in the laboratory frame of reference; $\alpha_2 = \beta_2 = 1, \epsilon = \kappa = 0.5$. (Reprinted with permission from Morton, D. S., Subramanian, R. S., and Balasubramaniam, R. The migration of a compound drop due to thermocapillarity. *Phys. Fluids A* 2, No. 12, 2119–2133. Copyright 1990, American Institute of Physics.)

separated reverse flow region in the wake, analogous to that seen in Section 4.17. There is no such reverse flow region when the bubble is concentrically located within the drop because the situation is symmetric and only the P_1-mode of the temperature field on the surface is excited. The separated reverse flow region appears in the eccentric case, where the symmetry is broken, for reasons discussed in Section 4.17.

We close with an interesting observation regarding compound drop configurations that can be encountered when the interfacial tension gradients at the two interfaces are of opposite sign. This is not common because interfacial tension commonly decreases with increasing temperature. As we have noted in Chapter One, however, there are cases where the opposite is encountered in multicomponent systems over some range of temperatures. In this case, the circulation in the drop will cause the motion of the bubble in one direction, and the temperature gradient will have an opposite influence. Therefore, it is possible to arrange conditions such that the velocity of the bubble relative to the drop is zero. Such a stationary location along the symmetry axis can be either stable or unstable to small perturbations along the axis. In Figure 7.10.5, a sample case is illustrated that shows the velocity of the bubble changing sign. The slopes of all the curves are negative at the zero crossing. Therefore, a small displacement of the bubble in either direction from the stationary location will lead to its moving away from that location. So, this type of configuration is not stable. The same curves apply to the mirror image case where the bubble is located in the upper half of the drop in Figure 7.10.1. In that situation, the negative slope at the zero crossing implies stable stationary locations. Note that we made an incorrect statement in Morton et al. (1990) regarding the sign of the slope in the mirror image case but drew the correct conclusion. An axisymmetric analysis cannot deal with other perturbations that destroy the symmetry. Therefore, no conclusion can be reached about the local stability of the configuration in a general sense. It is worth mentioning that these ideas regarding stability are equally applicable to the motion of a compound drop driven by a body force. In fact, they were originally put forth in that context by Sadhal and Oğuz (1985).

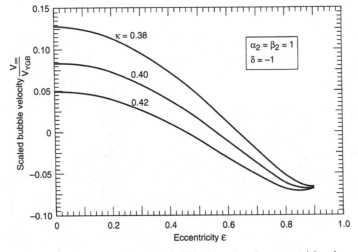

Figure 7.10.5 Scaled velocity of the bubble plotted against eccentricity, showing stationary locations. (Reprinted with permission from Morton, D. S., Subramanian, R. S., and Balasubramaniam, R. The migration of a compound drop due to thermocapillarity. *Phys. Fluids A* 2, No. 12, 2119–2133. Copyright 1990, American Institute of Physics.)

7.11   Interactions When Convective Transport Effects Are Not Negligible

7.11.1   Introduction

In this section, we discuss interaction problems in thermocapillary migration in which convective transport effects play a significant role. This implies that either the Reynolds number or the Marangoni number associated with the motion is nonnegligible. Analytical solutions would be difficult to obtain because of the nonlinearity of the governing equations, and the inherent time dependence that arises in this situation. As noted in Section 5.1.1, when the Reynolds and Marangoni numbers are both negligible, one can assume a quasi-steady situation to prevail. That is, the velocity and temperature fields in the fluids are assumed to be at steady states corresponding to the instantaneous geometric configuration of the system. In general, quasi-steady motion cannot be expected when either Re or Ma is nonzero. A numerical solution of the governing equations and boundary conditions must be used. Given this fact, it is not surprising that little has been done in this area. A discussion of the literature on interactions among bubbles in gravity-driven motion for nonzero values of the Reynolds number can be found in Harper (1997).

Lavrenteva, Leshansky, and Nir (1999) recently treated the influence of unsteady and convective energy transport effects on the Stokes motion of a pair of drops which is driven by interfacial tension gradients. This work extends the quasi-steady analysis of Golovin et al. (1995), which was mentioned in Section 7.3.5. The driving force for motion arises from the transport of surfactant from the drops to a medium in which the concentration of surfactant is uniform in the undisturbed state. The authors solved the problem by the method of matched asymptotic expansions, using a suitably defined Péclet number, Pe, as the small parameter. Solutions in bispherical coordinates are employed, along with solutions obtained by the method of reflections, which are useful for establishing the asymptotic behavior when the drops are far from each other. When the drops are close together, change in the geometric configuration with time takes on a primary role, leading to a history term that occurs in an $O(\sqrt{Pe})$ correction to the leading order concentration field. As a consequence, a time-dependent correction to the approach velocity is found at $O(\sqrt{Pe})$. The authors showed that when the initial separation distance is asymptotically large, the motion becomes quasi-steady. In this case, the quasi-steady correction to the approach velocity between the drops occurs at $O(Pe)$. The analogous situation in the thermocapillary case would occur when the drops are present in a continuous phase that is isothermal in the undisturbed state. To provide a thermocapillary driving force, the initial temperature of the drops must be different from that of the continuous phase or energy sources must be present within the drops. The interaction between the drops gives rise to a nonuniform flux of energy and a nonuniform temperature field over the surface of each drop, leading to the thermocapillary migration of both drops. In the present section, we only consider situations in which there is a gradient in temperature in the continuous phase in the undisturbed state that causes the motion of the drops.

The only available numerical work on the thermocapillary migration of interacting drops, in which nonzero values of the Reynolds and Marangoni numbers are considered, is that of Nas (1995), who permitted several drops to interact in fluid contained in a suitable volume. Nas also accommodated shape deformation of the drops. The test cases considered by Nas are nonaxisymmetric, and he reported results in representative situations both for a pair of drops and for a collection of several drops. He found that the drops tend to line up, after the passage of some time, in planes normal to the applied temperature gradient. When drops of different sizes are present, drops of any given size display a tendency to align themselves on a plane normal to the applied temperature gradient, and drops of a different size collect on a different plane, thus leading to a segregation by size. The only analysis available is an asymptotic treatment of the axisymmetric thermocapillary motion of two interacting bubbles, by Balasubramaniam and Subramanian (1999). This study was motivated by observations made in experiments aboard the United States space shuttle in 1994 and 1996. The results from these reduced gravity experiments, in the context of the motion of isolated bubbles and drops, have been described in Section 4.16. In the present section, we comment on results obtained when bubbles or drops interacted with each other in the same apparatus.

We begin with the highlights of the analysis of Balasubramaniam and Subramanian (1999), which applies in the gas bubble limit, wherein the viscosity and the thermal conductivity of the gas within the bubbles are both considered negligible when compared with the corresponding properties of the continuous phase. Therefore, only the transport problem in the continuous phase needs to be analyzed. We further restrict the analysis to axisymmetric fields.

7.11.2 Analysis

Consider the thermocapillary migration of a pair of bubbles of radii R_1 and R_2 respectively, spaced a distance d between their centers, as shown in Figure 7.11.1.

Bubble 1 is the leading bubble, and bubble 2 is the trailing bubble. Their motion occurs because of an undisturbed uniform temperature gradient ∇T_∞ in the continuous phase. All the relevant physical properties are assumed constant, as in the analyses presented in earlier sections. We assume both bubbles to retain the spherical shape, which requires that the Weber number be sufficiently small for each bubble. The physical velocities of the bubbles are designated $v_{\infty_1}^*$ and $v_{\infty_2}^*$, and the same symbols without the asterisk represent the scaled velocities of these bubbles. The velocities are scaled using reference velocities v_{0_1} and v_{0_2}. The reference velocity $v_{0_1} = -\dfrac{\sigma_T |\nabla T_\infty| R_1}{\mu}$, where as usual, σ_T is the rate of change of surface tension with temperature and is assumed negative. The symbol μ represents the viscosity of the continuous phase. The reference velocity v_{0_2} is defined in the same manner as v_{0_1}, using the radius R_2 of the trailing bubble as the length scale. We assume that v_{∞_1} and v_{∞_2} are independent of time. When the two bubbles move at unequal velocities, the distance d between them will be time dependent, and it would appear that the interaction between the bubbles would also depend on time. We show later that in the limiting case analyzed here, it is possible for the bubbles to move at steady velocities, even though d depends on time.

We define Reynolds and Marangoni numbers in the usual manner, using the radius of each bubble as the length scale.

$$Re_1 = \frac{R_1 v_{0_1}}{\nu}, \tag{7.11.1}$$

$$Ma_1 = \frac{R_1 v_{0_1}}{\kappa}, \tag{7.11.2}$$

$$Re_2 = \frac{R_2 v_{0_2}}{\nu}, \tag{7.11.3}$$

$$Ma_2 = \frac{R_2 v_{0_2}}{\kappa}. \tag{7.11.4}$$

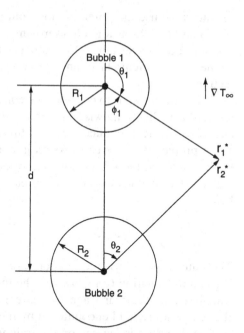

Figure 7.11.1 Sketch showing a pair of bubbles and the coordinate systems centered in each of them.

In the above, ν represents the kinematic viscosity of the liquid, and κ stands for its thermal diffusivity. The analysis is confined to the limit when Re_1, $Re_2 \to \infty$ and Ma_1, $Ma_2 \to \infty$. Therefore, we can use information from Section 4.13 in which we presented a similar asymptotic analysis for the migration of an isolated bubble. Scaled variables are employed in the analysis, which is performed individually on each bubble, for reasons that will be evident shortly. It is convenient to work with two spherical polar coordinate systems, each centered in one of the bubbles. The radial coordinates are designated r_1^* and r_2^* as shown in Figure 7.11.1 and are scaled in each case by the radius of the corresponding bubble. The same symbols, without the asterisks, are used to designate the scaled radial coordinates. The polar angles θ_1 and θ_2 are both measured from the direction of the temperature gradient, as shown in Figure 7.11.1. Velocities are scaled as usual by the reference velocities noted earlier. The temperature field, in each case, is scaled by first subtracting the temperature prevailing in the undisturbed continuous phase at the current location of a bubble and then dividing by the reference quantity $|\nabla T_\infty| R$, where a suitable subscript is used with the bubble radius R in each case.

The key idea in this asymptotic analysis is that convective transport of momentum and energy are dominant. First, consider the implication of the assumption of large values of each Reynolds number. The velocity field everywhere is given by the solution of the potential flow equations, with suitable boundary layers around the surface of each bubble and momentum wakes behind each bubble. Harper (1970, 1997) has analyzed the axisymmetric problem for the gravitational rise of a pair of identical bubbles in this situation. He showed that the interaction between them in that problem leads to a correction velocity on the surface of each bubble that is of $O(\delta^3)$ and a repulsion force between the two bubbles that is of $O(\delta^4)$, where $\delta = \frac{R}{d}$, and R is the radius of either bubble. It is evident that the potential flow interactions in that problem become negligible for sufficiently large distances of separation. In an analogous manner, we assume that we can omit such interactions in the present thermocapillary migration problem when d is sufficiently large compared with the radius of either bubble. This permits us to use the potential flow solution around each bubble, given in Equations (3.3.1) and (3.3.2), as a good approximation, ignoring the presence of the other bubble. As noted in Section 4.13, we neglect the correction to the velocity field in the momentum boundary layer and the momentum wake as well for reasons stated in that section.

The second step is to consider the consequences of assuming Ma_1, $Ma_2 \to \infty$. This means that in the streamwise direction, convective transport is the dominant mechanism, and we can neglect conduction in that direction. Thus, information on temperature fields is not propagated upstream. As an immediate consequence, we see that the leading bubble moves as though it is completely unaffected by the trailing bubble. Therefore, its scaled velocity, at leading order, is already known from Section 4.13, and is reproduced below.

$$v_{\infty 1} = \frac{1}{3} - \frac{1}{8} \log 3. \tag{7.11.5}$$

The outer solution for the temperature field in the wake behind the leading bubble is given in Equation (4.13.49), and the inner solution in the wake, which corrects the singularity on the rear stagnation streamline, is given in Equation (4.13.56). This is the temperature field encountered by the trailing bubble. The analysis in the present section therefore is focused on the trailing bubble, which encounters fluid in potential flow around it with this approaching temperature field.

As we can see from the above, the problem of interaction has been reduced to the problem of determining the solution for the motion of the trailing bubble, given a potential flow velocity distribution and a known temperature distribution in the "undisturbed" fluid approaching it. The only difference from the problem for the leading bubble is that this temperature distribution is not a simple linear field in distance but is given by the wake solution for the leading bubble mentioned earlier. We need to express that solution in terms of the coordinate system erected from the center of the trailing bubble as the origin. For this purpose, we use the following geometrical relationships:

$$\lambda^2 r_2^2 = r_1^2 + D^2 - 2r_1 D \cos \phi_1 \tag{7.11.6}$$

and

$$\theta_2 = \arctan\left(\frac{r_1 \sin \phi_1}{D - r_1 \cos \phi_1}\right). \tag{7.11.7}$$

In the above results, the radius ratio λ and the scaled separation distance D are defined as follows:

$$\lambda = \frac{R_2}{R_1} \tag{7.11.8}$$

and

$$D = \frac{d}{R_1}. \tag{7.11.9}$$

The angle ϕ_1 is related to θ_1 through

$$\phi_1 = \pi - \theta_1. \tag{7.11.10}$$

Because the analysis of the energy equation for the trailing bubble is performed in the limit $Ma_2 \to \infty$, it is necessary only to know the approaching temperature field in a narrow bundle of streamlines surrounding its forward stagnation streamline. The plan is to construct a leading order outer solution, neglecting conduction, in the vicinity of the surface of the trailing bubble and match it with a leading order inner solution valid near the surface. With this in mind, we approximate the geometric relations in Equations (7.11.6) and (7.11.7) for ϕ_1, $\theta_2 \ll 1$ as follows:

$$\lambda r_2 \approx D - r_1 \tag{7.11.11}$$

and

$$\theta_2 \approx \frac{r_1 \phi_1}{\lambda r_2}. \tag{7.11.12}$$

We can use the above relationships with the outer wake solution, given in Equation (4.13.49), to write that solution in the trailing bubble coordinates. We assume that the distance of separation between the two bubbles is large compared with the radius of either bubble. Therefore, the solution is approximated for large values of the scaled radial coordinate, r_1, as follows:

$$T_{\text{wake}} = A + r_2 + \frac{2}{3\lambda} \log r_2 \theta_2. \tag{7.11.13}$$

Here, the constant A is given by

$$A = \frac{1}{\lambda}\left(1.6727 + \frac{\pi}{3\sqrt{3}} - D(0) - \frac{1}{6}\log\frac{5832\,Ma_1 v_{\infty_1}}{\lambda^4}\right). \tag{7.11.14}$$

It is important to note that the scaled distance D depends on scaled time τ, which is defined as

$$\tau = \frac{v_{0_1} t^*}{R_1}, \tag{7.11.15}$$

where t^* is physical time. In writing Equation (7.11.14), we have used the fact that the time dependence of the scaled distance of separation is given by

$$D(\tau) = D(0) + \left(v_{\infty_1} - \lambda v_{\infty_2}\right)\tau. \tag{7.11.16}$$

It is remarkable that the outer thermal wake solution from the leading bubble, far away from it, is independent of time when expressed in terms of the trailing bubble coordinates. This is a consequence of the small angle approximation used in Equation (7.11.12). In this approximation of the wake solution, the correction to the applied linear temperature field, in the absence of conduction, depends only on the distance from the axis of symmetry, and hence is independent of time. The undisturbed linear temperature field, by itself, is independent of time when written in terms of the trailing bubble coordinates, without the need for the small angle approximation. We see that the effect of the leading bubble on the trailing bubble arises from the correction to the linear temperature field that is present in the wake. It will prove sufficient for our purposes to use the outer temperature field from the wake solution, without correcting for the singularity on the axis of symmetry. The singularity that arises, as a consequence, in the temperature field at the forward stagnation point of the trailing bubble does not influence the result for the migration velocity we obtain. Balasubramaniam and Subramanian (1999) provide a solution accommodating the inner wake solution from the leading bubble. They show that when it is used as the approaching temperature field in the vicinity of the forward stagnation streamline of the trailing bubble, the singularity is relieved.

The governing equation for the steady temperature field around the trailing bubble, in a reference frame attached to it, is given below.

$$v_{\infty_2} + \mathbf{v} \bullet \nabla T = \epsilon^2 \nabla^2 T. \tag{7.11.17}$$

The symbol $\mathbf{v}$ represents the scaled velocity vector, with components in spherical polar coordinates given in Equations (3.3.1) and (3.3.2). The symbol $\epsilon = \frac{1}{\sqrt{Ma_2}}$. For convenience, we do not include the subscript 2 in the analysis, except when referring to Ma_2 and v_{∞_2}, but it is implied where appropriate. Far from the bubble, the temperature field in the continuous phase must approach that in the wake of the leading bubble.

$$T(r \to \infty, s) \to T_{\text{wake}}. \tag{7.11.18}$$

In the above equation, as usual, $s = \cos \theta$. The assumption of negligible thermal conductivity of the gas inside the bubble leads to an adiabatic boundary condition at the surface of the bubble.

$$\frac{\partial T}{\partial r}(1, s) = 0. \tag{7.11.19}$$

Since we are seeking a solution in the limit $\epsilon \to 0$, the mathematical problem is virtually identical to that considered in Section 4.13 in the limit $Re \to \infty$. The only difference is in the undisturbed temperature field. This field includes two additional terms beyond the linear term that appears in Equation (4.13.2), one of which is just a constant. The solution procedure is identical, and therefore the details are omitted. The leading order outer solution, in a region near the surface of the bubble, is written as

$$T_0 = \frac{1}{3} \log(r - 1) + \frac{2}{3} \log(1 + \cos \theta) + G(\psi), \tag{7.11.20}$$

where $\psi = \frac{3}{2}(r - 1) \sin^2 \theta$ is the streamfunction near the surface of the bubble. The function $G(\psi)$ is determined by a patching procedure described in Section 4.13 as

$$G(\psi) = 1 + A + \frac{\pi}{6\sqrt{3}} - \frac{1}{6} \log 48 + \frac{1}{3\lambda} \log 2\psi. \tag{7.11.21}$$

The corresponding leading order inner solution, $t_0(x, s)$, where $x = \sqrt{Ma_2}(r - 1)$, can be written as follows:

$$t_0 = B + \frac{2}{3} \log \left(\frac{1 + \cos \theta}{\sin \theta} \right) + \frac{(1 + \lambda)}{\lambda} \left[\frac{1}{6} \log \xi + \frac{2}{3} F(\zeta) \right]. \tag{7.11.22}$$

In the above result, the variables ξ and ζ, as well as the function $F(\zeta)$, are those defined in Equations (4.13.27) through (4.13.29) with the understanding that the definitions apply to the trailing bubble. The constant B is defined below.

$$B = 1 + A + \frac{\pi}{6\sqrt{3}} - \frac{1}{6} \log 27 + \frac{1}{3\lambda} \log 4 - \frac{(1 + \lambda)}{\lambda} \left[0.3273 + \frac{1}{6} \log Ma_2 v_{\infty_2}^2 \right]. \tag{7.11.23}$$

The scaled migration velocity of the trailing bubble can be determined in the same way as in Section 4.13, using Equation (4.13.19). The result is

$$v_{\infty_2} = \frac{1}{2} \int_{-1}^{1} t_0(0, s) P_1(s) \, ds = \frac{1}{3} - \frac{1 + \lambda}{8\lambda} \log 3. \tag{7.11.24}$$

It is evident that the scaled velocity of the trailing bubble is smaller than that of the leading bubble, and this is a consequence of its immersion in the thermal wake field of the leading bubble. When the bubbles are of equal size, the velocity of the trailing bubble is reduced by approximately 70% compared with its velocity when isolated. The radius of the trailing bubble has to be approximately 1.7 times that of the leading bubble for the two bubbles to move at the same velocity, which will maintain a fixed separation distance between them. Also, we find that when $\lambda = 0.7006$, the velocity of the trailing bubble is zero. Of course, before this limit is reached, the analysis will break down because convective transport of energy will not be predominant over conduction.

We also note a remarkable feature discussed by Balasubramaniam and Subramanian (1999) regarding the temperature distribution on the surface of the trailing bubble. It is not monotonic, as is the case for an isolated bubble. On the forward portion of the surface of the trailing bubble, $\frac{\partial t_0}{\partial \theta}$ is positive. A maximum occurs at θ slightly less than $\frac{\pi}{2}$. Over the rear half of the surface, the temperature decreases with increasing θ. The increase in temperature with θ over the forward portion occurs because the thermal wake of the leading bubble, which is relatively cold, influences the temperature on the surface of the trailing bubble in this region. The reader will find it useful to review the discussion in Section 4.13, which occurs immediately following Equation (4.13.32). There, we discuss the way in which the temperature of fluid elements in a bundle of streamlines changes as they approach the forward stagnation point of a bubble in a reference frame that is attached to the bubble. It is the outer solution in this bundle of streamlines that determines conditions at the edge of the radial thermal boundary layer on the surface of the trailing bubble. In the case of the leading bubble, the temperature is uniform in a plane normal to the forward stagnation streamline, when the plane is located far from that stagnation point. The fluid elements move at comparable velocities in this bundle of streamlines, and therefore the temperature variations across the streamlines in this bundle are not large. All of this fluid is relatively cold, which leads to the logarithmic leading order term in the inner solution in Equation (4.13.32). In contrast, we see from the wake solution in Equation (7.11.13) that the temperature decreases logarithmically as the distance from the symmetry axis decreases. This means that the fluid approaching the trailing bubble in the immediate neighborhood of the forward stagnation point is much colder than the fluid a little distance away. This information is passed on to the region near the surface of the bubble by conduction in the inner solution. As a result, the temperature increases with θ, at least near the forward stagnation point. Eventually, the sink term in the energy equation leads to a decrease of temperature with distance in the θ direction along the bubble surface. In addition to the logarithmic singularity in the temperature field near the forward stagnation point mentioned earlier, there is another logarithmic singularity in the surface temperature distribution in the vicinity of the rear stagnation point. This must be relieved by accommodating conduction in the θ direction, in the same manner as was done in Section 4.13 in developing the wake solution for an isolated bubble.

We must add a word of caution. The analysis assumes that the orientation of two bubbles in line is stable. It is possible that when a bubble moves slightly away from the axis because of a disturbance, it may continue to move away in the same direction. We mention some experimental evidence in the following subsection that this may indeed be the case. Nonetheless, the analysis still serves a useful purpose in predicting the qualitative nature of the effect of interactions on the velocities of trailing bubbles as well as drops in our reduced gravity experiments. We now proceed to discuss the experimental observations.

7.11.3 Experimental Observations

We mentioned in the introduction that experiments on interacting drops and bubbles were performed aboard the space shuttle in summer 1994 and again in summer 1996. The apparatus and procedure have been described in Section 4.16, wherein two references are given, in which more details can be found. Here, we describe interesting observations from these experiments on interacting pairs of drops and pairs of bubbles.

In the International Microgravity Laboratory (IML-2) mission in summer 1994, we performed reduced gravity experiments on air bubbles, as well as Fluorinert FC-75 drops, migrating in a temperature gradient established in a DC-200 silicone oil of nominal viscosity 50 centistokes. In some of these experiments, two Fluorinert drops were injected one after the other, with a small space between them. We first introduced the smaller of the two drops and followed it with the larger drop. The expectation was that the larger drop, moving more rapidly, would catch up to the smaller drop and perhaps coalesce with it. To our surprise, the smaller drop, which was the leader, moved slightly more rapidly than the larger drop. Figure 7.11.2 shows a typical set of trajectories from an experiment. Both the drops moved in the direction of the temperature gradient, and their transverse position in the plane of the image did not change appreciably during the experiment. We could not monitor motion normal to that plane. Assuming that there was negligible movement in that direction, we can surmise that the interaction was axisymmetric.

The solid curves in Figure 7.11.2 represent the trajectories that would have resulted if each drop was moving at the velocity it would have had if isolated. For this purpose,

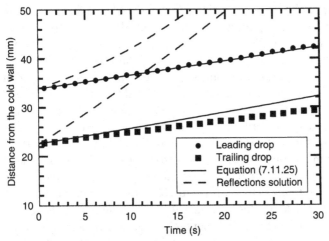

Figure 7.11.2 Trajectories of a pair of FC-75 Fluorinert drops migrating in a DC-200 silicone oil, of nominal viscosity 50 centistokes, along the center line of the test cell in the IML-2 flight experiment. The ordinate is the distance measured from the cold wall in the direction of the applied temperature gradient. The radius of the leading drop is 2.40 mm and that of the trailing drop is 4.72 mm; the temperature gradient is 0.75 K/mm. The solid curve was obtained by integrating the result for the velocity of an isolated drop from Equation (7.11.25), and the dashed curves represent the reflections solution. (Reprinted with permission from Balasubramaniam, R., Lacy, C. E., Wozniak, G., and Subramanian, R. S. Thermocapillary migration of bubbles and drops at moderate values of the Marangoni number in reduced gravity. *Phys. Fluids* 8, No. 4, 872–880. Copyright 1996, American Institute of Physics.)

we have used an empirical fit of the measured velocities of isolated drops in the same fluid. The equation of the fit is

$$\frac{v_\infty}{v_{YGB}} = 1.49 \, Ma^{-0.328}. \tag{7.11.25}$$

Here, v_{YGB} stands for the velocity of a drop in the limit $Re \to 0$ and $Ma \to 0$. It was necessary to integrate the result for the instantaneous velocity to calculate the trajectory of each drop. The dashed curves in the figure are predictions from the reflections solution, given in Section 7.3, which is applicable to interacting drops in the limit $Re \to 0$ and $Ma \to 0$. At the start of the run, the Marangoni number corresponding to the leading drop was 31.2, and that for the trailing drop was 102. At the end of the run, the Marangoni numbers were 35.3 and 113, respectively. The Prandtl number was at least 370 or larger, so that the Reynolds numbers corresponding to both drops were small. Similar results were noted for other pairs of liquid drops, both in this series of experiments and in those performed later in summer 1996 in a DC-200 series silicone oil of nominal viscosity 10 centistokes aboard the Life and Microgravity Spacelab (LMS) mission of the space shuttle.

It is clear that the reflections solution is not adequate to describe the behavior of the data in Figure 7.11.2. Convective transport of energy is evidently important in these experiments. We see that the leading drop moves virtually as though it is isolated. The trailing drop moves more slowly than when it is isolated. Both these facts are consistent with the predictions of the analysis in the gas bubble limit presented in the previous subsection. Remarkably, the analysis predicts the extent of retardation of the trailing drop as approximately 36%, a value close to the observed retardation of about 28%. The viscosity of Fluorinert FC-75 is indeed small compared with that of the silicone oil. Therefore, the assumption made in the analysis that $\alpha \to 0$ is reasonable. The thermal conductivity ratio is not negligible, however. The heat transport process within the drops is important in this situation, and the gas bubble limit does not accommodate it. Furthermore, the Reynolds number was small in the experiments, and therefore, we must regard the closeness of the prediction from the simplified theoretical model to the observed result as being fortuitous, and not an adequate test of the predictions.

We were unable to observe pairs of gas bubbles moving along the axis of the cell in an axisymmetric situation. Even though several pairs of bubbles were introduced both in the IML-2 experiments, and later in the LMS experiments, the trailing bubble showed a tendency to move off the axis. Its velocity toward the hot wall would increase, and in some cases, the trailing bubble would pass the smaller leading bubble. In the case of liquid drops, in all the experiments in the IML-2 mission, and in some of the experiments in the LMS mission, we did observe interactions that appeared to occur under nearly axisymmetric conditions. Nonetheless, we also saw the trailing drop behave in the same manner as the trailing bubble in some instances in the LMS experiments. We conjecture that the source of the instability is the temperature distribution in the thermal wake of the leading bubble or drop. From the sample isotherms in Figures 4.15.3(a) and 4.15.8, we can see that if a natural disturbance leads to a slight displacement of the trailing drop (or bubble) in a direction perpendicular to the applied temperature gradient, it would lead to a transverse contribution to the temperature gradient experienced by the drop, propelling it away from the axis. When the trailing drop has moved a sufficient distance away from the axis, the interaction with the thermal wake of the leading drop would be weak. The hydrodynamic interaction with the leading drop is likely to be weak as well, as long as the separation distance is at least three times the radius of the leading drop. Uninfluenced by the leading drop, the trailing drop would move more rapidly, as though it were isolated. Because the velocity of isolated drops increases with their size, the trailing drop can pass the leading drop when there is sufficient time during the traverse to do so. This would explain the movement of the trailing bubbles and drops, away from the axis, in the IML-2 and the LMS experiments. From the above arguments, we see that the instability should have been observed in the case of drops in the IML-2 experiments as well. Yet in those experiments, the trailing drop did not show a tendency to move substantially away from the axis. Of course, the Marangoni numbers for the leading drop were considerably smaller than 400, which is the value used in Figure 4.15.8. This should not alter our conclusion in a qualitative sense, however. Therefore, the situation must be more complex than implied by the above picture. Unfortunately, this phenomenon could not be explored systematically in the LMS experiments, and it will need to be studied in detail in a future flight experiment.

Another interesting observation we made in a small number of experiments involved a chain of drops. These were typically introduced into the test cell by accident, usually during the first attempt to inject a drop. These drops executed a three-dimensional trajectory through the cell. Unfortunately, only one view could be captured on video. In this view, the trajectory of each drop across the cell appeared as an undulating pattern, not quite sinusoidal, but qualitatively similar. Also, these spatial oscillations in position take the drop away from the axis. Interferometry images recorded from an orthogonal direction revealed a similar pattern, even though making precise measurements from this view was not possible. Therefore, we are led to conjecture that the drops did not proceed in a straight path along the axis of the cell, but rather followed a spiral path. This phenomenon remains unexplained and unexplored at this time. The reader interested in examining the data from these interaction experiments aboard the LMS mission of the space shuttle may consult a report submitted to NASA by Subramanian (1997).

PART FOUR

RELATED TOPICS

CHAPTER EIGHT

Mass Transfer between a Bubble or Drop and a Continuous Phase

8.1 Introduction

In this chapter, we discuss mass transfer issues that can be expected to arise in reduced gravity situations involving bubbles and drops. We shall not particularly emphasize heat transfer because the problems considered in earlier chapters deal with a variety of heat transfer issues already. New situations in which there is an analogy between mass and heat transfer will be noted.

The subject of mass transfer between a drop and a continuous phase in reduced gravity can be organized into two classes of problems. In the first, a drop is stationary in a continuous phase. One example is that of bubbles accidentally formed or deliberately introduced into an isothermal liquid in a space environment. Another is that of a drop of liquid held fixed without a container in a gaseous medium such as air. Combustion experiments and those on the containerless processing of materials in the liquid state fall into this category. In the second class of problems, the drop moves in the continuous phase, likely because of gradients in interfacial tension. The motion of the drop modifies the diffusive flux at the interface between the drop and the continuous phase. Practical situations involving mass transport under reduced gravity conditions can arise in separation processes. Examples would be the processes used for separating life-sustaining materials such as water and oxygen from waste products so that they can be recycled.

The reason for the occurrence of mass transfer is simple to understand. Consider a gas bubble of a single species placed in a liquid with a negligible vapor pressure at the prevailing temperature. The interfacial layer of liquid in contact with the gas will rapidly achieve equilibrium with the gas in the bubble at the interface, leading to a dissolved gas concentration that makes the chemical potentials of the species in the gas the same on both sides of the interface. The dissolved gas concentration in the liquid far from the bubble, and therefore its chemical potential, will be different from that prevailing at the interface, however. This chemical potential gradient will drive a diffusive flux of species whose sense will be such as to achieve equilibrium everywhere in the liquid. Therefore, gas will either leave or enter the bubble, depending on the direction of the chemical potential gradient at the interface. This will cause the bubble to change in size with time. If a limited quantity of a gas and a liquid are brought into contact with each other, one might expect that thermodynamic equilibrium would eventually be reached between the two phases. When a bubble is present in a large liquid body, however, it must either dissolve completely or grow indefinitely. The implicit assumption in the following argument is that the liquid is unbounded in extent. Presuming the bubble and the liquid are at thermodynamic equilibrium at some instant, slight departures caused by a disturbance will cause the system to move away from that state. This is because of the action of surface tension in elevating the pressure within a bubble compared with that prevailing outside. For instance, if the bubble becomes slightly smaller, the pressure inside the bubble will rise. This will increase the concentration of the dissolved species at the interface, leading to diffusion of species away from the interface and into the liquid. The consequence is shrinkage of the bubble, leading to even higher pressures within the bubble. This will further increase the driving force for dissolution. The process ultimately will result in the complete disappearance of the bubble. A similar argument applies if the radius of the bubble becomes slightly larger than that at equilibrium. In this case, the process will lead to continued growth in size. Therefore, in all cases, the bubble will either dissolve or grow indefinitely, depending on the direction of the diffusive flux. The same idea holds if the bubble contains more than one species, such as when it is present in a volatile liquid or when multicomponent gas bubbles are introduced in a liquid. In this case, the composition at the interface on the gas side need not be the same as that in the gas away from the interface. Therefore, in principle, one must consider the mass transport problem within the bubble as well. As a practical matter, diffusivities in gases are typically 3 orders of magnitude larger than those in a liquid. Hence, composition gradients within the bubble can be considered negligible.

In the case of liquid drops, mass transport to a continuous phase, or from it, is again driven by chemical potential gradients of the transferring species. Unlike the case of a gas bubble, however, the equilibrium conditions at the interface are not sensitive to pressure at ordinary pressures. Therefore, as a practical matter, a drop can reach an equilibrium state with the surrounding liquid and retain that state indefinitely. We have assumed thermodynamic equilibrium between the layers of fluid that straddle the interface in all cases. Bird et al. (1960) have noted that at high rates of mass transfer, this assumption of equilibrium need not hold, but for most common situations, it is a good assumption.

When a bubble or drop moves in a fluid in which the temperature is not uniform, it is clear that compositional equilibrium cannot be expected at all locations. For example, consider the experiments of Hardy (1979), mentioned in Section 4.16. Hardy introduced air bubbles into a silicone oil, in which a temperature gradient had been established. The silicone oil had a certain amount of air dissolved in it. For the purpose of this discussion, we shall treat air as though it is a single species even though, in reality, one must take into account the fact that it contains numerous gaseous species. The uniform concentration of dissolved air present in the bulk liquid would likely correspond to equilibrium with air at some temperature and pressure. At other temperatures or pressures, such equilibrium cannot be preserved, however. At virtually every location in the cell, the concentration of air in the liquid at the interface would be different from that far from the bubble. The resulting gradient of chemical potential would lead to a diffusive flux toward the interface or away from it. As a consequence, air would either enter or leave the bubble, leading to a change in the mass of the bubble and therefore its volume. The role of convection is to modify the gradient of concentration at the interface, and thereby affect the diffusive flux, and to assist in the transport of the dissolved species through the liquid. An added complication in mass transfer is that the diffusive transport of species, by itself, causes mass average motion in general, leading to convective transport. In dilute systems, this effect is relatively small and can be ignored, as we do in some problems considered here. Furthermore, when multiple species are present, the diffusive fluxes are generally coupled through the Maxwell–Stefan equations. This aspect has been discussed in depth by Taylor and Krishna (1993), and we ignore such coupling when considering the transfer of more than one species.

There is a vast literature on mass transfer between a drop and its surroundings, which already is summarized in available books. Clift et al. (1978) provided a substantial amount of material on mass transfer between fluid objects and a continuous phase, particularly when the object is in motion with respect to the fluid because of the action of a body force, but also when the object is stationary. The book by Sadhal et al. (1997) contains a comprehensive discussion of heat and mass transfer problems involving moving and stationary drops. These authors provided numerous references to articles that have been published since the appearance of the book by Clift et al. Furthermore, Sadhal et al. included a detailed discussion of the evaporation of a liquid drop into a gas and related combustion issues. Mass transfer problems involving motion in non-Newtonian fluids are discussed in a book by Chhabra (1993). The literature on transport from an isolated stationary object to a surrounding fluid was also reviewed by Davis, Ravindran, and Ray (1980, 1981).

In this chapter, attention is focused on problems relevant to reduced gravity conditions that are not discussed in the books by Clift et al. or Sadhal et al., or that are only mentioned briefly in those sources. In Section 8.2, we present the equations governing the dissolution or growth of a stationary sphere in a liquid, accounting for density variation in the liquid due to variation in the composition. Also, we provide the simpler form of these equations for dissolution and growth of gas bubbles, wherein such density variations are negligible, and the density of the gas within the bubble is negligible compared with that of the liquid. Then, solutions of these problems are given in Sections 8.3 to 8.5. Finally, in Sections 8.6 and 8.7, we treat the dissolution or growth of a drop that moves because of thermocapillarity in the two limiting situations when convective transport plays a small role and when it is a dominant influence, respectively.

8.2 Governing Equations for the Dissolution or Growth of a Stationary Sphere in a Large Liquid Body

8.2.1 Introduction

In this section, we provide the governing equations for the dissolution or growth of an isolated stationary sphere containing a single species in a liquid that can be considered infinite in extent. The principal objective is to predict the rate of change of the radius of the sphere with time. This requires a knowledge of the gradient of the concentration field at the surface of the sphere. It is therefore necessary to solve for the concentration field of the dissolved solute in the continuous phase. The sphere can be a solid or a fluid object. We use this model problem to illustrate how to accomodate density variations in the continuous phase that arise from composition variations.

We assume that the sphere consists of a pure species A, and the surrounding fluid contains solute A and solvent B of mass concentrations ρ_A and ρ_B, respectively. There are no chemical reactions, and enthalpy changes due to dissolution are considered negligible. This permits us to consider the system as being isothermal. Because we only treat spheres of a single component, the density ρ' is spatially uniform within the sphere and does not vary with time. The density of the continuous phase ρ will depend on position and time because of the variation of the solute concentration. Readey and

Cooper (1966) provided a mathematical model for problems of this type, assuming that the partial specific volumes of the solute and the solvent are independent of solute concentration. The development in this section is based on their work. In the following, we presume the sphere to be fluid in general, but the equations are equally applicable for a solid sphere.

8.2.2 The Case When Density Variations Are Permitted in the Continuous Phase

At the outset, we need to recognize that the normal components of the velocities of the two fluids straddling the interface need not match each other or that of the interface, when there is mass transport. The kinematic condition in Equation (2.3.6), which has been used in Parts Two and Three, implies such equality. A more general version, known as the overall jump mass balance, is used in the present chapter:

$$\rho(\mathbf{v} - \mathbf{v}_i) \bullet \mathbf{n} = \rho'(\mathbf{v}' - \mathbf{v}_i) \bullet \mathbf{n}. \tag{8.2.1}$$

Here, all the fields are evaluated at the interface. Physical variables are used in this section; to avoid clutter, we do not use the asterisk to distinguish them as in earlier chapters. In Equation (8.2.1), $\mathbf{n}$ is the unit normal to the interface, $\mathbf{v}_i$ is the velocity of the interface normal to itself, and the prime refers to quantities within the sphere. In the present spherically symmetric problem, the only nonzero velocity component is the radial velocity field $v_r(t, r)$ where t is time and r is the radial coordinate, measured from an origin at the center of the sphere. Due to symmetry, v_r does not depend on the polar or azimuthal angular coordinates. The continuity equation within the sphere yields the result that $r^2 v_r'$ must be a constant. Because it must be zero at the center, we find that $v_r' = 0$ everywhere within the sphere. For a sphere of instantaneous radius $R(t)$, the velocity of the interface normal to itself is $\mathbf{v}_i = \frac{dR}{dt}\mathbf{i}_r$. Therefore, we obtain the following result from the kinematic condition in Equation (8.2.1):

$$v_r(t, R) = \frac{\rho'}{\rho(t, R)} v_r'(t, R) + \left[1 - \frac{\rho'}{\rho(t, R)}\right] \frac{dR}{dt} = \left[1 - \frac{\rho'}{\rho(t, R)}\right] \frac{dR}{dt}. \tag{8.2.2}$$

The continuity equation can be written as follows:

$$\frac{\partial \rho}{\partial t} + \frac{1}{r^2}\frac{\partial}{\partial r}(\rho r^2 v_r) = 0. \tag{8.2.3}$$

The partial specific volumes of the solute and solvent are designated as $\overline{V}_A$ and $\overline{V}_B$, respectively. The definition of partial quantities in a multicomponent mixture can be found in Guggenheim (1967). The density of the liquid can be written as

$$\rho = \frac{1 - \rho_A(\overline{V}_A - \overline{V}_B)}{\overline{V}_B} = \frac{1 - \Gamma \rho_A}{\overline{V}_B}, \tag{8.2.4}$$

where $\Gamma = (\overline{V}_A - \overline{V}_B)$. Substitution of the expression for the density field into the continuity equation leads to the following result.

$$\frac{1}{r^2}\frac{\partial}{\partial r}(r^2 v_r) = \Gamma \left[\frac{\partial \rho_A}{\partial t} + \frac{1}{r^2}\frac{\partial}{\partial r}(r^2 v_r \rho_A)\right]. \tag{8.2.5}$$

When Fick's law for a binary system is substituted into the equation of conservation of mass and no assumptions are made about the spatial variations of the density or the diffusivity, the following simple form is obtained in the spherically symmetric case:

$$\frac{\partial \rho_A}{\partial t} + \frac{1}{r^2} \frac{\partial}{\partial r} (r^2 v_r \rho_A) = \frac{1}{r^2} \frac{\partial}{\partial r} \left[\mathcal{D} \rho r^2 \frac{\partial}{\partial r} \left(\frac{\rho_A}{\rho} \right) \right]. \tag{8.2.6}$$

The symbol $\mathcal{D}$ represents the binary diffusivity $\mathcal{D}_{AB}$. Substitution of Equation (8.2.6) into Equation (8.2.5) permits the continuity equation to be integrated from the surface of the sphere to an arbitrary radial location in the liquid. The result for the radial velocity field is given below.

$$v_r(t, r) = \frac{R^2}{r^2} v_r(t, R) + \Gamma \left\{ \frac{\mathcal{D}}{1 - \Gamma \rho_A} \frac{\partial \rho_A}{\partial r} - \frac{R^2}{r^2 [1 - \Gamma \rho_A(t, R)]} \left[\mathcal{D} \frac{\partial \rho_A}{\partial r} \right] (t, R) \right\}. \tag{8.2.7}$$

In writing the above, we have used the fact that $\rho \frac{\partial}{\partial r} \left(\frac{\rho_A}{\rho} \right) = \frac{1}{1 - \Gamma \rho_A} \frac{\partial \rho_A}{\partial r}$, and the notation $[\cdots](t, R)$ implies that the entire quantity within the square brackets is evaluated at $r = R$. Now, the jump mass balance across the interface for species A, given in Equation (2.3.16), can be used. Under the present conditions, it can be written as follows:

$$\rho' \left[v_r'(t, R) - \frac{dR}{dt} \right] = \rho_A(t, R) \left[v_r(t, R) - \frac{dR}{dt} \right] - \left[\mathcal{D} \rho \frac{\partial}{\partial r} \left(\frac{\rho_A}{\rho} \right) \right] (t, R). \tag{8.2.8}$$

On setting $v_r'(t, R) = 0$ and rearranging, Equation (8.2.8) becomes

$$\frac{dR}{dt} = \frac{1}{\rho_A(t, R) - \rho'} \left\{ \rho_A(t, R) v_r(t, R) - \frac{[\mathcal{D} \frac{\partial \rho_A}{\partial r}](t, R)}{1 - \Gamma \rho_A(t, R)} \right\}. \tag{8.2.9}$$

Using Equations (8.2.2), (8.2.4), and (8.2.9) yields the following result for the rate of change of the radius of the sphere:

$$\frac{dR}{dt} = \frac{[\mathcal{D} \frac{\partial \rho_A}{\partial r}](t, R)}{\rho' [1 - \overline{V}_A \rho_A(t, R)]}. \tag{8.2.10}$$

Now we can write the radial velocity at the interface in the continuous phase as

$$v_r(t, R) = (1 - \rho' \overline{V}_A) \frac{dR}{dt} + \frac{\Gamma [\mathcal{D} \frac{\partial \rho_A}{\partial r}](t, R)}{1 - \Gamma \rho_A(t, R)} \tag{8.2.11}$$

and the radial velocity distribution as

$$v_r(t, r) = \frac{R^2}{r^2} (1 - \rho' \overline{V}_A) \frac{dR}{dt} + \frac{\Gamma \mathcal{D} \frac{\partial \rho_A}{\partial r}}{1 - \Gamma \rho_A}. \tag{8.2.12}$$

Using Equation (8.2.12), the equation of conservation of species A can be recast in the following form:

$$\frac{\partial \rho_A}{\partial t} + \left[\frac{R^2}{r^2} (1 - \rho' \overline{V}_A) \frac{dR}{dt} \right] \frac{\partial \rho_A}{\partial r} = \frac{1}{r^2} \frac{\partial}{\partial r} \left(r^2 \mathcal{D} \frac{\partial \rho_A}{\partial r} \right). \tag{8.2.13}$$

Equation (8.2.13), for the distribution of the mass concentration of species A in the continuous phase, must be solved along with Equation (8.2.10) and the associated initial and boundary conditions to obtain the radius as a function of time. A similar set of equations is obtained in an analogous heat transport problem associated with a phase change process that generates or consumes latent heat at the interface. By making suitable assumptions in the conduction-controlled growth or collapse of a vapor bubble, it is possible to reduce the heat transport problem to the same mathematical form as Equations (8.2.10) and (8.2.13). The interested reader may wish to consult Duda and Vrentas (1971) for further information.

8.2.3 The Limiting Case of a Gas Bubble

It is common practice, when considering gas bubble dissolution, to ignore the slight contribution made by the dissolved gas to the density of the liquid. In this case, the density in the continuous phase can be regarded as a constant. Furthermore, the density of a gas can be neglected when compared with that of a liquid. Following a development similar to that given above, we can write the result given below for the velocity of the interface, which is equal to the velocity in the liquid at the interface:

$$v_r(t, R) = \frac{dR}{dt} = \frac{\left[\mathcal{D} \frac{\partial \rho_A}{\partial r} \right](t, R)}{\rho'}.$$

(8.2.14)

The distribution of radial velocity is written as

$$v_r(t, r) = \frac{R^2}{r^2} v_r(t, R) = \frac{R^2}{r^2} \frac{dR}{dt},$$

(8.2.15)

and the governing conservation equation for the dissolving gaseous species A reduces to

$$\frac{\partial \rho_A}{\partial t} + \left[\frac{R^2}{r^2} \frac{dR}{dt} \right] \frac{\partial \rho_A}{\partial r} = \frac{1}{r^2} \frac{\partial}{\partial r} \left(r^2 \mathcal{D} \frac{\partial \rho_A}{\partial r} \right).$$

(8.2.16)

In the case of a gas bubble, we need to know the pressure within the bubble to determine the thermodynamic equilibrium concentration that would exist in the liquid at the interface. The pressure will vary with radial position in the liquid because of inertial and viscous effects. This variation can be inferred by integration of the Navier–Stokes equation in the liquid. Because only the pressure in the bubble is of interest, the integration in the liquid can be performed from the interface to the point at infinity. Then the jump across the interface due to surface tension can be added to yield the result for the pressure within the bubble, which is uniform. In the Newtonian case, this yields

$$p - p_\infty = \frac{2\sigma}{R} + \rho \left[R \frac{d^2 R}{dt^2} + \frac{3}{2} \left(\frac{dR}{dt} \right)^2 + 4 \frac{\mu}{\rho R} \frac{dR}{dt} \right].$$

(8.2.17)

Here, p_∞ is the pressure in the undisturbed liquid. We have assumed zero gravity conditions so that there is no hydrostatic variation of pressure. In the right side of Equation (8.2.17), the first two terms in the square bracket arise from inertia, and the last arises from the viscous stress. In problems of bubble dissolution and growth controlled by diffusion, the contributions of inertial and viscous effects are small, and the entire quantity within the square bracket can be ignored with little error. Also, except for small bubbles, the increase in pressure across the interface due to surface tension effects is negligible as well. For example, even when the surface tension is as large as 300 mN/m, for a bubble of radius 0.6 mm, the increase in pressure is less than 1% of an atmosphere. In orbiting laboratories such as the space shuttle and the international space station, the ambient pressure is maintained at the same order of magnitude as the atmospheric pressure at the surface of Earth. Therefore, in the reduced gravity conditions aboard these vehicles, it is reasonable to assume that the pressure is uniform in the liquid and within the bubble and equal to p_∞. The effects being neglected here were considered in some detail in Venerus and Yala (1997) in the Newtonian case. Venerus, Yala, and Bernstein (1998) provided similar calculations in the case of a viscoelastic liquid.

When a bubble contains more than one species that can dissolve in the liquid, we have a multicomponent diffusion problem. As noted in Section 8.1, diffusion within the bubble will be rapid compared with that in the surrounding liquid because gas phase diffusivities are typically at least three orders of magnitude larger than those in liquids. Hence, we can treat the contents of the bubble as being of uniform composition. Now, considering diffusion in the liquid, for most gaseous solutes the solutions are very dilute; therefore, we can ignore the coupling among the diffusion fluxes and treat each species as though it alone is diffusing through the liquid according to Fick's law. In this case, we can use Equation (8.2.16) for the mass fraction ρ_i of each of the species. In addition, a mass balance is written for each of the N species within the bubble, equating the rate of change of the mass of the species i in the bubble to the rate of influx of species i through the interface.

$$\frac{dX_i'}{dt} = \frac{3}{R} \left[\frac{[\mathcal{D}_i \frac{\partial \rho_i}{\partial r}](t, R)}{C' M_i} - X_i' \frac{dR}{dt} \right] \quad (i = 1, \ldots, N). \tag{8.2.18}$$

When species of different molecular weights diffuse into or out of the bubble at different rates, it is the total molar concentration that remains constant when the temperature and pressure remain constant. The mass density of the bubble can change under these conditions. Therefore, it is necessary to use mole fractions X_i' in the gas phase, along with the total molar concentration of the gas in the bubble C'. The molecular weight of species i is M_i, and the symbol $\mathcal{D}_i$ stands for the binary diffusivity of species i in the liquid. Summing the individual mass balances in Equation (8.2.18) yields a differential equation for the radius:

$$\frac{dR}{dt} = \frac{1}{C'} \sum_{i=1}^{N} \frac{1}{M_i} \left[\mathcal{D}_i \frac{\partial \rho_i}{\partial r} \right](t, R). \tag{8.2.19}$$

In this chapter, we do not provide solutions in the case of multicomponent gas bubbles. The reader interested in such problems can consult the articles by Weinberg and Subramanian (1980) and Ramos (1986) for some examples of the use of the above model equations.

8.3 Stationary Sphere in a Large Liquid Body – The Quasi-Stationary Solution

8.3.1 Introduction

The quasi-stationary solution, provided by Epstein and Plesset (1950), has been superseded by other solutions that are easier to use. Nonetheless, it has some historical significance. It is an approximate solution for the slow dissolution or growth of a stationary sphere in a fluid body that can be considered infinite in extent. In the case of mass transport, the concentration disturbance field due to the sphere decays rapidly away from the sphere surface; therefore a container that is larger in size than several sphere radii is sufficient for considering the fluid as being infinite in extent. As seen from Equation (8.2.10), we must know the gradient of the concentration field in the continuous phase at the interface to obtain the time history of the sphere radius. Yet, the rate of change of the radius with time appears in the convective transport term in the governing conservation equation for the species entering or leaving the sphere. The mathematical problem contains two complications. The first is that the boundary condition of equilibrium concentration is applied at a time-dependent boundary, and the second is the appearance of the nonlinear convective transport term in the conservation equation because of the movement of the boundary. Early work on the problem before the advent of computers was directed at approximating it further so that analytical results could be written. The physical approximation made in the quasi-stationary model is that the driving force for dissolution is not large. In this case, the rate of change of the radius with time is so small that the concentration field can be approximated as that around a sphere of constant radius. Of course, the resulting concentration gradient at the surface of the sphere will display a dependence on this radius, and hence the time dependence of the radius is accommodated through its influence on this gradient. Consistent with the assumption that the radius is held constant in calculating the concentration field, the motion in the continuous phase caused by the transport process also is neglected. The result is known as the *quasi-stationary model*. It also is assumed that the diffusivity is constant. It is straightforward to write an analytical solution in this case.

8.3.2 Quasi-Stationary Concentration Field

With the above assumptions, Equation (8.2.13) for the mass concentration $\rho_A(t, r)$ in the continuous phase reduces to the following diffusion equation:

$$\frac{\partial \rho_A}{\partial t} = \frac{D}{r^2} \frac{\partial}{\partial r} \left(r^2 \frac{\partial \rho_A}{\partial r} \right). \tag{8.3.1}$$

Here, t is time, and r is the radial coordinate measured from the center of the stationary sphere. Next, the initial and boundary conditions on the concentration field are written. The initial concentration is taken to be $\rho_{A,\infty}$, and this concentration prevails at all times in the limit $r \to \infty$.

$$\rho_A(0, r) = \rho_{A,\infty}, \tag{8.3.2}$$

$$\rho_A(t, \infty) = \rho_{A,\infty}. \tag{8.3.3}$$

In writing the boundary condition in the limit $r \to \infty$, we have used the simple expedient of substituting the symbol ∞ for r, but the limit process is implied here and in other places where similar notation is used. At the surface of the sphere,

$$\rho_A(t, R) = \rho_{A,i}. \tag{8.3.4}$$

Note that the radius of the sphere R that appears in Equation (8.3.4) is treated as a constant parameter in the present approximation, even though it depends on time. The solution of this diffusion problem can be found in Carslaw and Jaeger (1959) or Crank (1975).

$$\frac{\rho_A - \rho_{A,\infty}}{\rho_{A,i} - \rho_{A,\infty}} = \frac{R}{r} \, \text{erfc} \left[\frac{r - R}{2\sqrt{Dt}} \right]. \tag{8.3.5}$$

Differentiation of the concentration field and evaluation of the gradient at the surface of the sphere leads to

$$\frac{\partial \rho_A}{\partial r}(t, R) = -(\rho_{A,i} - \rho_{A,\infty}) \left[\frac{1}{\sqrt{\pi Dt}} + \frac{1}{R} \right]. \tag{8.3.6}$$

If we had assumed a steady concentration distribution, ignoring the initial transient, we would have obtained a quasi-steady result for this gradient:

$$\frac{\partial \rho_A}{\partial r}(t, R) = -\frac{\rho_{A,i} - \rho_{A,\infty}}{R}. \tag{8.3.7}$$

8.3.3 Time Evolution of the Radius of the Sphere

The governing equation for the radius is Equation (8.2.10). It is reproduced here as the starting point:

$$\frac{dR}{dt} = \frac{\mathcal{D}}{\rho'(1 - \overline{V}_A \rho_{A,i})} \frac{\partial \rho_A}{\partial r}(t, R). \tag{8.3.8}$$

Substitution of the quasi-stationary solution for the concentration gradient at the interface into Equation (8.3.8) leads to the following differential equation describing the time evolution of the radius:

$$\frac{dR}{dt} = -\frac{\mathcal{D}(\rho_{A,i} - \rho_{A,\infty})}{\rho'(1 - \overline{V}_A \rho_{A,i})} \left[\frac{1}{\sqrt{\pi \mathcal{D} t}} + \frac{1}{R} \right]. \tag{8.3.9}$$

The solution, with $R(0) = R_0$, was given by Epstein and Plesset (1950) in parametric form. Subsequently, Weinberg, Onorato, and Uhlmann (1980) provided an equivalent result, which we give below in slightly modified form.

$$\log \frac{R_0}{R} = \frac{1}{2} \log [1 + \Omega p(1 + p)] + \sqrt{\frac{\Omega}{4 - \Omega}} \left[\arctan \left\{ \sqrt{\frac{\Omega}{4 - \Omega}}(1 + 2p) \right\} \right.$$

$$\left. - \arctan \left\{ \sqrt{\frac{\Omega}{4 - \Omega}} \right\} \right], \quad 0 \leq \Omega < 4, \tag{8.3.10}$$

$$\log \frac{R_0}{R} = \frac{1}{2} \log [1 + \Omega p(1 + p)] - \sqrt{\frac{\Omega}{\Omega - 4}} \left[\operatorname{arctanh} \left\{ \sqrt{\frac{\Omega}{\Omega - 4}}(1 + 2p) \right\} \right.$$

$$\left. - \operatorname{arctanh} \left\{ \sqrt{\frac{\Omega}{\Omega - 4}} \right\} \right], \quad \Omega > 4 \text{ or } \Omega < 0, \tag{8.3.11}$$

$$\log \frac{R_0}{R} = \frac{2p}{1 + 2p} + \log(1 + 2p), \quad \Omega = 4. \tag{8.3.12}$$

Here,

$$\Omega = \frac{2(\rho_{A,i} - \rho_{A,\infty})}{\pi \rho'(1 - \overline{V}_A \rho_{A,i})} \tag{8.3.13}$$

and

$$p = \frac{\sqrt{\pi \mathcal{D} t}}{R}, \tag{8.3.14}$$

so that the solution is implicit for the radius. Equations (8.3.10) and (8.3.11) are equivalent and provided as separate results for ease in calculation with real-valued quantities. When the parameter Ω is positive, dissolution occurs; when Ω is negative, the sphere grows in size.

If the quasi-steady gradient of concentration at the surface is used in Equation (8.3.8), integration is simple, leading to the following result, which applies to either dissolution or growth:

$$\frac{R^2}{R_0^2} = 1 - \pi\Omega\left(\frac{\mathcal{D}t}{R_0^2}\right). \tag{8.3.15}$$

This approximation is not as good as the quasi-stationary result. It is analogous to the d^2-law used in droplet growth or shrinkage by evaporation or combustion, when the process is assumed to be diffusion-limited and quasi-steady. The symbol d represents the diameter of the droplet in that case. Information regarding the applicability of the d^2-law can be found in Sadhal et al. (1997).

Sample predictions from the quasi-stationary solution are provided in Section 8.5, where results from various approximations are compared with those from a numerical solution.

8.4 Stationary Sphere in a Large Liquid Body – Similarity Solution for Growth from Zero Initial Size

8.4.1 Introduction

When describing the growth of a sphere due to diffusion of a species to it, an exact solution can be written in the special case of growth from zero initial size. In principle, there must be a tiny cluster of atoms, called a nucleus, from which such a sphere will grow, but for practical purposes, this can be regarded as zero initial size. In fact, such a solution is useful for describing growth from a finite initial size after the sphere has grown to more than about five times its initial radius. The solution is obtained by using the idea that the concentration profiles at various times are self-similar, varying only because difussion causes the region of concentration variation to expand with time. When the radial coordinate is scaled with a reference length that grows as the square root of time, the dependence of the concentration field on radial position and time collapses into a dependence on a single similarity coordinate. This solution was presented and discussed by Scriven (1959).

8.4.2 Analysis

The governing differential equations for this problem are given in Section 8.2 as Equation (8.2.13) for the mass concentration ρ_A of the diffusing species in the continuous phase and Equation (8.2.10) for the rate of change of the radius R of the sphere with time. All the assumptions stated in that section apply here; in addition, it is assumed that the diffusivity $\mathcal{D}$ is constant and that the sphere is present in an infinite extent of fluid. A normalized mass concentration $C(t, r)$ is defined as $C = \frac{\rho_{A,\infty} - \rho_A}{\rho_{A,\infty} - \rho_{A,i}}$, where the constant $\rho_{A,\infty}$ is the mass concentration of A prevailing in the undisturbed continuous phase and the constant $\rho_{A,i}$ is the equilibrium concentration in the continuous phase at the surface of the sphere. The governing equations are given below, along with the initial and boundary conditions.

$$\frac{\partial C}{\partial t} + \left[\frac{R^2}{r^2}(1 - \rho'\overline{V}_A)\frac{dR}{dt}\right]\frac{\partial C}{\partial r} = \frac{\mathcal{D}}{r^2}\frac{\partial}{\partial r}\left(r^2\frac{\partial C}{\partial r}\right), \tag{8.4.1}$$

$$C(0, r) = 0, \tag{8.4.2}$$

$$C(t, \infty) = 0, \tag{8.4.3}$$

$$C(t, R) = 1, \tag{8.4.4}$$

$$\frac{dR}{dt} = \frac{\mathcal{D}(\rho_{A,i} - \rho_{A,\infty})}{\rho'(1 - \overline{V}_A \rho_{A,i})} \frac{\partial C}{\partial r}(t, R), \tag{8.4.5}$$

$$R(0) = 0. \tag{8.4.6}$$

Shortly, it will become clear why we must assume that $R(0) = 0$ for this solution procedure to work. For a similarity solution, we require that $C(t, r) = F(\eta)$ where

$$\eta = \frac{r}{\delta(t)} \tag{8.4.7}$$

and trace through the consequences. When Equation (8.4.1) is transformed in this way, it becomes

$$\frac{d^2 F}{d\eta^2} + \left[\frac{2}{\eta} + \eta \left(\frac{1}{\mathcal{D}} \delta \frac{d\delta}{dt} \right) - \frac{1}{\eta^2} \left(\frac{\{1 - \rho' \overline{V}_A\}}{\mathcal{D}} \frac{R^2}{\delta} \frac{dR}{dt} \right) \right] \frac{dF}{d\eta} = 0, \tag{8.4.8}$$

along with

$$F\left(\frac{R(t)}{\delta(t)} \right) = 1 \tag{8.4.9}$$

and

$$F(\infty) = 0. \tag{8.4.10}$$

It is evident that certain conditions must apply for similarity to hold. We must require that time cannot explicitly appear in the coefficient of $\frac{dF}{d\eta}$ in Equation (8.4.8). This in turn implies that we should set the coefficients of η and $\frac{1}{\eta^2}$ to constants. Fortunately, this leads to proportional time dependencies of $\delta(t)$ and $R(t)$, which resolve the difficulty with the explicit appearance of time dependence in Equation (8.4.9). Both are proportional to the square root of time. Without loss of generality, we can write

$$\delta(t) = 2\sqrt{\mathcal{D}t} \tag{8.4.11}$$

and

$$R(t) = 2\beta\sqrt{\mathcal{D}t}, \tag{8.4.12}$$

where the dimensionless constant β is called the *growth constant*. This permits Equations (8.4.8) and (8.4.9) to be rewritten as

$$\frac{d^2 F}{d\eta^2} + 2 \left[\frac{1}{\eta} + \eta - \frac{\beta^3 (1 - \rho' \overline{V}_A)}{\eta^2} \right] \frac{dF}{d\eta} = 0 \tag{8.4.13}$$

and

$$F(\beta) = 1. \tag{8.4.14}$$

The solution is given below.

$$F(\eta) = \frac{I(\eta)}{I(\beta)}. \tag{8.4.15}$$

Here

$$I(\eta) = \int_{\eta}^{\infty} \exp\left[-\eta^2 - \frac{2\beta^3(1 - \rho'\overline{V}_A)}{\eta}\right]\frac{d\eta}{\eta^2}. \tag{8.4.16}$$

It only remains to determine the growth constant β. This is done from the equation for the time evolution of the radius of the sphere:

$$2\beta^3 I(\beta)\exp[\beta^2(3 - 2\rho'\overline{V}_A)] = \frac{(\rho_{A,\infty} - \rho_{A,i})}{\rho'(1 - \overline{V}_A\rho_{A,i})}. \tag{8.4.17}$$

On defining

$$N_a = \frac{\rho_{A,i} - \rho_{A,\infty}}{\rho'(1 - \overline{V}_A\rho_{A,i})} \tag{8.4.18}$$

and

$$\epsilon = 1 - \rho'\overline{V}_A, \tag{8.4.19}$$

Equation (8.4.17) can be rewritten as follows:

$$\phi(\epsilon, \beta) = 2\beta^3 I(\beta)\exp[\beta^2(1 + 2\epsilon)] = -N_a. \tag{8.4.20}$$

The negative constant N_a represents a dimensionless driving force parameter for growth. For any given set of values of N_a and ϵ, the growth constant can be obtained by solving Equation (8.4.20). Scriven (1959) calculated the function $\phi(\epsilon, \beta)$ for a wide range of values of the parameters ϵ and β, providing a table as well as a figure. For numerical evaluation, it is convenient to transform the variable of integration in $I(\eta)$ to $x = 1 - \frac{\beta}{\eta}$, so that $\phi(\epsilon, \beta)$ can be rewritten as

$$\phi(\epsilon, \beta) = 2\beta^2 \int_0^1 \exp\left[-\beta^2\left(\frac{1}{(1 - x)^2} - 2\epsilon x - 1\right)\right]dx. \tag{8.4.21}$$

It is seen that in the limit as $\beta \to 0$, $\phi \to 2\beta^2$, and this result is independent of the value of ϵ. Therefore, in this limit, the growth constant is simply related to N_a via $\beta = \sqrt{(-N_a/2)}$. Although this approximation becomes inadequate when $\beta \geq 0.01$, the integral can be numerically evaluated with ease for small to moderate β. When β becomes large in magnitude, the integral must be handled carefully. In the limit as $\beta \to \infty$, Scriven used the saddle point method to evaluate the integral and obtained the following asymptotic result:

$$\phi(\epsilon, \beta) \sim \sqrt{\frac{\pi}{3}}\,\beta\exp(y^2)\,\text{erfc}(y). \tag{8.4.22}$$

Here,

$$y = \frac{(1 - \epsilon)\beta}{\sqrt{3}}. \tag{8.4.23}$$

When $\epsilon = 1$, which corresponds to a sphere of negligible density, or a solute of negligible partial specific volume in the continuous phase, a simple asymptotic result can be written for large β:

$$\phi(1, \beta) \sim \sqrt{\frac{\pi}{3}} \left[\beta - \frac{4}{9} + O\left(\frac{1}{\beta}\right) \right]. \tag{8.4.24}$$

The result in Equation (8.4.22) serves as an adequate approximation for $\beta \geq 30$ and $\epsilon \leq 0.9$. When $\beta \geq 100$, it can be used with an error that is less than 1% over the entire range of values of ϵ. When $\epsilon = 1$, the error incurred by using Equation (8.4.24) for $\beta = 10$ is approximately 0.25%, and the approximation is even better at larger values of β. Finally, it is worthy of note that the function $\phi(\epsilon, \beta)$ is most sensitive to change in the value of β near $\epsilon = 1$. For any fixed $\epsilon < 1$, ϕ changes significantly with β for small β, but the sensitivity is greatly reduced at large β. This has an important implication because the growth constant is obtained by solving Equation (8.4.20). Minute changes in the value of N_a can lead to changes in the value of β by orders of magnitude. Hence, one must be precise in evaluating N_a.

8.5 Stationary Sphere in a Large Liquid Body – Asymptotic Expansions

8.5.1 Introduction

In this section, we provide results from perturbation solutions for the time evolution of the radius of a stationary sphere of a pure substance in a large body of fluid. Even though the analysis requires some effort, the final results for the radius of the sphere are simpler to calculate than those from the quasi-stationary solution and more accurate than that solution. The accuracy is established by comparison with a solution of the governing equations by the method of finite differences. Two types of asymptotic expansions have been used for solving this problem. In the traditional type of expansion, developed by Duda and Vrentas (1969) and generalized by Vrentas and Shin (1980a, 1980b), two small parameters are identified and expansions are developed in powers of these parameters. In the other technique, introduced in the gas bubble limit by Subramanian and Weinberg (1981) and subsequently generalized by Shankar, Wiltshire, and Subramanian (1984), an asymptotic expansion is developed in time, which is a variable in the problem. The expansion in time provides results that are complementary to those obtained by an expansion in the parameters. A discussion of various available solutions, and recommendations regarding their use, can be found in Vrentas, Vrentas, and Ling (1983).

8.5.2 Governing Equations

The governing differential equations for this problem are given in Section 8.2 as Equation (8.2.13) for the mass concentration ρ_A of the diffusing species in the continuous phase and Equation (8.2.10) for the rate of change of the radius R of the sphere with time. All the assumptions stated in that section apply here; in addition, it is assumed that the diffusivity $\mathcal{D}$ is constant and that the sphere is present in an infinite extent of fluid. The initial radius R_0 of the sphere is used as the length scale and $\frac{R_0^2}{\mathcal{D}}$ as the time scale. The scaled radial coordinate is r, and scaled time is T. A normalized mass concentration $C(T, r)$ is defined as $C = \frac{\rho_A - \rho_{A,\infty}}{\rho_{A,i} - \rho_{A,\infty}}$, where the constant $\rho_{A,\infty}$ is the mass concentration of A prevailing in the undisturbed continuous phase, and the constant $\rho_{A,i}$ is the equilibrium concentration in the continuous phase at the surface of the sphere. The scaled governing equations are given below, along with the initial and boundary conditions.

$$\frac{\partial C}{\partial T} + (1 - \Lambda)\frac{g^2}{r^2}\frac{dg}{dT}\frac{\partial C}{\partial r} = \frac{1}{r^2}\frac{\partial}{\partial r}\left(r^2\frac{\partial C}{\partial r}\right),$$ (8.5.1)

$$C(0, r) = 0,$$ (8.5.2)

$$C(T, \infty) = 0,$$ (8.5.3)

$$C(T, g(T)) = 1,$$ (8.5.4)

$$\frac{dg}{dT} = N_a\frac{\partial C}{\partial r}[T, g(T)],$$ (8.5.5)

$$g(0) = 1.$$ (8.5.6)

Here, the scaled radius of the sphere is $g(T)$. Two parameters appear in the model:

$$N_a = \frac{\rho_{A,i} - \rho_{A,\infty}}{\rho'(1 - \overline{V}_A\rho_{A,i})}$$ (8.5.7)

and

$$\Lambda = \rho'\overline{V}_A.$$ (8.5.8)

The partial specific volume $\overline{V}_A$ of the solute A in the continuous phase has been encountered in Section 8.2. The parameter N_a can be regarded as a scaled driving force for mass transfer. Positive values of N_a imply dissolution, whereas negative values correspond to growth. The parameter Λ represents the ratio of the partial specific volume of the diffusing species in the continuous phase to the specific volume of this species within the sphere. It is convenient to define another parameter N_b as follows:

$$N_b = \Lambda N_a.$$ (8.5.9)

8.5.3 Expansion in the Parameters N_a and N_b

For the dissolution of gaseous species from a bubble, or the dissolution of a soluble liquid or solid sphere into a liquid, the parameter N_a is small compared with unity when the solute is sparingly soluble. The parameter N_b can be much smaller than N_a in the case of diffusing gases and comparable to N_a when the dissolving object is a liquid or solid sphere. Therefore, it is reasonable to try to develop an asymptotic expansion in these two parameters. The location of the surface of the sphere depends on time and the parameters. We define a transformed radial coordinate ξ to remove this dependence:

$$\xi = \frac{r}{g(T)}.$$ (8.5.10)

Problems involving diffusion in a spherical geometry can be reduced to diffusion in a planar geometry by defining a new dependent variable as the product of the radial coordinate and the original dependent variable. To obtain this benefit, the variable $\Psi(T, \xi)$ is defined as follows:

$$\Psi(T, \xi) = \xi C(T, r).$$ (8.5.11)

After transformation to the new variables, Equations (8.5.1) to (8.5.6) can be rewritten as follows:

$$h\frac{\partial \Psi}{\partial T} + \frac{1}{2}\frac{dh}{dT}\left[\frac{1-\Lambda}{\xi^2} - \xi\right]\left[\frac{\partial \Psi}{\partial \xi} - \frac{\Psi}{\xi}\right] = \frac{\partial^2 \Psi}{\partial \xi^2},$$

(8.5.12)

$$\Psi(0, \xi) = 0,$$

(8.5.13)

$$\Psi(T, \infty) = 0,$$

(8.5.14)

$$\Psi(T, 1) = 1,$$

(8.5.15)

$$\frac{dh}{dT} = 2N_a\left[-1 + \frac{\partial \Psi}{\partial \xi}(T, 1)\right],$$

(8.5.16)

and

$$h(0) = 1.$$

(8.5.17)

Here,

$$h(T) = g^2(T).$$

(8.5.18)

The procedure now is to write a straightforward expansion of the form

$$\Psi = \Psi_0 + \Psi_{10} N_a + \Psi_{01} N_b + \Psi_{20} N_a^2 + \Psi_{11} N_a N_b + \Psi_{02} N_b^2 + \cdots$$

(8.5.19)

for the transformed concentration field. This expansion is substituted into Equations (8.5.12) through (8.5.17), leading to a set of linear equations satisfied by the coefficient functions that can be solved by standard techniques. Equation (8.5.16) can be integrated, along with the initial condition in Equation (8.5.17), to yield

$$h(T) = 1 - 2N_a T + 2N_a \int_0^T \frac{\partial \Psi}{\partial \xi}(T, 1)\, dT.$$

(8.5.20)

When the expansion in Equation (8.5.19) is used in Equation (8.5.20), an ordered result is obtained for the square of the scaled radius of the sphere. Algebraic complexity restricts the number of coefficient functions that can be determined conveniently. Duda and Vrentas (1969) obtained Ψ_0 and the coefficients of the linear terms in N_a and N_b in Equation (8.5.19). Using those coefficients, they provided a result for $h(T)$. Later, Vrentas and Shin (1980a) pointed out that the expansion in the parameter N_a is not uniformly valid for all values of time or at all spatial locations. This is a common occurrence when the range of variables is semi-infinite, which happens to be the case here for both time and the radial coordinate. The authors developed outer expansions in powers of N_a, holding suitable transformed variables such as $|N_a|T$ fixed and $\sqrt{|N_a|}(\xi - 1)$ fixed, as appropriate. The details are lengthy and therefore omitted here. The ultimate result is a uniformly valid composite solution provided by Vrentas and Shin and is reproduced here:

$$h(T) = 1 - 2N_a\left(T + \sqrt{\frac{T}{\pi}}\right) + N_a^2\left[\frac{8}{\pi}\sqrt{\frac{T}{\pi}} + 2T - 2I(T)\right]$$

$$+ N_a N_b[T + 2I(T)] + f(T).$$

(8.5.21)

The function $I(T)$ is given by Duda and Vrentas in the following convenient form for numerical evaluation:

$$I(T) = \sqrt{\frac{T}{\pi}} \left[2 - \int_0^1 \int_0^\infty \frac{\mathrm{erfc}\left(\frac{x}{2\sqrt{Ty}}\right) \mathrm{erfc}\left(\frac{x}{2\sqrt{T(1-y)}}\right)}{(1+x)^3 \sqrt{1-y}} \, dx \, dy \right]$$

$$- \frac{1}{\pi} \int_0^1 \int_0^\infty \frac{\exp\left(-\frac{x^2}{4Ty}\right) \mathrm{erfc}\left(\frac{x}{2\sqrt{T(1-y)}}\right)}{(1+x)^2 \sqrt{y(1-y)}} \, dx \, dy. \tag{8.5.22}$$

The other function $f(T)$ appearing in Equation (8.5.21) is defined by Vrentas and Shin as

$$f(T) = \sqrt{\frac{2N_a}{\pi}} (2N_a T - 1) \mathrm{arctanh}\left(\sqrt{2N_a T}\right), \quad N_a > 0 \tag{8.5.23}$$

and

$$f(T) = \sqrt{\frac{2(-N_a)}{\pi}} (1 - 2N_a T) \arctan\left(\sqrt{2(-N_a)T}\right), \quad N_a < 0. \tag{8.5.24}$$

The two definitions are equivalent when using the connection between the inverse trigonometric and inverse hyperbolic functions for imaginary arguments but provided individually for convenience. It is worthwhile mentioning here that in a second article, Vrentas and Shin (1980b) considered the opposite case of large values of $|N_a|$ when $N_b = 0$. They developed a perturbation solution in $\frac{1}{|N_a|}$, once again using the method of matched asymptotic expansions for this purpose. This process yields an accurate solution but involves cumbersome numerical integration. Here, we provide another approach for dealing with this situation that yields simpler results of comparable accuracy for describing the early stages of the dissolution process.

8.5.4 Expansion in the Square Root of Time

In a different approach to the problem, Subramanian and Weinberg (1981) obtained an ordered expansion in a variable proportional to $\sqrt{T}$. In physical terms, a change in concentration is made at the interface at time zero, and its influence gradually spreads to larger distances from the interface as time progresses. The quantity $\sqrt{T} = \frac{\sqrt{Dt}}{R_0}$ represents the order of magnitude of this distance from the interface at any given time, divided by the initial radius. A similarity solution can be found for the scaled concentration field in the limit as $T \to 0$. The expansion in $\sqrt{T}$ permits this solution to be improved in an orderly manner. Such an expansion in time would naturally be expected to be useful only for small values of T, breaking down as time is increased. The authors considered the case $N_b = 0$, and found that the expansion produced a result for the time evolution of the radius of the sphere that was simple enough to be calculated by hand. Yet this result was found to be more accurate than the prediction from the quasi-stationary solution, when compared against a numerical solution obtained by using the method of finite differences. The results from this expansion are useful over a significant portion of the lifetime of a dissolving sphere and over comparable times for a growing sphere. Later, the method was extended to the case $N_b \neq 0$ by Shankar et al. (1984). Here, we provide details of the technique and the results for the case $N_b = 0$, which corresponds to $\Lambda = 0$. The following additional transformations are introduced into Equations (8.5.12) through (8.5.17):

$$X = 2\sqrt{T} \tag{8.5.25}$$

and

$$Y = \frac{\xi - 1}{2\sqrt{T}}. \tag{8.5.26}$$

Because the independent variables have been changed, we define

$$F(X, Y) = \Psi(T, \xi) \tag{8.5.27}$$

and

$$H(X) = h(T) \tag{8.5.28}$$

and write asymptotic power series for $F(X, Y)$ and $H(X)$ as follows:

$$F(X, Y) \sim \sum_{j=0}^{\infty} F_j(Y) X^j \tag{8.5.29}$$

and

$$H(X) \sim \sum_{j=0}^{\infty} H_j X^j. \tag{8.5.30}$$

These series are substituted into the governing equations for $F(X, Y)$, and by matching coefficients of X^j for each value of j, a series of ordinary differential equations and associated boundary conditions are obtained for $F_j(Y)$. In a similar way, the use of the expansion for $H(X)$ in Equation (8.5.30), in conjunction with Equations (8.5.16), (8.5.17) and (8.5.25) through (8.5.28), yields a set of algebraic equations for the coefficients H_j.

The governing differential equation satisfied by the function $F_j(Y)$ is given below.

$$
\begin{aligned}
F_j'' &+ 2Y F_j' - 2j F_j \\
&= \sum_{\ell=0}^{j-1} \Big[F_\ell \{ 2\ell H_{j-\ell} + 3(j+\ell-1)Y H_{j-\ell-1} + 3(j+\ell-2)Y^2 H_{j-\ell-2} \\
&\quad + (j+\ell-3)Y^3 H_{j-\ell-3} \} - F_\ell' \{ (3j-3\ell+2)Y H_{j-\ell} \\
&\quad + 6(j-\ell)Y^2 H_{j-\ell-1} + 2(2j-2\ell-1)Y^3 H_{j-\ell-2} \\
&\quad + (j-\ell-1)Y^4 H_{j-\ell-3} \} \Big] - Y(3F_{j-1}'' + 3Y F_{j-2}'' + Y^2 F_{j-3}''), \\
&\hspace{6cm} (j = 0, 1, 2, \ldots). \tag{8.5.31}
\end{aligned}
$$

Here the convention applies that functions and constants with a negative subscript such as F_{-1} and H_{-1} are identically zero. The boundary conditions on these functions are

$$F_j(0) = \delta_{j0} \tag{8.5.32}$$

and

$$F_j(\infty) = 0. \tag{8.5.33}$$

The coefficients H_j are obtained from

$$H_{j+1} = \frac{N_a}{j+1} [F_j'(0) - \delta_{1j}], \quad (j = 0, 1, 2 \ldots), \tag{8.5.34}$$

with Equation (8.5.17) providing the result that

$$H_0 = 1. \tag{8.5.35}$$

The solutions of the above equations for the first few functions and the corresponding coefficients are reported below.

$$F_0(Y) = \text{erfc}(Y), \tag{8.5.36}$$

$$H_1 = N_a F_0'(0) = -\frac{2}{\sqrt{\pi}} N_a, \tag{8.5.37}$$

$$F_1(Y) = \frac{10}{3\pi} N_a Y e^{-Y^2}, \tag{8.5.38}$$

$$H_2 = \frac{N_a}{2}(F_1'(0) - 1) = N_a\left(\frac{5}{3\pi} N_a - \frac{1}{2}\right), \tag{8.5.39}$$

$$F_2(Y) = \frac{N_a}{\sqrt{\pi}}\left[\frac{48}{5\sqrt{\pi}} i^2 \text{erfc}(Y) + 3Y \text{erfc}(Y)\right.$$
$$\left. - e^{-Y^2}\left\{\frac{12}{5\sqrt{\pi}} - \left(1 - \frac{5}{3\pi} N_a\right)Y + \frac{6}{5\sqrt{\pi}} Y^2 - \frac{50}{9\pi} N_a Y^3\right\}\right], \tag{8.5.40}$$

$$H_3 = \frac{N_a}{3} F_2'(0) = \frac{N_a^2}{\sqrt{\pi}}\left[\frac{4}{3} - \frac{16}{5\pi} - \frac{5}{9\pi} N_a\right]. \tag{8.5.41}$$

In Equation (8.5.40), $i^2\text{erfc}(Y)$ stands for a repeated integral of the complementary error function defined in Abramowitz and Stegun (1965). The following final result can be written for $h = g^2$:

$$h(T) = 1 - 4N_a\sqrt{\frac{T}{\pi}} + \left(\frac{10}{3\pi} N_a - 1\right) 2N_a T$$
$$+ \left(\frac{4}{3} - \frac{16}{5\pi} - \frac{5}{9\pi} N_a\right) 8N_a^2 T\sqrt{\frac{T}{\pi}} + O(T^2). \tag{8.5.42}$$

Note that the square of the radius appears as a natural dependent variable in both the parameter expansion mentioned earlier and the expansion in time. Subramanian and Weinberg (1981) provide results for $F_3(Y)$ and H_4, but indicate that the resulting $O(T^2)$ contribution in Equation (8.5.42) offers little improvement.

8.5.5 Comparison with a Numerical Solution

If an accurate answer is desired in the general case, a numerical solution must be used. Duda and Vrentas (1969) described a coordinate transformation to render the interface stationary so that the concentration gradients could be accurately represented near the interface in a solution by the method of finite differences. These authors also established the utility of their asymptotic solution by comparison with their numerical results. Later, Subramanian and Weinberg (1981) used the same approach for constructing a numerical solution as a standard against which the solutions by asymptotic expansions could be tested. The reader may wish to consult those articles to obtain more details. Here we present two sample drawings in the case of dissolution in Figures 8.5.1 and 8.5.2, for values of $N_a = 0.2$ and 1.5 respectively. In these figures, results from the two expansions given in the present section, as well as those from the quasi-stationary solution, are

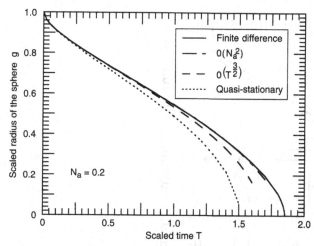

Figure 8.5.1 Scaled radius of the sphere plotted against scaled time. Results from the approximate solutions are compared with those from a finite-difference solution; $N_a = 0.2$.

compared with predictions from a numerical solution. We have calculated the square root of the results from Equations (8.5.21) and (8.5.42) to plot the scaled radius of the sphere against scaled time. It is evident that the expansion for small N_a is superior to the expansion in time for $N_a = 0.2$, and the reverse is true for $N_a = 1.5$. At larger values of N_a, the expansion in time continues to do better than the small N_a expansion, as illustrated in Subramanian and Weinberg (1981). It is useful for describing the radius-time history until the sphere has dissolved to approximately one-half its initial radius. It also is seen from Figures 8.5.1 and 8.5.2 that the quasi-stationary solution is not as good as either of the two expansions.

In the case of growth, illustrative results are shown in Figure 8.5.3 for $N_a = -1.5$. Here, we have plotted the square of the scaled radius of the sphere against scaled time. Based on the prediction from the similarity solution for growth from zero initial size, one would expect h to be asymptotically linear in time when the sphere has grown to

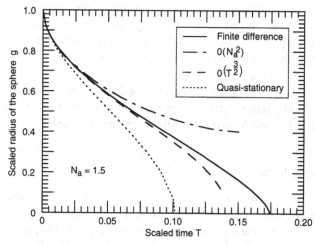

Figure 8.5.2 Scaled radius of the sphere plotted against scaled time. Results from the approximate solutions are compared with those from a finite-difference solution; $N_a = 1.5$.

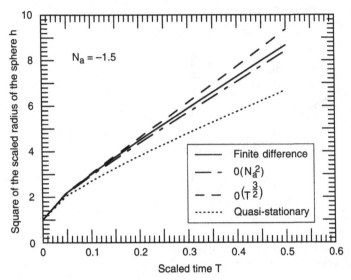

Figure 8.5.3 Square of the scaled radius of the sphere plotted against scaled time. Results from the approximate solutions are compared with those from a finite-difference solution; $N_a = -1.5$.

a new radius that is large compared with its initial radius. This is the rationale for the choice of the variable plotted as the ordinate. Even though Figure 8.5.3 depicts only the initial stages of growth, the trend for the square of the radius to become linear with time is clear. For this value of N_a, the growth constant β in Equation (8.4.12) is 1.8306. Therefore, the asymptotic trend should be $h \sim 4\beta^2 T \approx 13.40T$. The slope estimated from the numerical solution for h, between $T = 0.45$ and 0.5, is 13.79, which is already close to the asymptotic value even though the sphere has grown to less than three times its initial radius. For growth from finite initial size, Vrentas et al. (1983) indicated that the similarity solution is adequate for describing the dependence of the radius on time after the sphere has grown to a radius larger than five times its initial radius.

It is seen from Figure 8.5.3 that the expansion for small $|N_a|$ does better than the expansion in time, and both are superior to the quasi-stationary approximation. Subramanian and Weinberg (1981) displayed a similar comparison for the case of $N_a = -10$, demonstrating that it takes relatively large values of $|N_a|$ for the small time expansion to perform better than the expansion for small $|N_a|$. Eventually, the small time expansion can be expected to diverge away from the correct solution.

As noted earlier, we have only shown results in this section for the case $N_b = 0$. Analytical results, as well as graphical comparisons with a numerical solution, similar to those in Figures 8.5.1, 8.5.2, and 8.5.3 for the case $N_b \neq 0$, can be found in Shankar et al. (1984).

8.5.6 Extensions to Include Other Effects

The present section brings us to the conclusion of our discussion of the dissolution and growth of a stationary sphere. We have not considered the consequences of including the effects of surface tension in increasing the pressure within a gas bubble and, therefore, the equilibrium concentration in the liquid at the interface. This effect is discussed in detail in Weinberg (1981) and Subramanian and Weinberg (1983). Viscous effects in gas bubble dissolution and growth, in the context of polymer processing, are considered by Venerus and Yala (1997). These authors use a numerical solution, obtained by the method of finite differences, to establish the ranges of utility of approximate solutions. Numerical solutions in model problems involving stationary multicomponent gas bubbles are provided in Weinberg and Subramanian (1980), Ramos (1986), Cable and Frade (1987), and Yung et al. (1989). A noteworthy feature of such problems is that the bubble can initially grow because of a rapid influx of a new gas but subsequently dissolve because of the slower efflux of a gas initially present. The reverse situation, involving initial dissolution followed by subsequent growth, also can occur. This implies the existence of an extremum in the radius versus time curve. Weinberg (1986) performed a qualitative analysis of the governing equations in the two-component case using the quasi-stationary approximation. He noted that although the quasi-stationary approximation is not accurate in a quantitative sense, it retains the qualitative features of the problem adequately for this purpose. Weinberg concluded that for a two-component bubble, zero, one, or two extrema can occur in the bubble radius as a function of time. Another effect of great practical importance is that of chemical reactions in the liquid surrounding the bubble. If solute is removed by reaction, it is possible to use the quasi-steady state approximation for the concentration field, provided the reaction is sufficiently rapid. The quasi-steady approximation is similar to the quasi-stationary approximation, with the additional assumption that the time derivative of the concentration field can be neglected. For the case of a first-order irreversible reaction that consumes the solute, a numerical solution was used to establish the range of utility of the approximate solutions in Subramanian and Chi (1980), and Weinberg and Subramanian (1981). Weinberg (1982) performed a similar analysis for a reversible reaction with first order kinetics and a rapid forward reaction rate constant. When the reaction mechanisms and kinetic descriptions become complex, the quasi-steady approximation is commonly employed as an alternative to a full numerical solution. A numerical solution of the governing equations must be used to obtain accurate answers, however, because the range of utility of the quasi-steady approximation cannot be established a priori in the general case.

8.6 Mass Transport from a Drop in Thermocapillary Motion at Small Compositional Péclet Number

8.6.1 Introduction

Now, we consider the case of a sphere moving through a fluid. A variety of problems can be posed in this situation, but most require numerical means for the solution of the governing equations. This is because of additional complications that arise when the sphere moves through the continuous phase. A substantial literature is available on mass transport from a rigid or fluid object that moves through a continuous phase because of a body force. The problem is unsteady if the interior of the object contains more than one component because the temporal evolution of the composition distribution within the object must be considered. Also, if proper accounting of the change in size of the object is to be made, the problem once again is unsteady even in the case of a sphere of a pure component. In any case, most investigators assume the size to be fixed while solving the mass conservation equations. It also is typical to assume that the transferring solute is sparingly soluble. This permits one to neglect motion induced by diffusion. Clift et al. (1978) and Sadhal et al. (1997) have provided ample discussion of the existing work on mass and heat transport from rigid or fluid objects moving through a fluid because of a body force, and we refer the reader to them for further information. Here, we only consider mass transport from a spherical drop that undergoes thermocapillary migration in a continuous phase.

In Sections 8.6.2 and 8.6.3, we present the governing equations and the definitions of the local and average mass transfer coefficients. These equations are equally applicable to the subsequent development in Section 8.6.4 and to that in Section 8.7.

8.6.2 Governing Equations for the Concentration Field

To keep the problem tractable, the drop, of radius R, is assumed to be of a pure substance A, which can dissolve in the continuous phase containing solvent B and solute A. The size of the drop is assumed to change so slowly with time that it is treated as a constant in analyzing the thermocapillary migration and mass transport problems. In addition, the solution is assumed to be sufficiently dilute that convective transport arising from diffusion is negligible. We also set the partial specific volume of the solute A in the solution, $\overline{V}_A$, equal to zero. The symbol $\mathcal{D}$ is used to designate the binary diffusivity of A in B. Thermocapillary migration is assumed to occur under conditions wherein the velocity field obtained in Section 4.2 is applicable. The physical migration velocity of the drop is v_∞^*, which is given in Equation (4.2.28). As noted in Section 4.3, Equation (4.2.28), and the associated velocity field given in Equations (4.2.19) and (4.2.20) in scaled variables, can be used when convective transport of energy can be neglected. This implies that the Marangoni number in the continuous phase, Ma, which is defined in Section 4.1, is negligibly small. In this case, the result applies for all values of the Reynolds number, Re, also defined in Section 4.1. In addition, in the limiting situation when $Re \to \infty$, the same velocity field is applicable, with asymptotically small corrections in a momentum boundary layer of thickness $O(\frac{1}{\sqrt{Re}})$, at any value of the Marangoni number. In this limiting situation, however, Equation (4.2.28) no longer holds, and other appropriate results must be used for the migration velocity.

First, consider the case when the Reynolds number is negligible. It is possible for the Péclet number for mass transport, defined as $Pe_M = \frac{2Rv_\infty^*}{\mathcal{D}}$, to take on a wide range of values. This is because the Péclet number is the product of the Reynolds number and the Schmidt number $Sc = \frac{\nu}{\mathcal{D}}$, where ν is the kinematic viscosity of the continuous phase. The Schmidt number for typical solutes in liquids can range in value from a few hundred to several thousand. Therefore, it is appropriate to consider situations wherein Pe_M is small, as well as when it is large. On the other hand, when the Reynolds number is large, the Péclet number for mass transport will necessarily be large as well. Note that we have used the diameter of the drop as the length scale in defining Pe_M to conform with standard practice in the literature.

The approach is to solve for the distribution of mass concentration of the dissolving solute ρ_A in the continuous phase and infer the rate of mass transport at the interface from the solution. A spherical polar coordinate system, with its origin at the center of the drop, is used. The radial coordinate, scaled with the radius R of the drop, is designated r, and the polar angle θ is measured from the forward stagnation streamline. The sketch given in Figure 3.1.1 can be used as a schematic of the system. Velocity is scaled using the natural reference v_∞^*. In a reference frame attached to the moving drop, after an initial transient, the mass concentration field will be steady because the size of the drop and the equilibrium concentration $\rho_{A,i}$ at the interface are constant with time. The concentration of A far from the drop remains at its undisturbed value of $\rho_{A,\infty}$. As in the earlier sections, a dimensionless concentration $C = \frac{\rho_A - \rho_{A,\infty}}{\rho_{A,i} - \rho_{A,\infty}}$ is used. The field $C(r, \theta)$ satisfies the following governing equation and boundary conditions:

$$\frac{1}{2} Pe_M \left[v_r \frac{\partial C}{\partial r} + \frac{v_\theta}{r} \frac{\partial C}{\partial \theta} \right] = \nabla^2 C, \tag{8.6.1}$$

$$C(1, \theta) = 1, \tag{8.6.2}$$

and

$$C \to 0 \quad \text{as} \quad r \to \infty. \tag{8.6.3}$$

In addition, the scaled concentration C must remain bounded at $\theta = 0, \pi$. The scaled velocity components are given below.

$$v_r = -\left(1 - \frac{1}{r^3}\right) s, \tag{8.6.4}$$

$$v_\theta = \left(1 + \frac{1}{2r^3}\right) \sqrt{1 - s^2}. \tag{8.6.5}$$

In the above results, $s = \cos\theta$ as usual. It is convenient to use (r, s) as the set of independent variables henceforth.

8.6.3 The Local and Average Mass Transfer Coefficients and the Governing Equation for the Evolution of the Radius of the Drop

It is customary to report the results for the rate of mass transport in terms of a local mass transfer coefficient K_ℓ, which is defined as follows:

$$K_\ell(s) = -\frac{\mathcal{D}}{\rho_{A,i} - \rho_{A,\infty}} \frac{\partial \rho_A}{\partial r^*}(R, s). \tag{8.6.6}$$

Here r^* is the physical radial coordinate, and $\rho_A = \rho_A(r^*, s)$. Note that K_ℓ is a function of the polar angle. The dimensionless version of the mass transfer coefficient is the Sherwood number $Sh_\ell = \frac{2K_\ell R}{\mathcal{D}}$. This can be related to the gradient of the scaled concentration of solute at the interface:

$$Sh_\ell(s) = -2\frac{\partial C}{\partial r}(1, s). \tag{8.6.7}$$

Using the rate of mass transport over the entire drop surface, an average Sherwood number Sh_{av} can be defined as follows:

$$Sh_{av} = -\int_{-1}^{+1} \frac{\partial C}{\partial r}(1, s) \, ds. \tag{8.6.8}$$

Up to this point, we have assumed that the radius of the drop is independent of time. Now, we can determine how that radius changes with time. A mass balance written for the drop, using this quasi-steady approach for estimating the mass transfer rate, yields

$$\frac{dR}{dt} = -N_a \, Sh_{av} \frac{\mathcal{D}}{2R}, \tag{8.6.9}$$

where t is time. Because $\overline{V}_A$ has been assumed to be zero, the parameter N_a is defined as follows:

$$N_a = \frac{\rho_{A,i} - \rho_{A,\infty}}{\rho'}. \tag{8.6.10}$$

The drop will dissolve when N_a is positive and grow when it is negative. Equation (8.6.9), along with the initial condition $R(0) = R_0$, can be used to obtain the time evolution of the radius of the drop when a suitable expression is available for the average Sherwood number as a function of the radius.

Analytical solutions for the quasi-steady concentration field can be obtained in the asymptotic cases when $Pe_M \to 0$ and $Pe_M \to \infty$. In Section 8.6.4, we present the solution in the limit when $Pe_M \to 0$. Because the Schmidt number is large, this analysis is applicable only in the limiting case when $Re \to 0$. In Section 8.7, the asymptotic problem in the opposite limit $Pe_M \to \infty$ is analyzed.

8.6.4 Solution for Small Compositional Péclet Number

Let $\epsilon = \frac{1}{2} Pe_M$. Consider the solution in the limit as $\epsilon \to 0$. Because ϵ multiplies the convective transport terms in Equation (8.6.1), the mathematical problem for the concentration field is similar to that for the temperature field considered in Section 4.12 but somewhat simpler in nature. An analogous mathematical problem was considered by Acrivos and Taylor (1962) in the context of heat or mass transfer from a rigid sphere in body force driven Stokes motion and extended to the case of fluid spheres by Brunn (1982). As discussed in Section 4.12, a solution by regular perturbation fails because the neglected convective transport terms become comparable to the included diffusion terms when $r \sim O(\frac{1}{\epsilon})$. As before, the remedy is to construct an outer asymptotic expansion with $\rho = \epsilon r$ fixed and match with the asymptotic expansion for small ϵ with r fixed. We use the symbol $c(r, s)$ to designate the inner field and $C(\rho, s)$ for the outer field. An expansion for $c(r, s)$ can be written as follows:

$$c(r, s) \sim \sum_j f_j(\epsilon) c_j(r, s). \tag{8.6.11}$$

Because the boundary condition at the surface of the drop must be satisfied by the inner field, $c(1, s) = 1$, and

$$f_0(\epsilon) = 1. \tag{8.6.12}$$

In a similar manner, the outer field is expanded in an asymptotic series:

$$C(\rho, s) \sim \sum_j F_j(\epsilon) C_j(\rho, s). \tag{8.6.13}$$

Here, $F_0(\epsilon)$ cannot be established from the only boundary condition to be satisfied by the outer field, which is Equation (8.6.3). Instead, it must be determined by matching requirements. A straightforward process yields the following results after matching at leading order:

$$c_0(r, s) = \frac{1}{r}, \tag{8.6.14}$$

$$F_0(\epsilon) = \epsilon, \tag{8.6.15}$$

and

$$C_0(\rho, s) = \frac{1}{\rho} \exp\left[-\frac{\rho}{2}(1 + s)\right]. \tag{8.6.16}$$

The inner solution at leading order is the quasi-steady result for a stationary sphere.

The leading order local and average Sherwood numbers are the same and are given by $Sh = Sh_{av} = 2$, corresponding to a pure diffusion field around the sphere.

Proceeding with the correction to the inner field, assuming

$$f_1(\epsilon) = \epsilon, \tag{8.6.17}$$

c_1 is found to be

$$c_1(r, s) = -\frac{1}{2}\left[\left(1 - \frac{1}{r}\right) + \left(1 - \frac{3}{2r^2} + \frac{1}{2r^3}\right)P_1(s)\right]. \tag{8.6.18}$$

Note that c_1 does not approach 0 as $r \to \infty$. This is the reason for developing the outer expansion. If a straightforward expansion is employed, c_1 will fail to meet the boundary condition as $r \to \infty$. In the method of matched asymptotic expansions, as $r \to \infty$, c_1 only needs to match the outer field as $\rho \to 0$, which it does. Matching confirms the correctness of the assumption that $f_1(\epsilon) = \epsilon$. It is possible to continue to find higher order terms, and the following results can be obtained without too much labor. Up to the order presented here, the asymptotic sequence proceeds in powers of ϵ in an orderly manner so that $f_2(\epsilon) = F_1(\epsilon) = \epsilon^2$, $f_3(\epsilon) = F_2(\epsilon) = \epsilon^3$, and $F_3(\epsilon) = \epsilon^4$.

$$C_1(\rho, s) = \frac{1}{2\rho}\exp\left[-\frac{\rho}{2}(1+s)\right], \tag{8.6.19}$$

$$c_2(r, s) = \frac{r}{6} - \frac{1}{4} + \frac{7}{80r} + \frac{1}{24r^2} - \frac{1}{16r^4} + \frac{1}{60r^5} - \left[\frac{1}{4} - \frac{r}{4} - \frac{1}{8r^2} + \frac{1}{8r^3}\right]P_1(s)$$

$$+ \left[\frac{r}{12} - \frac{1}{4r} + \frac{5}{24r^2} + \frac{3}{56r^3} - \frac{1}{8r^4} + \frac{5}{168r^5}\right]P_2(s), \tag{8.6.20}$$

$$C_2(\rho, s) = \frac{1}{8\rho}\exp\left[-\frac{\rho}{2}(1+s)\right]\left[\frac{17}{10} + 3\left(1 + \frac{2}{\rho}\right)P_1(s)\right], \tag{8.6.21}$$

$$c_3(r, s) = -\frac{r^2}{24} + \frac{r}{12} - \frac{17}{160} + \frac{11}{240r} + \frac{1}{48r^2} - \frac{1}{96r^4} + \frac{1}{120r^5}$$

$$- \left[\frac{3r^2}{40} - \frac{r}{8} + \frac{23}{160} + \frac{3}{40r} - \frac{247}{1120r^2} + \frac{11}{320r^3} + \frac{3}{112r^4}\right.$$

$$\left. + \frac{3}{560r^5} - \frac{3}{160r^6} + \frac{9}{2240r^7}\right]P_1(s) - \left[\frac{r^2}{24} - \frac{r}{24} + \frac{1}{12r}\right.$$

$$\left. - \frac{5}{48r^2} + \frac{5}{336r^3} + \frac{1}{48r^4} - \frac{5}{336r^5}\right]P_2(s) - \left[\frac{r^2}{120} - \frac{3}{80} + \frac{1}{20r}\right.$$

$$\left. + \frac{9}{560r^2} - \frac{3}{40r^3} + \frac{9}{224r^4} + \frac{9}{1120r^5} - \frac{1}{80r^6} + \frac{1}{420r^7}\right]P_3(s), \tag{8.6.22}$$

$$C_3(\rho, s) = \frac{1}{\rho}\exp\left[-\frac{\rho}{2}(1+s)\right]\left[\frac{1}{15} - \frac{1}{4\rho^2}P_1(s) + \frac{1}{48}\left(1 + \frac{6}{\rho}\right)P_2(s)\right]. \tag{8.6.23}$$

The average Sherwood number can be calculated using Equation (8.6.8) in conjunction with the inner solution. The result is

$$Sh_{av} = 2 + \frac{1}{2}Pe_M - \frac{13}{160}Pe_M^2 + \frac{7}{320}Pe_M^3 + \cdots. \tag{8.6.24}$$

In contrast, for the equivalent problem in which convective transport of momentum is neglected and the drop moves because of a body force such as gravity, Brunn (1982) provided the following result, which extends that of Acrivos and Taylor (1962) for a rigid sphere:

$$Sh_{av} = 2 + \frac{1}{2}Pe_M + \frac{1}{4}APe_M^2 \log Pe_M - \frac{1}{4}$$

$$\times \left[\frac{13}{40} - \left(\gamma - \log 2 + \frac{37}{120}\right)A - \frac{43}{160}A^2\right]Pe_M^2 + \cdots. \tag{8.6.25}$$

Here, to conform with the usual mathematical convention, we use the symbol γ to designate Euler's constant, $0.5772\ldots$ instead of the ratio of densities, as we have done elsewhere. The other constant A, appearing in Equation (8.6.25), is defined as

$$A = \frac{2 + 3\alpha}{3(1 + \alpha)}, \tag{8.6.26}$$

where α is the ratio of the viscosity of the drop to that of the continuous phase.

It is noteworthy that logarithmic terms begin to appear in the asymptotic sequence in the body force driven motion problem at $O(Pe_M^2 \log Pe_M)$. This is not the case in the thermocapillary migration problem. We have checked the governing equation for $c_4(r, s)$ and determined that the particular solution for c_4 does not necessitate adding a term $O(Pe_M^4 \log Pe_M)$ to the sequence that would appear before the $O(Pe_M^4)$ term. It is not evident at which point such logarithmic terms would arise in the present problem, and the question must be left open at this time.

Using the quasi-steady result for the Sherwood number given in Equation (8.6.24), in conjunction with Equation (8.6.9) for the time evolution of the radius, we can calculate the radius of the drop as a function of time when it migrates due to thermocapillarity. For this, we note that it is possible to write $v_\infty^* = \Omega R$ from Equation (4.2.28), where Ω is a constant that has dimensions of $\frac{1}{time}$. Using just the leading term in Equation (8.6.24) corresponds to setting $Pe_M = 0$, which reduces the problem to that of mass transport between a stationary drop and a continuous phase. The quasi-steady solution in this case already is given in Equation (8.3.15). Note that the symbol Ω has a different meaning in that equation, and is defined in Equation (8.3.13). Truncation of Equation (8.6.24) at $O(Pe_M)$ yields the following result for $R^2(t)$:

$$R^2 = R_0^2 \exp(-N_a \Omega t) - \frac{2D}{\Omega}[1 - \exp(-N_a \Omega t)]. \tag{8.6.27}$$

If we include the $O(Pe_M^2)$ term in the result for the Sherwood number, we obtain the improved result:

$$R^2 = \frac{20D}{13\Omega}\left[1 + \frac{6}{\sqrt{10}}\tanh\left(d - \frac{3}{\sqrt{10}}N_a \Omega t\right)\right]. \tag{8.6.28}$$

In Equation (8.6.28), the constant d can be evaluated by applying the initial condition on the radius. It is possible to carry out the integration analytically when the $O(Pe_M^3)$ term in the result for the Sherwood number is included. The result is unwieldy, however, and therefore omitted.

We note that in the case of dissolution, if Pe_M is initially small, it will stay small throughout the process; therefore, the results given here can be used to describe the entire time evolution of the radius. For growth, caution is necessary when using Equation (8.6.24), because Pe_M will increase with time as the drop grows in size. Therefore, the expansion for small Pe_M will not be applicable beyond a certain point.

In the analogous body force driven migration problem, the velocity of the drop would be proportional to the square of the radius of the drop, and therefore, we can write $v_\infty^* = \Lambda R^2$, where the constant Λ has dimensions of $\frac{1}{\text{length} \cdot \text{time}}$. Here, because of the appearance of the logarithmic term in Equation (8.6.25) for the Sherwood number, it is possible to carry out the integration analytically only if we truncate the result at $O(Pe_M)$. This leads to the following solution in implicit form for the radius:

$$\log \frac{R^2 - pR + p^2}{(R + p)^2} + 2\sqrt{3} \arctan \left(\frac{\sqrt{3}\, R}{2p - R} \right) = d - 3pN_a \Lambda t. \tag{8.6.29}$$

Here,

$$p = \left(\frac{2D}{\Lambda} \right)^{\frac{1}{3}}, \tag{8.6.30}$$

and d is a constant to be determined by applying the initial condition on the radius.

8.7 Mass Transport from a Drop in Thermocapillary Motion at Large Compositional Péclet Number

8.7.1 Introduction

Here we analyze the problem posed in Section 8.6 in the opposite limit as $Pe_M \to \infty$. All the assumptions made in that section apply, except the one that Pe_M is small. As noted in Section 8.6, it is possible to use the scaled velocity distribution given in Equations (4.2.19) and (4.2.20) in two limiting situations. The first is one where the Marangoni number is negligible. In this case, for any value of the Reynolds number, the physical migration velocity is given by the result in Equation (4.2.28). Also, for any value of the Marangoni number when the Reynolds number is very large, the velocity distribution still is applicable as long as an appropriate result for the migration velocity for the drop is used.

We retain the same notation and scalings that were used in Section 8.6. Spherical polar coordinates are used, and we define $s = \cos\theta$ as before. We can begin with Equations (8.6.1) through (8.6.5) as the starting point. In the limit $Pe_M \to \infty$, convective transport of A dominates over diffusive transport. It is possible to neglect diffusion altogether, were it not for the need to satisfy the crucial boundary condition on the concentration field at the drop surface. In the absence of diffusion, the solution $C = 0$ is one of uniform mass concentration $\rho_A = \rho_{A,\infty}$ and does not satisfy the requirement that $C(1, s) = 1$. Thus, a boundary layer of sharp concentration variation adjacent to the drop surface, in which diffusive terms are comparable to convective transport terms, needs to be resolved. The details of the solution in this boundary layer and the result for the Sherwood number are given in this section. The treatment is similar to that of Levich (1962), who considered the same problem for body-force-driven motion.

8.7.2 Analysis

By requiring that the diffusive term in the radial direction in the governing differential equation for C be comparable to the convective term in the direction tangential to the drop surface, it is possible to establish that the scaled thickness of the boundary layer adjacent to the drop surface is proportional to $\frac{1}{\sqrt{Pe_M}}$. With this knowledge, we define $\epsilon = \sqrt{\frac{2}{Pe_M}}$. It is evident that a straightforward expansion in ϵ yields a leading order problem that does not include diffusive transport. The solution is $C = 0$ as mentioned earlier. We now define a transformed radial coordinate as

$$y = \frac{r-1}{\epsilon}, \tag{8.7.1}$$

which is a magnified distance from the interface. Let the solution be $c(y, s)$ in this new inner variable. After making the transformation given in Equation (8.7.1) in Equations (8.6.1) through (8.6.5), we expand the solution in the limit $\epsilon \to 0$ as

$$c(y, s) \sim \sum_j f_j(\epsilon) c_j(y, s). \tag{8.7.2}$$

We recognize that $f_0(\epsilon) = 1$ and find that the leading order field c_0 satisfies the following differential equation:

$$-3ys\frac{\partial c_0}{\partial y} - \frac{3}{2}(1 - s^2)\frac{\partial c_0}{\partial s} = \frac{\partial^2 c_0}{\partial y^2}. \tag{8.7.3}$$

This field should satisfy the boundary condition at the surface of the drop, which is

$$c_0(0, s) = 1. \tag{8.7.4}$$

Symmetry at $\theta = 0$ requires that $\frac{\partial c_0}{\partial s}(y, 1)$ be bounded. In addition, the field $c_0(y, s)$ must satisfy matching conditions with the outer field in the limit $y \to \infty$. Because the leading order outer result is simply that $C_0 = 0$, this leads to the boundary condition

$$c_0(y, s) \to 0 \quad \text{as} \quad y \to \infty. \tag{8.7.5}$$

A solution can be obtained by postulating similarity in the concentration profiles. Let $c_0(y, s) = F(\eta)$ where $\eta = \frac{y}{\delta(s)}$. We find that

$$c_0(y, s) = F(\eta) = \text{erfc}(\eta) \tag{8.7.6}$$

and

$$\delta(s) = \frac{\sqrt{8(2 - 3s + s^3)}}{3(1 - s^2)}. \tag{8.7.7}$$

Note that $\delta \to \sqrt{\frac{2}{3}}$ as $s \to 1$ but that δ becomes unbounded in the limit as $s \to -1$. This poses no difficulty in obtaining the local and average Sherwood numbers using the definitions given in Section 8.6.

$$Sh_l = \sqrt{\frac{8Pe_M}{\pi}}\frac{1}{\delta(s)}, \tag{8.7.8}$$

$$Sh_{av} = 2\sqrt{\frac{Pe_M}{\pi}}. \tag{8.7.9}$$

When the result for the average Sherwood number from Equation (8.7.9) is used in Equation (8.6.9), we obtain the following result for the rate of change of the radius of the drop with time in the case when the Marangoni number Ma, defined in Section 4.1, is negligible, for any value of the Reynolds number:

$$\frac{dR}{dt} = -N_a\sqrt{\frac{2D\Omega}{\pi}}. \tag{8.7.10}$$

Here, the parameter Ω is the constant of proportionality in the linear relationship between the migration velocity of the drop and its radius, which was introduced in Section 8.6.4. The parameter N_a is positive for dissolution and negative for growth. With the initial condition $R(0) = R_0$, we can integrate Equation (8.7.10) to write

$$R(t) = R_0 - \sqrt{\frac{2D\Omega}{\pi}} N_a t. \tag{8.7.11}$$

In the case $Re \to \infty$, Equation (8.7.9) is applicable for any value of the Marangoni number. We already have considered the limiting case of $Ma \to 0$ above. The only other case wherein analytical results for the migration velocity are available is that of $Ma \to \infty$. In the gas bubble limit, considered in Section 4.13, the physical migration velocity continues to be proportional to the radius R of the bubble when $Ma \to \infty$. Therefore, Equations (8.7.10) and (8.7.11) continue to be applicable in this situation, when the proper value of Ω is used. In the general case of a drop, in the limit $Ma \to \infty$, we must use the result given in Equation (4.14.67). This can be written as $v_{\infty,0}^* = \tilde{\Omega} R^3$, where the constant $\tilde{\Omega}$ has dimensions of $\frac{1}{\text{length}^2 \cdot \text{time}}$. In this case, we can write the following result for the rate of change of the radius of the drop with time:

$$\frac{dR}{dt} = -N_a \sqrt{\frac{2D\tilde{\Omega}}{\pi}} R. \tag{8.7.12}$$

Integration leads to

$$R(t) = R_0 \exp\left(-\sqrt{\frac{2D\tilde{\Omega}}{\pi}} N_a t\right). \tag{8.7.13}$$

In the case of dissolution, even if the Reynolds number is very large in the beginning, it will decrease as the radius of the drop decreases. One must keep this in mind when using the result in Equation (8.7.13). Similarly, in the case of growth, the result in Equation (8.7.11) will break down as the Marangoni number increases with increasing size of the drop. The above results for the radius in the limit $Pe_M \to \infty$ may be contrasted with the results given in Equations (8.6.27) and (8.6.28) in the case wherein Pe_M is small. In that problem, even when we set $Pe_M = 0$, it is the square of the radius that is linear in time. When Pe_M is not zero, the results are more involved.

In the analogous body force driven motion case, for low Reynolds number motion, the average Sherwood number can be written as

$$Sh_{av} = 2\sqrt{\frac{Pe_M}{3\pi(1+\alpha)}}. \tag{8.7.14}$$

Here, α is the ratio of the viscosity of the drop to that of the continuous phase. If we write $v_\infty = \Lambda R^2$, as in Section 8.6.4, where Λ is independent of the radius of the drop, we can integrate Equation (8.7.12), along with the initial condition on the radius, to yield

$$\sqrt{R} = \sqrt{R_0} - \sqrt{\frac{D\Lambda}{6\pi(1+\alpha)}} N_a t. \tag{8.7.15}$$

Therefore, for low Reynolds number body-force-driven motion, it is the square root of the radius that is linear in time.

If the value of the Péclet number Pe_M is such that it can be considered neither large nor small, one must resort to numerical methods of solution of the governing equations, along with the associated boundary conditions. This has not been done to date in the thermocapillary migration problem. Also, note that we have used a simple result for the velocity field that applies in certain limiting situations. If it becomes necessary to use a less restrictive result for the velocity field, the asymptotic mass transfer problem for $Pe_M \to \infty$ still can be solved in the same manner. In a general way, the results for $\delta(s)$ and the average Sherwood number can be written as follows:

$$\delta(s) = \frac{2}{g(s)} \sqrt{\int_s^1 g(\tilde{s}) \, d\tilde{s}} \tag{8.7.16}$$

and

$$Sh_{av} = \sqrt{\frac{2}{\pi} Pe_M \int_{-1}^1 g(s) \, ds}. \tag{8.7.17}$$

Here, the function $g(s)$ is related to the scaled tangential velocity distribution at the surface of the drop via

$$g(s) = v_\theta(1, s) \sqrt{1 - s^2}. \tag{8.7.18}$$

The velocity scale is the migration velocity of the drop. This leading order result is valid regardless of the detailed nature of the velocity distribution or the origin of the motion of the drop. For the case of body-force-driven motion, more details regarding this and other related results can be found in Clift et al. (1978) and Sadhal et al. (1997).

CHAPTER NINE

Motion Driven by the Interface in a Body of Fluid

9.1 Introduction

In Chapter One, we introduced the phenomenon of thermocapillarity, and in Chapter Four, it was shown how the tangential stress discontinuity it induces at interfaces leads to the migration of bubbles and drops. The subject of the present chapter is the flow induced by the variation of the interfacial tension in a body of liquid with a free surface. Specifically, we consider thermocapillary flows in rectangular and cylindrical liquid pools, cylindrical liquid bridges, and a stationary spherical drop. In all these instances, the liquid mass itself is stationary as a whole, unlike the drops and bubbles that moved because of thermocapillarity in the problems considered earlier. The containing boundaries of the liquid prevent the motion of its center of mass, and when the liquid is not contained, we assume it is held in position by suitable means, such as an acoustic field.

In addition to the mere containment of the mass of liquid, the presence of boundaries can introduce additional features. Consider a pool of liquid in an open rectangular trough. The liquid is in contact with rigid boundaries on all sides, with the exception of the free surface where it is in contact with the surrounding gas. The thermal conditions at the container walls can be such that the temperature gradient in the liquid near the free surface could have components parallel or perpendicular to the free surface. The variation of temperature parallel to the free surface induces a surface tension gradient that leads to motion in the liquid because of the tangential stresses exerted at the free surface. This is precisely the mechanism that causes the migration of drops discussed in earlier chapters. We shall refer to this motion in a fluid as *thermocapillary convection*. On the other hand, if temperature varies only in a direction perpendicular to the free surface, the surface tension would be uniform, and the above mechanism for flow would be absent. The ensuing quiescent state of the liquid can become unstable to disturbances, however, spontaneously leading to organized motion. For liquids with a negative value of the rate of change of surface tension with temperature, σ_T, this happens when the surface is cooler than the liquid in its vicinity and the Marangoni number exceeds a critical value. The resulting motion is called Bénard convection. When the liquid layer is sufficiently thin, the flow is in the form of repeating hexagonal patterns that are called Bénard cells, named after the French scientist Henre Bénard (1900, 1901), who observed it experimentally. There is a vast literature on the subject of Bénard convection. After the pioneering work by Pearson (1958), who performed a linear stability analysis of the problem assuming the liquid-air free surface to be flat, much theoretical and experimental research has been conducted. We shall not discuss Bénard convection any further. The interested reader can consult the book by Koschmieder (1993) for additional information and references on this topic. A related instability is the Rayleigh–Bénard instability, where flow in the form of hexagonal cells occurs in a mass of fluid that is heated from below in a gravitational field. An introductory treatment can be found in Chandrasekhar (1961).

In this chapter, we consider thermocapillary flows in single and multiple layers of liquids. The thermal boundary conditions are such that there is a variation of temperature along liquid-gas and liquid-liquid interfaces. These interfaces are assumed to be clean, so that the only response of the fluid to the thermocapillary stress is a state of motion. This is true regardless of the magnitude of this stress, no matter how small. We pose problems in liquid pools that are rectangular or cylindrical and in a stationary spherical mass of liquid. The results for the velocity and temperature fields in these systems and the free surface shape, which we obtain in the rectangular geometry, may be used in a variety of applications, such as welding, crystal growth, burning of liquid fuel pools, and the measurement of physical properties in levitated samples of material in reduced gravity.

We focus here on thermocapillary flows, and neglect motion induced by buoyancy. The latter topic is discussed in the book by Gebhart et al. (1988). We can determine the conditions when buoyant convection can be neglected in comparison with thermocapillary convection by comparing the order of magnitude of the velocity induced by the buoyant force with that induced by interfacial tension gradients. For simplicity, we neglect the role of inertia in the following discussion. For a specific example, consider a rectangular pool of liquid, of length L, depth H, and of large width, so that the problem is two-dimensional. Let us impose a temperature difference ΔT^* between the end walls in the direction in which the length is measured. Buoyant convection will ensue in this pool. When the density differences are small compared with the value of the density, they can be neglected everywhere in the Navier–Stokes equation, except in the body-force term. This is known as the Boussinesq approximation. After making this assumption, by balancing the order of magnitude of the viscous term in the Navier–Stokes equation against the buoyancy term, it is possible to infer that the steady characteristic velocity due to buoyancy is given by

$$u_b \sim \frac{\rho g \zeta \Delta T^* H^3}{\mu L}, \tag{9.1.1}$$

where ρ is the density of the liquid, g represents the magnitude of the acceleration due to gravity, μ is the viscosity of the liquid, and the coefficient of thermal expansion of the liquid ζ is defined as follows:

$$\zeta = -\frac{1}{\rho} \frac{d\rho}{dT^*}. \tag{9.1.2}$$

In the same liquid pool, we can estimate the steady characteristic thermocapillary velocity u_t, by balancing the thermocapillary stress against the viscous tangential stress at the free surface. This leads to the following result for u_t:

$$u_t \sim \frac{|\sigma_T| \Delta T^* H}{\mu L}. \tag{9.1.3}$$

The ratio of the two characteristic velocities can be written as

$$\frac{u_b}{u_t} = \frac{\rho g \zeta H^2}{|\sigma_T|}. \tag{9.1.4}$$

Therefore, we can conclude that, even in liquid pools of substantial length scale L over which a temperature difference is applied, thermocapillary motion will dominate over that induced by buoyancy, so long as the depth of the pool is sufficiently small. Similar results can be developed in cylindrical and spherical geometries. In the case of a cylindrical liquid column, when a temperature difference is applied over the length of the cylinder, thermocapillarity will dominate when the radius of the cylinder is sufficiently small. In a like manner, in a spherical drop of sufficiently small radius, motion driven by a temperature difference applied across the drop will predominantly be of thermocapillary origin. Arguments similar to the above can be developed when inertia is important as well. In this chapter, however, we mostly analyze situations wherein inertial effects are unimportant.

Throughout this chapter, we assume an incompressible flow of a Newtonian fluid with spatially and temporally constant physical properties, except the interfacial tension, which is assumed to depend linearly on temperature. In the first six sections, flow in a rectangular pool is analyzed. For ease of mathematical analysis, we assume that the depth of the liquid pool is small compared with its length, except in Section 9.6. First, steady flow and the resulting deformation of the free surface, assumed small, are considered. Next we analyze transients in the velocity field and the deformation. A treatment of motion in multiple fluid layers is then presented. This is followed by an analysis of the flow in the vicinity of the end walls. Finally, we consider the effects of inertia and convective heat transfer in a deep liquid layer, with a specified distribution of the heat flux at the free surface. Next, we treat the flow in a cylindrical liquid pool, the free surface of which is subjected to a heat flux, and follow with a discussion of thermocapillary convection in floating zones. We conclude with an analysis of the flow generated inside a stationary spherical drop due to an arbitrary distribution of the temperature on its surface and a brief discussion of the literature on flow driven by a nominally hemispherical bubble attached to a plane wall.

9.2 Flow in a Rectangular Liquid Layer

9.2.1 Introduction

In this section, we introduce the problem of thermocapillary convection in a rectangular geometry and the relevant notation. We also provide solutions for the steady velocity and temperature fields and the shape of the free surface. As noted in Section 9.1, additional aspects of this problem are considered in several subsequent sections. In all cases, we assume that the layer is shallow, implying that the depth is small compared with its length. In treating the deformation of the free surface, we include the role of gravity through a hydrostatic head, as well as surface tension effects. It is assumed that this deformation is small so that an analytical approach can be used for solving the problem. In our discussion of rectangular pools, we occasionally use the terms *up*, *down*, *top*, and *bottom*. In the absence of a gravitational field, these terms are used with reference to the orientation of Figure 9.2.1, which displays the geometry and the coordinate system. When we include a gravitational force in the problem, it is in the context of

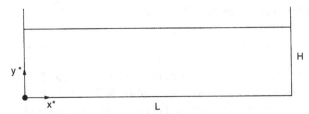

Figure 9.2.1 Geometry and the coordinate system used in the analysis of flow in a thin rectangular liquid layer.

determining the shape of the free surface. We always neglect buoyant convection as noted in Section 9.1.

Figure 9.2.1 shows an open rectangular container of length L filled with a liquid of viscosity μ to a depth H. We assume it is sufficiently wide in the third dimension to consider the velocity and temperature fields two-dimensional. This implies that the velocity component in the third direction is zero and that there is no variation of the fields in that direction. We further assume that the aspect ratio $\epsilon = \frac{H}{L} \ll 1$. Distances in the x^* direction are scaled by L, and those in the y^* direction by H. The scaled coordinates are $x = \frac{x^*}{L}$ and $y = \frac{y^*}{H}$.

We consider three problems that differ in the thermal conditions at the boundaries. In the first problem, labeled problem I, the left wall of the container, at $x = 0$, is maintained at a temperature T_a^*. The right wall at $x = 1$ is likewise at a temperature T_b^*. We assume that $T_a^* > T_b^*$, and that all the other surfaces of the container are thermally insulated. The symbol ΔT^* represents the temperature difference $T_a^* - T_b^*$. In the quiescent state, the free surface is flat and is located at $y = 1$. There is no loss of heat from the free surface as well. In problem II, we assume that the left and right walls are maintained at the same temperature and that the bottom boundary is thermally insulated. The free liquid surface is subjected to a heat flux, such as that from a laser beam or a radiant heater. Problem III is similar to problem II, except that all the rigid boundaries are maintained at the same temperature. Our goal is to determine the velocity field in the liquid pool due to the thermocapillary stress exerted at the free surface and the shape of the free surface itself, which is influenced by the flow. The analysis of the flow field is common to all the three problems. Levich (1962) performed a steady-state analysis of this problem, which has subsequently been extended by Birikh (1966), Yih (1968, 1969), and Adler and Sowerby (1970). A treatment of the problem also appears in the book by Probstein (1994). Pimputkar and Ostrach (1980) and Lai (1984) have included unsteady effects in their analyses. Our development parallels that in the latter two studies. In the present section, we consider only the steady state. Unsteady effects will be examined in Section 9.3.

In this chapter, when employing rectangular coordinates, we use u^* and v^* to designate the physical velocity components in the x^* and y^* directions, respectively. The corresponding scaled velocity components are denoted by u and v and are obtained by dividing the physical velocities by U_R and ϵU_R, respectively. Here $U_R = \frac{(-\sigma_T)\Delta T^* \epsilon}{\mu}$ is the thermocapillary reference velocity and is obtained by balancing a characteristic tangential viscous stress and the thermocapillary stress at the free surface. The symbol σ_T represents the rate of change of the interfacial tension with temperature at the free surface. The scaled pressure p is obtained by dividing the hydrodynamic pressure by the reference pressure $\frac{\mu U_R L}{H^2}$. The scaled temperature is T and is obtained in problem I by subtracting T_b^* from the physical temperature and dividing by ΔT^*. The scaled temperature T in problems II and III is defined in Section 9.2.4. When the free surface is deformed, the scaled depth of the liquid varies with x and is denoted by $h(x)$. For ease of analysis, we assume that there are only slight changes to the shape. The implication is that $h(x) = 1 + h_1(x)$, where $h_1(x) \ll 1$. The conditions under which this holds will be established later.

When the continuity, momentum, and energy equations given in Chapter Two are nondimensionalized as indicated above, the following equations for the steady scaled velocity and temperature fields are obtained:

$$\frac{\partial u}{\partial x} + \frac{\partial v}{\partial y} = 0, \tag{9.2.1}$$

$$\epsilon^2 Re\left(u\frac{\partial u}{\partial x} + v\frac{\partial u}{\partial y}\right) = -\frac{\partial p}{\partial x} + \epsilon^2 \frac{\partial^2 u}{\partial x^2} + \frac{\partial^2 u}{\partial y^2}, \tag{9.2.2}$$

$$\epsilon^2 Re\left(u\frac{\partial v}{\partial x} + v\frac{\partial v}{\partial y}\right) = -\frac{1}{\epsilon^2}\frac{\partial p}{\partial y} + \epsilon^2 \frac{\partial^2 v}{\partial x^2} + \frac{\partial^2 v}{\partial y^2}, \tag{9.2.3}$$

and

$$\epsilon^2 Ma\left(u\frac{\partial T}{\partial x} + v\frac{\partial T}{\partial y}\right) = \epsilon^2 \frac{\partial^2 T}{\partial x^2} + \frac{\partial^2 T}{\partial y^2}. \tag{9.2.4}$$

Here $Re = \frac{U_R L}{\nu}$ is the Reynolds number, and $Ma = \frac{U_R L}{\kappa}$ is the Marangoni number. As in earlier chapters, ν represents the kinematic viscosity of the liquid, and κ represents its thermal diffusivity. When the Reynolds and Marangoni numbers are of $O(1)$, in the limit $\epsilon \to 0$, the momentum and energy equations are decoupled. In this limit, it is possible to obtain the velocity field for an arbitrary temperature distribution $T(x, h)$ at the free surface. Therefore, we first determine this velocity field, which is common to the three problems mentioned earlier.

9.2.2 Velocity Field

For $\epsilon \to 0$ the momentum equations reduce to

$$\frac{\partial p}{\partial x} = \frac{\partial^2 u}{\partial y^2} \tag{9.2.5}$$

and

$$\frac{\partial p}{\partial y} = 0. \tag{9.2.6}$$

For convenience, we have not used the subscript 0 to denote leading order quantities in the above equations, but it is implied.

From Equations (9.2.1), (9.2.5), and (9.2.6), we infer that the flow in the liquid layer is nearly parallel and is dominated by viscous forces. The turning flows near the end walls at $x = 0$ and $x = 1$ cannot be accommodated by Equations (9.2.5) and (9.2.6), which apply at leading order in the limit $\epsilon \to 0$. In this straightforward perturbation procedure, the derivatives in the x direction are absent in the equations at leading order. Thus, the ability to satisfy boundary conditions at the end walls is lost. The remedy is to rescale the x-coordinate near the end walls so that these terms are retained. An inner expansion should be developed near each end wall and matched to the expansion away from the end walls. Such an analysis is performed in Section 9.5. Here, we proceed to obtain a solution that is valid away from the end walls.

Next, we give the boundary conditions associated with the above equations. At the bottom boundary, the no slip and kinematic conditions are used.

$$u(x, 0) = 0, \tag{9.2.7}$$

$$v(x, 0) = 0. \tag{9.2.8}$$

The kinematic condition and the tangential stress balance at the free surface are written as follows:

$$v(x, h) = 0 \tag{9.2.9}$$

and

$$\frac{\partial u}{\partial y}(x, h) = -\frac{\partial T}{\partial x}(x, h). \tag{9.2.10}$$

In writing the above conditions at the free surface, as well as the normal stress balance to be given subsequently, we assume that $\frac{dh}{dx} \ll 1$. Because the liquid pool is shallow and its total volume is fixed, this is equivalent to the assumption of small changes to the free surface shape. The location of the free surface, $h(x)$, is unknown and needs to be determined as part of the solution. Problems of this type are known as free boundary problems. The function $h(x)$ is determined by using the normal stress balance at the free surface. We have discussed this boundary condition in Section 2.3.3. It states that the jump in the normal stress across the interface is equal to twice the product of the surface tension and the mean curvature. The normal stress consists of a contribution from pressure and one from viscous stress. The pressure in the surrounding gas at the free surface is the uniform ambient pressure. The viscous stresses in the gas are negligible compared with those in the liquid because the dynamic viscosity of a gas is negligible compared with that of a typical liquid. The pressure in the liquid is measured relative to the ambient gas pressure and is the total pressure that includes the hydrodynamic pressure $p(x, y)$ and a hydrostatic contribution from gravity. Therefore, the normal stress balance can be written as follows:

$$Bo_s h - Ca\, p(x, h) + 2\epsilon^2\, Ca \frac{\partial v}{\partial y}(x, h) = \epsilon^2 (1 + Ca\, T)\frac{d^2 h}{dx^2}. \tag{9.2.11}$$

Here $Bo_s = \frac{\rho g H^2}{\sigma_0}$ is the static Bond number, and $Ca = \frac{(-\sigma_T)\Delta T^*}{\sigma_0}$ is the Capillary number, where σ_0 is a reference value for the surface tension. For a pool of silicone oil of depth $H \approx 1$ mm, in Earth's gravitational field, $Bo_s \approx 0.5$. Because $Bo_s \propto H^2$, it rapidly becomes small compared with unity for thinner layers. In low gravity conditions, it is likely to be small in most systems. In the limit $Bo_s \to 0$, the surface tension force plays a dominant role in determining the shape of the free surface. In the opposite extreme, $Bo_s \gg 1$, the hydrostatic head will be the governing force in controlling the shape of the free surface.

Because Equation (9.2.11) contains the second derivative of $h(x)$, it must be supplemented by conditions prescribed at the boundaries $x = 0$ and 1, where the free surface meets these boundaries. Two types of conditions are commonly used. In the first, it is assumed that the free surface is pinned to the wall and unable to move. This implies prescribing the value of h at the boundary. In the second, the free surface is permitted to move, but the contact angle is assumed to be fixed. This means that the first derivative

$\frac{dh}{dx}$ is fixed at the boundary. The contact angle is a physical property determined by the nature of the materials in contact. The above two conditions represent the ideal cases of a free surface that sticks to the wall and a free surface that freely slips at the wall. A gamut of conditions that lie in between these limiting cases has been used in the literature dealing with the motion of the contact line, as noted by Satterlee and Reynolds (1964). Wherever necessary, we shall assume the contact angle to be 90°, which yields $\frac{dh}{dx} = 0$.

In the subsequent development, we shall have occasion to use an integrated form of the continuity equation. It provides a condition to be used in conjunction with the remaining governing equations. To obtain it, we multiply both sides of the continuity equation by an area element $dxdy$, and integrate with respect to x from 0 to x, and with respect to y from 0 to h. The result is

$$\int_0^h u(x, y) \, dy = 0. \tag{9.2.12}$$

The physical implication of this condition is that the volumetric flow rate in the liquid pool across any location at a fixed value of x is zero.

We have now completed the specification of the problem to determine the steady velocity distribution and the free surface shape in the liquid pool for a known steady temperature distribution at the free surface. Because of the free boundary nature of the problem, the solution of the equations along with the boundary conditions is, in general, formidable. Indeed, such problems are typically solved numerically. It is possible to obtain an analytical solution via a perturbation approach when the free surface is only slightly deformed. This is the method we pursue below. The assumption of small deformation is valid when the Capillary number is small. The idea is that because $h(x) = 1 + h_1(x)$, and $h_1(x) \ll 1$, at leading order, we can use $h(x) = 1$ in the boundary conditions. This means that the boundary location is no longer unknown in the leading order problem for the field variables. Therefore, the solutions for the velocity and temperature fields can be obtained with the boundary conditions specified at fixed boundaries. Equation (9.2.11) is then used to determine an approximation to $h_1(x)$, thus providing a first correction to the shape of the free surface. If desired, the process can be repeated with conditions specified at the new location of this boundary. The technique is analogous to that used in Section 4.3 for determining small corrections to the spherical shape of a drop due to inertia. The leading order solution of the equations and boundary conditions with $h(x) = 1$ is presented in this subsection. The perturbation solution for $h_1(x)$ is calculated in Section 9.2.3, where the order of magnitude of h_1 is established in the limiting cases of small and large values of the Bond number.

The velocity field $u(x, y)$ is determined from the x-momentum equation. Because $\frac{\partial p}{\partial y}$ is zero, the pressure depends only on x. Equation (9.2.5) then may be integrated to yield

$$u(x, y) = \frac{dp}{dx}\left(\frac{y^2}{2} - y\right) - y\frac{\partial T}{\partial x}(x, 1). \tag{9.2.13}$$

We see that velocity field in the liquid layer consists of a superposition of two flows. The first is a pressure gradient driven flow that is analogous to the flow in a wide channel of rectangular cross section. The second contribution is that driven by the tangential stress at the free surface, which is similar to a Couette flow. Using the constraint of zero volumetric flow rate across the cross-section at any x, given in Equation (9.2.12), in the result for the velocity field in Equation (9.2.13), the pressure gradient is determined to be

$$\frac{dp}{dx} = -\frac{3}{2}\frac{\partial T}{\partial x}(x, 1). \tag{9.2.14}$$

Thus, the constraint on the volumetric flow rate determines the pressure gradient so that it induces just the right flow rate to cancel that induced by the thermocapillary stress. Using this result for $\frac{dp}{dx}$, the velocity field can be rewritten as

$$u(x, y) = -\frac{\partial T}{\partial x}(x, 1)\left[\frac{3}{4}y^2 - \frac{1}{2}y\right]. \tag{9.2.15}$$

Using Equation (9.2.1), we can determine the velocity field $v(x, y)$ as

$$v(x, y) = -\frac{\partial}{\partial x}\int_0^y u(x, \tilde{y})\,d\tilde{y} = \frac{1}{4}\frac{\partial^2 T}{\partial x^2}(x, 1)y^2(y - 1). \tag{9.2.16}$$

We have used the kinematic condition in Equation (9.2.8) in obtaining this result. Note that the reference velocity used for defining v is of $O(\epsilon)$, whereas that used in defining u is $O(1)$. Therefore, the velocities in the y-direction are small in magnitude compared with those in the x-direction. If the temperature distribution at the free surface is linear in x, then $v(x, y) = 0$. Thus the leading order velocity field is strictly parallel in this case.

9.2.3 Shape of the Free Surface

The flow in the liquid layer alters the shape of its free surface. We saw how pressure variations naturally arise from the constraint of vanishing volumetric flow rate. This nonuniform pressure field drives a recirculation that opposes the flow induced by the thermocapillary stress. The pressure in the gas just above the free surface is constant, however. Therefore, from the normal stress balance in Equation (9.2.11), we see that the free surface must be deformed. For $\epsilon \ll 1$, the viscous normal stresses are negligible compared with the pressure in determining the free surface shape. The free surface deformation and the pressure gradient in the liquid evolve together in time, as we shall see in Section 9.3. Therefore, viewing Equation (9.2.11) as a description of the sources that produce the required pressure variations in the liquid, we see that both the gravitational force and the surface tension force contribute to these variations. Depending on the value of the static Bond number, we can distinguish between two limiting cases in which one or the other dominates.

When Bo_s is large, the pressure at any location in the pool is determined predominantly by gravity. Because the hydrodynamic pressure does not depend on y, the total pressure at a given location in the liquid is seen to be merely the hydrostatic head. Neglecting the viscous normal stress, we can differentiate Equation (9.2.11) with respect to x to yield

$$\frac{dp}{dx} = \frac{Bo_s}{Ca}\frac{dh}{dx} = Bo_e\frac{dh}{dx} \tag{9.2.17}$$

in the limit $\epsilon \to 0$. The group $\frac{Bo_s}{Ca} = \frac{\rho g H^2}{(-\sigma_T)\Delta T^*}$ is assigned the symbol Bo_e, which is termed the elevation Bond number. Equation (9.2.14) can be combined with Equation (9.2.17) to yield the following differential equation for $h(x)$:

$$\frac{dh}{dx} = -\frac{3}{2}\frac{1}{Bo_e}\frac{\partial T}{\partial x}(x, 1). \tag{9.2.18}$$

Because we have assumed that $\frac{dh}{dx} \ll 1$, the value of Bo_e must be large compared with unity. Writing $h(x) = 1 + (Bo_e)^{-1} f(x)$ and using the constraint that the total volume of liquid in the pool must be constant when the free surface deforms, which can be shown to yield $\int_0^1 f(x)\,dx = 0$, $h(x)$ may be determined to be

$$h(x) = 1 + \frac{3}{2}\frac{1}{Bo_e}\left[\int_0^1 T(\tilde{x}, 1)\,d\tilde{x} - T(x, 1)\right]. \tag{9.2.19}$$

If the temperature distribution at the free surface is linear, we obtain $h(x) = 1 + \frac{3}{2}\frac{1}{Bo_e}(x - \frac{1}{2})$.

When the static Bond number is small, surface tension forces are important in determining the shape of the free surface. Unlike the case when the static Bond number is large, the free surface at equilibrium, even in the absence of the thermocapillary stress, is a surface of constant curvature, the shape of which is determined by the conditions at the contact line. We have implicitly assumed that $h(x) = 1$ is the static equilibrium shape of the free surface. This is valid when the contact line is pinned or when the contact angle is 90°. In the case when $Bo_s \ll 1$, by differentiating Equation (9.2.11) and substituting for the pressure gradient from Equation (9.2.14), we can write

$$\frac{d^3 h}{dx^3} = \frac{3}{2}\frac{Ca}{\epsilon^2}\frac{\partial T}{\partial x}(x, 1). \tag{9.2.20}$$

Here, we have used the condition that $\frac{Ca}{\epsilon^2} \ll 1$ to restrict the free surface deformation to be small. Using the constraint on the liquid volume and assuming that the contact angles between the liquid and the rigid boundaries at $x = 0$ and 1 are both 90°, h may be obtained as

$$h(x) = 1 + \frac{3Ca}{2\epsilon^2}\left[\int_0^x \int_0^{\tilde{x}} T(\hat{x}, 1)\,d\hat{x}\,d\tilde{x} + \left(\frac{1}{6} - \frac{x^2}{2}\right)\int_0^1 T(x, 1)\,dx\right.$$
$$\left. - \int_0^1 \int_0^x \int_0^{\tilde{x}} T(\hat{x}, 1)\,d\hat{x}\,d\tilde{x}\,dx\right]. \tag{9.2.21}$$

If the temperature distribution at the free surface is linear, the required integrations can be performed analytically to yield

$$h(x) = 1 - \frac{Ca}{\epsilon^2}\left(\frac{1}{4}x^3 - \frac{3}{8}x^2 + \frac{1}{16}\right). \tag{9.2.22}$$

When the static Bond number is large, from Equation (9.2.19), we see that the free surface deformation is controlled by the magnitude of $\frac{1}{Bo_e}$, which must be small compared with unity, so that the free surface is only slightly deformed. Likewise when Bo_s is small, surface tension effects are predominant, and the extent of the free surface deformation is determined by $\frac{Ca}{\epsilon^2}$, the magnitude of which must be small compared with unity for this analysis to be valid.

9.2.4 Temperature Field

In the previous subsections, we obtained the velocity field in the liquid layer and the shape of the free surface, assuming that the temperature distribution at the free surface is given. Here, we determine the temperature field in the layer for the three specific thermal problems mentioned in the introduction. From these fields, the surface temperature distribution can be evaluated and used in the expressions for the velocity fields and the shape of the free surface obtained earlier. First, we enumerate the boundary conditions applicable to each of the three thermal problems. Then, we specialize the energy equation in the limit $\epsilon \to 0$, assuming that Ma is $O(1)$.

The adiabatic boundary conditions for problem I for a differentially heated liquid layer are

$$\frac{\partial T}{\partial y}(x, 0) = 0 \tag{9.2.23}$$

and

$$\frac{\partial T}{\partial y}(x, 1) = 0. \tag{9.2.24}$$

The boundary conditions at the remaining rigid boundaries are

$$T(0, y) = 1 \tag{9.2.25}$$

and

$$T(1, y) = 0. \tag{9.2.26}$$

In problem II, where the free surface is subjected to a heat flux, we can write the following boundary conditions:

$$\frac{\partial T}{\partial y}(x, 1) = q(x), \tag{9.2.27}$$

$$T(0, y) = T(1, y) = 0, \tag{9.2.28}$$

and

$$\frac{\partial T}{\partial y}(x, 0) = 0. \tag{9.2.29}$$

In problem III, the condition at the bottom boundary is altered:

$$T(x, 0) = 0. \tag{9.2.30}$$

The remaining boundary conditions are the same as in problem II. In problems II and III, the characteristic temperature difference ΔT^*, which is used to scale the temperature differences from a reference value, is defined to be $\frac{\hat{q}H}{k}$, where $\hat{q}$ is a reference value for the surface heat flux and k is the thermal conductivity of the liquid. The reference temperature is taken to be the temperature of the walls at $x = 0$ and 1. In problem III, this is the temperature of the bottom boundary as well.

We now consider the specialization of Equation (9.2.4), which is the energy equation governing the temperature field, for problems I and II. From the boundary conditions, it can be discerned that in these two problems, the essential variation of temperature occurs in the x-direction. This variation can be determined by a perturbation procedure discussed by Kevorkian and Cole (1981). We expand the temperature field in the liquid in problem I in an asymptotic series as follows:

$$T = T_0(x) + \epsilon^2 T_1(x, y) + \cdots. \tag{9.2.31}$$

To obtain the solution for T_0, we need to consider the problem for T_1. Substitution of Equation (9.2.31) into Equation (9.2.4), followed by taking the limit as $\epsilon \to 0$, yields the following governing equation for T_1:

$$\frac{\partial^2 T_1}{\partial y^2} = -\frac{d^2 T_0}{dx^2} + Ma\, u \frac{d T_0}{dx}. \tag{9.2.32}$$

Next we integrate Equation (9.2.32) once with respect to y and use adiabatic boundary conditions on T_1 at $y = 0$ and 1. These can be obtained by substituting the asymptotic expansion for $T(x, y)$ into Equations (9.2.23) and (9.2.24). Then, using the zero volumetric flow rate condition given in Equation (9.2.12), we obtain the following governing equation for $T_0(x)$:

$$\frac{d^2 T_0}{dx^2} = 0. \tag{9.2.33}$$

Substitution of the asymptotic expansion for T into Equations (9.2.25) and (9.2.26) shows that T_0 must satisfy the same boundary conditions. Therefore, the leading order solution for the temperature field in problem I for a differentially heated liquid layer is

$$T_0(x) = 1 - x. \tag{9.2.34}$$

A similar technique can be used in problem II. Here, we expand the temperature field as

$$T = \frac{1}{\epsilon^2} T_0(x) + T_1(x, y) + \cdots. \tag{9.2.35}$$

Following the steps in problem I leading to Equation (9.2.33), the equation for T_0 in problem II may be obtained:

$$\frac{d^2 T_0}{dx^2} = -q(x). \tag{9.2.36}$$

This equation can be integrated immediately, along with the boundary conditions on T_0. These are obtained by substituting the asymptotic expansion for T into Equation (9.2.28). The solution is

$$T_0(x) = x \int_0^1 \int_0^{\tilde{x}} q(\hat{x})\, d\hat{x}\, d\tilde{x} - \int_0^x \int_0^{\tilde{x}} q(\hat{x})\, d\hat{x}\, d\tilde{x}. \tag{9.2.37}$$

In problem III for a liquid pool with a specified surface heat flux and a scaled temperature of zero at the rigid boundaries, we can use the expansion given in Equation (9.2.31) with the understanding that, in this problem, $T_0 = T_0(x, y)$. At leading order, Equation (9.2.4) yields

$$\frac{\partial^2 T_0}{\partial y^2} = 0, \tag{9.2.38}$$

and the solution that satisfies boundary conditions obtained from Equations (9.2.27) and (9.2.30), after substitution of the asymptotic expansion for T, is

$$T_0(x, y) = q(x)y. \tag{9.2.39}$$

This solution does not satisfy the conditions at the end walls at $x = 0$ and $x = 1$, given in Equation (9.2.28), except in the special case when $q(x)$ is zero at these boundaries. Thus the temperature field given above is only valid away from these end walls.

9.3   Transient Development of the Flow and Free Surface Shape in a Rectangular Liquid Layer

9.3.1   Introduction

In the previous section, we considered only the steady-state velocity fields and the steady free surface shape in a rectangular liquid layer. We saw that a nonuniform temperature field at the free surface induces a recirculating flow in the liquid. This recirculation is caused by the presence of rigid boundaries at the two ends and leads to the establishment of a pressure gradient in the liquid. We continue to use Figure 9.2.1 and the notation introduced in Section 9.2 in the present section and analyze the time evolution of the flow in the layer. It will be shown that recirculation is not present in the initial stages but develops naturally with time. The normal stress balance at the free surface links the pressure within the layer at any given x with that in the gas, which is uniform. Therefore, the pressure and the surface shape evolve together. Let us consider an initially quiescent layer with a uniform depth and let a thermocapillary stress be imposed at time zero such that the flow at the free surface is from left to right. As the flow develops from rest, conservation of mass, in conjunction with the condition that fluid cannot pass through the end walls, leads to depletion of material from the left and accumulation of material on the right. This naturally leads to a pressure gradient, which opposes the flow. In a gravity-dominated situation, this is a consequence of increasing hydrostatic pressure from left to right due to the increasing thickness of the layer. When gravitational effects are unimportant and surface tension plays a dominant role, a curvature gradient develops along the layer from left to right. This again leads to a pressure gradient opposing the flow. This pressure gradient drives a reverse flow in the region near the bottom, overcoming the viscous forces there. As time progresses, a balance is reached wherein the surface shape becomes steady, and therefore there is no net flow across any cross section. This is the steady situation that was analyzed in Section 9.2.

9.3.2   Governing Equations and Time Scales

We analyze two cases here. In the first, the gravitational force dominates in determining the shape of the free surface. This is the limit of large values of the static Bond number Bo_s. In the opposite limit of small values of the static Bond number, the surface tension force dominates. Note that although we accommodate the hydrostatic effect of gravity, we continue to neglect buoyant convection in the liquid. All the assumptions made in Section 9.2 are applicable here, except that of steady state. In particular, we continue to assume that the aspect ratio ϵ is small compared with unity. The free surface under static equilibrium is assumed to be flat, and the deformation induced by the flow is assumed to be small. Furthermore, we assume that the Reynolds and Marangoni numbers are no larger than $O(1)$. The time constants associated with the evolution of the velocity and temperature fields, and that associated with the deformation of the free surface, can be quite different. We consider several situations with varying ratios of these time constants. The symbol t is used to denote physical time scaled by $\tau_f = \frac{H^2}{\nu}$. The scaled continuity and Navier–Stokes equations for transient flow in the liquid layer at leading order, in the limit $\epsilon \to 0$, are given below.

$$\frac{\partial u}{\partial x} + \frac{\partial v}{\partial y} = 0, \tag{9.3.1}$$

$$\frac{\partial u}{\partial t} = -\frac{\partial p}{\partial x} + \frac{\partial^2 u}{\partial y^2}, \tag{9.3.2}$$

$$\frac{\partial p}{\partial y} = 0. \tag{9.3.3}$$

We have dropped the subscript 0 in the above equations for convenience. We can write the energy equation as

$$Pr \frac{\partial T}{\partial t} = \epsilon^2 \frac{\partial^2 T}{\partial x^2} + \frac{\partial^2 T}{\partial y^2}. \tag{9.3.4}$$

Asymptotic analysis of this equation, in the limit $\epsilon \to 0$, must be performed carefully, as noted in Section 9.2. The boundary condition that $v = 0$ at the free surface and the constraint that the volumetric flow rate in the layer is zero are not valid when the surface deformation and the flow are time dependent. The more general kinematic condition at a fluid-fluid interface, given in Equation (2.3.5), connects the rate at which the free surface moves to the normal velocity of the elements of fluid at the free surface. In the present context, this condition may be written as

$$\frac{1}{\epsilon^2 Re} \frac{\partial h}{\partial t} = v(x, 1) = -\frac{\partial}{\partial x} \int_0^1 u \, dy. \tag{9.3.5}$$

The other boundary conditions are the same as in Section 9.2. The initial conditions are that the liquid layer is quiescent with a uniform temperature and is undeformed at $t = 0$.

Now, we discuss the important time scales. The time scale for the development of the velocity field is $\tau_f = \frac{H^2}{\nu}$. This represents the characteristic time for the thermocapillary stress at the interface to be communicated to the bulk of the liquid by viscous means in the y-direction. When the liquid layer is subjected to a heat flux at its surface, the time scale for the temperature changes to be communicated across the layer by conduction is likewise $\frac{H^2}{\kappa} = Pr\tau_f$. For a differentially heated layer that is thermally insulated at both the free surface and the bottom rigid boundary, the transfer of heat by conduction occurs in the x-direction, and L is the appropriate length scale. Hence the thermal time scale is $\frac{Pr\tau_f}{\epsilon^2}$. The characteristic time for surface deformation may be obtained from the dimensional form of Equation (9.3.5) to be $\tau_s = \frac{h_R}{\epsilon U_R}$, where h_R is a reference value for the deformation of the free surface. Note that the symbol h_R has dimensions of length, and we continue to use the symbol h for the scaled deformation as in Section 9.2. Using the steady state values for h_R obtained from Equations (9.2.19) and (9.2.21), τ_s may be determined to be $\frac{1}{\epsilon^2} \frac{\nu}{gH}$ when $Bo_s \gg 1$, and $\frac{1}{\epsilon^4} \frac{\mu H}{\sigma_0}$ when $Bo_s \ll 1$.

The ratio of the thermal time scale to the viscous time scale is proportional to the Prandtl number. Two limiting situations arise when this ratio is either very small or very large compared with unity, which permit a simplification in the analysis of the unsteady problem. First, let us consider the case of large values of the Prandtl number. When $Pr \gg 1$, the thermal time scale is much larger than the viscous time scale. It is given by $Pr\tau_f$ or $Pr\frac{\tau_f}{\epsilon^2}$, depending on the thermal boundary conditions. This implies that the velocity field develops more rapidly than the temperature field. Because the velocity field depends on the temperature distribution at the free surface through the tangential stress balance given in Equation (9.2.10), the thermal time scale is imposed on the flow by the thermocapillary stress. At any instant in time, the velocity field rapidly adjusts to comply with the prevailing temperature variation at the interface. When the thermal time scale is also much larger than the time scale for the surface deformation, the results of Section 9.2.1 for the velocity field are applicable in the unsteady problem. The only change is that the temperature distribution at the interface appearing in the expressions for the velocities should be treated as a function of time. The unsteady temperature field is determined by solving Equation (9.3.4) along with an initial condition and the associated thermal boundary conditions in each of the thermal problems mentioned in Section 9.2.1, at leading order. Here, we provide the solution for the unsteady temperature field at leading order for problem I for a differentially heated layer with the initial condition $T(0, x) = 0$.

$$T(t, x) = 1 - x - \sum_{n=1}^{\infty} \frac{2 \sin n\pi x}{n\pi} \exp\left(-\frac{n^2\pi^2\epsilon^2 t}{Pr}\right). \tag{9.3.6}$$

Analogous results for the unsteady temperature fields can be constructed in problems II and III as well. Using such temperature fields, the instantaneous velocity and surface deformation fields can be constructed from the solutions given in Section 9.2.

The analyses in the remainder of Section 9.3 pertain to the situation when $Pr \ll 1$. Physically, this would be the case for liquid metals. In this situation, the temperature field attains a steady state much more rapidly than the velocity and the surface deformation fields. We only treat problem I in a differentially heated liquid layer in the remainder of Section 9.3 to illustrate the transient development of the velocity and surface deformation fields. Similar analyses can be performed in the other two problems as well, but we do not pursue them here. In problem I, the temperature gradient at the free surface is given by $\frac{\partial T}{\partial x} = -1$. The thermocapillary stress is independent of x, which permits simplification in the mathematical analysis. Away from the end walls, the flow at steady state is fully developed and independent of x as well. Several cases arise, depending on the values of two parameters. The first parameter is the static Bond number, and the second parameter is the ratio of the time scale for surface deformation to that for the viscous diffusion of momentum. This ratio of the time scales is $\frac{\tau_s}{\tau_f} = \frac{h_R}{\epsilon^2 Re H}$. When $Bo_s \gg 1$, it can be specialized to $\frac{\tau_s}{\tau_f} = \frac{1}{\epsilon^2 Re Bo_e} = \frac{\nu^2}{\epsilon^2 g H^3}$. When $Bo_s \ll 1$, it becomes $\frac{\tau_s}{\tau_f} = \frac{Ca}{\epsilon^4 Re} = \frac{\mu\nu}{\epsilon^4 \sigma_0 H}$.

The material that follows in Sections 9.3.3 to 9.3.7 is organized as follows. First, in Section 9.3.3, we consider the situation when the time scale for viscous diffusion of momentum in the y-direction is small compared with the time scale for surface deformation. In this case, for times that are much smaller than the time scale τ_s required for surface deformation, the surface can be treated as flat, and the velocity field evolves into a Couette flow over a time scale of the order of τ_f. We provide a solution for this transient velocity field. The results are valid regardless of the value of the static Bond number Bo_s. We separate the subsequent development into that in Sections 9.3.4 and 9.3.5, which deal with the case when $Bo_s \gg 1$, and that in Sections 9.3.6 and 9.3.7, where the analogous problems are treated in the opposite limit when $Bo_s \ll 1$. Physically, the former describes the situation when the deformation is controlled by gravity, whereas the deformation is controlled by surface tension in the latter case. In each case, in the first subsection, we consider a simple situation wherein the time scale for deformation is large compared with the viscous time scale across the depth of the layer. This permits one to decouple the evolution of the velocity field from that of the surface deformation field. Therefore, the velocity field is treated as being quasi-steady and is given by Equation (9.2.13), with the understanding that the pressure gradient appearing in that equation is evolving with time. Then, in the second subsection, we relax this assumption and provide results for the case when either the time scales τ_s and τ_f are comparable, or when they are disparate, for intermediate values of time lying between these two time scales. It will be necessary to use different reference quantities for scaling physical time in each subsection. To avoid clutter, we use the same symbol t to designate scaled time in all five subsections. Therefore, the reader should check the reference quantity used for scaling time, mentioned near the beginning of each subsection, before interpreting the results in terms of physical time.

9.3.3 Evolution of the Velocity Field for Small Values of Time

When $\tau_s \gg \tau_f$, for times much smaller than τ_s, the flow in the liquid layer is analogous to a developing Couette flow. There is no recirculation in the layer until times are comparable to τ_s. With time scaled by τ_f, the unsteady velocity field in the layer $u(t, y)$ satisfies the following governing equation:

$$\frac{\partial u}{\partial t} = \frac{\partial^2 u}{\partial y^2}. \tag{9.3.7}$$

The initial and boundary conditions are

$$u(0, y) = 0, \tag{9.3.8}$$

$$u(t, 0) = 0, \tag{9.3.9}$$

and

$$\frac{\partial u}{\partial y}(t, 1) = -\frac{\partial T}{\partial x}(t, x, 1) = 1. \tag{9.3.10}$$

The solution can be obtained by using the method of Laplace transforms, followed by inversion, or by employing the method of separation of variables:

$$u(t, y) = y - 8 \sum_{n=0}^{\infty} (-1)^n \frac{\sin\left[\frac{(2n+1)}{2}\pi y\right]}{(2n+1)^2\pi^2} \exp\left[-\frac{(2n+1)^2\pi^2 t}{4}\right]. \tag{9.3.11}$$

An alternative result, particularly useful for small t, can be written from the solution obtained by using Laplace transforms:

$$u(t, y) = \sum_{n=0}^{\infty} (-1)^n \left\{ 2\sqrt{\frac{t}{\pi}} \left[\exp\left(-\frac{(2n+1-y)^2}{4t}\right) - \exp\left(-\frac{(2n+1+y)^2}{4t}\right) \right] \right.$$
$$\left. + (2n+1+y)\,\text{erfc}\left(\frac{2n+1+y}{2\sqrt{t}}\right) - (2n+1-y)\,\text{erfc}\left(\frac{2n+1-y}{2\sqrt{t}}\right) \right\}. \tag{9.3.12}$$

Carslaw and Jaeger (1959) have provided details of the steps involved in obtaining solutions such as the above. From Equation (9.3.11), we see that the slowest transient mode decays as $\exp(-\pi^2 t/4)$. As noted earlier, the results in Equations (9.3.11) and (9.3.12) apply regardless of the value of Bo_s.

The similarity to Couette flow does not hold for relatively large values of time. Eventually, a return flow is generated in the liquid layer. When time is on the order of τ_s, the velocity field is almost steady and responds rapidly to changes in the shape of the free surface. We analyze several cases for large and intermediate times in the subsections that follow and classify them based on the relative magnitude of the static Bond number.

9.3.4 Unsteady Deformation of the Free Surface for Large Bond Number

Consider the case when the static Bond number Bo_s is large. This implies that the deformation of the free surface and the pressure gradient in the liquid are controlled by gravity. For times much larger than τ_f, the deformation of the free surface and the recirculating flow are important. The velocity field is assumed to be at a quasi-steady state with respect to the prevailing pressure gradient in the liquid at any given instant. This pressure gradient evolves with time and is related to the deformation of the free surface. In the following analysis, time is scaled by $\tau_s = \frac{1}{\epsilon^2}\frac{\nu}{gH}$. The quasi-steady velocity field may be written from Equations (9.2.13) and (9.2.34) as

$$u(t, x, y) = \frac{\partial p}{\partial x}(t, x)\left(\frac{y^2}{2} - y\right) + y. \tag{9.3.13}$$

Let the shape of the free surface be described by

$$h(t, x) = 1 + (Bo_e)^{-1} f(t, x). \tag{9.3.14}$$

Here, Bo_e is the elevation Bond number, related to the static Bond number Bo_s through $Bo_e = \frac{Bo_s}{Ca}$. From Equations (9.3.5), (9.3.13), and (9.3.14), we obtain the following equation for $f(t, x)$, upon using Equation (9.2.17), which relates the pressure gradient to the shape of the free surface.

$$\frac{\partial f}{\partial t} = \frac{1}{3}\frac{\partial^2 f}{\partial x^2}. \tag{9.3.15}$$

The velocity field given by Equation (9.3.13) is not valid in a region that is of scaled extent $O(\epsilon)$ from the end walls because the turning flow in this region is not described by the parallel flow theory used to obtain Equation (9.3.13). Therefore, the result we obtain here for the shape of the free surface cannot be expected to be valid in the region near the end walls. But we still need to use boundary conditions at $x = 0$ and 1 on the shape function $f(t, x)$. We write conditions on $\frac{\partial f}{\partial x}$ at these two end walls by requiring the volumetric flow rate to vanish at these boundaries, which is a consequence of the kinematic conditions. Although they are analogous to the contact angle conditions mentioned in the discussion following Equation (9.2.11), the origin is different.

$$\frac{\partial f}{\partial x}(t, 0) = \frac{3}{2}, \tag{9.3.16}$$

$$\frac{\partial f}{\partial x}(t, 1) = \frac{3}{2}. \tag{9.3.17}$$

The initial condition states that the free surface is undeformed at $t = 0$.

$$f(0, x) = 0. \tag{9.3.18}$$

If $F(x; q)$ is the Laplace transform of $f(t, x)$, we can write the solution for F as

$$F(x; q) = \frac{\sqrt{3}}{2q\sqrt{q}} \left[\frac{1 - \cosh\sqrt{3q}}{\sinh\sqrt{3q}} \cosh\sqrt{3q}\, x + \sinh\sqrt{3q}\, x \right]$$

$$= \frac{\sqrt{3}}{2q\sqrt{q}} \left[\sum_{n=0}^{\infty} (-1)^n \{\exp\left(-\sqrt{3q}(1 + n - x)\right) - \exp\left(-\sqrt{3q}(n + x)\right)\} \right].$$

$$(9.3.19)$$

It may be verified that $F(x; q)$ satisfies the constraint that the total volume of liquid in the layer is a constant, which can be shown to yield $\int_0^1 F\, dx = 0$. The volume constraint was used as a condition in determining the shape of the free surface at steady state given in Equation (9.2.19). In this transient analysis, it is not imposed but satisfied naturally. The inverse transform is

$$f(t, x) = \frac{\sqrt{3}}{2} \left\{ \sum_{n=0}^{\infty} (-1)^n \left[\sqrt{\frac{4t}{\pi}} \left\{ \exp\left(-\frac{3}{4t}(1 + n - x)^2\right) - \exp\left(-\frac{3}{4t}(n + x)^2\right) \right\} \right. \right.$$

$$+ \sqrt{3}(n + x)\, \mathrm{erfc}\left(\frac{\sqrt{3}(n + x)}{2\sqrt{t}}\right)$$

$$\left. \left. - \sqrt{3}(1 + n - x)\, \mathrm{erfc}\left(\frac{\sqrt{3}(1 + n - x)}{2\sqrt{t}}\right) \right] \right\}.$$

$$(9.3.20)$$

We can infer from either Equation (9.3.19) or Equation (9.3.20) that the free surface is undeformed at $x = \frac{1}{2}$ and that the deformation is antisymmetric about $x = \frac{1}{2}$, which implies that $f(\frac{1}{2} - x, t) = -f(\frac{1}{2} + x, t)$. These properties of the shape of the free surface hold for all the unsteady solutions presented here and in the subsequent subsections.

We display the evolution of the shape of the free surface in Figures 9.3.1 and 9.3.2. Figure 9.3.1 shows that material is depleted near the wall at $x = 0$, reducing the height of the free surface there. This material accumulates near the wall at $x = 1$, thereby increasing the height of the free surface at that boundary. The interface shape evolves monotonically with time and is nearly linear at $t = 5$. The temporal history of the surface

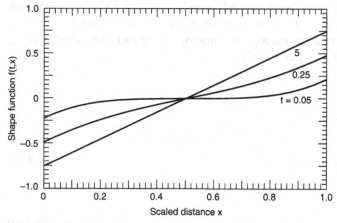

Figure 9.3.1 The shape of the free surface at selected values of scaled time for large Bond number and a quasi-steady velocity field.

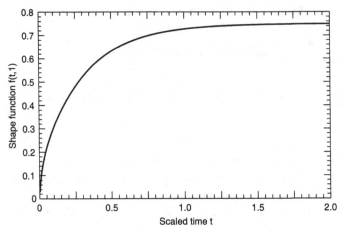

Figure 9.3.2 The evolution of the shape of the free surface at the cold wall $x = 1$ for large Bond number.

shape at the cold end wall ($x = 1$) is displayed in Figure 9.3.2. The monotonic nature of the evolution of the deformation is evident from this figure. We see that the deformation has almost attained steady state when the scaled time $t \approx 1.5$. The reader will recall from Equations (9.3.14) and (9.2.17) that the pressure distribution is simply related to the surface deformation f through an additive constant. Therefore, the evolution of the pressure field is also described by Figures 9.3.1 and 9.3.2.

9.3.5 Simultaneous Evolution of the Flow and Free Surface Deformation for Large Bond Number

In Section 9.3.2, we found that when the Bond number Bo_s is large, the time scale for viscous diffusion of momentum across the depth of the layer is related to the time scale for deformation through $\tau_f = \epsilon^2 Re\, Bo_e \tau_s$. Therefore, it is possible in the limit $\epsilon \to 0$ and $Bo_e^{-1} \to 0$ for the two time scales τ_f and τ_s to be comparable to each other. In this case, the velocity and surface deformation fields evolve together. The development in Sections 9.3.3 and 9.3.4 is no longer applicable. Even when τ_f and τ_s are disparate, for intermediate times, the transients in both the velocity and surface deformation fields must be considered. The analyses presented below and in Section 9.3.7 for small values of the Bond number are general and may be specialized to the limiting cases that are separately analyzed. Nondimensionalizing time by τ_f, we can write the following governing equations for $u(t, x, y)$ and $f(t, x)$:

$$\frac{\partial u}{\partial t} = -\frac{\partial f}{\partial x} + \frac{\partial^2 u}{\partial y^2} \tag{9.3.21}$$

and

$$\frac{1}{\epsilon^2 Re\, Bo_e} \frac{\partial f}{\partial t} = -\frac{\partial}{\partial x} \int_0^1 u\, dy. \tag{9.3.22}$$

The initial conditions are

$$u(0, x, y) = 0 \tag{9.3.23}$$

and

$$f(0, x) = 0,$$
(9.3.24)

and the boundary conditions are

$$u(t, x, 0) = 0,$$
(9.3.25)

$$\frac{\partial u}{\partial y}(t, x, 1) = 1,$$
(9.3.26)

$$\int_0^1 u(t, 0, y)\, dy = 0,$$
(9.3.27)

and

$$\int_0^1 u(t, 1, y)\, dy = 0.$$
(9.3.28)

Equation (9.3.25) represents the no-slip boundary condition at the bottom surface, and Equation (9.3.26) is the tangential stress balance at the free surface. Equations (9.3.27) and (9.3.28) arise from the impenetrability conditions at the end walls. The solutions for $U(x, y; q)$ and $F(x; q)$, which are the Laplace transforms of $u(t, x, y)$ and $f(t, x)$, respectively, can be written as follows:

$$U(x, y; q) = \frac{\sinh(\sqrt{q}\, y)}{q\sqrt{q}\cosh\sqrt{q}} + \frac{1}{q}\frac{\partial F}{\partial x}[\cosh(\sqrt{q}\, y) - 1 - \tanh\sqrt{q}\sinh(\sqrt{q}\, y)]$$
(9.3.29)

and

$$F(x; q) = \frac{\cosh\sqrt{q} - 1}{q\sqrt{l}\cosh\sqrt{q}(1 - \frac{\tanh\sqrt{q}}{\sqrt{q}})}\left[\frac{1 - \cosh\sqrt{l}}{\sinh\sqrt{l}}\cosh(\sqrt{l}\, x) + \sinh(\sqrt{l}\, x)\right].$$
(9.3.30)

Here, $l = \frac{q^2\sqrt{q}}{\epsilon^2\, Re\, Bo_e(\sqrt{q} - \tanh\sqrt{q})}$. It can be shown that $\frac{l}{q} \rightarrow 3\frac{\tau_s}{\tau_f}$ as $q \rightarrow 0$. The results given in Section 9.3.4, for the evolution of the shape of the free surface when the flow is quasi-steady, are obtained in the limit $q \rightarrow 0$, for fixed l. The opposite limit $l \rightarrow 0$, for fixed q, is interesting. It represents the physical situation when $\tau_s \ll \tau_f$. The deformation of the free surface is rapid compared with the time scale for the flow to evolve. At every instant, the shape of the free surface is quasi-steady and corresponds to the prevailing pressure gradient in the liquid. This pressure gradient is independent of x and y and is determined by the condition that the flow rate across the cross-section in the liquid must vanish. Using Equations (9.3.14) and (9.2.17), it can be established that, in the limit $l \rightarrow 0$, to within an additive constant, the pressure field is given by the inverse Laplace transform of $F(x; q)$. We note that the results given in Section 9.3.3 can be recovered from Equation (9.3.29) when the free surface deformation is neglected. In this case, only the first term in the right hand side survives, and its inverse is precisely the result in Equation (9.3.11).

The inverse Laplace transforms for the fields $u(t, x, y)$ and $f(t, x)$ can be determined numerically. Sample results that are obtained in the time domain are displayed in Figures 9.3.3 to 9.3.6 when $\frac{\tau_s}{\tau_f} = 1$. Figure 9.3.3 shows the evolution of the free surface shape from an initial state of no deformation to a final shape that is almost linear in x,

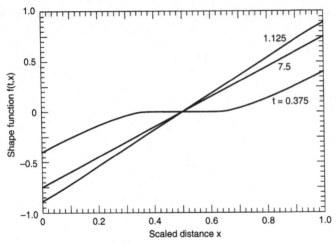

Figure 9.3.3 The shape of the free surface at selected values of scaled time for large Bond number; $\frac{\tau_s}{\tau_f} = 1$.

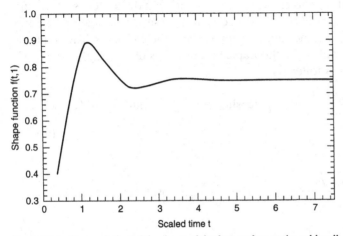

Figure 9.3.4 The evolution of the shape of the free surface at the cold wall $x = 1$ for large Bond number; $\frac{\tau_s}{\tau_f} = 1$.

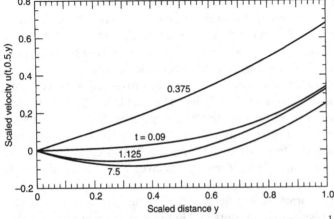

Figure 9.3.5 The velocity component in the x-direction in the plane $x = \frac{1}{2}$, plotted against y at selected values of scaled time, for large Bond number; $\frac{\tau_s}{\tau_f} = 1$.

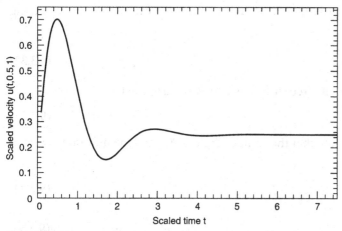

Figure 9.3.6 The evolution of the velocity component in the x-direction at the interface in the plane $x = \frac{1}{2}$, for large Bond number; $\frac{\tau_s}{\tau_f} = 1$.

at a scaled time $t = 7.5$. Note that at intermediate times, the deformation overshoots the final state. This is seen more clearly in Figure 9.3.4, which depicts the evolution of the deformation at the right boundary as a function of scaled time. Figure 9.3.5 shows the evolution of the velocity field from an initially quiescent state to the steady distribution given by Equation (9.2.15). We display the scaled velocity in the x-direction in the plane $x = \frac{1}{2}$, which is midway between the end walls, as a function of y at selected values of the scaled time. As discussed earlier, the flow begins as a Couette flow with no recirculation. At a later stage, a return flow naturally evolves with the pressure gradient. The distribution reaches steady state by $t = 7.5$. In Figure 9.3.6, the scaled velocity at the free surface $y = 1$, in the same plane $x = \frac{1}{2}$, is displayed as a function of scaled time. The figure shows a substantial overshoot of the scaled velocity at the interface before it settles into a steady state through an oscillatory approach.

9.3.6 Unsteady Deformation of the Free Surface for Small Bond Number

Next, we analyze the deformation of the free surface when the static Bond number $Bo_s \ll 1$. This case can be physically realized on Earth in thin layers or when the surface tension is large. It also is the appropriate limiting case under reduced gravity conditions in most applications. Capillary forces determine the extent of the deformation, as well as the time scale τ_s, for the response of the free surface. We assume that the contact angle between the liquid and the container wall is 90°. In this subsection, we consider the case when the surface deformation occurs over a large time scale compared with the time it takes for the velocity field to achieve steady state for a given surface shape. This is analogous to the problem considered in Section 9.3.4. When $\tau_s \gg \tau_f$, the velocity field can be treated as being quasi-steady. It is given by Equation (9.3.13). In this subsection, time is scaled by $\tau_s = \frac{1}{\epsilon^4} \frac{\mu H}{\sigma_0}$.

When the static Bond number is small, the scale for the deformation is obtained from Equation (9.2.21). This permits us to write the time-dependent shape function as

$$h(t, x) = 1 + \frac{Ca}{\epsilon^2} f(t, x). \tag{9.3.31}$$

From Equations (9.2.11), (9.3.5), (9.3.13), and (9.3.31), we can obtain the governing equation for $f(t, x)$:

$$\frac{\partial f}{\partial t} = -\frac{1}{3}\frac{\partial^4 f}{\partial x^4}. \tag{9.3.32}$$

The initial condition is that the free surface is undeformed at $t = 0$.

$$f(0, x) = 0 \tag{9.3.33}$$

The boundary conditions state that the contact angle is 90°, and the flow rate is zero at $x = 0$ and 1.

$$\frac{\partial f}{\partial x}(t, 0) = 0, \tag{9.3.34}$$

$$\frac{\partial^3 f}{\partial x^3}(t, 0) = -\frac{3}{2}, \tag{9.3.35}$$

$$\frac{\partial f}{\partial x}(t, 1) = 0, \tag{9.3.36}$$

$$\frac{\partial^3 f}{\partial x^3}(t, 1) = -\frac{3}{2}. \tag{9.3.37}$$

The solution for $F(x; q)$, which is the Laplace transform of $f(t, x)$, is given below.

$$F(x; q) = \frac{3}{8l^3 q}\left[\frac{(\sin l - \sinh l)\cos lx \cosh lx}{(\cosh l + \cos l)} + \cos lx \sinh lx \right.$$

$$\left. + \frac{(\sin l + \sinh l)\sin lx \sinh lx}{(\cosh l + \cos l)} - \sin lx \cosh lx \right]. \tag{9.3.38}$$

In Equation (38), $l = \left(\frac{3q}{4}\right)^{\frac{1}{4}}$.

Representative results in the time domain, obtained numerically, are shown in Figures 9.3.7 and 9.3.8. The results in these figures are analogous to those in Figures 9.3.1 and 9.3.2, respectively. The buildup of fluid toward the cold wall is similar, even though the driving force that determines the shape of the free surface is different here. In both situations, the evolution is monotonic, as is clearly illustrated in Figures 9.3.2

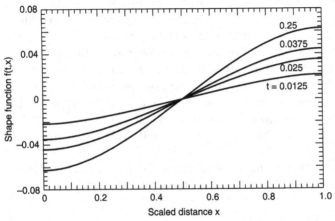

Figure 9.3.7 The shape of the free surface at selected values of scaled time for small Bond number and a quasi-steady velocity field.

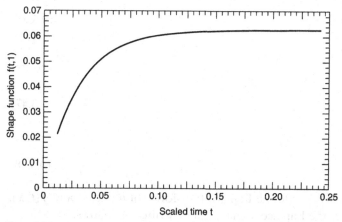

Figure 9.3.8 The evolution of the shape of the free surface at the cold wall $x = 1$ for small Bond number.

and 9.3.8. Note that the steady interface shape, which is attained in the present case for scaled time $t > 0.2$, is not linear, however.

9.3.7 Simultaneous Evolution of the Flow and Free Surface Deformation for Small Bond Number

Now, we provide results for $Bo_s \ll 1$ in the more general case wherein the surface deformation and the velocity fields are evolving simultaneously. This situation is analogous to that treated in Section 9.3.5 for large values of the static Bond number. The extent of the deformation of the free surface and the time scale are both controlled by surface tension forces. Such an analysis is necessary when $\frac{\tau_s}{\tau_f} (= \frac{Ca}{\epsilon^4 Re})$ is on the order of unity or for any values of τ_s and τ_f at intermediate times which lie between the two. The analysis also covers the limiting cases considered earlier in Sections 9.3.3 and 9.3.6. With time scaled by τ_f, we can write the following governing equations for $u(t, x, y)$ and $f(t, x)$:

$$\frac{\partial u}{\partial t} = \frac{\partial^3 f}{\partial x^3} + \frac{\partial^2 u}{\partial y^2} \tag{9.3.39}$$

and

$$\frac{Ca}{\epsilon^4 Re} \frac{\partial f}{\partial t} = -\frac{\partial}{\partial x} \int_0^1 u \, dy. \tag{9.3.40}$$

The initial conditions are

$$u(0, x, y) = 0 \tag{9.3.41}$$

and

$$f(0, x) = 0, \tag{9.3.42}$$

and the boundary conditions, the physical basis of which has been discussed earlier, are written as follows:

$$u(t, x, 0) = 0, \tag{9.3.43}$$

$$\frac{\partial u}{\partial y}(t, x, 1) = 1, \tag{9.3.44}$$

$$\int_0^1 u(t, 0, y)\, dy = 0, \tag{9.3.45}$$

$$\int_0^1 u(t, 1, y)\, dy = 0, \tag{9.3.46}$$

$$\frac{\partial f}{\partial x}(t, 0) = 0, \tag{9.3.47}$$

and

$$\frac{\partial f}{\partial x}(t, 1) = 0. \tag{9.3.48}$$

Let $U(x, y; q)$ and $F(x; q)$ denote the Laplace transforms of $u(t, x, y)$ and $f(t, x)$, respectively. The solution in the Laplace domain can be written as follows:

$$U(x, y; q) = \frac{\sinh(\sqrt{q}\, y)}{q\sqrt{q}\cosh\sqrt{q}} - \frac{1}{q}\frac{\partial^3 F}{\partial x^3}[\cosh(\sqrt{q}\, y) - 1 - \tanh\sqrt{q}\sinh(\sqrt{q}\, y)] \tag{9.3.49}$$

and

$$F(x; q) = \frac{\cosh\sqrt{q} - 1}{4l^3\sqrt{q}\cosh\sqrt{q}(\sqrt{q} - \tanh\sqrt{q})}\left\{\frac{\sin l - \sinh l}{\cos l + \cosh l}\cos lx\cosh lx \right.$$
$$\left. + \cos lx\sinh lx + \frac{\sin l + \sinh l}{\cos l + \cosh l}\sin lx\sinh lx - \sin lx\cosh lx\right\}. \tag{9.3.50}$$

Here, $l = [\frac{Caq^2\sqrt{q}}{4\epsilon^4 Re(\sqrt{q} - \tanh\sqrt{q})}]^{\frac{1}{4}}$. Results for the cases $\tau_s \gg \tau_f$ and $\tau_s \ll \tau_f$ can be recovered in the limits l fixed, $q \to 0$, and q fixed, $l \to 0$, respectively. In the limit $q \to 0$, $\frac{l^4}{q} \to \frac{3\tau_s}{4\tau_f}$. The limit $l \to 0$ is similar to that discussed in Section 9.3.5, and the pressure field is equal to $f(t, x)$, to within an additive constant. We have obtained the inverse Laplace transforms numerically for $\frac{\tau_s}{\tau_f} = 5$, and the results are displayed in Figures 9.3.9 through 9.3.12. These figures are analogous to Figures 9.3.3 through 9.3.6 in Section 9.3.5. We

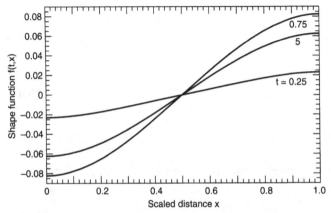

Figure 9.3.9 The shape of the free surface at selected values of scaled time for small Bond number; $\frac{\tau_s}{\tau_f} = 5$.

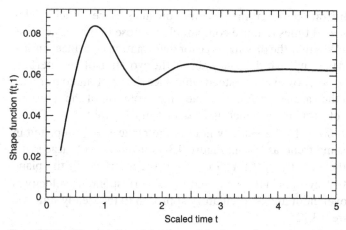

Figure 9.3.10 The evolution of the shape of the free surface at the cold wall $x = 1$ for small Bond number; $\frac{\tau_s}{\tau_f} = 5$.

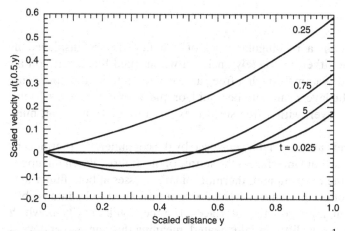

Figure 9.3.11 The velocity component in the x-direction in the plane $x = \frac{1}{2}$, plotted against y at selected values of scaled time, for small Bond number; $\frac{\tau_s}{\tau_f} = 5$.

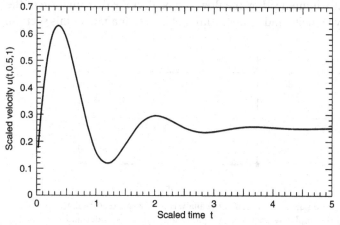

Figure 9.3.12 The evolution of the velocity component in the x-direction at the interface in the plane $x = \frac{1}{2}$ for small Bond number; $\frac{\tau_s}{\tau_f} = 5$.

have selected $\frac{\tau_s}{\tau_f} = 5$ for illustration so that the deformation and surface velocity fields achieve a steady state in scaled times that are comparable to those in the earlier case considered in Section 9.3.5. Whether the driving force for deformation is surface tension or gravity, we see that liquid buildup near the cold wall and the evolution of the interface shape are similar. When $Bo_s \ll 1$, however, the steady interface shape, attained by $t \simeq 5$, is nonlinear, in contrast to the case when $Bo_s \gg 1$. The time evolution of the interface shape is not monotonic, but oscillatory, which is evident from Figure 9.3.10. This is similar to the behavior for $Bo_s \gg 1$. The velocity field in the plane $x = \frac{1}{2}$, depicted in Figure 9.3.11, follows the same trend as that in Figure 9.3.5 for the case of large static Bond number. Similarly, the velocity in the x-direction at the interface, in the plane $x = \frac{1}{2}$, shown in Figure 9.3.12, evolves in the same way as the corresponding velocity in the earlier case. A strong overshoot and an oscillatory approach to the steady state are important features in Figure 9.3.12.

9.4 Flow in Multiple Rectangular Layers

9.4.1 Introduction

Next we consider motion in a rectangular pool of two fluid layers, displayed in Figure 9.4.1, which can be either completely enclosed within rigid boundaries on all sides or exposed to a gas or vapor through a free surface, chosen to be the upper surface. In the former case, the layers can both be liquid, or one layer can be a gas. In the latter situation, the upper layer, with a free surface exposed to a gas or vapor, must necessarily be a liquid.

When the two fluid layers are completely enclosed by the container, the thermocapillary stress at the single fluid-fluid interface is the source of the motion. When the upper liquid layer is exposed to the environment, thermocapillary stresses at both fluid interfaces contribute to the motion in the layers. The fluids are assumed to be immiscible, with no mass transfer occurring across the interfaces. The rectangular cavity, in which the fluid layers are contained, is differentially heated, meaning that the two end walls are maintained at different temperatures. We assume that the depths of both layers are small compared with their length. The equilibrium shapes of the interfaces, in the absence of motion in the liquids, are assumed to be planar. Furthermore, we assume the shapes to be only slightly influenced by the flow and neglect the deformation of the interfaces. Only the steady velocity and temperature fields are obtained in this section.

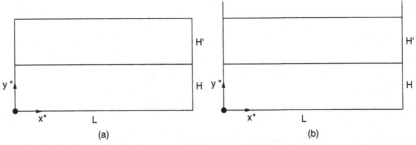

Figure 9.4.1 Geometry and the coordinate system used in the analysis of flow in a rectangular pool of two thin fluid layers: (a) completely enclosed by rigid boundaries on all sides and (b) exposed to a surrounding gas at the upper surface.

9.4.2 Motion Driven by a Single Fluid-Fluid Interface

When the fluids are completely enclosed by rigid surfaces, the thermocapillary stress at the interface between the two layers induces a flow in both the fluids. This is a generalization of the motion considered in Section 9.2. Now, transport in the fluids on both sides of the interface becomes important. The coordinates x^* and y^* are shown in Figure 9.4.1a. A prime is used to denote quantities in the upper fluid. The length of the cavity is L, and the depths of the lower and the upper layers are H and H', respectively. The scaled coordinates are defined as $x = \frac{x^*}{L}$ and $y = \frac{y^*}{H+H'}$, and the interface is located at $y = l = \frac{H}{H+H'}$. The hot wall at $x = 0$ is maintained at a temperature T_a^*, and the temperature at the cold wall at $x = 1$ is T_b^*. The velocity scale is chosen as $U_R = \frac{(-\sigma_T)\Delta T^* \epsilon}{\mu}$, where σ_T represents the rate of change of the interfacial tension with temperature at the interface between the two layers, $\Delta T^* = T_a^* - T_b^*$, μ is the viscosity of the fluid in the lower layer, and $\epsilon = \frac{(H+H')}{L} \ll 1$. Temperatures are scaled by first subtracting T_b^*, and then dividing by ΔT^*. The scaled velocity components in the x-direction in the two fluids are represented by u and u'. Likewise, the symbols T and T' represent the scaled temperature fields. The pressure in the lower layer is scaled by $\frac{\mu U_R L}{(H+H')^2}$, and that in the upper layer is scaled by $\frac{\mu' U_R L}{(H+H')^2}$. Here, μ' stands for the viscosity of the fluid in the upper layer. We define a viscosity ratio $\alpha = \frac{\mu'}{\mu}$.

The boundaries at $y = 0$ and $y = 1$ are assumed to be thermally insulated. The temperature variation in the layers is then one-dimensional at steady state, and the scaled temperature distribution is given by $T = T' = 1 - x$. The governing equations and boundary conditions for the leading order velocity fields $u(x, y)$ and $u'(x, y)$, in the limit $\epsilon \to 0$, are given below, and are similar to those used in Section 9.2.

$$\frac{dp}{dx} = \frac{\partial^2 u}{\partial y^2}, \tag{9.4.1}$$

$$\frac{dp'}{dx} = \frac{\partial^2 u'}{\partial y^2}. \tag{9.4.2}$$

We have not used a subscript 0 in the above equations to avoid clutter. At leading order, the motion in the liquid layers is parallel and is dominated by viscous forces. The velocity components only depend on the coordinate y and henceforth are designated $u(y)$ and $u'(y)$. Of course, the parallel flow solution does not apply near the end walls, for reasons already mentioned in Section 9.2. At the lower and upper rigid boundaries, the no-slip condition yields

$$u(0) = 0 \tag{9.4.3}$$

and

$$u'(1) = 0. \tag{9.4.4}$$

Because the velocity field is continuous across the interface,

$$u(l) = u'(l). \tag{9.4.5}$$

The tangential stress balance leads to

$$\frac{du}{dy}(l) - \alpha \frac{du'}{dy}(l) = 1. \tag{9.4.6}$$

Because the volumetric flow rate across the cross-section vanishes at any x in each fluid,

$$\int_0^l u(y)\,dy = \int_l^1 u'(y)\,dy = 0. \tag{9.4.7}$$

The solution can be obtained in a straightforward manner. It is useful to replace the independent variable y by a new variable y_+, where $y_+ = \frac{y}{l}$ in the lower layer, and $y_+ = \frac{1-y}{1-l}$ in the upper layer. This permits us to display the symmetry in the results for the two fields explicitly. Even though the independent variable has been changed, we have retained the same symbols u and u' for convenience in writing the solution below.

$$u = u' = \frac{l(1-l)}{2[1 + l(\alpha - 1)]}\left(\frac{3}{2}y_+^2 - y_+\right). \tag{9.4.8}$$

The scaled pressure gradient in each layer can be evaluated from the scaled velocity profile in that layer, using Equations (9.4.1) or (9.4.2) as appropriate. It is twice the coefficient of the y^2 term in the velocity profile in the layer, after it is rewritten in terms of y. The reader can see that the velocity profiles given in Equation (9.4.8) are identical to that for a single fluid layer given in Equation (9.2.15). The interfacial velocity at $y = l$ can be obtained as $u(l) = u'(l) = \frac{l(l-1)}{4(1+\alpha l-l)}$. When α is of order unity, it is seen that the motion in both fluids is reduced significantly when l is near 0 or 1. This is because of the increased resistance to flow caused by the viscous forces when one of the layers is very thin. Although the mechanism is different, a thin film of surfactant at the surface of a liquid also can retard the flow in the liquid, as one might infer from the discussion in Section 4.6.

9.4.3 Motion Driven by Two Interfaces

The analysis of the problem with two interfaces appears in Doi and Koster (1993). An example can be given from crystal growth. Molten boron oxide is used as an encapsulation layer in the growth of gallium arsenide crystals. This is necessary to prevent one of the components, which has a high vapor pressure, from evaporating and thereby changing the composition of the crystal as it grows from the melt. Of course the actual geometries are different from the idealized one used here. Nevertheless, the analysis can be used as a guide regarding the influence of the various parameters in the problem.

The geometry and the coordinate system are shown in Figure 9.4.1(b). In this problem, the upper layer of the two-layer system is a liquid, and the boundary at $y^* = H + H'$ is a free surface that is exposed to a gas. Thermocapillary stresses are exerted at both the interface between the layers and the free surface. The symbol σ'_T represents the rate of change of the interfacial tension with temperature at the free surface, and σ_T represents a similar quantity at the interface between the two layers. The ratio of these two quantities is denoted by $\delta = \frac{\sigma'_T}{\sigma_T}$. Typically, δ is greater than unity.

We can continue to use all the scalings introduced in Section 9.4.2, including the definition of the reference velocity based on σ_T. We assume heat transfer from the upper liquid to the surrounding gas to be negligible. Thus, the same one-dimensional temperature field, given in Section 9.4.2, prevails in the layers. The velocity fields satisfy Equations (9.4.1) and (9.4.2), along with the conditions in Equations (9.4.3) and (9.4.5) through (9.4.7). The boundary condition in Equation (9.4.4) is replaced by the tangential stress balance at the free surface. Neglecting the viscous stress in the gas, this yields

$$\frac{du'}{dy}(1) = \frac{\delta}{\alpha}. \tag{9.4.9}$$

The solution is

$$u = \frac{l(1-l)(2-\delta)}{(4-4l+3\alpha l)}\left(\frac{3}{2}y_+^2 - y_+\right) \tag{9.4.10}$$

and

$$u' = \frac{(1-l)}{2\alpha(4-4l+3\alpha l)}\{[6\delta(1-l)+3\alpha l(1+\delta)]y_+^2$$
$$- [8\delta(1-l)+6\alpha\delta l]y_+ + 2\delta(1-l)+\alpha l(2\delta-1)\}, \tag{9.4.11}$$

where y_+ in each layer is defined as in Section 9.4.2. The velocity profile in the lower layer is the same in both problems. The presence of the free surface alters the magnitudes of the velocities in the layers and the velocity profile in the upper layer, however. The velocity at the fluid-fluid interface is $u(l) = \frac{l(1-l)(2-\delta)}{2(4-4l+3\alpha l)}$. The velocity at the free surface is $u'(1) = \frac{(1-l)(2\delta-2\delta l+2\alpha\delta l-\alpha l)}{2\alpha(4-4l+3\alpha l)}$. As before, the scaled pressure gradients in the two layers are given by twice the coefficient of the y^2 term in each scaled velocity profile, written in terms of y.

A remarkable prediction is that the bottom layer is motionless when $\delta = 2$. The return flow in the upper layer opposes the flow that would be generated by the thermocapillary stress in the lower layer, and the two effects precisely counterbalance each other when $\delta = 2$. In this case, the flow in the upper layer assumes the form of a vortex with a clockwise circulation. When $\delta > 2$, a vortex with a counterclockwise circulation is induced in the lower layer. The free surface in the upper layer is motionless when $\delta = \frac{\alpha l}{2(1-l+\alpha l)}$. This occurs in the range $0 \le \delta \le 0.5$, depending on the values of α and l. For $\frac{\alpha l}{2(1-l+\alpha l)} \le \delta \le 2$, a vortex with a clockwise circulation occurs in the lower layer. The upper layer now has two vortices, one with a clockwise circulation near the free surface and the other with a counterclockwise circulation near the fluid-fluid interface. When $-\infty < \delta < \frac{\alpha l}{2(1-l+\alpha l)}$, the circulation everywhere in the upper layer is counterclockwise, whereas that in the lower layer is clockwise. In this discussion, we have assumed that when δ is negative, σ'_T is positive and σ_T is negative, so that the reference velocity U_R remains positive. When U_R is negative, the directions of all the circulations will be opposite to those given above.

9.5 Flow Near an End Wall in a Thin Rectangular Liquid Layer

9.5.1 Introduction

We have described the flow induced in a thin rectangular liquid layer by the thermocapillary stress at its free surface in Sections 9.1 to 9.4. In that problem, we defined an aspect ratio $\epsilon = \frac{H}{L}$ where H is the depth of the layer and L is its length. In the limit when $\epsilon \to 0$, we saw that the leading order flow is a recirculating parallel flow. The boundary conditions of no slip and no penetration were not imposed at the end walls. Therefore, the parallel flow solution is valid only in regions away from the end walls. Our objective in this section is to obtain the velocity field, subject to the conditions at the end walls. This will be done by using the method of matched asymptotic expansions. An inner solution will be determined that is matched to the parallel flow away from the end walls. This problem has been analyzed by Sen and Davis (1982), who accounted for the deformation of the free surface, and calculated the streamfunction to $O(\epsilon^2)$. At leading order, the free surface is undeformed. Here, the leading order outer and inner solutions for the streamfunction are determined. Whereas Sen and Davis computed the leading order inner solution numerically, we obtain an analytical solution. de Socio (1979) analytically solved the problem of combined thermocapillary and buoyant convection in a rectangular cavity in the Stokes limit by using the method of biorthogonal series described by Joseph (1977) and Joseph and Sturges (1978). It is not convenient to obtain the structure of the flow as a recirculating parallel flow with turning flows at the end walls from the limit $\epsilon \to 0$ in de Socio's analysis. Therefore, we have based the material in this section on the analysis provided by Balasubramaniam (1998b), who used the biorthogonal series method in a more natural way for the boundary conditions that are applicable to the present problem.

9.5.2 Analysis

We assume that the Reynolds and Marangoni numbers, defined in Section 9.2, are negligibly small. The analysis is performed in a rectangular coordinate system shown in Figure 9.2.1. The scaled coordinates are $x = \frac{x^*}{L}$ and $y = \frac{y^*}{H}$. The free surface is assumed to be undeformed and located at $y = 1$. Temperature in the layer is scaled by first subtracting T_b^*, and then dividing by $\Delta T^* = T_a^* - T_b^*$, where T_a^* and T_b^* are the temperatures at the hot and cold walls, located at $x = 0$ and 1, respectively. We assume the thermal conditions at the boundaries to be such that the scaled temperature field is one-dimensional and is given by $T(x, y) = 1 - x$. Therefore, the thermocapillary stress at the free surface is uniform. It is convenient to work with the streamfunction in this two-dimensional flow problem. The streamfunction, scaled by the reference quantity $\frac{(-\sigma_T)\Delta T^* \epsilon H}{\mu}$, is labeled ψ and is related to the velocity vector $\mathbf{v}$, scaled by $\frac{(-\sigma_T)\Delta T^* \epsilon}{\mu}$, through $\mathbf{v} = \nabla \psi \times \mathbf{k}$. Here, as usual, σ_T is the rate of change of the surface tension with temperature, and μ is the dynamic viscosity of the liquid.

The scaled streamfunction satisfies the following equation:

$$\epsilon^4 \frac{\partial^4 \psi}{\partial x^4} + 2\epsilon^2 \frac{\partial^4 \psi}{\partial x^2 \partial y^2} + \frac{\partial^4 \psi}{\partial y^4} = 0. \tag{9.5.1}$$

In the limit $\epsilon \to 0$, the governing equation and boundary conditions for the leading order outer streamfunction ψ_o can be written as follows:

$$\frac{\partial^4 \psi_o}{\partial y^4} = 0, \tag{9.5.2}$$

$$\psi_o(x, 0) = \frac{\partial \psi_o}{\partial y}(x, 0) = 0, \tag{9.5.3}$$

$$\psi_o(x, 1) = 0, \tag{9.5.4}$$

and

$$\frac{\partial^2 \psi_o}{\partial y^2}(x, 1) = 1. \tag{9.5.5}$$

We have omitted the subscript 0, corresponding to leading order, for convenience and used the subscript o, which identifies this solution as the outer solution. Equation (9.5.3) represents the kinematic and no-slip boundary conditions at the rigid bottom boundary. Equation (9.5.4) results from the kinematic condition at the free surface. Equation (9.5.5) represents the tangential stress balance at the free surface. The solution for ψ_o is

$$\psi_o(x, y) = \frac{1}{4}y^2(y - 1). \tag{9.5.6}$$

Thus, ψ_o is a function of y alone and represents the recirculating parallel flow solution, given in Equation (9.2.15). The conditions at the walls $x = 0$ and 1 cannot be imposed on ψ_o. To obtain a solution valid near these walls, an inner streamfunction ψ_i must be introduced near each end wall. Because of the symmetry in the problem, it is necessary to consider only the flow near one end wall, and we choose this as the wall at $x = 0$. Then, the inner variable is $\eta = \frac{x}{\epsilon}$. The inner streamfunction $\psi_i(\eta, y)$ must be matched, in the limit $\eta \to \infty$, with the outer streamfunction $\psi_o(x, y)$ in the limit $x \to 0$. We restrict the analysis to determining the inner solution at leading order, which satisfies the following governing equation:

$$\nabla^4 \psi_i(\eta, y) = \frac{\partial^4 \psi_i}{\partial \eta^4} + 2\frac{\partial^4 \psi_i}{\partial \eta^2 \partial y^2} + \frac{\partial^4 \psi_i}{\partial y^4} = 0. \tag{9.5.7}$$

Again, the subscript 0 has been deleted from ψ_i. Equation (9.5.7) is known as the biharmonic equation, and it is frequently encountered in the theory of elasticity. The boundary conditions on ψ_i are as follows:

$$\psi_i(0, y) = \frac{\partial \psi_i}{\partial \eta}(0, y) = 0, \tag{9.5.8}$$

$$\psi_i(\eta, 0) = \frac{\partial \psi_i}{\partial y}(\eta, 0) = 0, \tag{9.5.9}$$

$$\psi_i(\eta, 1) = 0, \tag{9.5.10}$$

$$\frac{\partial^2 \psi_i}{\partial y^2}(\eta, 1) = 1, \tag{9.5.11}$$

and

$$\psi_i(\eta \to \infty, y) \to \frac{1}{4}y^2(y - 1). \tag{9.5.12}$$

Equations (9.5.8) and (9.5.9) represent the kinematic and no-slip conditions at the left wall and the bottom boundary, respectively. Equation (9.5.10) is the kinematic condition at the free surface, and Equation (9.5.11) arises from the tangential stress balance. Equation (9.5.12) represents the matching condition between the inner and outer solutions. It is convenient to pose the inner problem slightly differently. Let

$$F(\eta, y) = \psi_i(\eta, y) - \frac{1}{4}y^2(y - 1). \tag{9.5.13}$$

$F(\eta, y)$ satisfies the biharmonic equation as well. The boundary conditions that are altered are

$$F(0, y) = -\frac{1}{4}y^2(y - 1),$$ (9.5.14)

$$\frac{\partial^2 F}{\partial y^2}(\eta, 1) = 0,$$ (9.5.15)

and

$$F(\eta \to \infty, y) \to 0.$$ (9.5.16)

The remaining homogeneous boundary conditions in Equations (9.5.8), (9.5.9), and (9.5.10) continue to apply to the function $F(\eta, y)$.

It is known that if $\phi(x, y)$ is a harmonic function satisfying Laplace's equation $\nabla^2\phi = 0$, then ϕ, $x\phi$, $y\phi$, and $(x^2 + y^2)\phi$ are solutions of the biharmonic equation. For cavities with aspect ratios of the order of unity, de Socio (1979) constructed solutions for $\psi(x, y)$ of the form $\{\phi, x\phi\}$ where ϕ is $e^{\pm ky} \cos kx$ and $e^{\pm ky} \sin kx$, and $-\epsilon \leq y \leq 0$. As noted earlier, it is not convenient to obtain the structure of the solution for a thin layer from the formulation used by de Socio. For the type of boundary conditions in the present problem, the appropriate form of the solution for $F(\eta, y)$ is $\{\phi, y\phi\}$, where ϕ is $e^{-k\eta} \cos ky$ or $e^{-k\eta} \sin ky$. We write the solution as

$$F(\eta, y) = \sum_{n=1}^{\infty} H_n \frac{e^{-k_n\eta}}{k_n^2} f_n(y),$$ (9.5.17)

where

$$f_n(y) = \sin(k_n y) - y \sin k_n \cos[k_n(1 - y)]$$ (9.5.18)

and k_n are the roots of

$$2k_n = \sin(2k_n); \quad k_n \neq 0; \quad Re[k_n] > 0.$$ (9.5.19)

The function $f_n(y)$ is known as the Papkovich–Fadle eigenfunction. It satisfies a fourth order differential equation, known as the reduced biharmonic equation, which is given below.

$$\frac{d^4 f_n}{dy^4} + 2k_n^2 \frac{d^2 f_n}{dy^2} + k_n^4 f_n = 0.$$ (9.5.20)

The boundary conditions on $f_n(y)$ are

$$f_n(0) = f_n'(0) = 0$$ (9.5.21)

and

$$f_n(1) = f_n''(1) = 0.$$ (9.5.22)

The Papkovich–Fadle eigenfunctions considered by Joseph (1977) and de Socio (1979) are slightly different and satisfy $f_n = f_n' = 0$ at both end points. These eigenfunctions are also used to describe the Moffat eddies that are encountered in many Stokes flow problems in corners. This topic is discussed in Moffat (1964), Pan and Acrivos (1967), Joseph and Sturges (1978), and Sherman (1990). It is evident that the values of k_n that satisfy Equation (9.5.19) must be complex numbers. The only real root $k = 0$ leads to a trivial solution. Roots of Equation (9.5.19) occur in all the four quadrants in the complex plane, in pairs as complex conjugates. The need to satisfy Equation (9.5.16) restricts acceptable roots to the right half-plane. The eigenvalues k_n are arranged as complex conjugates in the order of increasing magnitudes; k_1 and k_2 correspond to the first set of complex conjugates, k_3 and k_4 to the next set which is larger in magnitude, and so on. Joseph and Sturges (1978) provided an asymptotic result for k_n for large n:

$$\{k_{2n-1}, k_{2n}\} = \left[\left(n + \frac{1}{4}\right)\pi - \frac{\log[(4n+1)\pi]}{(4n+1)\pi} + o\left(\frac{\log n}{n}\right)\right]$$

$$\mp i\left[\frac{1}{2}\log[(4n+1)\pi] + \left(\frac{\log[(4n+1)\pi]}{(4n+1)\pi}\right)^2 + o\left(\frac{\log n}{n}\right)^2\right].$$

$$(9.5.23)$$

In general, H_n also is a complex number whose real and imaginary parts must be determined so that the boundary conditions on $F(0, y)$ and $\frac{\partial F}{\partial \eta}(0, y)$ are satisfied. Following Joseph (1977), we obtain a biorthogonality condition satisfied by the eigenfunctions $f_n(y)$, which must be used to determine H_n. Let

$$p_{n_1}(y) = f_n(y) \qquad (9.5.24)$$

and

$$p_{n_2}(y) = \frac{f_n''(y)}{k_n^2} = \frac{p_{n_1}''(y)}{k_n^2}. \qquad (9.5.25)$$

The adjoint system is defined as

$$q_{n_2}(y) = p_{n_1}(y) \qquad (9.5.26)$$

and

$$q_{n_1}''(y) = -k_n^2 q_{n_2}(y). \qquad (9.5.27)$$

Equation (9.5.20) may be written in vector form as

$$\frac{d^2\mathbf{p}_n}{dy^2} + k_n^2 \mathbf{A} \bullet \mathbf{p}_n = 0 \qquad (9.5.28)$$

and

$$\frac{d^2\mathbf{q}_n}{dy^2} + k_n^2 \mathbf{A}^T \bullet \mathbf{q}_n = 0, \qquad (9.5.29)$$

where

$$\mathbf{p}_n = \begin{pmatrix} p_{n_1}(y) \\ p_{n_2}(y) \end{pmatrix}; \quad \mathbf{q}_n = \begin{pmatrix} q_{n_1}(y) \\ q_{n_2}(y) \end{pmatrix}; \quad \mathbf{A} = \begin{pmatrix} 0 & -1 \\ 1 & 2 \end{pmatrix}. \qquad (9.5.30)$$

The following biorthogonality relation can be derived in a straightforward manner:

$$\int_0^1 \mathbf{q}_m \bullet \mathbf{A} \bullet \mathbf{p}_n \, dy = 0, \quad m \neq n. \qquad (9.5.31)$$

Note that the biorthogonality relation is obtained for boundary conditions of the form given in Equations (9.5.21) and (9.5.22), as well as those considered by Joseph (1977).

When $m = n$, it can be verified by direct integration that

$$\int_0^1 \mathbf{q}_n \bullet \mathbf{A} \bullet \mathbf{p}_n \, dy = -2 \tan^2 k_n. \tag{9.5.32}$$

We are now in a position to determine H_n. The boundary conditions to be satisfied at $\eta = 0$ are

$$F(0, y) = \sum_{n=1}^\infty \frac{H_n}{k_n^2} p_{n_1}(y) = -\frac{1}{4} y^2(y-1) \tag{9.5.33}$$

and

$$\frac{\partial F}{\partial \eta}(0, y) = -\sum_{n=1}^\infty \frac{H_n}{k_n} p_{n_1}(y) = 0. \tag{9.5.34}$$

Differentiating Equation (9.5.33) twice with respect to y, and using the relation $p''_{n_1} = k_n^2 p_{n_2}$, the above equations may be written as

$$\begin{pmatrix} 0 \\ -\frac{1}{2}(3y-1) \end{pmatrix} = \sum_{n=1}^\infty \begin{pmatrix} \frac{H_n}{k_n} p_{n_1}(y) \\ H_n p_{n_2}(y) \end{pmatrix} = \sum_{n=1}^\infty H_n \begin{pmatrix} p_{n_1}(y) \\ p_{n_2}(y) \end{pmatrix}$$

$$+ \sum_{n=1}^\infty \left(-H_n + \frac{H_n}{k_n} \right) \begin{pmatrix} p_{n_1}(y) \\ 0 \end{pmatrix}. \tag{9.5.35}$$

The operator $\int_0^1 \begin{pmatrix} q_{m_1}(y) \\ q_{m_2}(y) \end{pmatrix} \begin{pmatrix} 0 & -1 \\ 1 & 2 \end{pmatrix} (\,\cdot\,) \, dy$ is applied to both sides. This yields

$$\int_0^1 (3y-1) \left(\frac{1}{2} q_{m_1}(y) - q_{m_2}(y) \right) dy$$

$$= -2 H_m \tan^2 k_m + \sum_{n=1}^\infty H_n \left(\frac{1}{k_n} - 1 \right) \int_0^1 q_{m_2}(y) p_{n_1}(y) \, dy, \tag{9.5.36}$$

which can be simplified to

$$-\sec k_m \tan k_m = -2 H_m \tan^2 k_m + \sum_{n=1}^\infty H_n \left(\frac{1}{k_n} - 1 \right) B_{nm}, \tag{9.5.37}$$

where

$$B_{nm} = \frac{1}{2(k_m - k_n)^2} [(\sin k_n - \sin k_m)^2 + 2 \sin k_m k_n]$$

$$- \frac{1}{2(k_m + k_n)^2} [(\sin k_n + \sin k_m)^2 - 2 \sin k_m k_n]$$

$$- \frac{\sin k_m \sin k_n \sin(k_m + k_n)}{(k_m + k_n)^3} - \frac{\sin k_m \sin k_n \sin(k_m - k_n)}{(k_m - k_n)^3}. \tag{9.5.38}$$

When $m = n$, B_{nm} reduces to

$$B_{mm} = \frac{1}{2} - \frac{1}{3} \sin^2 k_m - \frac{\sin^2 k_m}{2 k_m^2}. \tag{9.5.39}$$

9.5.3 Results

Balasubramaniam (1998b) determined the unknown coefficients H_m numerically from Equation (9.5.37) after truncation of the infinite series. Sixty terms were retained in the series, corresponding to thirty pairs of values for k that are complex conjugates satisfying Equation (9.5.19). He verified the convergence of the series solution to the boundary conditions given in Equations (9.5.33) and (9.5.34). The deviations were found to be of the order 10^{-6} and 10^{-3}, respectively. In Table 9.5.1, we provide the values of k_m and H_m for $1 \leq m \leq 59$. Only the results for odd values of m are included. The companion set of (k_m, H_m), for even values of m, is obtained simply as the complex conjugate of (k_{m-1}, H_{m-1}). The exact values of k_m are in good agreement with those predicted by the asymptotic result in Equation (9.5.23) for all values of m. Even for $m = 1$, the difference is only 0.08% in the real part and 1.7% in the imaginary part of k; when $m = 9$, the

Table 9.5.1 The Eigenvalues k_m and the Coefficients H_m for $1 \leq m \leq 59$ for Odd Values of m; the Companion Set of (k_m, H_m) for Even Values of m is Obtained as the Complex Conjugate of (k_{m-1}, H_{m-1}).

m	k_m	H_m
1	$3.74884 - 1.38434i$	$-0.40335 - 0.26681i$
3	$6.94998 - 1.67610i$	$0.30755 + 0.27779i$
5	$10.1193 - 1.85838i$	$-0.22297 - 0.21431i$
7	$13.2773 - 1.99157i$	$0.20344 + 0.19328i$
9	$16.4299 - 2.09663i$	$-0.17534 - 0.17325i$
11	$19.5794 - 2.18340i$	$0.16236 + 0.15930i$
13	$22.7270 - 2.25732i$	$-0.14967 - 0.14757i$
15	$25.8734 - 2.32171i$	$0.13896 + 0.14007i$
17	$29.0188 - 2.37876i$	$-0.13305 - 0.12899i$
19	$32.1636 - 2.42996i$	$0.12337 + 0.12852i$
21	$35.3079 - 2.47640i$	$-0.12090 - 0.11336i$
23	$38.4518 - 2.51890i$	$0.11257 + 0.12252i$
25	$41.5954 - 2.55807i$	$-0.11059 - 0.09827i$
27	$44.7387 - 2.59439i$	$0.10598 + 0.12142i$
29	$47.8819 - 2.62825i$	$-0.09970 - 0.08218i$
31	$51.0248 - 2.65997i$	$0.10468 + 0.12488i$
33	$54.1677 - 2.68979i$	$-0.08508 - 0.06460i$
35	$57.3104 - 2.71794i$	$0.11154 + 0.13155i$
37	$60.4530 - 2.74459i$	$-0.06214 - 0.04760i$
39	$63.5955 - 2.76988i$	$0.13121 + 0.13610i$
41	$66.7379 - 2.79396i$	$-0.02561 - 0.04043i$
43	$69.8803 - 2.81694i$	$0.16667 + 0.12172i$
45	$73.0226 - 2.83890i$	$0.02121 - 0.07021i$
47	$76.1649 - 2.85994i$	$0.19694 + 0.04732i$
49	$79.3071 - 2.88014i$	$0.01869 - 0.18791i$
51	$82.4492 - 2.89954i$	$0.08806 - 0.11799i$
53	$85.5914 - 2.91823i$	$-0.25887 - 0.29340i$
55	$88.7335 - 2.93624i$	$-0.27367 + 0.09574i$
57	$91.8755 - 2.95362i$	$0.11478 + 0.23571i$
59	$95.0176 - 2.97042i$	$0.03644 - 0.05029i$

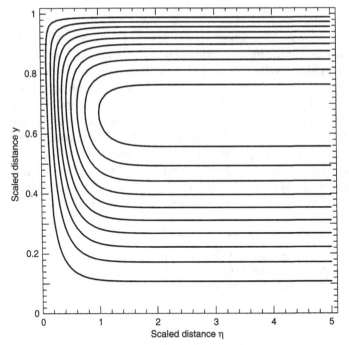

Figure 9.5.1 Streamlines near the end wall. $\psi_{min} = -0.037$ at $\eta = 5$, $y = \frac{2}{3}$; $\psi_{max} = 0$ at the lower, upper, and left boundaries; $\Delta\psi = 0.003364$. Distances in the y-direction are magnified by a factor of 5 for clearity.

real part is accurate to four significant figures, and the deviation in the imaginary part is 0.1%.

The streamlines corresponding to the flow near the end wall are displayed in Figure 9.5.1. The coordinates used are (η, y) so that both the distance variables are scaled by H, but we have magnified the scale for the y-axis by a factor of 5 for clarity. We see that when $\eta \approx 1.5$, which corresponds to a physical distance from the end wall that is 1.5 times the depth of the layer, the flow becomes predominantly parallel. This is in agreement with the results of Sen and Davis (1982). The shape of the streamlines near the corner is slightly different from that obtained by Sen and Davis. The present results are similar to the results of de Socio in this respect. Results for the flow near the wall at $x = 1$ can be obtained immediately by symmetry. It is then possible to construct a composite streamfunction, which is uniformly valid everywhere in the liquid layer, if desired. Also, the velocity field can be determined by using the relationship $\mathbf{v} = \nabla\psi \times \mathbf{k}$, which was given earlier.

9.6 Effects of Inertia and Convective Energy Transport on the Flow in a Deep Liquid Layer

9.6.1 Introduction and Governing Equations

In the past four sections, we analyzed problems involving the flow driven by a thermocapillary stress in a thin liquid layer. Neither the influence of inertia nor that of convective transport of energy was included in those problems. In the present section, we consider a liquid mass with a free surface, which is infinite in depth as well as in its lateral extent. Temperature gradients are introduced at the free surface by imposing a nonuniform heat flux. In this situation, both inertia and convective transport of energy play important roles. Boundary layers in both the velocity and temperature fields will form near the free surface, and their thicknesses will depend on the magnitudes of a suitably defined Reynolds number and the Prandtl number. The present model can be considered a reasonable approximation for a contained body of liquid, away from the walls of the container, provided the momentum and thermal boundary layers are thin compared with the depth of the liquid. We shall establish the scales for the thicknesses of these boundary layers during the course of the analysis. All the variation in the velocity and temperature fields will occur within the momentum and thermal boundary layers respectively. The velocity component in the direction parallel to the free surface will be zero outside the momentum boundary layer. The other velocity component is not zero, however. Far from the surface, there is a uniform flow of liquid toward the surface to replenish liquid that is drawn away by the surface tension gradient. In the unbounded body of fluid considered here, there will be no return flow. When containing boundaries are present, however, there will be such a recirculation. Therefore, the analysis is applicable only to situations where such return flow is weak.

We demonstrate here that an exact solution can be obtained when the heat flux quadratically depends on distance along the free surface. This holds for both the two-dimensional and axisymmetric configurations at steady state. We consider the two-dimensional case here. The analysis was first presented by Chan, Chen, and Mazumder (1988) and independently by Sanochkin (1989). It was later extended to include flows caused by rotation and buoyant convection in the axisymmetric case by Vedha-Nayagam and Balasubramaniam (1993). These authors showed that an exact solution also is obtained when the liquid is subjected simultaneously to a thermocapillary stress, rotation, and buoyant convection. The analysis provided below closely follows the development by Sanochkin (1989). The effects of buoyant convection and rotation are not considered in the analysis.

Consider a deep pool of liquid occupying the space $y^* > 0$ in rectangular coordinates (x^*, y^*, z^*). The free surface is located at $y^* = 0$ and is assumed to be undeformed. It can be shown that the solution we obtain is indeed consistent with an undeformed free surface. The liquid is heated by an imposed heat flux at the free surface of the form $Q(x^*, z^*) = Q_0 - \frac{Q_1}{L^2} x^{*2}$ in terms of physical quantities. As noted earlier, we restrict the analysis to a quadratic variation of the imposed surface heat flux. When this heat flux is not quadratic, analytical solution is not possible, and the problem must be solved numerically. The length scale L is associated with the variation of the heat flux on the free surface. It is used to define scaled variables $x = \frac{x^*}{L}$, and $y = \frac{y^*}{L}$. A typical source of heat might be a laser beam with an intensity that is uniform in the z^* direction, or a radiant heater such as a long hot filament aligned in the same direction. Although the actual heat flux profile from such sources can be complex (as for example, a Gaussian distribution), it is assumed that the profile can be fitted by a quadratic distribution by appropriate selection of Q_0, Q_1, and L such that x can be assumed to extend to infinity. The heat flux and the resulting flow are assumed to be independent of z^*. Scaling the heat flux by a reference value Q_R, the dimensionless heat flux $q = \frac{Q}{Q_R}$ may be written as $q(x) = q_0 - q_1 x^2$. Velocities are scaled by a reference velocity U_R that will be determined from a scaling analysis. The symbols u and v are used to represent the scaled velocity components in the x- and y-directions, respectively. The symbol p designates scaled hydrodynamic pressure, as usual. Pressure is measured relative to its value far away from the free surface, as $y \to \infty$ and is scaled by ρU_R^2. Temperature is nondimensionalized by subtracting the physical temperature as $y \to \infty$ and then dividing by the quantity $\frac{Q_R L}{k}$, where k is the thermal conductivity of the liquid. We use the symbol T to designate the scaled temperature. The governing equations are

$$\frac{\partial u}{\partial x} + \frac{\partial v}{\partial y} = 0, \tag{9.6.1}$$

$$u\frac{\partial u}{\partial x} + v\frac{\partial u}{\partial y} = -\frac{\partial p}{\partial x} + \frac{1}{Re}\nabla^2 u, \tag{9.6.2}$$

$$u\frac{\partial v}{\partial x} + v\frac{\partial v}{\partial y} = -\frac{\partial p}{\partial y} + \frac{1}{Re}\nabla^2 v, \tag{9.6.3}$$

and

$$u\frac{\partial T}{\partial x} + v\frac{\partial T}{\partial y} = \frac{1}{Ma}\nabla^2 T. \tag{9.6.4}$$

Here, $Re = \frac{U_R L}{\nu}$ is the Reynolds number, $Ma = Pr\,Re$ is the Marangoni number, and $Pr = \frac{\nu}{\kappa}$ is the Prandtl number. As usual, the symbols ν and κ represent the kinematic viscosity and the thermal diffusivity of the liquid, respectively. The boundary conditions are

$$\frac{\partial u}{\partial y}(x, 0) = \frac{(-\sigma_T)Q_R L}{\mu k U_R}\frac{\partial T}{\partial x}(x, 0), \tag{9.6.5}$$

$$v(x, 0) = 0, \tag{9.6.6}$$

$$-\frac{\partial T}{\partial y}(x, 0) = q(x) = q_0 - q_1 x^2, \tag{9.6.7}$$

and

$$u \to 0, \quad p \to 0, \quad T \to 0 \quad \text{as} \quad y \to \infty. \tag{9.6.8}$$

Equations (9.6.5) and (9.6.6) represent the tangential stress balance and the kinematic condition at the free surface, respectively. Equation (9.6.7) describes the imposed heat flux distribution at the surface, and Equation (9.6.8) represents conditions in the undisturbed fluid far from the free surface. In Equation (9.6.5), σ_T represents the rate of change of surface tension with temperature, and μ is the viscosity of the liquid.

Note that a boundary condition has not been specified on v as $y \to \infty$, even though the governing equations contain the term $\nabla^2 v$. For a liquid of infinite extent in x and y, imposing a condition on v as $y \to \infty$ leads to overspecification of the problem. As mentioned earlier, the flow in the boundary layer tangential to the free surface is replenished by an axial flow far away from the free surface. Thus, v is nonzero as $y \to \infty$ and is determined as a part of the solution. A similar situation is encountered in other fluid mechanical problems in a liquid of infinite extent, such as the flow due to a rotating disk and the flow in the vicinity of a stagnation point.

9.6.2 Scaling Analysis

We now perform a scaling analysis to determine the unknown reference velocity and the dependence of the boundary layer thicknesses on the parameters. We scale the boundary layer thicknesses using the reference length L. Let δ and δ_T denote the scaled thicknesses of the momentum and thermal boundary layers, respectively, and ΔT^* denote the reference value for temperature differences at the free surface. The unknown scales U_R, δ, δ_T, and ΔT^* are obtained by using the following balances:

(1) the tangential stress at the free surface balances the thermocapillary stress,
(2) viscous forces and inertia must be of the same order of magnitude in the momentum boundary layer,
(3) the temperature gradient in the thermal boundary layer is proportional to the heat flux at the free surface, and
(4) the ratio of the thickness of the momentum boundary layer to that of the thermal boundary layer is a function of the Prandtl number. As is typical for forced convection heat transfer, this ratio is proportional to Pr for low values of Pr, and it is proportional to $\sqrt{Pr}$ for large values of Pr. These order estimates of the ratios are subject to confirmation from the solution at a later stage.

The above balances may be written symbolically as follows:

$$\frac{1}{\delta} \sim \frac{(-\sigma_T)Q_R L}{\mu k U_R} \Delta T^*; \quad \delta \sim \frac{1}{\sqrt{Re}}; \quad \Delta T^* \sim q_1 \delta_T.$$

$$\frac{\delta}{\delta_T} \sim Pr, \quad Pr \ll 1; \quad \frac{\delta}{\delta_T} \sim \sqrt{Pr}, \quad Pr \gg 1.$$

(9.6.9)

From Equation (9.6.9), we can establish the velocity scale U_R as

$$U_R = Pr^{-\frac{1}{2}} \left(\frac{(-\sigma_T)q_1 Q_R}{\rho k} \right)^{\frac{1}{2}}, \quad Pr \ll 1;$$

$$U_R = Pr^{-\frac{1}{4}} \left(\frac{(-\sigma_T)q_1 Q_R}{\rho k} \right)^{\frac{1}{2}}, \quad Pr \gg 1.$$

(9.6.10)

Rivas and Ostrach (1992) have obtained the above scaling in the case when $Pr \ll 1$. Because the scaling analysis shows that for intermediate values of the Prandtl number, the reference velocity can depend on it in a complex manner, we simply choose the following reference velocity omitting the dependence on the Prandtl number:

$$U_R = \left(\frac{2(-\sigma_T)q_1 Q_R}{\rho k} \right)^{\frac{1}{2}}.$$

(9.6.11)

We shall find that the scaled velocity in our results depends on the Prandtl number through a multiplicative constant that appears naturally in the analysis. A multiplicative factor $\sqrt{2}$ has been introduced in defining the reference velocity for convenience in the analysis. The result for the reference velocity in Equation (9.6.11) is different from that obtained by Ostrach (1979, 1982) for a differentially heated liquid layer at large Reynolds number. The difference arises because, in the present problem, the reference value for the temperature variation along the free surface, which is responsible for the flow, is in turn influenced by the velocity field.

9.6.3 Solution for the Scaled Streamfunction and Temperature Fields

In the general case, analytical solution is not possible as noted in the introduction, and numerical methods must be used. For a quadratic variation of the heat flux on the free

surface, however, the governing equations admit an exact solution. First, we transform
the variables as follows:

$$\eta = Re^{\frac{1}{2}} y, \quad u = x F'(\eta), \quad v = -\frac{1}{Re^{\frac{1}{2}}} F(\eta),$$

$$p = \frac{1}{Re} K(\eta), \quad T = \frac{1}{Re^{\frac{1}{2}}} [q_0 G_0(\eta) - q_1 x^2 G_1(\eta)]$$

(9.6.12)

The variable η is independent of x. This implies that the thickness of the momentum
boundary layer does not depend on x. This also is the case in the analogous problems
of stagnation point flow and flow induced by a rotating disk. The function $x F(\eta)$ is
the scaled streamfunction for this flow. The function $G_0(\eta)$ is related to the scaled
temperature distribution in the liquid along the plane of symmetry $x = 0$. At any location
x, the function $x^2 G_1(\eta)$ is a measure of the temperature of the liquid relative to that at
the plane of symmetry. The transformed equations and boundary conditions are written
below.

$$F''' + F F'' - F'^2 = 0,$$

(9.6.13)

$$K' = -(F'' + F F'),$$

(9.6.14)

$$\frac{1}{Pr} G_1'' + F G_1' - 2 F' G_1 = 0,$$

(9.6.15)

$$\frac{1}{Pr} G_0'' + F G_0' = \frac{2}{Ma} \frac{q_1}{q_0} G_1,$$

(9.6.16)

$$F(0) = 0,$$

(9.6.17)

$$G_0'(0) = G_1'(0) = -1,$$

(9.6.18)

$$F''(0) = -G_1(0),$$

(9.6.19)

$$F', G_0, G_1, K \to 0 \quad \text{as} \quad \eta \to \infty.$$

(9.6.20)

The solutions for the functions $F(\eta)$ and $K(\eta)$ are

$$F = \chi(1 - e^{-\chi\eta})$$

(9.6.21)

and

$$K = -\frac{\chi^2}{2} e^{-2\chi\eta},$$

(9.6.22)

where

$$\chi = \chi(Pr) = [G_1(0)]^{\frac{1}{3}}.$$

(9.6.23)

We see that the velocity and pressure fields are now known to within an unknown
constant χ. This constant determines the thickness of the momentum boundary layer
and its dependence on the Prandtl number. It can be obtained once the solution for $G_1(\eta)$
is known. We follow Sanochkin (1989) and construct the following series solution:

$$G_1 = B e^{-\chi Pr \eta} \left(1 + \sum_{n=1}^{\infty} C_n e^{-\chi\eta}\right),$$

(9.6.24)

where the constant B can be written as

$$B = \left[\chi \left(Pr + \sum_{n=1}^{\infty} (Pr + n)C_n \right) \right]^{-1}$$

(9.6.25)

and the constants C_n are obtained as follows:

$$C_n = (-1)^n \frac{Pr^n}{n!} \frac{(Pr - 2)(Pr - 1) \cdots (Pr + n - 3)}{(Pr + 1)(Pr + 2) \cdots (Pr + n)}.$$

(9.6.26)

This permits us to determine χ as

$$\chi = \left(\frac{1 + \sum_{n=1}^{\infty} C_n}{Pr + \sum_{n=1}^{\infty} (Pr + n)C_n} \right)^{1/4}.$$

(9.6.27)

In the limit $Pr \to 0$, it can be shown that $\chi \to (3Pr)^{-1/4}$. Therefore, in this limit, the momentum boundary layer thickness scales as $Pr^{1/4}$. From Equations (9.6.21) and (9.6.24), we see that the ratio of the thickness of the momentum boundary layer to that of the thermal boundary layer scales as Pr in this limit, as anticipated in Equation (9.6.9).

When $Pr \gg 1$, the thermal boundary layer is much thinner than the momentum boundary layer. At leading order, the streamfunction field within the thermal boundary layer is linear in η. It can be obtained by expanding the right side of Equation (9.6.21) in a Taylor series about $\eta = 0$. Introducing the scaled thermal boundary layer coordinate $\zeta = \chi \sqrt{Pr}\, \eta$, Equation (9.6.15) may be written as

$$G_1'' + \zeta G_1' - 2G_1 = 0.$$

(9.6.28)

In the above equation, for economy in notation, we have retained the same symbol G_1 to designate the same function when expressed in terms of ζ, and the primes refer to differentiation with respect to ζ. The solution of Equation (9.6.28) that satisfies the boundary conditions on G_1, given in Equations (9.6.18) and (9.6.20), is

$$G_1(\zeta) = \frac{1}{\chi \sqrt{Pr}} (1 + \zeta^2) \int_{\zeta}^{\infty} \frac{\exp(-\tilde{\zeta}^2/2)}{(1 + \tilde{\zeta}^2)^2} \, d\tilde{\zeta}.$$

(9.6.29)

Setting $\zeta = 0$ in Equation (9.6.29), and using Equation (9.6.23), we obtain $\chi^4 = \frac{0.625}{\sqrt{Pr}}$. Therefore, the momentum boundary layer thickness scales as $Pr^{1/8}$ when $Pr \to \infty$. The ratio of the thickness of the momentum boundary layer to that of the thermal boundary layer is proportional to $\sqrt{Pr}$ in this limit, confirming the order estimate made in Equation (9.6.9).

The result for the function $G_0(\eta)$ can be expressed as a homogeneous solution that involves the incomplete gamma function and a particular solution that accommodates the inhomogeneity in Equation (9.6.16). When the Marangoni number is large, the inhomogeneity can be neglected. The homogeneous solution is

$$G_0 = \frac{1}{\chi} e^{Pr(1 - \log Pr)} \gamma(Pr, Pr e^{-\chi \eta}),$$

(9.6.30)

where

$$\gamma(a, z) = \int_0^z e^{-t} t^{a-1} \, dt$$

(9.6.31)

is the incomplete gamma function, the properties of which can be found in Abramowitz and Stegun (1965). For $Pr \gg 1$, in a manner analogous to that described above in the case of G_1, G_0 may be obtained as

$$G_0(\zeta) = \frac{1}{\chi}\sqrt{\frac{\pi}{2Pr}}\,\mathrm{erfc}\left(\frac{\zeta}{\sqrt{2}}\right),$$
(9.6.32)

where, once again, we have retained the same symbol G_0 for the function of ζ. Now, we can infer the asymptotic dependence of $G_0(0)$ on the Prandtl number when Pr is either very small or very large. From Equation (9.6.30), we find that $G_0(0) = 1.316\,Pr^{-3/4}$ when $Pr \ll 1$, and from Equation (9.6.32), it can be shown that $G_0(0) = 1.410\,Pr^{-3/8}$ when $Pr \gg 1$.

9.6.4 Results

Now, we display representative results for values of $Pr = 0.1, 1$, and 10. In Figure 9.6.1, we have plotted the function $F(\eta)$ from Equation (9.6.21). Note that $xF(\eta)$ is the scaled streamfunction and that the scaled velocity component parallel to the free surface is equal to $xF'(\eta)$, and the component normal to the free surface is proportional to $F(\eta)$. Figure 9.6.2 is a corresponding plot of the function $K(\eta)$, from Equation (9.6.22), which is proportional to the scaled pressure. We see that the thickness of the momentum boundary layer increases weakly with increasing values of the Prandtl number. In the range of Pr values covered in the figures, the edge of the boundary layer lies between $\eta = 3$ and 6. Figure 9.6.1 shows that the magnitude of the scaled velocity perpendicular to the free surface decreases with increasing values of the Prandtl number. Because this flow replenishes the fluid drawn away by the thermocapillary stress, it follows that the velocity in the direction parallel to the free surface also must decrease with increasing values of the Prandtl number. This behavior is consistent with the scalings obtained in Equation (9.6.10).

Figures 9.6.3 and 9.6.4 show the function G_0, which is proportional to the scaled temperature distribution on the plane of symmetry, plotted against η. In Figure 9.6.3, results calculated from Equation (9.6.30) are displayed for values of $Pr = 0.1, 1$, and 10, and the prediction from the asymptotic solution for $Pr \gg 1$ from Equation (9.6.32)

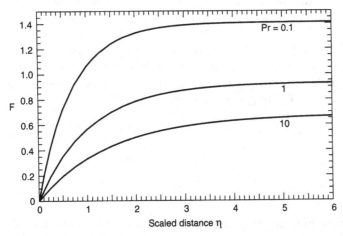

Figure 9.6.1 The function $F(\eta)$ for selected values of the Prandtl number.

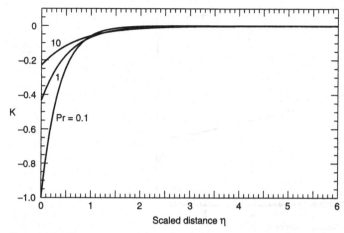

Figure 9.6.2 The function $K(\eta)$ for selected values of the Prandtl number.

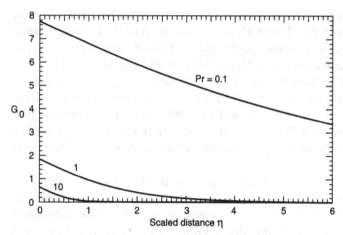

Figure 9.6.3 The function $G_0(\eta)$ for selected values of the Prandtl number.

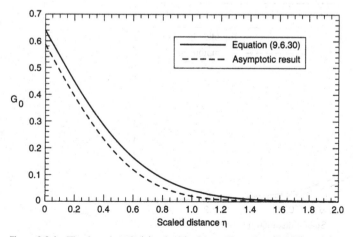

Figure 9.6.4 The function $G_0(\eta)$ calculated from Equation (9.6.30) and compared with the asymptotic result from Equation (9.6.32), for $Pr = 10$.

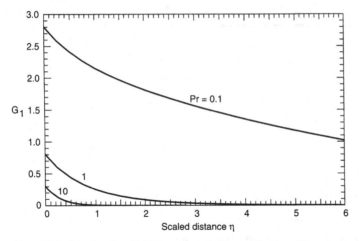

Figure 9.6.5 The function $G_1(\eta)$ for selected values of the Prandtl number.

is compared with the result from Equation (9.6.30) for $Pr = 10$ in Figure 9.6.4. From Figure 9.6.3, we see that both the magnitude of G_0 and the thickness of the thermal boundary layer are strongly dependent on the value of the Prandtl number, especially when the Prandtl number is small. Figure 9.6.4 shows that the asymptotic result is close to that from Equation (9.6.30) at $Pr = 10$, illustrating its utility. Similar drawings are provided in Figures 9.6.5 and 9.6.6 for the function G_1. Results calculated from Equation (9.6.24) are displayed in Figure 9.6.5, and the asymptotic prediction for $Pr \gg 1$ from Equation (9.6.29) is compared with the result from Equation (9.6.24) in Figure 9.6.6. The behavior of the results for G_1 parallels that of the results for G_0.

An analogous solution can be obtained for thermocapillary flow in a deep liquid layer when the heat flux at the surface is symmetric about an axis normal to the free surface and varies quadratically with radial location from this axis. Unlike the case of the two-dimensional flow presented here, the solution for the streamfunction is not

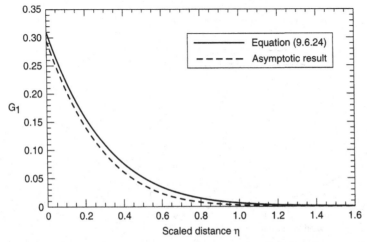

Figure 9.6.6 The function $G_1(\eta)$ calculated from Equation (9.6.24) and compared with the asymptotic result from Equation (9.6.29) for $Pr = 10$.

simple and must be constructed in a series of exponential functions. Sanochkin (1989) has discussed the solution for the axisymmetric situation in detail. We note that the qualitative features of the velocity and temperature fields are virtually the same as in the two-dimensional case considered here.

A problem closely related to the one analyzed in this section is that of the flow driven by a thermocapillary stress at the free surface of a liquid contained in a differentially heated cavity of finite depth when the Reynolds and Marangoni numbers are large. Ostrach (1982) presented a scaling analysis of this problem and showed that the characteristic velocity in the liquid scales as $[\frac{(\sigma_T)^2 \Delta T^{*2}}{\rho \mu L}]^{\frac{1}{3}}$ when the Reynolds number is large. Here, ΔT^* is the temperature difference imposed across the cavity, and L is its length. Zebib, Homsy, and Meiburg (1985) and Carpenter and Homsy (1990) have performed numerical calculations to obtain the two-dimensional velocity and temperature fields in a liquid with a free surface, contained in an enclosure of square cross-section that is infinitely long in the third dimension. Their results confirm the scaling obtained by Ostrach (1982) when the Prandtl number is of order unity. Perhaps the most important consequence of increasing the Prandtl number above unity, when the Reynolds number is large, is to weaken the temperature gradient over most of the free surface, away from the end walls, as confirmed by Carpenter and Homsy (1990). For Prandtl numbers larger than or equal to 10, these authors found a nonmonotonic dependence of the scaled free surface velocity at the midpoint between the end walls on the Reynolds number.

9.7 Flow in Cylindrical Layers and Float Zones

9.7.1 Introduction

Next, we consider thermocapillary convection in a liquid in the cylindrical geometry. Broadly speaking, there are two possibilities for the static shape of the free surface. In the first case, the free surface is nominally flat. An example is a pool of liquid in a cylindrical container. Such a pool is encountered in the Czochralski technique for growing crystals. In this method, crystals are formed by touching the free surface of the cylindrical pool with a seed crystal, often with rotation of both the crystal and the crucible, in a thermally controlled environment. The crystal is pulled out gradually as it grows. Alternatively, the free surface can be nominally cylindrical in shape. This occurs when the gap between two cylindrical rods, separated coaxially, is bridged by a column of liquid. The liquid region is termed a *float zone* in crystal growth applications. A float zone is sometimes used in the growth and purification of semiconductor crystals. The zone is formed by traversing the cylindrical crystal to be purified through a heated region. A cylindrical free surface is also encountered when a thin layer of liquid wets the interior or the exterior of a cylinder. This might occur, for example, in the process of condensation of a saturated vapor on a cylindrical surface.

In this section, we first analyze the thermocapillary flow induced by temperature variations at the free surface of a shallow cylindrical pool. This is followed by a brief description of the literature on flow in liquid bridges. We do not present an analysis of thermocapillary flow in this configuration because the problem must be solved numerically. The problem of flow in a thin layer of liquid formed by condensation on a cylindrical surface will not be considered. The interested reader will find a treatment of that problem in Jacobi and Goldschmidt (1989).

9.7.2 Flow in Cylindrical Layers

Consider a thin layer of liquid of uniform depth H and a planar free surface, present in a cylindrical container of radius R. We assume that $\epsilon = \frac{H}{R}$ is small compared with unity. The free surface is subjected to a steady, nonuniform heat flux that leads to temperature variations on it. In the case of Czochralski crystal growth, the heat flux arises from a heat loss to the crystal and the surrounding medium. The problem is similar to that considered in Section 9.2 for a rectangular pool. The difference is that the heat flux is assumed to be symmetric about the axis of the cylinder. This leads to a steady axisymmetric flow field in the liquid. We present a simplified version of the analysis performed by Balasubramaniam and Ostrach (1984), who also included the effect of buoyant convection. Here, we neglect the buoyant contribution to motion in the liquid. Only a leading order outer solution, valid away from the cylindrical boundary of the container, is given here. We briefly mention how an inner solution, valid near the wall, can be obtained using the method employed in Section 9.5.

Cylindrical polar coordinates are used in the analysis. Distances in the radial direction are scaled by the radius of the container and those in the axial direction by the depth of the liquid. The scaled radial and axial coordinates are designated by r and z, respectively. It is assumed that the rigid bottom boundary at $z = 0$ is thermally insulated. The rigid boundary at $r = 1$ is maintained at a uniform temperature T_c^*. Temperatures in the liquid are nondimensionalized by subtracting T_c^* from the physical temperature and dividing by the reference temperature difference $\Delta T^* = \frac{Q_R H}{k}$, where Q_R is a reference value for the heat flux at the free surface, and k is the thermal conductivity of the liquid. The scaled temperature field is designated $T(r, z)$. We neglect deformation of the free surface, assuming the Capillary number to be negligibly small. It would be possible to calculate a first approximation for the deformation for $Ca \ll 1$ using the same techniques that were employed in Section 9.2. The heat flux, scaled by Q_R, is labeled q. In general q can be represented in a Fourier–Bessel series as

$$q(r) = \sum_{n=1}^{\infty} C_n J_0 (\alpha_n r), \tag{9.7.1}$$

where α_n are the zeroes of $J_0(x)$, which is the Bessel function of the first kind of order zero. We use the symbol $\psi(r, z)$ to designate the Stokes streamfunction, scaled by $\frac{(-\sigma_T) \Delta T^* H^2}{\mu}$, where σ_T is the rate of change of surface tension with temperature and μ is the dynamic viscosity of the liquid. The velocity vector is related to the Stokes streamfunction through Equation (3.1.3).

For small ϵ, the scaled streamfunction and temperature fields are expanded as follows:

$$\psi(r, z) = \frac{1}{\epsilon^2} \psi_0(r, z) + \cdots \tag{9.7.2}$$

and

$$T(r, z) = \frac{1}{\epsilon^2} T_0(r) + T_1(r, z) \cdots . \tag{9.7.3}$$

Substitution of the above expansions into the scaled Navier–Stokes and energy equations yields the following governing equations for ψ_0 and T_1.

$$\frac{\partial^4 \psi_0}{\partial z^4} = 0 \tag{9.7.4}$$

$$\frac{1}{r} \frac{d}{dr} \left(r \frac{dT_0}{dr} \right) + \frac{\partial^2 T_1}{\partial z^2} = 0 \tag{9.7.5}$$

The governing equation for T_0 will be obtained shortly, using the technique introduced in Section 9.2 in the thermal problem of Type II. The boundary conditions satisfied by ψ_0 and T_1 are given below.

$$\psi_0(r, 0) = 0, \tag{9.7.6}$$

$$\frac{\partial \psi_0}{\partial z}(r, 0) = 0, \tag{9.7.7}$$

$$\frac{\partial T_1}{\partial z}(r, 0) = 0, \tag{9.7.8}$$

$$\psi_0(r, 1) = 0, \tag{9.7.9}$$

$$\frac{\partial^2 \psi_0}{\partial z^2}(r, 1) = r \frac{\partial T_0}{\partial r}(r, 1), \tag{9.7.10}$$

$$\frac{\partial T_1}{\partial z}(r, 1) = -q(r), \tag{9.7.11}$$

$$T_1(1, z) = 0. \tag{9.7.12}$$

Equations (9.7.6) and (9.7.7) represent the kinematic and no-slip conditions, respectively, at the rigid bottom surface. Equation (9.7.8) implies that the bottom boundary is adiabatic. Equations (9.7.9) and (9.7.10) arise from the kinematic condition and the tangential stress balance, respectively, applied at the free surface. Equation (9.7.11) describes the imposed heat flux at the same boundary. Equation (9.7.12) represents the isothermal condition at $r = 1$. Integrating Equation (9.7.5) once with respect to z and using the boundary conditions given in Equations (9.7.8) and (9.7.11) yields the following equation for $T_0(r)$:

$$\frac{1}{r} \frac{d}{dr} \left(r \frac{dT_0}{dr} \right) = q(r) = \sum_{n=1}^{\infty} C_n J_0(\alpha_n r). \tag{9.7.13}$$

The isothermal condition at $r = 1$ yields

$$T_0(1) = 0. \tag{9.7.14}$$

The solution for T_0 that satisfies Equation (9.7.14), and the requirement that it be bounded, can be written as

$$T_0(r) = -\sum_{n=1}^{\infty} \frac{C_n}{\alpha_n^2} J_0(\alpha_n r). \tag{9.7.15}$$

Using this temperature field in the tangential stress balance, we can solve for the leading order streamfunction:

$$\psi_0(r, z) = -\frac{1}{4}(z^2 - z^3) \sum_{n=1}^{\infty} \frac{C_n}{\alpha_n} r J_1(\alpha_n r). \tag{9.7.16}$$

We see that the streamfunction field depends on both r and z. The variation of the streamfunction with r is forced parametrically by the heat flux profile at the free surface. The flow field is a superposition of pure modes driven by each term in the Fourier–Bessel series of the surface heat flux. The first mode of the surface heat flux profile, and the corresponding streamline structure, are displayed in Figure 9.7.1, for $\epsilon = 0.2$. Likewise, we have illustrated the second pure mode of the heat flux, suitably normalized, in Figure 9.7.2, along with the corresponding streamline pattern. The mode driven by the

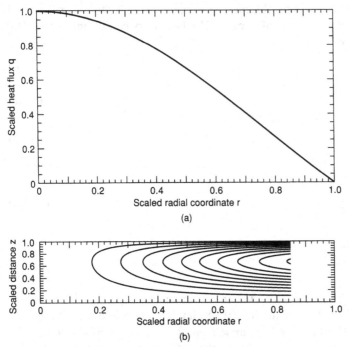

Figure 9.7.1 The mode $n = 1$: (a) Scaled heat flux q as a function of the scaled radial coordinate r, (b) streamlines for this mode; $\psi_{max} = 0$ at the boundaries, and $\Delta\psi = -0.00073$.

heat flux variation of the form $J_0(\alpha_n r)$ will contain n cells. We note that multiple cells can be obtained also in the rectangular liquid layer that was analyzed in Section 9.2. An arbitrary surface heat flux in Equation (9.2.27) can be expanded in a Fourier series. The flow field then can be decomposed into various pure modes in a manner analogous to that shown in Equation (9.7.16).

The no-slip condition at $r = 1$ could not be imposed on the leading order stream-function because derivatives in r are absent in Equation (9.7.4). Therefore, the stream-lines are not closed near $r = 1$ in Figures 9.7.1 and 9.7.2, and we have terminated them at $r = 0.85$. To obtain the proper flow structure near $r = 1$, an inner analysis must be performed in a region of $O(\epsilon)$ near $r = 1$. Such an analysis is analogous to that presented in Section 9.5. It is worthy of note that, to within a multiplicative factor, the expression for the outer streamfunction at $r = 1$ in Equation (9.7.16) is the same as that obtained in the rectangular geometry, which is given in Equation (9.5.6). Also, in the limit $\epsilon \to 0$, the curvature of the cylindrical boundary at $r = 1$ is unimportant at leading order, and it can be shown that the governing equations and boundary conditions in the inner region are virtually the same as those in Section 9.5. Therefore, the inner solution given in that section can be adapted to the present problem in a straightforward manner.

Kamotani, Ostrach, and Pline (1994) and Kamotani, Ostrach, and Masud (1998) have performed reduced gravity experiments on the thermocapillary flow induced in a cylindrical pool of a silicone oil that was heated at its surface by a laser beam. Kamotani et al. (1994) observed steady flow in the form of a single toroidal cell in the layer. They measured the velocities in the liquid by following the trajectories of tracer particles. The authors noted that above a critical value of the imposed heat flux, the flow and

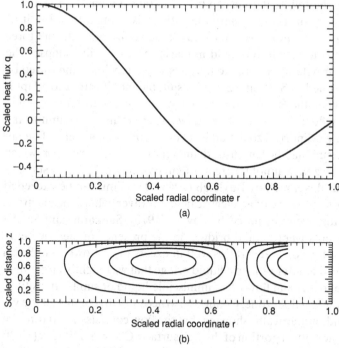

Figure 9.7.2 The mode $n = 2$: (a) Scaled heat flux q as a function of the scaled radial coordinate r, (b) streamlines for this mode; $\psi_{min} = -0.0015$ at $r = 0.436$, $z = \frac{2}{3}$ and $\Delta\psi = 0.00035$.

surface temperature distribution, obtained by an infrared camera, underwent temporal oscillations. They suggested that deformation of the free surface plays an important role in inducing the oscillations.

9.7.3 Flow in Float Zones

As mentioned earlier, a float zone is a nominally cylindrical column of liquid held between two coaxial cylindrical rods by the force of surface tension. In crystal growth, the zone is formed by traversing a solid sample of the material to be purified through a heated region. In reality, the float zone is never exactly cylindrical and is deformed principally due to the effects of gravity. The method is used to grow single crystals of silicon of high purity for certain applications. As noted earlier, solutions of governing equations applicable to models of float zone processes must be obtained by numerical means. Therefore, we only present a brief review of relevant work on this topic.

The subject of thermocapillary flow in float zones has been reviewed by Schwabe (1988), who mentioned that Chang and Wilcox (1975) were the first to point out the importance of thermocapillary flow during the purification of silicon. A simplified experimental model of a float zone is called the half zone. This appears to have been originally used by Schwabe et al. (1978) and Chun and Wuest (1979). In it, the zone is formed by a liquid bridge in the gap between two coaxial solid rods maintained at different temperatures. For stable stratification on Earth, the half zone is heated from above. Such a half zone therefore forms a model for the flow in the lower half of an actual float zone, albeit with altered boundary conditions.

Both the half zone and the full float zone can become statically unstable. A liquid bridge held in any orientation on Earth is unstable if there is enough liquid in it. If the force of gravity acting on the liquid is sufficiently large compared with the force of surface tension that tends to retain the liquid in place, the zone will collapse. The relative importance of the gravitational force to that of surface tension is measured by the static Bond number defined in Section 9.2.1, with suitable modification to adapt it to the present geometry. When the Bond number is large, the zone is unstable. Even in the absence of gravity, long float zones will be unstable. There is an upper limit to the length-to-diameter ratio of a cylindrical zone that is set by a capillary instability. Plateau (1863-1866) obtained the result that the length-to-diameter ratio cannot exceed π, which implies that the length of the zone cannot be greater than its circumference. A stability theory for this problem was developed by Rayleigh (1945). This limit can be exceeded under some conditions in the presence of an electric field, however, where the electrical force can act as a stabilizing agent, as noted by Saville (1970). Sankaran and Saville (1993) performed experiments where liquid bridges, in the presence of electric fields, were maintained with length-to-diameter ratios well in excess of π. Lowry and Steen (1997) achieved a length-to-diameter ratio in a liquid bridge that was slightly larger than π by subjecting the bridge to an axial flow of an immiscible liquid of similar density.

For thermocapillary flow in a float zone, Chang and Wilcox (1976) computed the velocity and temperature fields numerically, showing that the flow consists of two toroidal vortices above and below the hottest portion of the free surface. Clark and Wilcox (1980) corrected some errors in the numerical formulation of Chang and Wilcox and obtained similar results. Schwabe et al. (1978) performed experiments in molten sodium nitrate and obtained the streamline pattern at steady state by flow visualization using tracer particles. They qualitatively confirmed the findings of Chang and Wilcox (1976). In the case of a half zone, a flow structure with a single toroidal vortex has been predicted from numerical solutions by McNeil, Cole, and Subramanian (1984), Fu and Ostrach (1985), and Shen et al. (1990). Such a flow pattern has also been observed experimentally by Schwabe and Scharmann (1979) and Preisser, Schwabe, and Scharmann (1983) in half zones of sodium nitrate. McNeil et al. (1984) performed experiments in half-zones of silicone oils over a range of viscosities and found the measured surface velocities, as well as the velocity as a function of radial distance from the surface, to be in reasonable agreement with their predictions from a numerical solution. Subsequently, McNeil, Cole, and Subramanian (1985) made similar measurements in a sodium borate melt at temperatures in the range 865 to 950°C and found the observed velocities to be consistent with their predictions. Hyer, Jankowski, and Neitzel (1991) analyzed the problem accounting for deformation of the free surface in both the half and the full float zone.

It is difficult to determine the relative importance of buoyant convection in a float zone in experiments conducted on Earth. As discussed in Section 9.1, it will change depending on the parameters. Preisser et al. (1983) attempted to suppress thermocapillary convection in a sodium nitrate zone by using an oil film of 50 μm thickness at the free surface. They reported that under otherwise identical conditions, the peak velocity at the free surface was reduced by a factor of approximately 5 when the oil film was used. It is not possible to conclude that this is a measure of the relative magnitude of the contribution of buoyant convection in their experiments, because thermocapillary convection is not entirely absent. As noted in Section 9.4, a thin (second) layer of liquid can lead to reduction in the magnitude of the thermocapillary velocities in both liquids because of the relativity large viscous stress needed to maintain circulation in a thin layer. Also, the interfacial tension gradient between sodium nitrate and the surrounding gaseous environment in the experiments would likely have been different from the two interfacial tension gradients at the new fluid-fluid interfaces when an oil film was added. To avoid buoyant convection, Chun (1984) performed experiments to obtain the steady thermocapillary flow pattern in a liquid bridge under reduced gravity conditions on board a sounding rocket, and similar experiments were conducted aboard a space shuttle flight by Napolitano, Monti, and Russo (1986). Chun observed satisfactory agreement between the experimentally observed flow and that predicted theoretically. Napolitano et al. indicated that though the Marangoni number in their experiments was large, no oscillatory flow was observed. We briefly mention some references to oscillatory flows below.

Under both normal and reduced gravity conditions, thermocapillary flow in a liquid bridge is observed to be time dependent for large values of the Marangoni number. On Earth, velocity and temperature field oscillations have been observed in a variety of liquids in both half and full float zones by Schwabe et al. (1978), Chun (1980), Preisser et al. (1983), Kamotani, Ostrach, and Vargas (1984), Kamotani and Lee (1989), Cröll, Müller-Sebert, and Nitsche (1990), Velten, Schwabe, and Scharmann (1991), and Masud, Kamotani, and Ostrach (1997). Similar oscillations have been reported in experiments conducted in reduced gravity by Schwabe, Preisser, and Scharmann (1982), Chun (1984), and Schwabe and Scharmann (1984, 1985). Eyer, Leiste, and Nitsche (1985) grew crystals of silicon under reduced gravity conditions and concluded that oscillatory thermocapillary flow produces undesirable striations of dopant in the crystal. Eyer and Leiste (1985) suggested that even on Earth, thermocapillary flow is responsible for striations in silicon crystals. These authors were able to grow striation-free crystals from melts when the free surface was coated by an oxide skin. Stability analyses of the flow in float zones have been performed by Xu and Davis (1984), Shen et al. (1990), Neitzel et al. (1993), Kuhlmann and Rath (1993), and Wanschura et al. (1995), with the objective of predicting the conditions for the onset of time-dependent flow.

We have provided a brief discussion of thermocapillary flow in a cylindrical geometry in this section. This subject, and especially the origin of the onset of the transition to oscillatory fields and that of higher order transitions that ultimately lead to turbulent flow, is one of active inquiry at this time.

9.8 Flow in the Spherical Geometry

9.8.1 Introduction

In this section, we consider two problems in the spherical geometry. The first is that of motion generated within a stationary liquid drop by an arbitrary temperature distribution imposed at its surface. Most of this section is devoted to an analysis of this problem subject to simplifying assumptions. At the end of the section, we provide a brief discussion of available results on thermocapillary motion caused by a bubble attached to a wall.

9.8.2 Flow in a Stationary Spherical Liquid Drop

Temperature gradients can be induced in a drop, for example, by applying a nonuniform heat flux at the surface using a laser or other suitable device. We considered the axisymmetric version of this problem in Section 4.10 and obtained a result for the migration velocity of the drop. There are two important differences between the present problem and the one considered in the earlier section. Here, we assume the temperature on the drop surface to be nonaxisymmetric and known. Also, in the present case, the drop is stationary and surrounded by a gas, such as air. As mentioned in Section 4.2.2, when the drop is much more viscous than the continuous phase, its thermocapillary migration velocity will be small. In the absence of gravity, it is possible to arrest this motion and hold the drop fixed by using an acoustic field. The acoustic field can induce some flow in the drop, which we neglect here. When gravity is present, we assume that the drop is levitated by appropriate means and therefore stationary. We neglect buoyant convection that will arise in that case, however, assuming the length scale of the drop to be sufficiently small, as noted in Section 9.1.

The present problem was first analyzed by Dragoo (1974), who also permitted the drop to deform. We construct the solution assuming that the drop is undeformed. The analysis is performed in spherical polar coordinates, using spherical harmonics that arise in the solution of Laplace's equation in these coordinates. The tools used here to obtain the velocity and vorticity fields are analogous to those developed and used by Chandrasekhar (1961) in the context of instabilities in fluid spheres.

We assume that the Reynolds and Marangoni numbers associated with the flow are small and therefore neglect convective transport effects. The flow is considered to be steady and driven by the thermocapillary stress resulting from the prescribed temperature distribution on the surface of the drop. Temperatures are scaled by first subtracting a reference temperature T_0^*, and then dividing by ΔT^*, which is a characteristic temperature difference on the surface of the drop. The scaled temperature is designated as $T(r, \theta, \phi)$, where r is the radial coordinate, scaled by the radius R of the drop, and θ and ϕ are the polar and azimuthal coordinates in a spherical coordinate system with its origin at the center of the drop. The scaled temperature distribution prescribed at the surface is represented by the symbol $T_S(\theta, \phi)$. The scaled temperature field inside the drop satisfies Laplace's equation, $\nabla^2 T = 0$, and its solution can be written as a superposition of solid spherical harmonics. This solution, which is bounded everywhere inside the drop and which is equal to $T_S(\theta, \phi)$ at the surface, is given by

$$T(r, \theta, \phi) = \sum_{n=0}^{\infty} \sum_{m=-n}^{n} T_{mn} r^n Y_n^m(\theta, \phi). \tag{9.8.1}$$

In the above result, the functions $Y_n^m(\theta, \phi)$ are known as surface spherical harmonics. They are defined as

$$Y_n^m(\theta, \phi) = e^{im\phi} P_n^{|m|}(\cos \theta), \tag{9.8.2}$$

where $P_n^m(\cos \theta)$ is the associated Legendre function of the first kind, the properties of which may be found in MacRobert (1967). For the temperature field to be a real-valued function, the coefficients T_{mn} must assume complex values and also must occur as a set of complex conjugate pairs. From the orthogonality relation satisfied by the surface harmonics, these coefficients can be obtained as

$$T_{mn} = T_{(-m)(n)}^* = \frac{2n+1}{4\pi} \frac{(n-m)!}{(n+m)!} \int_0^{2\pi} \int_0^{\pi} T_S(\theta, \phi) e^{im\phi} P_n^m(\cos \theta) \sin \theta \, d\theta \, d\phi,$$

$$n = 0, \cdots, \infty; \quad m = 0, \cdots, n. \tag{9.8.3}$$

Here, an asterisk (*) denotes the complex conjugate. Equations (9.8.1) and (9.8.3) show that each pure surface harmonic mode of the prescribed temperature field excites a corresponding solid spherical harmonic mode in the temperature field within the drop.

The continuity and Navier–Stokes equations for the scaled velocity field **v** and the scaled hydrodynamic pressure field p can be written as

$$\nabla \bullet \mathbf{v} = 0 \tag{9.8.4}$$

and

$$\nabla^2 \mathbf{v} = \nabla p. \tag{9.8.5}$$

The velocity is scaled by the thermocapillary reference velocity $U_R = \frac{(-\sigma_T)\Delta T^*}{\mu}$ and the hydrodynamic pressure by $p_R = \frac{\mu U_R}{R}$. The symbol σ_T represents the rate of change of surface tension with temperature, and μ is the viscosity of the liquid. The velocity field satisfies the following boundary conditions:

$$v_r(1, \theta, \phi) = 0, \tag{9.8.6}$$

$$\frac{\partial}{\partial r}\left(\frac{v_\theta}{r}\right)(1, \theta, \phi) = -\frac{\partial T_S}{\partial \theta}, \tag{9.8.7}$$

$$\frac{\partial}{\partial r}\left(\frac{v_\phi}{r}\right)(1, \theta, \phi) = -\frac{1}{\sin\theta}\frac{\partial T_S}{\partial \phi}, \tag{9.8.8}$$

and

$$|\mathbf{v}(0, \theta, \phi)| < \infty. \tag{9.8.9}$$

Here v_r, v_θ, and v_ϕ are the components of the velocity vector. Equation (9.8.6) represents the kinematic condition, and Equations (9.8.7) and (9.8.8) arise from the tangential stress balance at the surface. Because the pressure can be determined only to within an arbitrary additive constant, we shall set the scaled pressure equal to zero at the center of the drop. Taking the curl of Equation (9.8.5) and the curl of the resulting equation, we obtain

$$\nabla^2 \mathbf{\Omega} = 0 \tag{9.8.10}$$

and

$$\nabla^4 \mathbf{v} = 0, \tag{9.8.11}$$

where $\mathbf{\Omega} = \nabla \times \mathbf{v}$ is the vorticity. Chandrasekhar (1961) showed that because both **v** and $\mathbf{\Omega}$ are solenoidal, a term implying that they are divergence free, $\mathbf{x} \cdot \mathbf{v}$ and $\mathbf{x} \cdot \mathbf{\Omega}$ satisfy the same equations. Here, **x** represents the scaled position vector. Chandrasekhar also showed that, in spherical polar coordinates, once the r-components of these two vectors are known, the complete vectors **v** and $\mathbf{\Omega}$ can be constructed. This is achieved by making use of a general representation of any solenoidal vector field in a basis set of toroidal and poloidal vector fields. Definitions of the terms *toroidal* and *poloidal* can be found in Chandrasekhar (1961). Members of the basis set involve the radial components of the vector field and the surface harmonics $Y_n^m(\theta, \phi)$. Sani (1963) provided the following convenient relation for constructing the solution for the nth pure mode:

$$\mathbf{v} = v_r\mathbf{i}_r + \frac{r^2}{n(n+1)}\left[\nabla_s\left(\frac{1}{r^2}\frac{\partial}{\partial r}(r^2 v_r)\right) - \mathbf{i}_r \times \nabla_s \Omega_r\right]. \tag{9.8.12}$$

Here $\mathbf{i}_r$ is the unit vector in the radial direction, and $\nabla_s = \nabla - \mathbf{i}_r \frac{\partial}{\partial r}$ is the surface gradient operator. Let

$$r v_r(r, \theta, \phi) = \sum_{n=1}^{\infty} \sum_{m=-n}^{n} W_{mn}(r) Y_n^m(\theta, \phi)$$

(9.8.13)

and

$$r \Omega_r(r, \theta, \phi) = \sum_{n=1}^{\infty} \sum_{m=-n}^{n} Z_{mn}(r) Y_n^m(\theta, \phi).$$

(9.8.14)

Chandrasekhar derived the identity

$$\nabla^2 Y_n^m(\theta, \phi) f(r) \equiv Y_n^m(\theta, \phi) \mathcal{D}_n f(r),$$

(9.8.15)

where the operator $\mathcal{D}_n$ is defined as

$$\mathcal{D}_n \equiv \frac{d^2}{dr^2} + \frac{2}{r} \frac{d}{dr} - \frac{n(n+1)}{r^2}.$$

(9.8.16)

The equations satisfied by $W_{mn}(r)$ and $Z_{mn}(r)$ are

$$\mathcal{D}_n^2 W_{mn} = 0$$

(9.8.17)

and

$$\mathcal{D}_n Z_{mn} = 0.$$

(9.8.18)

The boundary conditions on W_{mn} and Z_{mn} are derived by applying the operator $r \frac{\partial}{\partial r}$ to both sides of the continuity equation and to the vector Ω and evaluating the results at $r = 1$, together with the use of Equations (9.8.7) and (9.8.8). This yields

$$\frac{d^2 W_{mn}}{dr^2}(1) = n(n+1) T_{mn}$$

(9.8.19)

and

$$\frac{d Z_{mn}}{dr}(1) - Z_{mn}(1) = 0.$$

(9.8.20)

The solutions that satisfy the above boundary conditions, along with the requirement that W_{mn} and Z_{mn} be bounded at $r = 0$, are

$$W_{mn}(r) = -\frac{n(n+1)}{2(2n+1)} T_{mn} r^n (1 - r^2)$$

(9.8.21)

and

$$Z_{mn}(r) = 0.$$

(9.8.22)

It is therefore evident that each pure mode of the surface temperature field excites a corresponding pure mode of the flow. The complete solutions for the velocity and vorticity fields can be constructed using Equation (9.8.12). The pressure field can be obtained from the following relation, which is derived from Equation (9.8.5):

$$r \frac{\partial p}{\partial r} = \nabla^2 (r v_r).$$
(9.8.23)

The solutions are

$$v_r = -\sum_{n=1}^{\infty} \sum_{m=-n}^{n} \frac{n(n+1)}{2(2n+1)} T_{mn} r^{n-1} (1 - r^2) Y_n^m(\theta, \phi),$$
(9.8.24)

$$v_\theta = -\sum_{n=1}^{\infty} \sum_{m=-n}^{n} \frac{n+1}{2(2n+1)} T_{mn} r^{n-1} \left(1 - \frac{n+3}{n+1} r^2\right) \frac{\partial Y_n^m}{\partial \theta}(\theta, \phi),$$
(9.8.25)

$$v_\phi = -\sum_{n=1}^{\infty} \sum_{m=-n}^{n} \frac{n+1}{2(2n+1)} T_{mn} r^{n-1} \left(1 - \frac{n+3}{n+1} r^2\right) \frac{1}{\sin \theta} \frac{\partial Y_n^m}{\partial \phi}(\theta, \phi),$$
(9.8.26)

$$\Omega_\theta = -\sum_{n=1}^{\infty} \sum_{m=-n}^{n} \frac{2n+3}{2n+1} T_{mn} r^n \frac{1}{\sin \theta} \frac{\partial Y_n^m}{\partial \phi}(\theta, \phi),$$
(9.8.27)

$$\Omega_\phi = \sum_{n=1}^{\infty} \sum_{m=-n}^{n} \frac{2n+3}{2n+1} T_{mn} r^n \frac{\partial Y_n^m}{\partial \theta}(\theta, \phi),$$
(9.8.28)

and

$$p = \sum_{n=1}^{\infty} \sum_{m=-n}^{n} \frac{(n+1)(2n+3)}{2n+1} T_{mn} r^n Y_n^m(\theta, \phi).$$
(9.8.29)

We note that the radial component of the vorticity is identically zero. The vorticity field is thus purely toroidal and the velocity field is purely poloidal. The pressure field is harmonic, which is the case in any Stokes flow. Dragoo (1974) discussed the flow structures of the modes that are axisymmetric and therefore independent of the azimuthal coordinate ϕ. These correspond to setting $m = 0$, in which case the associated Legendre function $P_n^0(\cos \theta)$ is simply the Legendre Polynomial $P_n(\cos \theta)$. The axisymmetric mode with $n = 0$ corresponds to a uniform surface temperature and does not induce flow in the drop. Therefore, this mode is not present in Equations (9.8.24) through (9.8.29). The mode $n = 1$ is Hill's spherical vortex which was first mentioned in Section 3.1.2. All the higher modes have a stagnation point at $r = 0$. In addition to this, the nth mode has n internal stagnation rings about the axis of symmetry. These rings are described by $r^2 = \frac{k+1}{k+3}$, $k = 1, 2, 3, \ldots, n$, and $P_n^0(\cos \theta) = 0$. All the axisymmetric modes have two additional stagnation points at the poles of the drop.

The streamlines corresponding to the first three modes, distorted by the presence of a bubble within the drop, already have been displayed in Figures 7.7.2(b), 7.7.3(b), and 7.7.4(b). Despite the distortion, the symmetries associated with these flows are evident from these figures. Therefore, we have not included additional drawings in the present section. The interested reader will find the streamline structures for the first few modes in the article by Dragoo (1974) or in Jayaraj, Cole, and Subramanian (1982). The latter authors analyzed axisymmetric flow in a drop, which is driven by the combined action of thermocapillarity and buoyancy, subject to the same assumptions we have made here.

9.8.3   Flow Induced by a Bubble Attached to a Wall

Now, we briefly discuss the literature on thermocapillary flow that is induced by a bubble which is attached to a wall. This situation is common in applications in boiling, where the bubble is that of the vapor of the liquid and may contain other gaseous components as well. Typically, convective transport effects are significant in applications to boiling heat transfer. McGrew, Bamford, and Rehm (1966) suggested that thermocapillary flow may be of some importance as a heat transfer mechanism in boiling. In a pure liquid in contact with its own vapor, as long as the pressure within the bubble is uniform, the temperature also must be uniform along the interface, if equilibrium conditions are assumed at the interface. Therefore, thermocapillary flow can arise only if one or more of these conditions is relaxed. For example, the presence of other species in a vapor bubble or in the liquid can lead to a variation of the equilibrium temperature along the surface of the vapor bubble if the composition varies along the interface. The theoretical analyses, as well as the experimental work that we review, are confined to the case of a gas bubble, and not a vapor bubble, in a liquid, because this issue does not arise in the former case.

Larkin (1970) appears to have been the first to analyze the thermocapillary flow caused by a bubble, assumed hemispherical in shape, attached to a rigid plane wall. Larkin was aware of potential applications in reduced gravity and specifically mentioned heat transfer to cryogenic fluids in long-term storage under weightless conditions as a motivation for the work. A uniform heat flux is supplied from the wall to the fluid everywhere except underneath the bubble, where the heat flux is taken to be zero. The viscosity and the thermal conductivity of the gas were assumed negligible, compared with the corresponding properties in the liquid. The governing equations in the liquid were solved numerically by the method of finite differences. Larkin considered the transient problem, wherein the flow initially evolved rapidly and then began to decay slowly. The Reynolds and Marangoni numbers are defined by using the bubble radius as the length scale and the velocity v_0, defined in Equation (4.1.1), as the characteristic velocity. When the heat flux q^* is prescribed, $\frac{q^*}{k}$, where k is the thermal conductivity of the liquid, is used instead of $|\nabla T_\infty|$ in the definition of v_0. Larkins's numerical calculations extend to $Ma = 100,000$, with the Prandtl number $Pr = 1$ and 5. The liquid is drawn along the surface of the bubble to the apex and forms a jet extending several bubble diameters into the liquid, with a return flow toward the surface, to satisfy conservation of mass.

Kao and Kenning (1972) performed experiments on air bubbles on a heated surface facing downward in water. The authors chose this orientation to achieve a stable stratification of the liquid. They took special precautions to make sure that the water was pure. They solved the steady flow problem for a uniform heat flux applied to the base of the solid surface. The continuity, Navier–Stokes, and energy equations in the liquid and the conduction equation in the solid were discretized and solved by the method of finite differences. The influence of buoyancy was accommodated in the model through the Boussinesq approximation; the authors also included the influence of thermocapillarity. The observed flow patterns were consistent with those predicted, and the magnitude of the maximum value of the streamfunction was in good agreement with predictions from the model, extrapolated to the values of the Marangoni number in the experiments, which were of the order of 10^4. Kao and Kenning provided photographs of the flow pattern in a meridian section, obtained by using tracers, which illustrate not only the flow in the case of a clean surface, but also the role of surface active contaminants in forming a stagnant cap in another experiment. The single vortex produced by the jet at the apex of the bubble in thermocapillary flow extends over a large region in the liquid, and the effect of buoyancy is to moderate the thermocapillary jet and substantially reduce the size of this vortex. Also, buoyancy produces weaker secondary and tertiary vortices, in addition to the primary vortex.

Subsequently, flow driven by the combined action of gravity and thermocapillarity in this geometry in silicone oils was investigated experimentally by Raake, Siekmann, and Chun (1989). The bubbles were typically of radius 4 mm and were not hemispherical in shape. One can expect the shapes to be deformed primarily because of the hydrostatic variation of pressure, but there also would have been a contribution from hydrodynamic effects, which would depend on the value of the Weber number. Electrical heating was used to keep the top surface at a desired temperature, and the bottom surface of the cell was maintained at a different temperature by circulating fluid. Raake et al. used Schlieren interferometry to map temperature variations and hollow glass spheres as tracers to investigate flow patterns. The interferograms were used to plot isotherms around the bubble for a range of values of the Marangoni number. The definition of the Marangoni number is unusual in that the authors used the temperature gradient along the bubble surface instead of the applied uniform gradient far from the bubble. The authors also plotted radial temperature profiles and the temperature distribution along the bubble surface. As expected, the temperature variations are distributed over the entire surface at relatively small values of Ma, whereas at large values of Ma, the temperature is virtually uniform over a large portion of the surface, changing only near the region where the bubble contacts the wall. The authors reported that for a silicone oil with a relatively small value of the Prandtl number, at sufficiently large applied temperature gradients, velocity and temperature field oscillations were observed. This is reminiscent of the oscillatory instability mentioned in the context of nominally cylindrical liquid bridges in Section 9.7. After the onset of oscillations occured at a critical value of Ma, Raake et al. found that the frequency of the oscillations stayed nearly constant with further increase of Ma until a second critical range of values was reached. In this range, the frequency more than doubled and kept increasing with subsequent increase in the Marangoni number. The velocity fields, mapped by tracers, are similar to those observed by Kao and Kenning.

Wozniak, Wozniak, and Bergelt (1996) performed similar experiments, both on Earth and in a European Texus rocket flight experiment that provided approximately six minutes of reduced gravity conditions. They used liquid crystal tracers that permitted them to map the temperature and the velocity fields simultaneously because the tracers changed color depending on their temperature. Color images were recorded by using multiple exposures at constant time intervals. In the reduced gravity experiment, several bubbles, instead of a single bubble, were introduced because of a malfunction in the system. These bubbles migrated to the heated wall and collected there. Some of these bubbles eventually coalesced, forming a single large bubble attached to the hot wall, with another bubble nearly touching it on the opposite side. Differences are clearly evident in the flow patterns and isotherms obtained in the Earth-based experiment, when compared to those from the reduced gravity experiment. The latter display the thermocapillary jet predicted by the models. Wozniak and Wozniak (1998) extended the technique to two bubbles on a wall and indicated that the experiment would be conducted under reduced gravity conditions in the future.

New theoretical work has been published since the appearance of the articles of Larkin (1970) and Kao and Kenning (1972). It is not our intent to review the literature in depth. However, there are two articles worthy of mention. Lee and Chen (1990) solved for the flow generated by a nominally hemispherical bubble placed on a wall kept at a uniform temperature in an unbounded fluid in the presence of a uniform temperature gradient in the undisturbed fluid far from the bubble. The direction of the temperature gradient was toward the wall. The authors permitted the shape of the bubble to be determined as part of the solution. The method of finite differences was employed, and boundary-fitted coordinates were used. Lee and Chen first presented results from calculations made in the case of an undeformed bubble. They displayed sample streamlines and isotherms, as well as the distribution of the tangential velocity along the bubble surface and the variation of the radial velocity with distance from the bubble when gravitational effects are neglected. The Capillary number, defined in the usual manner, was varied from 0 to 0.2, and representative results were displayed for the velocity distributions and the shape of the bubble, which was an oblate spheroid. The authors suggested that this shape was consistent with that observed by Raake et al. (1989). The calculations appear to be for a case where gravitational effects are neglected, however, which is not explicitly stated. Therefore, a direct comparison is inappropriate, because hydrostatic effects must have played a large role in determining the shapes of the bubbles in those Earth-based experiments. Finally, the authors reported sample results for streamlines and isotherms including a buoyant contribution. When buoyancy is accommodated, the flow structure changes sharply in character, as noted earlier by Kao and Kenning (1972).

Recently, Arlabosse et al. (1999) used the method of finite elements to solve the governing equations both within the bubble and in the liquid, in the same problem as that modeled by Kao and Kenning, with the exception that the shape of the bubble was taken to be that obtained from experiments. The authors also accommodated the variation of the viscosity of the liquid with temperature and made the Boussinesq approximation in handling the buoyant contribution. The temperatures of the hot and cold walls were assumed to be fixed, and adiabatic boundary conditions were employed at the side walls. The Marangoni number was varied from 0 to 600, and the Prandtl number was varied from 100 to 1000. Therefore, the flows were very viscous compared with those considered in the earlier studies. Buoyant contributions were characterized by a suitably defined Rayleigh number, which was varied from 0 to 220. The authors found good agreement of their predicted maximum velocities with those observed in their experiments. The principal conclusion of the authors is that increasing Ma leads to an increase in the strength of the flow, and gradually, in the number of vortices in the cell. Arlabosse et al. also noted that the flow structure changes qualitatively when buoyancy is accommodated. Finally, they noted that the strength of the primary vortex is controlled by thermocapillarity, and the Rayleigh number has only a weak influence on it in the range considered.

References

Abramowitz, M., and Stegun, I. A. 1965. *Handbook of Mathematical Functions.* New York: Dover.

Acrivos, A., Jeffrey, D.J., and Saville, D.A. 1990. Particle migration in suspensions by thermo-capillary or electrophoretic motion. *J. Fluid Mech.* **212,** 95–110.

Acrivos, A., and Taylor, T.D. 1962. Heat and mass transfer from single spheres in Stokes flow. *Phys. Fluids* **5,** No. 4, 387–394.

Adamson, A.W. 1976. *Physical Chemistry of Surfaces.* New York: Wiley.

Adler, J., and Sowerby, L. 1970. Shallow three-dimensional flows with variable surface tension. *J. Fluid Mech.* **42,** Pt. 3, 549–559.

Alexander, J.I.D. 1990. Low-gravity experiment sensitivity to residual acceleration: A review. *Microgravity Sci. Technol.* **III,** No. 2, 52–68.

Anderson, J.L. 1985. Droplet interactions in thermocapillary motion. *Int. J. Multiphase Flow* **11,** No. 6, 813–824.

Anderson, J.L. 1989. Colloid transport by interfacial forces. In *Ann. Rev. Fluid Mech.,* ed. J.L. Lumley, M. Van Dyke, and H.L. Reed (vol. 21, pp. 61–99). Palo Alto, CA: Annual Reviews, Inc.

Annamalai, P., and Cole, R. 1983. Drop motion in a rotating immiscible liquid body. In *Advances in Space Research,* ed. Y. Malmejac (vol. 3, No. 5, pp. 165–168). England: Pergamon Press.

Annamalai, P., and Cole, R. 1986. Particle migration in rotating liquids. *Phys. Fluids* **29,** No. 3, 647–649.

Annamalai, P., Shankar, N., Cole, R., and Subramanian, R.S. 1982a. Bubble migration inside a liquid drop in a space laboratory. In *Mechanics and Physics of Bubbles in Liquids,* ed. L. van Wijngaardan (pp. 179–186). The Hague, The Netherlands: Martinus Nijhoff. (Reprinted in 1982, *Appl. Sci. Res.* **38,** 179–186).

Annamalai, P., Subramanian, R.S., and Cole, R. 1982b. Bubble migration in a rotating, liquid-filled sphere. *Phys. Fluids* **25,** No. 7, 1121–1126.

Antar, B.N., and Nuotio-Antar, V.S. 1993. *Fundamentals of Low Gravity Fluid Dynamics and Heat Transfer.* Boca Raton, FL: CRC Press.

Anthony, T.R., and Cline, H.E. 1972. The thermomigration of biphase vapor-liquid droplets in solids. *Acta Metallurgica* **20,** 247–255.

Aris, R. 1962. *Vectors, Tensors, and the Basic Equations of Fluid Mechanics.* Englewood Cliffs, NJ: Prentice-Hall.

Arlabosse, P., Lock, N., Medale, M., and Jaeger, M. 1999. Numerical investigation of thermo-capillary flow around a bubble. *Phys. Fluids* **11,** No. 1, 18–29.

Ascoli, E.P., and Leal, L.G. 1990. Thermocapillary motion of a deformable drop toward a planar wall. *J. Colloid Interface Sci.* **138,** No. 1, 220–230.

Balasubramaniam, R. 1995. Thermocapillary bubble migration – solution of the energy equation for potential-flow approximated velocity fields. *Computational Fluid Dynamics J.* **3,** No. 4, 407–414.

Balasubramaniam, R. 1998a. Thermocapillary and buoyant bubble motion with variable viscosity. *Int. J. Multiphase Flow* **24,** No. 4, 679–683.

Balasubramaniam, R. 1998b. Thermocapillary flow near a corner in a thin liquid layer. CP420 Space Technology and Applications International Forum 1998, American Institute of Physics, pp. 429–434, 25–29 January 1998, Albuquerque, New Mexico.

Balasubramaniam, R., and Chai, A. 1987. Thermocapillary migration of droplets: An exact solution for small Marangoni numbers. *J. Colloid Interface Sci.* **119,** No. 2, 531–538.

Balasubramaniam, R., Lacy, C.E., Wozniak, G., and Subramanian, R.S. 1996. Thermocapillary migration of bubbles and drops at moderate values of the Marangoni number in reduced gravity. *Phys. Fluids* **8,** No. 4, 872–880.

Balasubramaniam, R., and Lavery, J.E. 1989. Numerical simulation of thermocapillary bubble migration under microgravity for large Reynolds and Marangoni numbers. *Num. Heat Transfer A* **16,** 175–187.

Balasubramaniam, R., and Ostrach, S. 1984. Fluid motion in the Czochralski method of crystal growth. *Physicochemical Hydrodynamics* **5,** No. 1, 3–18.

Balasubramaniam, R., and Subramanian, R.S. 1996. Thermocapillary bubble migration – thermal boundary layers for large Marangoni numbers. *Int. J. Multiphase Flow* **22,** No. 3, 593–612.

Balasubramaniam, R., and Subramanian, R.S. 1999. Axisymmetric thermal wake interaction of two bubbles in a uniform temperature gradient at large Reynolds and Marangoni numbers. *Phys. Fluids* **11,** No. 10, 2856–2864.

Balasubramaniam, R., and Subramanian, R.S. 2000. The migration of a drop in a uniform temperature gradient at large Marangoni numbers. *Physics of Fluids* **12,** No. 4, 733–743.

Barnocky, G., and Davis, R.H. 1989. The lubrication force between spherical drops, bubbles and rigid particles in a viscous fluid. *Int. J. Multiphase Flow* **15,** No. 4, 627–638.

Bart, E. 1968. The slow unsteady settling of a fluid sphere toward a flat fluid interface. *Chem. Eng. Sci.* **23,** 193–210.

Barton, K.D. 1990. Thermocapillary Migration of Drops. Ph.D. diss., in chemical engineering, Clarkson University: Potsdam, New York.

Barton, K.D., and Subramanian, R.S. 1989. The migration of liquid drops in a vertical temperature gradient. *J. Colloid Interface Sci.* **133,** No. 1, 211–222.

Barton, K.D., and Subramanian, R.S. 1990. Thermocapillary migration of a liquid drop normal to a plane surface. *J. Colloid Interface Sci.* **137,** No. 1, 170–182.

Barton, K.D., and Subramanian, R.S. 1991. Migration of liquid drops in a vertical temperature gradient – interaction effects near a horizontal surface. *J. Colloid Interface Sci.* **141,** No. 1, 146–156.

Batchelor, G.K. 1967. *An Introduction to Fluid Dynamics.* Cambridge, England: Cambridge University Press.

Batchelor, G.K. 1972. Sedimentation in a dilute suspension of spheres. *J. Fluid Mech.* **52,** Pt. 2, 245–268.

Bénard, H. 1900. Les tourbillons cellulaires dans une nappe liquide. *Rev. Générale Sci. Pures Appl.* **11,** 1261–1271, 1309–1328.

Bénard, H. 1901. Les tourbillons cellulaires dans une nappe liquide transportant de la chaleur par convection en régime permanent. *Annales de Chimie et de Physique* **23,** 62–144.

Bhaga, D., and Weber, M.E. 1981. Bubbles in viscous liquids: shapes, wakes and velocities. *J. Fluid Mech.* **105,** 61–85.

Bird, R.B., Stewart, W.E., and Lightfoot, E.N. 1960. *Transport Phenomena.* New York: Wiley.

Birikh, R.V. 1966. Thermocapillary convection in a horizontal layer of liquid. *J. Appl. Mech. Tech. Phys.* **7,** 43–44.

Bond, W.N. 1927. Bubbles and drops and Stokes' law. *Phil. Mag.* Ser. 7, **4,** 889–898.

Bond, W.N., and Newton, D.A. 1928. Bubbles, drops, and Stokes' law. *Phil. Mag.* Ser. 7, **5,** 794–800.

Borhan, A., Haj-Hariri, H., and Nadim, A. 1992. Effect of surfactants on the thermocapillary migration of a concentric compound drop. *J. Colloid Interface Sci.* **149,** No. 2, 553–560.

Brabston, D.C., and Keller, H.B. 1975. Viscous flows past spherical gas bubbles. *J. Fluid Mech.* **69,** Pt. 1, 179–189.

Bratukhin, Yu K. 1975. Termokapillyarnyy dreyf kapel'ki vyazkoy zhidkosti. *Izv. Akad. Nauk*

SSSR, Mekh. Zhidk. Gaza **5,** 156–161; Original in Russian, NASA Technical Translation 17093, June 1976.

Bratukhin, Yu K., Briskman, V.A., and Zuev, A.L. 1984. Brief review of thermocapillary migration experiments. Personal communication providing an English summary of Bratukhin Yu K., Briskman, V.A., Zuev, A.L., Pshenichnikov, A.F., Rivkind, V. Ya. 1982. Experimental investigation of thermocapillary migration of gas bubbles in liquid. *Hydromechanics and Heat/Mass Transfer in Microgravity,* pp. 98–109. Moscow: Nauka.

Braun, B., Ikier, C., Klein, H., and Woermann, D. 1993. Thermocapillary migration of droplets in a binary mixture with miscibility gap during liquid/liquid phase separation under reduced gravity. *J. Colloid Interface Sci.* **159,** 515–516.

Brenner, H. 1961. The slow motion of a sphere through a viscous fluid towards a plane surface. *Chem. Eng. Sci.* **16,** 242–251.

Brenner, H. 1964. The Stokes resistance of an arbitrary particle – IV Arbitrary fields of flow. *Chem. Eng. Sci.* **19,** 703–727.

Brignell, A.S. 1973. The deformation of a liquid drop at small Reynolds number. *Quart. J. Mech. and Applied Math.* **XXVI,** Pt. 1, 99–107.

Brignell, A.S. 1975. Solute extraction from an internally circulating spherical liquid drop. *Int. J. Heat Mass Transfer* **18,** 61–68.

Brunn, P.O. 1982. Heat or mass transfer from single spheres in a low Reynolds number flow. *Int. J. Eng. Sci.* **20,** No. 7, 817–822.

Buff, F.P. 1960. The theory of capillarity. In *Handbuch der Physik, Band X: Struktur der Flüssigkeiten,* ed. S. Flügge (pp. 281–304). Berlin: Springer-Verlag.

Burkersroda, C.V., Prakash, A., and Koster, J.N. 1994. Interfacial tension between fluorinert liquids and silicone oils. *Microgravity Q.* **4,** No. 2, 93–99.

Cable, M., and Frade, J.R. 1987. Diffusion-controlled growth of multi-component gas bubbles. *J. Materials Sci.* **22,** 919–924.

Carpenter, B.M., and Homsy, G.M. 1990. High Marangoni number convection in a square cavity: Part II. *Phys. Fluids A* **2,** No. 2, 137–149.

Carslaw, H.S., and Jaeger, J.C. 1959. *Conduction of Heat in Solids.* Oxford, England: Clarendon Press.

Chan, C.L., Chen, M.M., and Mazumder, J. 1988. Asymptotic solution for thermocapillary flow at high and low Prandtl numbers due to concentrated surface heating. *J. Heat Transfer* **110,** 140–146.

Chandrasekhar, S. 1961. *Hydrodynamic and Hydromagnetic Stability.* Oxford, England: Clarendon Press.

Chang, C.E., and Wilcox, W.R. 1975. Inhomogeneities due to thermocapillary flow in floating zone melting. *J. Crystal Growth* **28,** 8–12.

Chang, C.E., and Wilcox, W.R. 1976. Analysis of surface tension driven flow in floating zone melting. *Int. J. Heat Mass Transfer* **19,** 355–366. Erratum, Ibid, 1977, **20,** 891.

Chang, L.S., and Berg, J.C. 1985. The effect of interfacial tension gradients on the flow structure of single drops or bubbles translating in an electric field. *AIChE J.* **31,** No. 4, 551–557.

Chao, B.T. 1962. Motion of spherical gas bubbles in a viscous liquid at large Reynolds numbers. *Phys. Fluids* **5,** No. 1, 69–79.

Chen, J., Dagan, Z., and Maldarelli, C. 1991. The axisymmetric thermocapillary motion of a fluid particle in a tube. *J. Fluid Mech.* **233,** 405–437.

Chen, J., and Stebe, K.J. 1996. Marangoni retardation of the terminal velocity of a settling droplet: The role of surfactant physico-chemistry. *J. Colloid Interface Sci.* **178,** 144–155.

Chen, J., and Stebe, K.J. 1997. Surfactant-induced retardation of the thermocapillary migration of a droplet. *J. Fluid Mech.* **340,** 35–59.

Chen, J.C., and Lee, Y.T. 1992. Effect of surface deformation on thermocapillary bubble migration. *AIAA J.* **30,** No. 4, 993–998.

Chen, S.H. 1999a. Thermocapillary deposition of a fluid droplet normal to a planar surface. *Langmuir* **15,** 2674–2683.

Chen, S.H. 1999b. Thermocapillary migration of a fluid sphere parallel to an insulated plane. *Langmuir* **15**, 8618–8626.

Chen, S.H., and Keh, H.J. 1990. Thermocapillary motion of a fluid droplet normal to a plane surface. *J. Colloid Interface Sci.* **137**, No. 2, 550–562.

Chen, Y.S., Lu, Y.L., Yang, Y.M., and Maa, J.R. 1997. Surfactant effects on the motion of a droplet in thermocapillary migration. *Int. J. Multiphase Flow* **23**, No. 2, 325–335.

Chester, W., and Breach, D.R., with an appendix by Proudman, I. 1969. On the flow past a sphere at low Reynolds number. *J. Fluid Mech.* **37**, Pt. 4, 751–760.

Chhabra, R.P. 1993. *Bubbles, Drops, and Particles in Non-Newtonian Fluids.* Boca Raton, FL: CRC Press.

Chi, B.K., and Leal, L.G. 1989. A theoretical study of the motion of a viscous drop toward a fluid interface at low Reynolds number. *J. Fluid Mech.* **201**, 123–146.

Chisnell, R.F. 1987. The unsteady motion of a drop moving vertically under gravity. *J. Fluid Mech.* **176**, 443–464.

Chun, Ch.-H. 1980. Experiments on steady and oscillatory temperature distribution in a floating zone due to the Marangoni convection. *Acta Astronautica* **7**, No. 4–5, 479–488.

Chun, Ch.-H. 1984. Numerical study on the thermal Marangoni convection and comparison with experimental results from the Texus-rocket program. *Acta Astronautica* **11**, No. 3–4, 227–232.

Chun, Ch.-H., and Wuest, W. 1979. Experiments on the transition from the steady to the oscillatory Marangoni-convection of a floating zone under reduced gravity effect. *Acta Astronautica* **6**, No. 9, 1073–1082.

Churchill, S.W. 1988. *Viscous Flows.* Boston: Butterworths.

Clark, P.A., and Wilcox, W.R. 1980. Influence of gravity on thermocapillary convection in floating zone melting of silicon. *J. Crystal Growth* **50**, 461–469.

Clift, R., Grace, J.R., and Weber, M.E. 1978. *Bubbles, Drops and Particles.* New York: Academic Press.

Clifton, J.V., Hopkins, R.A., and Goodwin, D.W. 1969. Motion of a bubble in a rotating liquid. *J. Spacecraft* **6**, 215–217.

Collins, W.D. 1961. On some dual series equations and their application to electrostatic problems for spheroidal caps. *Proc. Cambridge Phil. Soc.* **57**, 367–384.

Cooley, M.D.A., and O'Neill, M.E. 1969. On the slow motion generated in a viscous fluid by the approach of a sphere to a plane wall or stationary sphere. *Mathematika* **16**, 37–49.

Cox, R.G., and Brenner, H. 1967. The slow motion of a sphere through a viscous fluid towards a plane surface – II Small gap widths, including inertial effects. *Chem. Eng. Sci.* **22**, 1753–1777.

Crank, J. 1975. *The Mathematics of Diffusion.* Oxford, England: Clarendon Press.

CRC Handbook of Chemistry and Physics. 1999. Boca Raton, FL: CRC Press LLC.

Crespo, A., and Jimenez-Fernandez, J. 1992a. Thermocapillary migration of bubbles at moderately large Reynolds numbers. In *Microgravity Fluid Mechanics IUTAM Symposium Bremen 1991*, ed. H.J. Rath (pp. 405–411). Berlin: Springer-Verlag.

Crespo, A., and Jimenez-Fernandez, J. 1992b. Thermocapillary migration of bubbles: A semi-analytical solution for large Marangoni numbers. *Proceedings of the 8th European Symposium on Materials and Fluid Sciences in Microgravity* (pp. 193–196), European Space Agency SP-333. 12–16 April 1992, Brussels, Belgium.

Crespo, A., and Manuel, F. 1983. Bubble motion under reduced gravity. *Proceedings of the 4th European Symposium on Materials Sciences under Microgravity* (pp. 45–49), European Space Agency SP-191. 5–8 April 1983, Madrid, Spain.

Crespo, A., Migoya, E., and Manuel, F. 1998. Thermocapillary migration of bubbles at large Reynolds numbers. *Int. J. Multiphase Flow* **24**, No. 4, 685–692.

Cröll, A., Müller-Sebert, W., and Nitsche, R. 1990. Transition from steady to time-dependent Marangoni convection in partially coated silicon melt zones. *Proceedings of the 7th European Symposium on Materials and Fluid Sciences in Microgravity* (pp. 263–269). European Space Agency SP-295. 10–15 September 1989, Oxford, UK.

Dandy, D.S., and Leal, L.G. 1989. Buoyancy-driven motion of a deformable drop through a quiescent liquid at intermediate Reynolds numbers. *J. Fluid Mech.* **208**, 161–192.

Davies, J.T., and Rideal, E.K. 1963. *Interfacial Phenomena*. New York: Academic Press.

Davies, R.M., and Taylor, G.I. 1950. The mechanics of large bubbles rising through extended liquids and through liquids in tubes. *Proc. Roy. Soc. Lond. A* **200**, 375–390.

Davis, E.J., Ravindran, P., and Ray, A.K. 1980. A review of theory and experiments on diffusion from submicroscopic particles. *Chem. Eng. Commun.* **5**, 251–268.

Davis, E.J., Ravindran, P., and Ray, A.K. 1981. Single aerosol particle studies. *Adv. Colloid Interface Sci.* **15**, 1–24.

Davis, M.H. 1969. The slow translation and rotation of two unequal spheres in a viscous fluid. *Chem. Eng. Sci.* **24**, 1769–1776.

Davis, R.E., and Acrivos, A. 1966. The influence of surfactants on the creeping motion of bubbles. *Chem. Eng. Sci.* **21**, 681–685.

Davis, R.H., Schonberg, J.A., and Rallison, J.M. 1989. The lubrication force between two viscous drops. *Phys. Fluids A* **1**, No. 1, 77–81.

de Socio, L.M. 1979. Convection driven by non-uniform surface tension. *Letters in Heat and Mass Transfer* **6**, 375–383.

Dean, W.R., and O'Neill, M.E. 1963. A slow motion of viscous liquid caused by the rotation of a solid sphere. *Mathematika* **10**, 13–24.

Defay, R., and Prigogine, I., with the collaboration of Bellemans, A. 1966. *Surface Tension and Adsorption*. Translated by D.H. Everett. London: Longmans, Green and Co.

Delitzsch, V., Eckelmann, H., and Wuest, W. 1984. The influence of thermocapillarity on the migration of droplets in a liquid possessing a uniform temperature gradient. *Proceedings of the 5th European Symposium on Material Sciences Under Microgravity* (pp. 245–249). European Space Agency SP-222. 5–7 November 1984, Schloss Elmau.

Dennis, S.C.R., and Walker, J.D.A. 1971. Calculation of the steady flow past a sphere at low and moderate Reynolds numbers. *J. Fluid Mech.* **48**, Pt. 4, 771–789.

Dill, L.H. 1991. On the thermocapillary migration of a growing or shrinking drop. *J. Colloid Interface Sci.* **146**, No. 2, 533–540.

Dill, L.H., and Balasubramaniam, R. 1992. Unsteady thermocapillary migration of isolated drops in creeping flow. *Int. J. Heat Fluid Flow* **13**, No. 1, 78–85.

Doi, T., and Koster, J.N. 1993. Thermocapillary convection in two immiscible liquid layers with free surface. *Phys. Fluids A* **5**, No. 8, 1914–1927.

Domb, C., and Sykes, M.F. 1957. On the susceptibility of a ferromagnetic above the Curie point. *Proc. Roy. Soc. Lond. A* **240**, 214–228.

Dragoo, A.L. 1974. Steady thermocapillary convection cells in liquid drops. *Proceedings of the International Colloquium on Drops and Bubbles*, ed. D.J. Collins, M.S. Plesset, and M.M. Saffren (vol. 1, pp. 208–226). Pasadena, CA: Cal Tech/JPL. 28–30 August 1974, Pasadena, California.

Duda, J.L., and Vrentas, J.S. 1969. Mathematical analysis of bubble dissolution. *AIChE J.* **15**, No. 3, 351–356.

Duda, J.L., and Vrentas, J.S. 1971. Heat or mass transfer-controlled dissolution of an isolated sphere. *Int. J. Heat Mass Transfer* **14**, 395–408.

Edwards, D.A., Brenner, H., and Wasan, D.T. 1991. *Interfacial Transport Processes and Rheology*. Boston: Butterworth-Heinemann.

Ehmann, M., Wozniak, G., and Siekmann, J. 1992. Numerical analysis of the thermocapillary migration of a fluid particle under zero-gravity. *Z. Angew. Math. Mech.* **72**, No. 8, 347–358.

Epstein, P.S., and Plesset, M.S. 1950. On the stability of gas bubbles in liquid-gas solutions. *J. Chem. Phys.* **18**, No. 11, 1505–1509.

Eyer, A., Leiste, H., and Nitsche, R. 1985. Floating zone growth of silicon under microgravity in a sounding rocket. *J. Crystal Growth* **71**, 173–182.

Eyer, A., and Leiste, H. 1985. Striation-free silicon crystals by float-zoning with surface-coated melt. *J. Crystal Growth* **71**, 249–252.

Feuillebois, F. 1989. Thermocapillary migration of two equal bubbles parallel to their line of centers. *J. Colloid Interface Sci.* **131**, No. 1, 267–274.

Fraenkel, L.E. 1969. On the method of matched asymptotic expansions. Part I: A matching principle. *Proc. Cambridge Phil. Soc.* **65**, 209–231.

Frumkin, A., and Levich, V. 1947. Effect of surface-active substances on movements at the boundaries of liquid phases (in Russian). *Zhur. Fiz. Khim.* **21,** 1183–1204.

Fu, B.-I., and Ostrach, S. 1985. Numerical solutions of thermocapillary flows in floating zones, in *Transport Phenomena in Materials Processing* (vol. 29, pp. 1–9). New York: The American Society of Mechanical Engineers, Heat Transfer Division.

Fuentes, Y.O., Kim, S., and Jeffrey, D.J. 1988. Mobility functions for two unequal viscous drops in Stokes flow. I. Axisymmetric motions. *Phys. Fluids* **31,** No. 9, 2445–2455.

Fuentes, Y.O., Kim, S., and Jeffrey, D.J. 1989. Mobility functions for two unequal viscous drops in Stokes flow. II. Asymmetric motions. *Phys. Fluids A* **1,** No. 1, 61–76.

Gaines, G.L. 1966. *Insoluble Monolayers at Liquid-Gas Interfaces.* New York: Wiley Interscience.

Galindo, V., Gerbeth, G., Langbein, D., and Treuner, M. 1994. Unsteady thermocapillary migration of isolated spherical drops in a uniform temperature gradient. *Microgravity Sci. Tech.* **VII/3,** 234–241.

Ganatos, P., Pfeffer, R., and Weinbaum, S. 1978. A numerical-solution technique for three-dimensional Stokes flows, with application to the motion of strongly interacting spheres in a plane. *J. Fluid Mech.* **84,** Pt. 1, 79–111.

Gebhart, B., Jaluria, Y., Mahajan, R.L., and Sammakia, B. 1988. *Buoyancy-Induced Flows and Transport.* New York: Hemisphere.

Goldman, A.J., Cox, R.G., and Brenner, H. 1967. Slow viscous motion of a sphere parallel to a plane wall – I Motion through a quiescent fluid. *Chem. Eng. Sci.* **22,** 637–651.

Goldstein, S. 1938. *Modern Developments in Fluid Dynamics, Volume II.* Oxford, England: Clarendon.

Golovin, A.A. 1995. Thermocapillary interaction between a solid particle and a gas bubble. *Int. J. Multiphase Flow* **21,** No. 4, 715–719.

Golovin, A.A., Gupalo, Yu P., and Ryazantsev, Yu S. 1986. Chemo-thermocapillary effect for the motion of a drop in a liquid (in Russian). *Dokl. Akad. Nauk SSSR* **290,** No. 1, 35–39. (English translation: 1986, *Sov. Phys. Dokl.* **31,** No. 9, 700–702.)

Golovin, A.A., Nir, A., and Pismen, L.M. 1995. Spontaneous motion of two droplets caused by mass transfer. *Ind. Eng. Chem. Res.* **34,** 3278–3288.

Greenspan, H.P. 1969. *The Theory of Rotating Fluids.* London, England: Cambridge University Press.

Guelcher, S.A., Solomentsev, Y.E., Sides, P.J., and Anderson, J.L. 1998. Thermocapillary phenomena and bubble coalescence during electrolytic gas evolution. *J. Electrochem. Soc.* **145,** No. 6, 1848–1855.

Guggenheim, E.A. 1967. *Thermodynamics.* Amsterdam: North-Holland.

Haber, S., Hetsroni, G., and Solan, A. 1973. On the low Reynolds number motion of two droplets. *Int. J. Multiphase Flow* **1,** 57–71.

Hadamard, J. 1911. Mouvement permanent lent d'une sphère liquide et visqueuse dans un liquide visqueux. *Compt. Rend. Acad. Sci. (Paris)* **152,** 1735–1738.

Hadland, P.H., Balasubramaniam, R., Wozniak, G., and Subramanian, R.S. 1999. Thermocapillary migration of bubbles and drops at moderate to large Marangoni number and moderate Reynolds number in reduced gravity. *Experiments in Fluids* **26,** 240–248.

Hähnel, M., Delitzsch, V., and Eckelmann, H. 1989. The motion of droplets in a vertical temperature gradient. *Phys. Fluids A* **1,** No. 9, 1460–1466.

Haj-Hariri, H., Nadim, A., and Borhan, A. 1990. Effect of inertia on the thermocapillary velocity of a drop. *J. Colloid Interface Sci.* **140,** No. 1, 277–286.

Haj-Hariri, H., Shi, Q., and Borhan, A. 1997. Thermocapillary motion of deformable drops at finite Reynolds and Marangoni numbers. *Phys. Fluids* **9,** No. 4, 845–855.

Hakimzadeh, R., Hrovat, K., McPherson, K.M., Moskowitz, M.E., and Rogers, M.J.B. 1997. Summary report of mission acceleration measurements for STS-78. *NASA Technical Memorandum 107401.*

Hamacher, H., Fitton, B., and Kingdon, J. 1987. The environment of Earth-orbiting systems. In *Fluid Sciences and Materials Science in Space,* ed. H.U. Walter (Chap. I: pp. 1–50). Berlin: Springer-Verlag.

Happel, J., and Brenner, H. 1965. *Low Reynolds Number Hydrodynamics.* Englewood Cliffs, New Jersey: Prentice-Hall.

Hardy, S.C. 1979. The motion of bubbles in a vertical temperature gradient. *J. Colloid Interface Sci.* **69**, No. 1, 157–162.

Hardy, S.C. 1984. The surface tension of liquid silicon. *J. Crystal Growth* **69**, 456–460.

Harper, J.F. 1970. On bubbles rising in line at large Reynolds numbers. *J. Fluid Mech.* **41**, Pt. 4, 751–758.

Harper, J.F. 1972. The motion of bubbles and drops through liquids. In *Advances in Applied Mechanics,* ed. C.S. Yih (vol. 12, pp. 59–129). New York: Academic Press.

Harper, J.F. 1997. Bubbles rising in line: why is the first approximation so bad? *J. Fluid Mech.* **351**, 289–300.

Harper, J.F., and Moore, D.W. 1968. The motion of a spherical liquid drop at high Reynolds number. *J. Fluid Mech.* **32**, Pt. 2, 367–391.

Harper, J.F., Moore, D.W., and Pearson, J.R.A. 1967. The effect of the variation of surface tension with temperature on the motion of bubbles and drops. *J. Fluid Mech.* **27**, Pt. 2, 361–366.

Hassonjee, Q., Ganatos, P., and Pfeffer, R. 1988. A strong-interaction theory for the motion of arbitrary three-dimensional clusters of spherical particles at low Reynolds number. *J. Fluid Mech.* **197**, 1–37.

He, Z., Maldarelli, C., and Dagan, Z. 1991. The size of stagnant caps of bulk soluble surfactant on the interfaces of translating fluid droplets. *J. Colloid Interface Sci.* **146**, No. 2, 442–451.

Herron, I.H., Davis, S.H., and Bretherton, F.P. 1975. On the sedimentation of a sphere in a centrifuge. *J. Fluid Mech.* **68**, Pt. 2, 209–234.

Hetsroni, G., and Haber, S. 1970. The flow in and around a droplet or bubble submerged in an unbound arbitrary velocity field. *Rheol. Acta* **9**, 488–496.

Hetsroni, G., and Haber, S. 1978. Low Reynolds number motion of two drops submerged in an unbounded arbitrary velocity field. *Int. J. Multiphase Flow* **4**, No. 1, 1–17.

Hill, T.L. 1963/1964. *Thermodynamics of small systems. Parts I and II.* New York: Benjamin.

Hinch, E.J. 1991. *Perturbation Methods.* Cambridge, England: Cambridge University Press.

Hnat, J.G., and Buckmaster, J.D. 1976. Spherical cap bubbles and skirt formation. *Phys. Fluids* **19**, No. 2, 182–194. Erratum. *Ibid.* **19**, No. 4, 611.

Holbrook, J.A., and LeVan, M.D. 1983a. Retardation of droplet motion by surfactant. Part 1. Theoretical development and asymptotic solutions. *Chem. Eng. Commun.* **20**, 191–207.

Holbrook, J.A., and LeVan, M.D. 1983b. Retardation of droplet motion by surfactant. Part 2. Numerical solutions for exterior diffusion, surface diffusion, and adsorption kinetics. *Chem. Eng. Commun.* **20**, 273–290.

Hyer, J.R., Jankowski, D.F., and Neitzel, G.P. 1991. Thermocapillary convection in a model float zone. *J. Thermophysics* **5**, No. 4, 577–582.

Jacobi, A.M., and Goldschmidt, V.W. 1989. The effect of surface tension variation on filmwise condensation and heat transfer on a cylinder in crossflow. *Int. J. Heat Mass Transfer* **32**, No. 8, 1483–1490.

Jasper, J.J. 1972. The surface tension of pure liquid compounds. *J. Phys. Chem. Ref. Data* **1**, No. 4, 841–1009.

Jayaraj, K., Cole, R., and Subramanian, R.S. 1982. Combined thermocapillary and buoyant flow in a drop in a space laboratory. *J. Colloid Interface Sci.* **85**, No. 1, 66–77.

Jaycock, M.J., and Parfitt, G.D. 1981. *Chemistry of Interfaces.* New York: Ellis Horwood.

Jeffery, G.B. 1912. On a form of the solution of Laplace's equation suitable for problems relating to two spheres. *Proc. Roy. Soc. Lond. A* **87**, 109–120.

Jeffrey, D.J. 1973. Conduction through a random suspension of spheres. *Proc. Roy. Soc. Lond. A* **335**, 355–367.

Jeffrey, D.J. 1982. Low-Reynolds-number flow between converging spheres. *Mathematika* **29**, 58–66.

Johnson, R.E., and Sadhal, S.S. 1985. Fluid mechanics of compound multiphase drops and bubbles. In *Ann. Rev. Fluid Mech.,* ed. M. Van Dyke, J.V. Wehausen, and J.L. Lumley (vol. 17, pp. 289–320). Palo Alto, CA: Annual Reviews, Inc.

Joseph, D.D. 1977. The convergence of biorthogonal series for biharmonic and Stokes flow edge problems. Part I. *SIAM J. Appl. Math.* **33,** No. 2, 337–347.

Joseph, D.D., and Sturges, L. 1978. The convergence of biorthogonal series for biharmonic and Stokes flow edge problems: Part II. *SIAM J. Appl. Math.* **34,** No. 1, 7–26.

Kamotani,Y., and Lee, K.J. 1989. Oscillatory thermocapillary flow in a liquid column heated by a ring heater. *Physicochemical Hydrodynamics* **11,** No. 5/6, 729–736.

Kamotani,Y., Ostrach, S., and Masud, J. 1999. Oscillatory thermocapillary flows in open cylindrical containers induced by CO_2 laser heating. *Int. J. Heat Mass Transfer* **42,** 555–564.

Kamotani,Y., Ostrach, S., and Pline, A. 1994. Analysis of velocity data taken in surface tension driven convection experiment in microgravity. *Phys. Fluids* **6,** No. 11, 3601–3609.

Kamotani, Y., Ostrach, S., and Vargas, M. 1984. Oscillatory thermocapillary convection in a simulated floating zone configuration. *J. Crystal Growth* **66,** 83–90.

Kang, I.S., and Leal, L.G. 1988. The drag coefficient for a spherical bubble in a uniform streaming flow. *Phys. Fluids* **31,** No. 2, 233–237.

Kao, Y.S., and Kenning, D.B.R. 1972. Thermocapillary flow near a hemispherical bubble on a heated wall. *J. Fluid Mech.* **53,** Pt. 4, 715–735.

Kaplun, S., and Lagerstrom, P.A. 1957. Asymptotic expansions of Navier–Stokes solutions for small Reynolds numbers. *J. Math. Mech.* **6,** No. 5, 585–593.

Keh, H.J., and Chen, L.S. 1992. Droplet interactions in axisymmetric thermocapillary motion. *J. Colloid Interface Sci.* **151,** No. 1, 1–16.

Keh, H.J., and Chen, L. S. 1993. Droplet interactions in thermocapillary migration. *Chem. Eng. Sci.* **48,** 3565–3582.

Keh, H.J., and Chen, S.H. 1990. The axisymmetric thermocapillary motion of two fluid droplets. *Int. J. Multiphase Flow* **16,** No. 3, 515–527.

Kenning, D.B.R. 1969. The effect of surface energy variations on the motion of bubbles and drops. *Chem. Eng. Sci.* **24,** 1385–1386.

Kevorkian, J., and Cole, J.D. 1981. *Perturbation Methods in Applied Mathematics.* New York: Springer-Verlag.

Kim, H.S. 1988. Surfactant Effects on the Thermocapillary Migration of a Droplet. Ph.D. diss., in chemical engineering, Clarkson University, Potsdam, NY.

Kim, H.S., and Subramanian, R.S. 1989a. Thermocapillary migration of a droplet with insoluble surfactant I. Surfactant cap. *J. Colloid Interface Sci.* **127,** No. 2, 417–428.

Kim, H.S., and Subramanian, R.S. 1989b. The thermocapillary migration of a droplet with insoluble surfactant II. General case. *J. Colloid Interface Sci.* **130,** No. 1, 112–129.

Kim, S., and Karrila, S.J. 1991. *Microhydrodynamics: Principles and Selected Applications.* Boston: Butterworth-Heinemann.

Koschmieder, E.L. 1993. *Bénard Cells and Taylor Vortices.* Cambridge, England: Cambridge University Press.

Kronig, R., and Brink, J.C. 1950. On the theory of extraction from falling droplets. *Appl. Sci. Res.* **A2,** 142–154.

Kuhlmann, H.C., and Rath, H.J. 1993. Hydrodynamic instabilities in cylindrical thermocapillary liquid bridges. *J. Fluid Mech.* **247,** 247–274.

Kurdyumov, V.N., Rednikov, A. Ye., and Ryazantsev, Yu S. 1994. Thermocapillary motion of a bubble with heat generation at the surface. *Microgravity Q.* **4,** No. 1, 5–8.

Kurosaki, Y., Satoh, I., Horiuchi, T., and Kashiwagi, T. 1989. Effect of Marangoni convection on the temperature profiles of a free surface subject to nonuniform radiative heating. *Experimental Thermal and Fluid Science* **2,** 365–373.

Lacy, L.L., Witherow, W.K., Facemire, B.R., and Nishioka, G.M. 1982. Optical studies of a model binary miscibility gap system. *NASA Technical Memorandum 82494.*

Lai, C.L. 1984. Studies of Thermocapillary Oscillation Phenomena. Ph.D. diss., in mechanical and aerospace engineering, Case Western Reserve University, Cleveland, Ohio.

Lamb, H. 1932. *Hydrodynamics.* Cambridge, England: Cambridge University Press.

Langbein, D., and Heide, W. 1984. The separation of liquids due to Marangoni convection. *Advances in Space Research* **4/5,** 27–36.

Langmuir, I. 1917. The constitution and fundamental properties of solids and liquids. II. Liquids. *J. Amer. Chem. Soc.* **39,** 1848–1906.

Larkin, B.K. 1970. Thermocapillary flow around hemispherical bubble. *AIChE J.* **16,** No. 1, 101–107.

Lavrenteva, O.M., Leshansky, A.M., and Nir, A. 1999. Spontaneous thermocapillary interaction of drops, bubbles, and particles: Unsteady convective effects at low Peclet numbers. *Phys. Fluids* **11,** No. 7, 1768–1780.

Leal, L.G. 1980. Particle motions in a viscous fluid. In *Ann. Rev. Fluid Mech.* ed. M. Van Dyke, J.V. Wehausen, and J.L. Lumley (vol. 12, pp. 435–476). Palo Alto, CA: Annual Reviews, Inc.

Leal, L.G. 1989. Computational studies of drop and bubble dynamics in a viscous fluid. In *Proceedings of Third International Colloquium on Drops and Bubbles,* Monterey, California, 1988. AIP Conference Proceedings 197, ed. T.G. Wang (pp. 147–168).

Leal, L.G. 1992. *Laminar Flow and Convective Transport Processes.* Boston: Butterworth-Heinemann.

Lee, Y.-T., and Chen, J.-C. 1990. Numerical investigation of the thermocapillary flow around a deformable bubble on the hot wall. *J. Chinese Soc. Mech. Engrs.* **11,** No. 3, 233–243.

Leshansky, A.M., Golovin, A.A., and Nir, A. 1997. Thermocapillary interaction between a solid particle and a liquid-gas interface. *Phys. Fluids* **9,** No. 10, 2818–2827.

LeVan, M.D. 1981. Motion of a droplet with a Newtonian interface. *J. Colloid Interface Sci.* **83,** No. 1, 11–17.

Levich, V.G. 1949. Motion of bubbles at large Reynolds numbers (in Russian). *Zhur. Eksp. i Teoret. Fiz.* **19,** 18–24.

Levich, V.G. 1962. *Physicochemical Hydrodynamics.* Englewood Cliffs, NJ: Prentice-Hall.

Loewenberg, M., and Davis, R.H. 1993a. Near-contact, thermocapillary migration of a non-conducting, viscous drop normal to a planar interface. *J. Colloid Interface Sci.* **160,** 265–274.

Loewenberg, M., and Davis, R.H. 1993b. Near-contact thermocapillary motion of two noncon-ducting drops. *J. Fluid Mech.* **256,** 107–131.

Lorentz, H.A. 1907. Ein allgemeiner satz, die bewegung einer reibenden flüssigkeit betreffend, nebst einigen anwendungen desselben. *Abhand. Theor. Phys., Leipzig* **1,** 23–42.

Lovalenti, P.M., and Brady, J.F. 1993. The hydrodynamic force on a rigid particle under-going arbitrary time-dependent motion at small Reynolds number. *J. Fluid Mech.* **256,** 561–605.

Lowry, B.J., and Steen, P.H. 1997. Stability of slender liquid bridges subjected to axial flows. *J. Fluid Mech.* **330,** 189–213.

Lyklema, J. 2000. *Fundamentals of Interface and Colloid Science, Volume III.* London, England: Academic Press.

Ma, X. 1998. Numerical Simulation and Experiments on Liquid Drops in a Vertical Temperature Gradient in a Liquid of Nearly the Same Density. Ph.D. diss., in chemical engineering, Clarkson University, Potsdam, New York.

Ma, X., Balasubramaniam, R., and Subramanian, R.S. 1999. Numerical simulation of thermo-capillary drop motion with internal circulation. *Num. Heat Transfer A* **35,** 291–309.

Mackay, G.D.M., Suzuki, M., and Mason, S.G. 1963. Approach of a solid sphere to a rigid plane interface. Part 2. *J. Colloid Interface Sci.* **18,** 103–104.

MacRobert, T.M. 1967. *Spherical Harmonics.* Oxford, England: Pergamon Press.

Manga, M., and Stone, H.A. 1993. Buoyancy-driven interactions between two deformable viscous drops. *J. Fluid Mech.* **256,** 647–683.

Manga, M., and Stone, H.A. 1995a. Low Reynolds number motion of bubbles, drops and rigid spheres through fluid-fluid interfaces. *J. Fluid Mech.* **287,** 279–298.

Manga, M., and Stone, H.A. 1995b. Collective hydrodynamics of deformable drops and bubbles in dilute low Reynolds number suspensions. *J. Fluid Mech.* **300,** 231–263.

Maris, H.J., Seidel, G.M., and Williams, F.I.B. 1987. Experiments with supercooled liquid hydrogen. *Phys. Rev. B* **36,** No. 13, 6799–6810.

Masliyah, J.H. 1994. *Electrokinetic Transport Phenomena.* Edmonton, Canada: Alberta Oil Sands Technology and Research Authority.

Masud, J., Kamotani, Y., and Ostrach, S. 1997. Oscillatory thermocapillary flow in cylindrical columns of high Prandtl number fluids. *J. Thermophysics and Heat Transfer* **11**, No. 1, 105–111.

Maude, A.D. 1961. End effects in a falling-sphere viscometer. *British J. Appl. Phys.* **12**, 293–295.

McGrew, J.L., Bamford, F.L., and Rehm, T.R. 1966. Marangoni flow: An additional mechanism in boiling heat transfer. *Science* **153**, 1106–1107.

McGrew, J.L., Rehm, T.R., and Griskey, R.G. 1974. The effect of temperature induced surface tension gradients on bubble mechanics. *Appl. Sci. Res.* **29**, 195–210.

McLaughlin, J.B. 1996. Numerical simulation of bubble motion in water. *J. Colloid Interface Sci.* **184**, 614–625.

McNeil, T.J., Cole, R., and Subramanian, R.S. 1984. Thermocapillary convection in a liquid bridge. *J. Colloid Interface Sci.* **98**, No. 1, 210–222.

McNeil, T.J., Cole, R., and Subramanian, R.S. 1985. Surface-tension-driven flow in a glass melt. *J. Amer. Ceram. Soc.* **68**, No. 5, 254–259.

Merritt, R.M. 1988. Bubble Migration and Interactions in a Vertical Temperature Gradient. Ph.D. diss., in chemical engineering, Clarkson University, Potsdam, New York.

Merritt, R.M., Morton, D.S., and Subramanian, R.S. 1993. Flow structures in bubble migration under the combined action of buoyancy and thermocapillarity. *J. Colloid Interface Sci.* **155**, 200–209.

Merritt, R.M., and Subramanian, R.S. 1988. The migration of isolated gas bubbles in a vertical temperature gradient. *J. Colloid Interface Sci.* **125**, No. 1, 333–339.

Merritt, R.M., and Subramanian, R.S. 1989. Migration of a gas bubble normal to a plane horizontal surface in a vertical temperature gradient. *J. Colloid Interface Sci.* **131**, No. 2, 514–525.

Merritt, R.M., and Subramanian, R.S. 1992. Bubble migration under the combined action of buoyancy and thermocapillarity. In *Microgravity Fluid Mechanics IUTAM Symposium Bremen 1991*, ed. H.J. Rath (pp. 237–244). Berlin: Springer-Verlag.

Meyyappan, M. 1984. Interaction Effects in Thermocapillary Bubble Migration. Ph.D. diss., in chemical engineering, Clarkson University, Potsdam, New York.

Meyyappan, M., and Subramanian, R.S. 1987. Thermocapillary migration of a gas bubble in an arbitrary direction with respect to a plane surface. *J. Colloid Interface Sci.* **115**, No. 1, 206–219.

Meyyappan, M., Subramanian, R.S., Wilcox, W.R., and Smith, H. 1982. Bubble behavior in molten glass in a temperature gradient. In *Materials Processing in the Reduced Gravity Environment of Space*, ed. G.E. Rindone (pp. 311–314). New York: North-Holland.

Meyyappan, M., Wilcox, W.R., and Subramanian, R.S. 1981. Thermocapillary migration of a bubble normal to a plane surface. *J. Colloid Interface Sci.* **83**, No. 1, 199–208.

Meyyappan, M., Wilcox, W.R., and Subramanian, R.S. 1983. The slow axisymmetric motion of two bubbles in a thermal gradient. *J. Colloid Interface Sci.* **94**, No. 1, 243–257.

Miller, C.A., and Neogi, P. 1985. *Interfacial Phenomena*. New York: Marcel Dekker.

Moffatt, H.K. 1964. Viscous and resistive eddies near a sharp corner. *J. Fluid Mech.* **18**, Pt. 1, 1–18.

Mok, L.S., and Kim, K. 1987. Motion of a gas bubble inside a spherical liquid container with a vertical temperature gradient. *J. Fluid Mech.* **176**, 521–531.

Moon, P., and Spencer, D.E. 1971. *Field Theory Handbook*. Berlin: Springer-Verlag.

Moore, D.W. 1959. The rise of a gas bubble in a viscous liquid. *J. Fluid Mech.* **6**, 113–130.

Moore, D.W. 1963. The boundary layer on a spherical gas bubble. *J. Fluid Mech.* **16**, 161–176.

Moore, D.W. 1965. The velocity of rise of distorted gas bubbles in a liquid of small viscosity. *J. Fluid Mech.* **23**, Pt. 4, 749–766.

Morick, F., and Woermann, D. 1993. Migration of air bubbles in silicone oil under the action of buoyancy and thermocapillarity. *Ber. Bunsenges. Phys. Chem.* **97**, No. 8, 961–969.

Morton, D.S. 1990. The Motion of Compound Drops. Ph.D. diss., in chemical engineering, Clarkson University, Potsdam, New York.

Morton, D.S., Subramanian, R.S., and Balasubramaniam, R. 1990. The migration of a compound drop due to thermocapillarity. *Phys. Fluids A* **2**, No. 12, 2119–2133.

Myshkis, A.D., Babskii, V.G., Kopachevskii, N.D., Slobozhanin, L.A., and Tyuptsov, A.D. 1987. *Low-Gravity Fluid Mechanics*. Berlin: Springer-Verlag. Originally published in Russian, 1976, *Gidromekhanika Nevesomosti*. Moscow: Nauka.

Nadim, A., and Borhan, A. 1989. The effects of surfactants on the motion and deformation of a droplet in thermocapillary migration. *Physicochemical Hydrodynamics* **11,** No. 5/6, 753–764.

Nadim, A., Haj-Hariri, H., and Borhan, A. 1990. Thermocapillary migration of slightly deformed droplets. *Particulate Sci. Tech.* **8,** 191–198.

Nähle, R., Neuhaus, D., Siekmann, J., Wozniak, G., and Srulijes, J. 1987. Separation of fluid phases and bubble dynamics in a temperature gradient – a Spacelab D1 experiment. *Z. Flugwiss. Weltraumforsch.* **11,** 211–213.

Nallani, M., and Subramanian, R.S. 1993. Migration of methanol drops in a vertical temperature gradient in a silicone oil. *J. Colloid Interface Sci.* **157,** 24–31.

Napolitano, L.G., Monti, R., and Russo, G. 1986. Marangoni convection in one- and two-liquids floating zones. *Naturwiss.* **73,** 352–355.

Nas, S. 1995. Computational Investigation of Thermocapillary Migration of Bubbles and Drops in Zero Gravity. Ph.D. diss., in aerospace engineering, University of Michigan, Ann Arbor.

Neilson, G.F., and Weinberg, M.C. 1977. Outer space formation of a laser host glass. *J. Non-Crystalline Solids* **23,** No. 1, 43–58.

Neitzel, G.P., Chang, K.-T., Jankowski, D.F., and Mittelmann, H.D. 1993. Linear-stability theory of thermocapillary convection in a model of the float-zone crystal growth process. *Phys. Fluids A* **5,** No. 1, 108–114.

Neuhaus, D., and Feuerbacher, B. 1987. Bubble motions induced by a temperature gradient. *Proceedings of the 6th European Symposium on Materials Sciences under Microgravity Conditions* (pp. 241–244). European Space Agency SP-256. 2–5 December 1986, Bordeaux, France.

Oğuz, H.N., and Sadhal, S.S. 1987. Growth and collapse of translating compound multiphase drops: Analysis of fluid mechanics and heat transfer. *J. Fluid Mech.* **179,** 105–136.

Oliver, D.L.R., and DeWitt, K.J. 1988. Surface tension driven flows for a droplet in a micro-gravity environment. *Int. J. Heat Mass Transfer* **31,** No. 7, 1534–1537.

Oliver, D.L.R., and DeWitt, K.J. 1994. Transient motion of a gas bubble in a thermal gradient in low gravity. *J. Colloid Interface Sci.* **164,** 263–268.

O'Neill, M.E. 1964. A slow motion of viscous liquid caused by a slowly moving solid sphere. *Mathematika* **11,** 67–74.

O'Neill, M.E. 1967. A slow motion of viscous liquid caused by a slowly moving solid sphere: An addendum. *Mathematika* **14,** 170–172.

O'Neill, M.E., and Majumdar, S.R. 1970a. Asymmetrical slow viscous fluid motions caused by the translation or rotation of two spheres. Part I: The determination of exact solutions for any values of the ratio of radii and separation parameters. *Z. Angew. Math. Phys.* **21,** No. 2, 164–179.

O'Neill, M.E., and Majumdar, S.R. 1970b. Asymmetrical slow viscous fluid motions caused by the translation or rotation of two spheres. Part II: Asymptotic forms of the solutions when the minimum clearance between the spheres approaches zero. *Z. Angew. Math. Phys.* **21,** No. 2, 180–187.

O'Neill, M.E., and Stewartson, K. 1967. On the slow motion of a sphere parallel to a nearby plane wall. *J. Fluid Mech.* **27,** Pt. 4, 705–724.

Ono, S., and Kondo, S. 1960. Molecular theory of surface tension in liquids. In *Handbuch der Physik, Band X: Struktur der Flüssigkeiten,* ed. S. Flügge (pp. 134–280). Berlin: Springer-Verlag.

Oseen, C.W. 1910. Über die Stokes'sche formel, und über eine verwandte aufgabe in der hydrodynamik. *Arkiv för matematik astronomi och fysik* Band 6, N:o 29, 1–20.

Ostrach, S. 1979. Convection due to surface-tension gradients. In *COSPAR Space Research,* ed. M.J. Rycroft (vol. 19, pp. 563–570). New York: Pergamon.

Ostrach, S. 1982. Low-gravity fluid flows. In *Ann. Rev. Fluid Mech.* ed. M. Van Dyke, J.V. Wehausen, and J.L. Lumley (vol. 14, pp. 313–345). Palo Alto, CA: Annual Reviews, Inc.

Pan, F., and Acrivos, A. 1967. Steady flows in rectangular cavities. *J. Fluid Mech.* **28,** Pt. 4, 643–655.

Pan, F.Y., and Acrivos, A. 1968. Shape of a drop or bubble at low Reynolds number. *Ind. Eng. Chem. Fund.* **7,** No. 2, 227–232.

Papazian, J.M., and Wilcox, W.R. 1978. Interaction of bubbles with solidification interfaces. *AIAA J.* **16,** 447–451.

Pearson, J.R.A. 1958. On convection cells induced by surface tension. *J. Fluid Mech.* **4,** 489–500.

Pimputkar, S.M., and Ostrach, S. 1980. Transient thermocapillary flows in thin layers. *Phys. Fluids* **23,** No. 7, 1281–1285.

Plateau, J. 1863–1866. Experimental and theoretical researches on the figures of equilibrium of a liquid mass withdrawn from the action of gravity. Translated in *The Annual Reports of the Board of Regents of the Smithsonian Institution,* Pt. 1, House of Representatives Misc. Doc. #83, 38th Congress, 1st Session, pp. 207–285; Pts. 2–4, Misc. Doc. #54, 38th Congress, 2nd Session, pp. 285–369; Pt. 5, Misc. Doc. #102, 39th Congress, 2nd Session, pp. 411–435; Pt. 6, Misc. Doc. #83, 39th Congress, 2nd Session, pp. 255–289.

Prandtl, L. 1905. Über flüssigkeitsbewegung bei sehr kleiner reibung. In *Proceedings of the 3rd International Mathematics Congress, Heidelberg,* 1904 (pp. 484–491). Reprinted as NACA TM 452 in 1928.

Preisser, F., Schwabe, D., and Scharmann, A. 1983. Steady and oscillatory thermocapillary convection in liquid columns with free cylindrical surface. *J. Fluid Mech.* **126,** 545–567.

Press, W.H., Teukolsky, S.A., Vetterling, W.T., and Flannery, B.P. 1992. *Numerical Recipes in Fortran.* Cambridge, England: Cambridge University Press.

Probstein, R.F. 1994. *Physicochemical Hydrodynamics.* New York: Wiley-Interscience.

Proudman, I., and Pearson, J.R.A. 1957. Expansions at small Reynolds numbers for the flow past a sphere and a circular cylinder. *J. Fluid Mech.* **2,** Pt. 3, 237–262.

Quintana, G.C. 1992. The effect of surfactants on flow and mass transport to drops and bubbles. In *Transport Processes in Bubbles, Drops, and Particles,* ed. R.P. Chhabra and D. De Kee (Chap 4: pp. 87–113). New York: Hemisphere.

Raake, D., Siekmann, J., and Chun Ch.-H. 1989. Temperature and velocity fields due to surface tension driven flow. *Experiments in Fluids* **7,** 164–172.

Ramos, J.I. 1986. Growth of multicomponent gas bubbles. *Chem. Eng. Commun.* **40,** 321–334.

Rashidnia, N., and Balasubramaniam, R. 1991. Thermocapillary migration of liquid droplets in a temperature gradient in a density matched system. *Experiments in Fluids* **11,** 167–174.

Rayleigh, Lord. 1945. *The Theory of Sound* (vol. 2, Chap. 20: pp. 343–375). New York: Dover.

Readey, D.W., and Cooper, A.R. 1966. Molecular diffusion with a moving boundary and spherical symmetry. *Chem. Eng. Sci.* **21,** 917–922.

Rednikov, A. Ye., Ryazantsev, Yu S., and Velarde, M.G. 1994a. On the development of translational subcritical Marangoni instability for a drop with uniform internal heat generation. *J. Colloid Interface Sci.* **164,** 168–180.

Rednikov, A. Ye., Ryazantsev, Yu S., and Velarde, M.G. 1994b. Active drops and drop motions due to nonequilibrium phenomena. *J. Non-Equilib. Thermodyn.* **19,** 95–113.

Reed, C.C., and Anderson, J.L. 1980. Hindered settling of a suspension at low Reynolds number. *AIChE J.* **26,** No. 5, 816–827.

Reid, R.C., Prausnitz, J.M., and Poling, B.E. 1987. *The Properties of Gases and Liquids.* New York: McGraw Hill.

Rippin, D.W.T., and Davidson, J.F. 1967. Free streamline theory for a large gas bubble in a liquid. *Chem. Eng. Sci.* **22,** 217–228.

Rivas, D., and Ostrach, S. 1992. Scaling of low-Prandtl-number thermocapillary flows. *Int. J. Heat Mass Transfer* **35,** No. 6, 1469–1479.

Rosner, D.E., Mackowski, D.W., Tassopoulos, M., Castillo, J., and Garcia-Ybarra, P. 1992. Effects of heat transfer on the dynamics and transport of small particles suspended in gases. *Ind. Eng. Chem. Res.* **31,** 760–769.

Ruggles, J.S., Cook, R.G., Annamalai, P., and Cole, R. 1988. Bubble and drop trajectories in rotating flows. *Experimental Thermal and Fluid Sci.* **1,** 293–301.

Ruggles, J.S., Cook, R.G., and Cole, R. 1990. Microgravity bubble migration in rotating flows. *J. Spacecraft and Rockets* **27,** No. 1, 43–47.

Rushton, E., and Davies, G.A. 1973. The slow unsteady settling of two fluid spheres along their line of centers. *Appl. Sci. Res.* **28,** 37–61.

Rushton, E., and Davies, G.A. 1974. The motion of liquid droplets in settling and coalescence. *Proceedings of the International Solvent Extraction Conference* (1, pp. 289–317). 8–14 September 1974, Lyon, France.

Rushton, E., and Davies, G.A. 1978. The slow motion of two spherical particles along their line of centers. *Int. J. Multiphase Flow* **4,** 357–381.

Russel, W.B., Saville, D.A., and Schowalter, W.R. 1989. *Colloidal Dispersions.* Cambridge, England: Cambridge University Press.

Ryazantsev, Yu S. 1985. Thermocapillary motion of a reacting droplet in a chemically active medium (in Russian). *Izv. Akad. Nauk SSSR, Mekh. Zhidk. Gaza* **3,** 180–183.

Rybczyński, W. 1911. Über die fortschreitende Bewegung einer flüssigen Kugel in einem zähen Medium. *Bull. Acad. Sci. Cracovie* ser. **A,** 40–46.

Ryskin, G., and Leal, L.G. 1984a. Numerical solution of free-boundary problems in fluid mechanics. Part 1. The finite-difference technique. *J. Fluid Mech.* **148,** 1–17.

Ryskin, G., and Leal, L.G. 1984b. Numerical solution of free-boundary problems in fluid mechanics. Part 2. Buoyancy-driven motion of a gas bubble through a quiescent liquid. *J. Fluid Mech.* **148,** 19–35.

Sadhal, S.S. 1983. A note on the thermocapillary migration of a bubble normal to a plane surface. *J. Colloid Interface Sci.* **95,** No. 1, 283–285.

Sadhal, S.S., Ayyaswamy, P.S., and Chung, J.N. 1997. *Transport Phenomena with Drops and Bubbles.* New York: Springer-Verlag.

Sadhal, S.S., and Johnson, R.E. 1983. Stokes flow past bubbles and drops partially coated with thin films. Part I. Stagnant cap of surfactant film – exact solution. *J. Fluid Mech.* **126,** 237–250.

Sadhal, S.S., and Oğuz, H.N. 1985. Stokes flow past compound multiphase drops: the case of completely engulfed drops/bubbles. *J. Fluid Mech.* **160,** 511–529.

Sampson, R.A. 1891. On Stokes's current function. *Phil. Trans. Roy. Soc.* A**182,** 449–518.

Sani, R.L. 1963. Convective Instability. Ph.D. diss., in chemical engineering, University of Minnesota, Minneapolis.

Sankaran, S., and Saville, D.A. 1993. Experiments on the stability of a liquid bridge in an axial electric field. *Phys. Fluids A* **5,** No. 4, 1081–1083.

Sanochkin, Yu. V. 1989. Thermocapillary convection associated with non-uniform heating of the free surface of a liquid (in Russian). *Izv. Akad. Nauk SSSR, Mekh. Zhidk. Gaza.* **1,** 136–142.

Satrape, J.V. 1992. Interactions and collisions of bubbles in thermocapillary motion. *Phys. Fluids A* **4,** No. 9, 1883–1900.

Satterlee, H.M., and Reynolds, W.C. 1964. The dynamics of the free liquid surface in cylindrical containers under strong capillary and weak gravity conditions. Department of Mechanical Engineering, Stanford University, Tech. Report LG-2, Stanford, California, U.S.A.

Savic, P. 1953. Circulation and distortion of liquid drops falling through a viscous medium. *National Research Laboratories, Ottawa, Canada, Division of Mechanical Engineering Report MT-22.*

Saville, D.A. 1970. Electrohydrodynamic stability: Fluid cylinders in longitudinal electric fields. *Phys. Fluids* **13,** No. 12, 2987–2994.

Schrage, D.L., and Perkins, H.C. 1972. Isothermal bubble motion through a rotating liquid. *Trans. ASME J. Basic Eng.* **94,** 187–192.

Schwabe, D. 1988. Surface-tension-driven flow in crystal growth melts. In *Crystals-11,* ed. H.C. Freyhart (pp. 75–112). Berlin-Heidelberg: Springer-Verlag.

Schwabe, D., Dupont, O., Queeckers, P., and Legros, J.C. 1990. Experiments on Marangoni-Benard instability problems under normal and microgravity conditions. *Proceedings of the 7ᵗʰ European Symposium on Materials and Fluid Sciences in Microgravity* (pp. 291–298), European Space Agency SP-295. 10–15 September 1989, Oxford, UK.

Schwabe, D., Preisser, F., and Scharmann, A. 1982. Verification of the oscillatory state of thermo-capillary convection in a floating zone under low gravity. *Acta Astronautica* **9,** No. 4, 265–273.

Schwabe, D., and Scharmann, A. 1979. Some evidence for the existence and magnitude of a critical Marangoni number for the onset of oscillatory flow in crystal growth melts. *J. Crystal Growth* **46,** 125–131.

Schwabe, D., and Scharmann, A. 1984. Measurements of the critical Marangoni number of the laminar-oscillatory transition of thermocapillary convection in floating zones. *Proceedings of the 5ᵗʰ European Symposium on Materials Sciences under Microgravity* (pp. 281–289), European Space Agency SP-222. 5–7 November 1984, Schloss Elmau.

Schwabe, D., and Scharmann, A. 1985. Messung der kritischen Marangonizahl für den übergang von stationärer zu oszillatorischer thermokapillarer konvektion unter mikrogravitation: Ergebnisse der experimente in den ballistischen raketen Texus 5 und Texus 8. *Z. Flugwiss. Weltraumforsch.* **9,** 21–28.

Schwabe, D., Scharmann, A., Preisser, F., and Oeder, R. 1978. Experiments on surface tension driven flow in floating zone melting. *J. Crystal Growth* **43,** 305–312.

Scriven, L.E. 1959. On the dynamics of phase growth. *Chem. Eng. Sci.* **10,** Nos. 1/2, 1–13.

Scriven, L.E. 1960. Dynamics of a fluid interface. *Chem. Eng. Sci.* **12,** 98–108.

Sen, A.K., and Davis, S.H. 1982. Steady thermocapillary flows in two-dimensional slots. *J. Fluid Mech.* **121,** 163–186.

Shankar, N. 1984. Motion of bubbles due to thermocapillary effects. Ph.D. diss., in chemical engineering, Clarkson University, Potsdam, New York.

Shankar, N., Cole, R., and Subramanian, R.S. 1981. Thermocapillary migration of a fluid droplet inside a drop in a space laboratory. *Int. J. Multiphase Flow* **7,** No. 6, 581–594.

Shankar, N., and Subramanian, R.S. 1983. The slow axisymmetric thermocapillary migration of an eccentrically placed bubble inside a drop in zero gravity. *J. Colloid Interface Sci.* **94,** No. 1, 258–275.

Shankar, N., and Subramanian, R.S. 1988. The Stokes motion of a gas bubble due to interfacial tension gradients at low to moderate Marangoni numbers. *J. Colloid Interface Sci.* **123,** No. 2, 512–522.

Shankar, N., Wiltshire, T.J., and Subramanian, R.S. 1984. The dissolution or growth of a sphere. *Chem. Eng. Commun.* **27,** 263–281.

Shaw, D.J. 1969. *Introduction to Colloid and Surface Chemistry.* London: Butterworths.

Shen, Y., Neitzel, G.P., Jankowski, D.F., and Mittelmann, H.D. 1990. Energy stability of thermo-capillary convection in a model of the float-zone crystal-growth process. *J. Fluid Mech.* **217,** 639–660.

Sherman, F.S. 1990. *Viscous Flow.* New York: McGraw-Hill.

Siekmann, J., and Dittrich, K. 1977. Note on bubble motion in a rotating liquid under residual gravity. *Int. J. Non-Linear Mechanics* **12,** 409–415.

Siekmann, J., and Johann, W. 1976. On bubble motion in a rotating liquid under simulated low and zero gravity. *Ingenieur-Archiv* **45,** 307–315.

Slattery, J.C. 1972. *Momentum, Energy, and Mass Transfer in Continua.* New York: McGraw-Hill.

Smith, H.D., Mattox, D.M., Wilcox, W.R., Subramanian, R.S., and Meyyappan, M. 1982. Experimental observation of the thermocapillary driven motion of bubbles in a molten glass under low gravity conditions. In *Materials Processing in the Reduced Gravity Environment of Space,* ed. G.E. Rindone (pp. 279–288). New York: North-Holland.

Smoluchowski, M. von. 1911. Über die wechselwirkung von kugeln, die sich in einer zähen flüssigkeit bewegen. *Bull. Int. Acad. Pol. Sci. Lett. Cl. Sci. Math. Nat. Ser. A* **1,** 28–39.

Smoluchowski, M. von. 1917. Versuch einer mathematischen theorie der koagulationkinetik kollider lösungen. *Z. Phys. Chem.* **92,** 129–168.

Srividya, C.V. 1993. Migration of Fluorinert FC-75 Drops in a Silicone Oil in a Vertical Temperature Gradient. Master's thesis in chemical engineering, Clarkson University, Potsdam, New York.

Stimson, M., and Jeffery, G.B. 1926. The motion of two spheres in a viscous fluid. *Proc. Roy. Soc. Lond. A* **111,** 110–116.

Stokes, G.G. 1851. On the effect of the internal friction of fluids on the motion of pendulums. *Trans. Camb. Phil. Soc.* **9,** Pt. II, 8–106.

Stone, H.A. 1993. An interpretation of the translation of drops and bubbles at high Reynolds numbers in terms of the vorticity field. *Phys. Fluids A* **5,** No. 10, 2567–2569.

Subramanian, R.S. 1981. Slow migration of a gas bubble in a thermal gradient. *AIChE J.* **27,** No. 4, 646–654.

Subramanian, R.S. 1983. Thermocapillary migration of bubbles and droplets. In *Advances in Space Research,* ed. Y. Malmejac (vol. 3, no. 5, pp. 145–153). Great Britain: Pergamon Press.

Subramanian, R.S. 1985. The Stokes force on a droplet in an unbounded fluid medium due to capillary effects. *J. Fluid Mech.* **153,** 389–400.

Subramanian, R.S. 1992. The motion of bubbles and drops in reduced gravity. In *Transport Processes in Bubbles, Drops, and Particles,* ed. R.P. Chhabra and D. De Kee (Chap. 1: pp. 1–42). New York: Hemisphere.

Subramanian, R.S. 1997. Data and Results from the LMS Flight Experiment – Thermocapillary Migration and Interactions of Bubbles and Drops. Report submitted to NASA Lewis Research Center.

Subramanian, R.S., Balasubramaniam, R., and Wozniak, G. 2001. Fluid mechanics of bubbles and drops. In *Physics of Fluids in Microgravity,* ed. R. Monti (Chap. 6). Amsterdam: Gordon and Breach (in press).

Subramanian, R.S., and Chi, B. 1980. Bubble dissolution with chemical reaction. *Chem. Eng. Sci.* **35,** 2185–2194.

Subramanian, R.S., and Weinberg, M.C. 1981. Asymptotic expansions for the description of gas bubble dissolution and growth. *AIChE J.* **27,** No. 5, 739–748.

Subramanian, R.S., and Weinberg, M.C. 1983. The role of surface tension in glass refining. *Glastech. Berichte* **56K,** Bd. 1, 76–81.

Sy, F., Taunton, J.W., and Lightfoot, E.N. 1970. Transient creeping flow around spheres. *AIChE J.* **16,** No. 3, 386–391.

Szymczyk, J., and Siekmann, J. 1988. Numerical calculation of the thermocapillary motion of a bubble under microgravity. *Chem. Eng. Commun.* **69,** 129–147.

Szymczyk, J.A., Wozniak, G., and Siekmann, J. 1987. On Marangoni bubble motion at higher Reynolds- and Marangoni-numbers under microgravity. *Appl. Microgravity Tech.* **1,** No. 1, 27–29.

Taylor, R., and Krishna, R. 1993. *Multicomponent Mass Transfer.* New York: Wiley.

Taylor, T.D., and Acrivos, A. 1964. On the deformation and drag of a falling viscous drop at low Reynolds number. *J. Fluid Mech.* **18,** Pt. 3, 466–476.

Thompson, R.L. 1979. Marangoni Bubble Motion in Zero Gravity. Ph.D. diss., in engineering science, University of Toledo, Ohio.

Thompson, R.L., DeWitt, K.J., and Labus, T.L. 1980. Marangoni bubble motion phenomenon in zero gravity. *Chem. Eng. Commun.* **5,** 299–314.

Torres, F.E., and Herbolzheimer, E. 1993. Temperature gradients and drag effects produced by convection of interfacial internal energy around bubbles. *Phys. Fluids A* **5,** No. 3, 537–549.

Trefethen, L.M. 1963. Surface tension in fluid mechanics, a color film by Encyclopaedia Brittanica Educational Corporation, Film No. 21610, Chicago, Illinois.

Treuner, M., Galindo, V., Gerbeth, G., Langbein, D., and Rath, H.J. 1996. Thermocapillary bubble migration at high Reynolds and Marangoni numbers under low gravity. *J. Colloid Interface Sci.* **179,** 114–127.

Unverdi, S.O., and Tryggvason, G. 1992. A front-tracking method for viscous, incompressible, multi-fluid flows. *J. Comput. Phys.* **100,** 25–37.

Van Dyke, M. 1975. *Perturbation Methods in Fluid Mechanics.* Stanford, CA: Parabolic Press.

Vedha-Nayagam, M., and Balasubramaniam, R. 1993. An exact solution for combined thermocapillary, rotation, and buoyancy induced flow in a semi-infinite liquid layer. HTD-Vol. 235, Heat Transfer in Microgravity Systems (pp. 25–32). New York: The American Society of Mechanical Engineers.

Velten, R., Schwabe, D., and Scharmann, A. 1991. The periodic instability of thermocapillary convection in cylindrical liquid bridges. *Phys. Fluids A* **3,** No. 2, 267–279.

Venerus, D.C., and Yala, N. 1997. Transport analysis of diffusion-induced bubble growth and collapse in viscous liquids. *AIChE J.* **43,** No. 11, 2948–2959.

Venerus, D.C., Yala, N., and Bernstein, B. 1998. Analysis of diffusion-induced bubble growth in viscoelastic liquids. *J. Non-Newtonian Fluid Mech.* **75,** 55–75.

Vrentas, J.S., and Shin, D. 1980a. Perturbation solutions of spherical moving boundary problems – I Slow growth or dissolution rates. *Chem. Eng. Sci.* **35,** 1687–1696.

Vrentas, J.S., and Shin, D. 1980b. Perturbation solutions of spherical moving boundary problems – II Rapid growth or dissolution rates. *Chem. Eng. Sci.* **35,** 1697–1705.

Vrentas, J.S., Vrentas, C.M., and Ling, H.-C. 1983. Equations for predicting growth or dissolution rates of spherical particles. *Chem. Eng. Sci.* **38,** No. 11, 1927–1934.

Wakiya, S. 1967. Slow motions of a viscous fluid around two spheres. *J. Phys. Soc. Japan* **22,** No. 4, 1101–1109.

Wang, T.G., Saffren, M.M., and Elleman, D.D. 1974. Drop dynamics in space. *Proceedings of the International Colloquium on Drops and Bubbles,* ed. D.J. Collins, M.S. Plesset, and M.M. Saffren (vol. 1, pp. 266–299), Cal Tech/JPL. 28–30 August 1974, Pasadena, California.

Wang, Y., Mauri, R., and Acrivos, A. 1994. Thermocapillary migration of a bidisperse suspension of bubbles. *J. Fluid Mech.* **261,** 47–64.

Wanschura, M., Shevtsova, V.M., Kuhlmann, H.C., and Rath, H.J. 1995. Convective instability mechanisms in thermocapillary liquid bridges. *Phys. Fluids* **7,** No. 5, 912–925.

Wei, H. 1994. Interactions of Bubbles in a Temperature Gradient. Ph.D. diss., in chemical engineering, Clarkson University, Potsdam, New York.

Wei, H., and Subramanian, R.S. 1993. Thermocapillary migration of a small chain of bubbles. *Phys. Fluids A* **5,** No. 7, 1583–1595.

Wei, H., and Subramanian, R.S. 1994. Interactions between two bubbles under isothermal conditions and in a downward temperature gradient. *Phys. Fluids* **6,** No. 9, 2971–2978.

Wei, H., and Subramanian, R.S. 1995. Migration of a pair of bubbles under the combined action of gravity and thermocapillarity. *J. Colloid Interface Sci.* **172,** 395–406.

Weinberg, M.C. 1981. Surface tension effects in gas bubble dissolution and growth. *Chem. Eng. Sci.* **36,** 137–141.

Weinberg, M.C. 1982. Dissolution of a stationary bubble in a glassmelt with a reversible chemical reaction: Rapid forward reaction rate constant. *J. Amer. Ceram. Soc.* **65,** No. 10, 479–485.

Weinberg, M.C. 1986. On the possibility of diffusionally driven oscillations in two component gas bubbles in fluids. *Chem. Eng. Sci.* **41,** No. 9, 2333–2340.

Weinberg, M.C., Onorato, P.I.K., and Uhlmann, D.R. 1980. Behavior of bubbles in glassmelts: I, Dissolution of a stationary bubble containing a single gas. *J. Amer. Ceram. Soc.* **63,** No. 3–4, 175–180.

Weinberg, M.C., and Subramanian, R.S. 1980. Dissolution of multicomponent bubbles. *J. Amer. Ceram. Soc.* **63,** No. 9–10, 527–531.

Weinberg, M.C., and Subramanian, R.S. 1981. The dissolution of a stationary bubble enhanced by chemical reaction. *Chem. Eng. Sci.* **36,** No. 12, 1955–1965.

Welch, S.W.J. 1998. Transient thermocapillary migration of deformable bubbles. *J. Colloid Interface Sci.* **208,** 500–508.

Whitehead, A.N. 1889. Second approximation to viscous fluid motion. *Quart. J. Math.* **23,** 143–152.

Wozniak, G. 1986. Experimentelle Untersuchung des Einflusses der Thermokapillarität auf die Bewegung von Tropfen und Blasen. Ph.D. diss., in mechanics, Universitat-GH-Essen, Germany.

Wozniak, G. 1991. On the thermocapillary motion of droplets under reduced gravity. *J. Colloid Interface Sci.* **141,** No. 1, 245–254.

Wozniak, G., Siekmann, J., and Srulijes, J. 1988. Thermocapillary bubble and drop dynamics under reduced gravity – survey and prospects. *Z. Flugwiss. Weltraumforsch.* **12,** 137–144.

Wozniak, G., Wozniak, K., and Bergelt, H. 1996. On the influence of buoyancy on the surface tension driven flow around a bubble on a heated wall. *Experiments in Fluids* **21,** 181–186.

Wozniak, K., and Wozniak, G. 1998. Temperature gradient driven flow experiments of two interacting bubbles on a hot wall. *Heat and Mass Transfer* **33,** 363–369.

Xu, J-J., and Davis, S.H. 1984. Convective thermocapillary instabilities in liquid bridges. *Phys. Fluids* **27,** No. 5, 1102–1107.

Yee, J.F., Lin, M-C., Sarma, K., and Wilcox, W.R. 1975. The influence of gravity on crystal defect formation in InSb-GaSb alloys. *J. Crystal Growth* **30,** 185–192.

Yiantsios, S.G., and Davis, R.H. 1990. On the buoyancy-driven motion of a drop towards a rigid surface or a deformable interface. *J. Fluid Mech.* **217,** 547–573.

Yiantsios, S.G., and Davis, R.H. 1991. Close approach and deformation of two viscous drops due to gravity and van der Waals forces. *J. Colloid Interface Sci.* **144,** No. 2, 412–433.

Yih, C-S. 1968. Fluid motion induced by surface-tension variation. *Phys. Fluids* **11,** No. 3, 477–480.

Yih, C-S. 1969. Three-dimensional motion of a liquid film induced by surface-tension variation or gravity. *Phys. Fluids* **12,** No. 10, 1982–1987.

Young, N.O., Goldstein, J.S., and Block, M.J. 1959. The motion of bubbles in a vertical temperature gradient. *J. Fluid Mech.* **6,** 350–356.

Yung, C.-N., DeWitt, K.J., Brockwell, J.L., McQuillen, J.B., and Chai, A.-T. 1989. A numerical study of parameters affecting gas bubble dissolution. *J. Colloid Interface Sci.* **127,** No. 2, 442–452.

Zebib, A., Homsy, G.M., and Meiburg, E. 1985. High Marangoni number convection in a square cavity. *Phys. Fluids* **28,** No. 12, 3467–3476.

Zhang, X., and Davis, R.H. 1991. The rate of collisions due to Brownian or gravitational motion of small drops. *J. Fluid Mech.* **230,** 479–504.

Zhang, X., and Davis, R.H. 1992. The collision rate of small drops undergoing thermocapillary migration. *J. Colloid Interface Sci.* **152,** No. 2, 548–561.

Zhang, X., Davis, R.H., and Ruth, M.F. 1993. Experimental study of two interacting drops in an immiscible fluid. *J. Fluid Mech.* **249,** 227–239.

Zhang, X., Wang, H., and Davis, R.H. 1993. Collective effects of temperature gradients and gravity on droplet coalescence. *Phys. Fluids A* **5,** No. 7, 1602–1613.

Zinchenko, A.Z. 1979. Calculation of hydrodynamic interaction between drops at low Reynolds numbers. *Prikl. Matem. Mekhan.* English translation **42,** 1046–1051; original in Russian, 1978, **42,** No. 5, 955–959.

Zinchenko, A.Z. 1981. The slow asymmetric motion of two drops in a viscous medium. *Prikl. Matem. Mekhan.* English translation **44,** 30–37; original in Russian, 1980, **44,** No. 1, 49–59.

Zinchenko, A.Z. 1982. Calculation of close interaction between drops, with internal circulation and slip effect taken into account. *Prikl. Matem. Mekhan.* English translation **45,** 564–567; original in Russian, 1981, **45,** No. 4, 759–763.

Zinchenko, A.Z. 1983. Calculation of the effectiveness of gravitational coagulation of drops with allowance for internal circulation. *Prikl. Matem. Mekhan.* English translation **46,** 58–65; original in Russian, 1982, **46,** No. 1, 72–82.

Index